T0181963

Lecture Notes in Artificial Intelligence 12855

Subseries of Lecture Notes in Computer Science

Series Editors

Randy Goebel
University of Alberta, Edmonton, Canada

Yuzuru Tanaka
Hokkaido University, Sapporo, Japan

Wolfgang Wahlster
DFKI and Saarland University, Saarbrücken, Germany

Founding Editor

Jörg Siekmann
DFKI and Saarland University, Saarbrücken, Germany

More information about this subseries at http://www.springer.com/series/1244

Leszek Rutkowski · Rafał Scherer ·
Marcin Korytkowski · Witold Pedrycz ·
Ryszard Tadeusiewicz ·
Jacek M. Zurada (Eds.)

Artificial Intelligence and Soft Computing

20th International Conference, ICAISC 2021
Virtual Event, June 21–23, 2021
Proceedings, Part II

 Springer

Editors
Leszek Rutkowski 🆔
Częstochowa University of Technology
Częstochowa, Poland

Rafał Scherer
Częstochowa University of Technology
Częstochowa, Poland

Marcin Korytkowski
Częstochowa University of Technology
Częstochowa, Poland

Witold Pedrycz
Edmonton, AB, Canada

Jacek M. Zurada
Electrical and Computer Engineering
University of Louisville
Louisville, KY, USA

Ryszard Tadeusiewicz
AGH University of Science and Technology
Krakow, Poland

ISSN 0302-9743   ISSN 1611-3349   (electronic)
Lecture Notes in Artificial Intelligence
ISBN 978-3-030-87896-2   ISBN 978-3-030-87897-9   (eBook)
https://doi.org/10.1007/978-3-030-87897-9

LNCS Sublibrary: SL7 – Artificial Intelligence

This Springer imprint is published by the registered company Springer Nature Switzerland AG
The registered company address is: Gewerbestrasse 11, 6330 Cham, Switzerland

Preface

This volume constitutes the proceedings of 20th International Conference on Artificial Intelligence and Soft Computing (ICAISC 2021) held in Zakopane, Poland, during June 21–23, 2021, which took place virtually due to the COVID-19 pandemic. The conference was organized by the Polish Neural Network Society in cooperation with the University of Social Sciences in Łódź, the Department of Intelligent Computer Systems at the Częstochowa University of Technology, and the IEEE Computational Intelligence Society, Poland Chapter. Previous conferences took place in Kule (1994), Szczyrk (1996), Kule (1997), and Zakopane (1999, 2000, 2002, 2004, 2006, 2008, 2010, and 2012–2020) and attracted a large number of papers and internationally recognized speakers: Lotfi A. Zadeh, Hojjat Adeli, Rafal Angryk, Igor Aizenberg, Cesare Alippi, Shun-ichi Amari, Daniel Amit, Plamen Angelov, Albert Bifet, Piero P. Bonissone, Jim Bezdek, Zdzisław Bubnicki, Jan Chorowski, Andrzej Cichocki, Swagatam Das, Ewa Dudek-Dyduch, Włodzisław Duch, Adel S. Elmaghraby, Pablo A. Estévez, João Gama, Erol Gelenbe, Jerzy Grzymala-Busse, Martin Hagan, Yoichi Hayashi, Akira Hirose, Kaoru Hirota, Adrian Horzyk, Eyke Hüllermeier, Hisao Ishibuchi, Er Meng Joo, Janusz Kacprzyk, Jim Keller, Laszlo T. Koczy, Tomasz Kopacz, Jacek Koronacki, Zdzisław Kowalczuk, Adam Krzyzak, Rudolf Kruse, James Tin-Yau Kwok, Soo-Young Lee, Derong Liu, Robert Marks, Ujjwal Maulik, Zbigniew Michalewicz, Evangelia Micheli-Tzanakou, Kaisa Miettinen, Krystian Mikołajczyk, Henning Müller, Ngoc Thanh Nguyen, Andrzej Obuchowicz, Erkki Oja, Nikhil R. Pal, Witold Pedrycz, Marios M. Polycarpou, José C. Príncipe, Jagath C. Rajapakse, Šarunas Raudys, Enrique Ruspini, Jörg Siekmann, Andrzej Skowron, Roman Słowiński, Igor Spiridonov, Boris Stilman, Ponnuthurai Nagaratnam Suganthan, Ryszard Tadeusiewicz, Ah-Hwee Tan, Dacheng Tao, Shiro Usui, Thomas Villmann, Fei-Yue Wang, Jun Wang, Bogdan M. Wilamowski, Ronald Y. Yager, Xin Yao, Syozo Yasui, Gary Yen, Ivan Zelinka, and Jacek Zurada. The aim of this conference is to build a bridge between traditional artificial intelligence techniques and so-called soft computing techniques. It was pointed out by Lotfi A. Zadeh that "soft computing (SC) is a coalition of methodologies which are oriented toward the conception and design of information/intelligent systems. The principal members of the coalition are: fuzzy logic (FL), neurocomputing (NC), evolutionary computing (EC), probabilistic computing (PC), chaotic computing (CC), and machine learning (ML). The constituent methodologies of SC are, for the most part, complementary and synergistic rather than competitive". These proceedings present both traditional artificial intelligence methods and soft computing techniques. Our goal is to bring together scientists representing both areas of research. This volume is divided into four parts:

- Computer Vision, Image and Speech Analysis
- Data Mining
- Various Problems of Artificial Intelligence
- Bioinformatics, Biometrics and Medical Applications

I would like to thank our participants, invited speakers, and reviewers of the papers for their scientific and personal contribution to the conference. Finally, I thank my co-workers, Łukasz Bartczuk, Piotr Dziwiński, Marcin Gabryel, Rafał, Grycuk, Marcin Korytkowski, and Rafał, Scherer, for their enormous efforts to make the conference a very successful event. Moreover, I appreciate the work of Marcin Korytkowski who was responsible for the Internet submission system.

June 2021   Leszek Rutkowski

Organization

ICAISC 2021 was organized by the Polish Neural Network Society in cooperation with the University of Social Sciences in Łódź and the Institute of Computational Intelligence at Częstochowa University of Technology.

Conference Chairs

General Chair

Leszek Rutkowski, Poland

Area Chairs

Fuzzy Systems

Witold Pedrycz, Canada

Evolutionary Algorithms

Zbigniew Michalewicz, Australia

Neural Networks

Jinde Cao, China

Computer Vision

Dacheng Tao, Australia

Machine Learning

Nikhil R. Pal, India

Artificial Intelligence with Applications

Janusz Kacprzyk, Poland

International Liaison

Jacek Żurada, USA

Program Committee

Hojjat Adeli, USA
Cesare Alippi, Italy
Shun-ichi Amari, Japan
Rafal A. Angryk, USA
Robert Babuska, The Netherlands
James C. Bezdek, Australia
Piero P. Bonissone, USA
Bernadette Bouchon-Meunier, France
Jinde Cao, China
Juan Luis Castro, Spain
Yen-Wei Chen, Japan
Andrzej Cichocki, Japan
Krzysztof Cios, USA
Ian Cloete, Germany
Oscar Cordón, Spain
Bernard De Baets, Belgium
Włodzisław Duch, Poland
Meng Joo Er, Singapore
Pablo Estevez, Chile
David B. Fogel, USA
Tom Gedeon, Australia
Erol Gelenbe, UK
Jerzy W. Grzymala-Busse, USA
Hani Hagras, UK
Saman Halgamuge, Australia
Yoichi Hayashi, Japan
Tim Hendtlass, Australia
Francisco Herrera, Spain
Kaoru Hirota, Japan
Tingwen Huang, USA
Hisao Ishibuchi, Japan
Mo Jamshidi, USA
Robert John, UK
Janusz Kacprzyk, Poland
Nikola Kasabov, New Zealand
Okyay Kaynak, Turkey
Vojislav Kecman, USA
James M. Keller, USA
Etienne Kerre, Belgium
Frank Klawonn, Germany
Robert Kozma, USA
László Kóczy, Hungary
Józef Korbicz, Poland

Rudolf Kruse, Germany
Adam Krzyzak, Canada
Věra Kůrková, Czech Republic
Soo-Young Lee, South Korea
Simon M. Lucas, UK
Luis Magdalena, Spain
Jerry M. Mendel, USA
Radko Mesiar, Slovakia
Zbigniew Michalewicz, Australia
Javier Montero, Spain
Eduard Montseny, Spain
Kazumi Nakamatsu, Japan
Detlef D. Nauck, Germany
Ngoc Thanh Nguyen, Poland
Erkki Oja, Finland
Nikhil R. Pal, India
Witold Pedrycz, Canada
Leonid Perlovsky, USA
Marios M. Polycarpou, Cyprus
Danil Prokhorov, USA
Vincenzo Piuri, Italy
Sarunas Raudys, Lithuania
Olga Rebrova, Russia
Vladimir Red'ko, Russia
Raúl Rojas, Germany
Imre J. Rudas, Hungary
Norihide Sano, Japan
Rudy Setiono, Singapore
Jennie Si, USA
Peter Sincak, Slovakia
Andrzej Skowron, Poland
Roman Słowiński, Poland
Pilar Sobrevilla, Spain
Janusz Starzyk, USA
Jerzy Stefanowski, Poland
Vitomir Štruc, Slovenia
Ron Sun, USA
Johan Suykens, Belgium
Ryszard Tadeusiewicz, Poland
Hideyuki Takagi, Japan
Dacheng Tao, Australia
Vicenç Torra, Spain
Burhan Turksen, Canada

Shiro Usui, Japan
Deliang Wang, USA
Jun Wang, Hong Kong
Lipo Wang, Singapore
Paul Werbos, USA
Bernard Widrow, USA
Kay C. Wiese, Canada

Bogdan M. Wilamowski, USA
Donald C. Wunsch, USA
Ronald R. Yager, USA
Xin-She Yang, UK
Gary Yen, USA
Sławomir Zadrożny, Poland
Jacek Zurada, USA

Organizing Committee

Rafał Scherer
Łukasz Bartczuk
Piotr Dziwiłski
Marcin Gabryel (Finance Chair)
Rafał Grycuk
Marcin Korytkowski (Databases and Internet Submissions)

Contents – Part II

Data Mining

Various Problems of Ariticial Intelligence

Bioinformatics, Biometrics and Medical Applications

Contents – Part I

Fuzzy Systems and Their Applications

Evolutionary Algorithms and Their Applications

Artificial Intelligence in Modeling and Simulation

Computer Vision, Image and Speech Analysis

Classification of Dermatological Asymmetry of the Skin Lesions Using Pretrained Convolutional Neural Networks

Michał Beczkowski[1] (ORCID), Norbert Borowski[2] (ORCID), and Piotr Milczarski[1(✉)] (ORCID)

[1] Faculty of Physics and Applied Informatics, Department of Computer Science, University of Lodz, Pomorska Street 149/153, 90-236 Lodz, Poland
{michal.beczkowski,piotr.milczarski}@uni.lodz.pl
[2] Faculty of Physics and Applied Informatics, Department of Nuclear Physics, and Radiation Safety, University of Lodz, Pomorska Street 149/153, 90-236 Lodz, Poland
norbert.borowski@uni.lodz.pl

Abstract. In dermatology, malignant melanoma is one the most deadly forms of skin cancer. It is extremely important to detect it at an early stage. One of the methods of detecting it is an evaluation based on dermoscopy combined with one of the criteria for assessing a skin lesion. Such an evaluation method is the Three-Point Checklist of Dermoscopy which is considered a sufficient screening method for the assessment of skin lesions. The proposed method, founded on the convolutional neural networks, is aimed at improving diagnostics and enabling the preliminary assessment of skin lesions by a family doctor. The current paper presents the results of the application of convolutional neural networks: VGG19, Xception, Inception-ResNet-v2, for the assessment of skin lesions asymmetry, along with various variations of the PH2 database. For the best CNN network, we achieved the following results: true positive rate for the asymmetry 92.31%, weighted accuracy 67.41%, F1 score 0.646 and Matthews correlation coefficient 0.533.

Keywords: CNN · Dermoscopy · Asymmetry · Pretrained convolutional neural networks · Invariant dataset augmentation

1 Introduction

1.1 Dermatological Asymmetry of Skin Lesions and Screening Methods

According to the statistics compiled by European Cancer Information System (ECIS) [1] and the American Cancer Society (ACS) [2] life-threatening melanoma can be completely cured if removed in the early stages [3, 4]. According to ECIS, the estimated risk for 2018 of melanoma varies from 38.2 new cases in Germany to 13.6 cases in Iceland per 100K age and gender standardized population [1]. Also, ACS reports in 2018 that the risk of Americans developing cancer over their lifetime is 37.6% for females and 39.7% for males, where the melanoma risk is 1 in 42 cases for females and 1 in 27

© Springer Nature Switzerland AG 2021
L. Rutkowski et al. (Eds.): ICAISC 2021, LNAI 12855, pp. 3–14, 2021.
https://doi.org/10.1007/978-3-030-87897-9_1

cases for males [2]. It is necessary to develop a quick and effective diagnostic method: to minimize the excision of benign lesions and increase the detection of melanoma. Dermatology experts use various screening methods such as the Three-Point Checklist of Dermoscopy (3PCLD) [5–7], The Seven-Point Checklist (7PCL) and the ABCD rule [9, 10]. All of them are considered effective in skin lesion assessments.

3PCLD methodology is based on the criteria of asymmetry in shape, hue, and structure distribution within the lesion defined and it can have a value of either 0 for symmetry in two axes, 1 for symmetry in one axis, or 2 for asymmetry. In this method, the pigmented network and blue-white veil are either present or absent. Another example of the screening method used in dermatology is the ABCD's of melanoma. In this rule ABCD stands for asymmetry, border (not well-defined, irregular), color (more than one shade), diameter (usually larger than 6 mm) and evolution (changing features over time). All those features are characteristics of melanoma that general physicians or dermatologists check while diagnosing. Like in 3PCLD methodology, this method focuses on the asymmetry of the lesion [11–13] which is one of the common characteristics of skin damage that can be noticed visually. These examples show the importance of symmetry/asymmetry in various screening methods of detecting melanoma. In the paper, we show the results of the CNN application to the problem of the asymmetry within the skin lesion in the dermoscopic images.

There are a few publications about the symmetry/asymmetry of the skin lesion using machine learning/AI methods. In the paper [16], there is only shape asymmetry discussed and the authors tested several ML methods on the PH2 dataset. The result showed 95.8% of accuracy, with true positive rates for the asymmetry 92.5%, 95.7% for the 1-axis symmetry and 100% for the symmetric lesions while using the SVM with the radial basis kernel function.

This research paper presents the results of the application of convolutional neural networks for the diagnosis of skin lesions asymmetry. The neural networks were based on available, pre-trained networks such as Xception (XN), VGG19 [14], and Inception-ResNet-v2 (IRN2). Those networks provide promising results even with a relatively small but well-described PH2 dataset [15].

1.2 Dermatological Datasets

From the available databases, we have chosen the PH2 dataset [15] to conduct our research. This database consists of dermoscopic images obtained at the Dermatology Service of Hospital Pedro Hispano (Matosinhos, Portugal) under the same conditions through the Tuebinger Mole Analyzer system using a magnification of 20 times. Images in the dataset are 8-bit RGB color images with a resolution of 768×560 pixels.

This image database contains a total of 200 dermoscopic images of melanocytic lesions, including 80 common nevi, 80 atypical nevi, and 40 melanomas. The PH2 database includes medical annotation of all the images namely medical segmentation of the lesion, clinical and histological diagnosis and the assessment of several dermoscopic criteria (colors; pigment network; dots/globules; streaks; regression areas; blue-whitish veil) [8, 15, 16].

One of the alternatives for the PH2 database is the ISIC Archive which contains the largest publicly available collection of quality controlled dermoscopic images of skin lesions [17]. The ISIC Archive contains over 24,000 dermoscopic images, which were collected from leading clinical centers internationally and acquired from a variety of devices within each center. The ISIC dataset metadata does not provide information about the asymmetry of lesions. The other examples of the dermatological datasets can be found in Interactive Atlas of Dermoscopy or An Atlas of Surface Microscopy of Pigmented Skin Lesions [18, 19].

1.3 Pretrained Convolutional Neural Network and Their Features

The pretrained Convolutional Neural Networks have different features that should be taken into account when choosing a network to apply to a given problem. The most important characteristics are network accuracy, true positive and negative rate, speed of classification, and size. While selecting a network these features should be taken into account. Currently, we can choose within several pretrained networks. The chosen three networks' characteristics are given in Table 1. The network depth is defined as the largest number of sequential convolutional or fully connected layers on a path from the input layer to the output layer. The inputs to all networks are RGB images.

Table 1. Pretrained convolutional neural networks parameters

Network	Depth	Size [MB]	Parameters [Millions]	Image input size	Average accuracy [%]
VGG19	19	535	144	224 × 224	70
Xception	71	85	22.9	299 × 299	80<
Inception-ResNet-v2	164	209	55.9	299 × 299	80

2 Data Preparation for the Research

2.1 Augmentation and Preparation of the Database

The PH2 database contains 117 fully symmetric, 31 symmetric in one axis and 52 fully asymmetric images of skin lesions. In order to use this database in our research, we had to increase the number of images while minimalizing possible influence on the pixel distribution. To create new images, various geometric transformations that do not change the asymmetry of shape, shade and structure distribution, as well as other features present in both 3PCLD and 7 PCL, were used. For the transformation of images, we chose three rotations by 90°, 180° and 270°, mirroring on the vertical and horizontal axis, and a 90° rotation of the images after mirroring (Fig. 1). In total, we got seven transformations for each image that did not change the pixels, shape, or color distribution. These transformations allowed us to increase the PH2 database from 200 to 1600 images.

To show the idea of the author method, to classify not only using the original image but as well its invariant copies we provide in Table 2 the classification probabilities of the exemplary image classification and its invariant copies (the image IMD168 from the PH2 dataset) by the chosen VGG19 CNN network trained on the images from PH2 and their seven copies. The probability of the classification for the asymmetry (column with value '0') networks varies from 0.013 to 0.94. The same variance of probability occurs for other CNN networks. It can be concluded that the same image and its invariant versions can provide us with opposite classification results due to convolutional network operations.

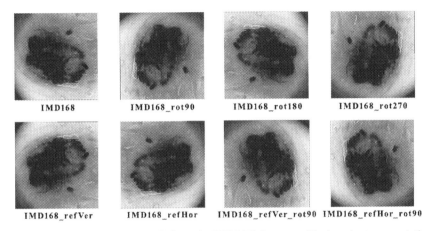

Fig. 1. A sample of image IMD168 from the PH2 [15] dataset and its invariant augmentation.

Table 2. VGG19 CNN classification probability of the image IMD168 from the PH2 dataset.

IMD168 image version	Classification probability for a class		
	0	1	2
original	0.0131	0.9593	0.0276
rot 90	0.5501	0.4017	0.0482
rot 180	0.9420	0.0407	0.0173
rot 270	0.3044	0.6713	0.0243
mir Vert	0.3781	0.6170	0.0050
mir Vert rot 90	0.4937	0.4964	0.0099
mir Hor	0.0543	0.9403	0.0054
mir Hor rot 90	0.6154	0.3441	0.0405

For example, during a convolution each of the eight images with the filters and their weights give a different output result due to the convolution properties that can be derived from the formula:

$$I(x, y) = \sum_{i=0}^{n} \sum_{j=0}^{m} k(i, j) I(x + i, y + j),$$

where the kernel k is of size n by m. The image is size NxM, where $N \geq n$ and $M \geq m$. As it is shown in the final section the probability of the classification of each of the invariant images can vary from 0.0 to 1.0.

The next step in preparing the database was to scale the images to the input sizes required by the selected networks Table 1. First, we scaled the shorter dimension of images (in our case, height) to the input size, e.g. 224 px (see Table 1) using the Bicubic Sharper algorithm in Photoshop. Then, all images were cropped to a square shape. As a result, we obtained a set of images scaled to the sizes required by each of the networks, e.g. 224 × 224 px, see Table 1. The dataset prepared in this way contains 936 fully symmetric, 248 symmetric in one axis and 416 fully asymmetric images of skin lesions and met the requirements of our research and could be used for network tests.

2.2   CNN Network Setting and Configuration

We used pre-trained networks in our research because they are trained on the ImageNet database [20]. Moreover, those networks use as starting point to learn a new task previous abilities to extract informative features from natural images. Since in each pre-trained network the last three layers are configured to classify 1000 classes, we separated all but the last three layers and replaced them so that the networks would classify images into 3 classes. Due to this method and 3PCLD, the networks classified the images as symmetrical, symmetrical in one axis, and asymmetrical.

To achieve the highest classification rates we have conducted the initial research testing the wide variety of the following parameters for all three networks:

- 30–60 epochs;
- learning rate from $1e - 4$ to $1e - 2$.

After the initial research we choose for:

- VGG19 – learning rate $1e - 4$ and 40 epoch;
- XN - learning rate $5e - 4$ and 30 epoch;
- IRN2 - learning rate $5e - 4$ and 30 epoch.

The time of training depends on the network and number of the training images and the machine specification. The times for the machine 1 specification have varied from around:

- 18 min for VGG19;
- 30 min for XN;
- 60 min for IRN2.

2.3   Hardware Description

To ensure the credibility of the results, the research was conducted independently on two computers with the same operating system (Microsoft Windows 10 Pro) and different configurations:

- Set 1. Processor: Intel(R) Core(TM) i7-8700K CPU @ 3.70 GHz (12 CPUs), Memory: 64 GB RAM, Graphics Card: NVIDIA GTX 1080Ti with 11 GB of Graphics RAM.
- Set 2. Processor: Intel(R) Core(TM) i7-9700K CPU @ 3.60 GHz (8 CPUs), Memory: 16 GB RAM, Graphics Card: NVIDIA GeForce RTX 2070 with 8 GB of Graphics RAM.

On both machines, the research was conducted using Matlab 2019b with up-to-date versions of Deep Learning Toolbox™ (v. 12). Deep Learning Toolbox allows to transfer learning with pretrained deep network models, see Table 1. The second hardware set was used to test the procedure to see whether the classification parameters depend on their hardware. Different configurations affected only the time of execution in CNN networks training and at the end training. When working on both machines the calculated average accuracy, as well as their maximum and minimum, showed results close to each other. It proved that the procedure was not hardware-dependent.

3   Research Method Description

The first step in our research method is database preparation. To selected networks, two databases were added: training and testing. Both databases were created by dividing the augmentation PH2 dataset into two sets in the following proportions 75% training and validation and 25% testing. The division was carried out so that the original images and their copies were in one set. The division was carried out 4 times so that each image was included in the test set. The training, validation and testing image cases for the three chosen networks were the same, although the image sizes were different. This allows us to assess and compare the results more thoroughly. Also, to check whether increasing the database with image copies obtained after rotations and mirroring gives better results, the tests were carried out on the original PH2 database file, which was also divided in the previously mentioned way into training and testing set, see Table 3. All steps were repeated on different image sizes to make it possible to research different networks, see Table 1.

Table 3. The number of the images in the original and augmentated PH2 set.

Number of images	Original PH2 dataset			Invariant dataset augmentation		
	Total	Train	Test	Total	Train	Test
Fully symmetric	117	88	29	936	704	232
Symmetric in 1 axes	31	23	8	248	184	64
Fully asymmetric	52	39	13	416	312	104
Total	200	150	50	1600	1200	400

The networks were tested 5 times on each pair of training, validation and testing sets. The resulting networks are saved for future testing and analysis of the results. For each CNN mentioned in Table 1 parameters such as accuracy, true positive rate were defined and calculated according to Eqs. (1)–(6). Next, their average values with the variance, minimum and maximum values were calculated for twenty-five (5 rounds ×5). Correct classification plus overestimation which is the Accuracy + Error Type I were considered best for purpose of our research: it is better if the screening method overestimates the diagnosis than the opposite (Underestimation Error Type II - False Negative) as the final diagnosis of malignant melanoma takes place after histopathological research.

The confusion matrix parameters are defined as follows:

$$ACC = (TP + TN)/N \tag{1}$$

$$TPR = TP/(TP + FN) \tag{2}$$

$$w.ACC = (TPR_0 + TPR_1 + TPR_2)/3 \tag{3}$$

$$FPR = FN/(FP + TN) \tag{4}$$

$$F1 = 2TP/(2TP + FP + FN) \tag{5}$$

$$MCC = (TP * TN - FP * FN)/\sqrt{(TP + FP)(TP + FN)(TN + FP)(TN + FN)} \tag{6}$$

where:

- N- a number of all cases;
- true positive, TP – number of positive results i.e. correctly classified cases;
- true negative, TN – number of negative results i.e. correctly classified cases;
- false positive, FP – number of negative results i.e. wrongly classified cases as positive ones;
- false negative, FN – number of positive results i.e. wrongly classified cases as
- negative ones, also called Type II error;
- accuracy, ACC; weighted accuracy, $w.\,ACC$;
- true positive rate, TPR, also called Recall; TPR_i – stands for true positive rate for the symmetry values $i = 0$, 1 and 2;
- false positive rate, FPR; FPR_i – stands for false positive rate for the symmetry values $i = 0$, 1 and 2;
- score test $F1$;
- Matthews correlation coefficient, MCC.

In Table 4 weighted F1 and MCC are calculated as for weighted accuracy, Eq. (3).

4   Results

The research method described above allowed us to obtain 60 neural networks. Results from those networks were recorded and analyzed. The results were analyzed in three ways:

1. T1 - networks tested on a subset of original images;
2. T8 - networks tested on the original set and its seven copies;
3. IDA - networks tested on the original set and its seven copies but in the worst-case scenario, i.e. if one of the 8 copies of the images has been recognized as asymmetric, all its copies have been classified as asymmetrical.

The advantage of the IDA procedure is the increased value of the true positive rate (TPR) for the positive cases i.e. asymmetric ones. Asymmetric lesions according to 3PCLD and ABCD rule are more prone to be melanocytic. On the other hand, this procedure increases the false-positive value (FPR) (see Table 4) which can be considered as its biggest disadvantage. However, this procedure finds more melanoma cases than the T1 or T8 methods. The procedure of IDA is also used in blue-white veil classification by CNN in [21].

When comparing the results of the networks, we also took into account such classification characteristics as weighted accuracy (ACC), F1 score and Matthews correlation coefficient (MCC), see Table 5. Within the network, the results were similar regardless of the method used (T1, T8, IDA). However, the results of each network differed. The best results for these classification characteristics were shown by the Xception (XN) network with an accuracy score of 78.9%.

Table 4. The classifications results for the asymmetry. The chosen confusion matrix factors true positive rate for full asymmetry (TPR_0), true positive rate for symmetry in one axis (TPR_1), true positive rate for full symmetry (TPR_2), false positive rate for full asymmetry (FPR_0), false positive rate for symmetry in one axis (FPR_1), false positive rate for full asymmetry full symmetry (FPR_2) with their average (AVG), variance (VAR), minimum (MIN) and maximum (MAX) values for the chosen CNN network.

CM factor		VGG19			Xception			Inception-ResNet-v2		
		T1	T8	IDA	T1	T8	IDA	T1	T8	IDA
TPR_0 [%]	AVG	58.8	60.2	71.9	65.4	67.1	80.4	53.1	55.9	70.4
	VAR	8.4	6.8	10.4	13.1	9.6	12.3	7.9	4.4	5.7
	Min	46.2	49.0	53.8	38.5	52.9	61.5	38.5	49.0	61.5
	Max	69.2	74.0	92.3	84.6	80.8	92.3	69.2	67.3	84.6
TPR_1 [%]	AVG	26.9	26.7	33.1	25.6	24.4	21.9	7.5	10.3	13.1
	VAR	15.3	12.9	9.3	22.0	16.3	12.7	8.5	10.0	9.5

(continued)

Table 4. (*continued*)

CM factor		VGG19			Xception			Inception-ResNet-v2		
		T1	T8	IDA	T1	T8	IDA	T1	T8	IDA
TPR_2 [%]	Min	12.5	12.5	12.5	0.0	6.3	0.0	0.0	0.0	0.0
	Max	62.5	57.8	50.0	62.5	53.1	37.5	25.0	31.3	25.0
	AVG	80.0	80.9	70.7	82.2	83.0	66.4	83.6	82.5	62.8
	VAR	11.8	10.6	13.0	7.1	5.7	7.2	5.7	4.7	9.9
FPR_0 [%]	Min	58.6	61.2	44.8	72.4	75.9	55.2	75.9	73.3	44.8
	Max	96.6	95.3	86.2	93.1	91.4	75.9	89.7	89.2	75.9
	AVG	9.2	8.9	15.9	11.6	11.1	25.0	10.8	10.9	23.4
	VAR	4.7	3.8	6.3	3.2	4.2	5.8	3.3	2.1	5.8
FPR_1 [%]	Min	0.0	2.0	2.7	5.4	6.1	16.2	5.4	8.1	13.5
	Max	16.2	14.5	2.4	18.9	17.2	2.4	16.2	15.5	4.8
	AVG	11.1	11.4	14.9	10.4	8.7	12.6	9.6	8.7	16.5
	VAR	9.8	8.7	11.6	7.8	6.4	9.3	7.3	5.3	9.9
FPR_2 [%]	Min	0.0	2.1	2.4	0.0	3.3	2.4	0.0	2.1	4.8
	Max	33.3	29.5	35.7	23.8	20.2	28.6	23.8	18.8	33.3
	AVG	42.6	40.5	25.5	33.1	35.7	19.0	48.6	49.0	28.6
	VAR	9.6	8.4	9.4	8.2	11.7	11.7	6.7	7.2	9.9
	Min	23.8	25.0	9.5	14.3	16.1	0.0	38.1	37.5	9.5
	Max	61.9	54.8	38.1	42.9	48.2	33.3	66.7	63.7	42.9

Additionally, to compare the networks, the area under curve (AUC) value was used. VGG19 turned out to be the best network and obtained a result of 0.9652. Figure 2 shows the best receiver operating characteristic curve (ROC) with the highest value of the area under curve (AUC).

From our research we have chosen the best CNN networks:

- VGG19 - true positive rate for the asymmetry 84.62%, weighted accuracy 68.29%, F1 score 0.682 and Matthews correlation coefficient 0.581;
- Xception - true positive rate for the asymmetry 92.31%, weighted accuracy 67.41%, F1 score 0.646 and Matthews correlation coefficient 0.533;

Inception-ResNet-v2 - true positive rate for the asymmetry 53.85%, weighted accuracy 51.57%, F1 score 0.528 and Matthews correlation coefficient 0.295.

Table 5. The classifications results for the asymmetry. The chosen confusion matrix factors weighted accuracy (w.ACC) with their average (AVG), variance (VAR), minimum (Min) and maximum (Max) values and weighted F1 score (w.F1), weighted Matthews correlation coefficient (w.MCC) for the chosen CNN network.

CM factor		VGG19			Xception			Inception-ResNet-v2		
		T1	T8	IDA	T1	T8	IDA	T1	T8	IDA
w.ACC [%]	AVG	55.2	56.0	58.6	57.8	58.1	56.2	48.1	49.6	48.8
	VAR	8.3	7.1	6.2	11.6	8.7	8.4	4.1	4.2	5.7
	Min	39.1	41.9	44.7	41.1	48.5	43.1	40.7	43.4	38.0
	Max	69.2	68.9	68.3	78.9	73.3	68.6	54.8	56.8	57.0
w.F1	AVG	0.543	0.548	0.555	0.560	0.560	0.511	0.446	0.465	0.439
	VAR	0.093	0.078	0.071	0.126	0.098	0.085	0.047	0.059	0.063
	Min	0.370	0.402	0.423	0.403	0.465	0.397	0.389	0.392	0.323
	Max	0.673	0.682	0.678	0.801	0.739	0.634	0.522	0.580	0.547
wMCC	AVG	0.400	0.324	0.438	0.463	0.363	0.424	0.318	0.249	0.309
	VAR	0.136	0.114	0.099	0.130	0.121	0.114	0.073	0.071	0.086
	Min	0.061	0.101	0.264	0.231	0.223	0.280	0.135	0.158	0.170
	Max	0.578	0.524	0.621	0.732	0.650	0.643	0.429	0.424	0.449

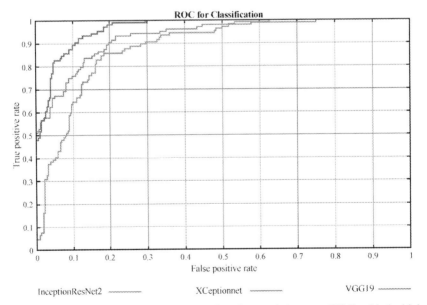

Fig. 2. The examples of the best receiver operating characteristic curve (ROC) with the highest value of the area under curve (AUC) for the three chosen CNNs.

5   Conclusions

Asymmetry plays an important role in the assessment of skin lesions, which is evident in dermatological diagnostic methods such as The Three-Point Checklist of Dermoscopy (3PCLD). Melanoma diagnosis based on asymmetry can also be made with the use of properly trained CNN networks. Such networks can serve as a helpful tool in the preliminary diagnosis of dangerous skin lesions.

In our research, we used three pretrained networks (Xception, VGG19, Inception-ResNet-v2) and trained them on our enlarged PH2 database. The method developed by us (using different forms of augmentation) turned out to be in many cases more effective than training the network only on the original images, even by 20% higher. In the studies we achieved a maximum of 68.56% weighted accuracy, 92.31% true positive rate, 66% false positive rate with tests F1 = 0.74, MCC = 0.58 and AUC = 0.97.

In our corresponding research [21, 22] in the field of a dermatological image processing using the Invariant Dataset of Augmentation is used with the PH2 [15] and the Atlas of Dermoscopy (Derm7pt) [8, 18] datasets to increase the classification rates e.g. true positive rate, test F1 and MCC using CNNs in comparison to the feature based methods used in [12, 16, 23].

References

1. European Cancer Information System (ECIS). https://ecis.jrc.ec.europa.eu. Accessed 05 Jan 2021
2. ACS – American Cancer Society. https://www.cancer.org/research/cancer-facts-statistics.html. Accessed 05 Jan 2021
3. Was, L., Milczarski, P., Stawska, Z., Wiak, S., Maslanka, P., Kot, M.: Verification of results in the acquiring knowledge process based on IBL methodology. In: Rutkowski, L., Scherer, R., Korytkowski, M., Pedrycz, W., Tadeusiewicz, R., Zurada, J.M. (eds.) ICAISC 2018. LNCS (LNAI), vol. 10841, pp. 750–760. Springer, Cham (2018). https://doi.org/10.1007/978-3-319-91253-0_69
4. Celebi, M.E., Kingravi, H.A., Uddin, B.: A methodological approach to the classification of dermoscopy images. Comput Med. Imaging Graph. 31(6), 362–373 (2007)
5. Soyer, H.P., Argenziano, G., Zalaudek, I., et al.: Three-point checklist of dermoscopy. A new screening method for early detection of melanoma. Dermatology 208(1), 27–31 (2004)
6. Argenziano, G., Soyer, H.P., et al.: Dermoscopy of pigmented skin lesions: results of a consensus meeting via the Internet. J. Am. Acad. Dermatol. 48(9), 679–693 (2003)
7. Milczarski, P.: Symmetry of hue distribution in the images. In: Rutkowski, L., Scherer, R., Korytkowski, M., Pedrycz, W., Tadeusiewicz, R., Zurada, J.M. (eds.) ICAISC 2018. LNCS (LNAI), vol. 10842, pp. 48–61. Springer, Cham (2018). https://doi.org/10.1007/978-3-319-91262-2_5
8. Kawahara, J., Daneshvar, S., Argenziano, G., Hamarneh, G.: Seven-point checklist and skin lesion classification using multitask multimodal neural nets. IEEE J. Biomed. Health Inform. 23(2), 538–546 (2019)
9. Argenziano, G., Fabbrocini, G., et al.: Epiluminescence microscopy for the diagnosis of doubtful melanocytic skin lesions. Comparison of the ABCD rule of dermatoscopy and a new 7-point checklist based on pattern analysis. Arch. Dermatol. 134, 1563–1570 (1998)

10. Carrera, C., Marchetti, M.A., Dusza, S.W., Argenziano, G., et al.: Validity and reliability of dermoscopic criteria used to differentiate nevi from melanoma: a web-based international dermoscopy society study. JAMA Dermatol. **152**(7), 798–806 (2016)
11. Nachbar, F., Stolz, W., Merkle, T., et al.: The ABCD rule of dermatoscopy. High prospective value in the diagnosis of doubtful melanocytic skin lesions. J. Am. Acad. Dermatol. **30**(4), 551–559 (1994)
12. Milczarski, P., Stawska, Z., Maslanka, P.: Skin lesions dermatological shape asymmetry measures. In: Proceedings of the IEEE 9th International Conference on Intelligent Data Acquisition and Advanced Computing Systems: Technology and Applications, IDAACS, pp. 1056–1062 (2017)
13. Menzies, S.W., Zalaudek, I.: Why perform Dermoscopy? The evidence for its role in the routine management of pigmented skin lesions. Arch Dermatol. **142**, 1211–1222 (2006)
14. Simonyan, K., Zisserman, A.: Very deep convolutional networks for large-scale image recognition. In: Conference Track Proceedings of 3rd International Conference on Learning Representations (ICRL), San Diego, USA (2015)
15. Mendoncca, T., Ferreira, P.M., Marques, J.S., Marcal, A.R.S., Rozeira, J.: PH2 – a dermoscopic image database for research and benchmarking. In: 35th Annual International Conference of the IEEE Engineering in Medicine and Biology Society (EMBC), Osaka, pp. 5437–5440 (2013)
16. Milczarski, P., Stawska, Z.: Classification of skin lesions shape asymmetry using machine learning methods. In: Barolli, L., Amato, F., Moscato, F., Enokido, T., Takizawa, M. (eds.) WAINA 2020. AISC, vol. 1150, pp. 1274–1286. Springer, Cham (2020). https://doi.org/10.1007/978-3-030-44038-1_116
17. The International Skin Imaging Collaboration: Melanoma Project. http://isdis.net/isic-project/. Accessed 21 Mar 2020
18. Argenziano, G., Soyer, H.P., De Giorgi, V., et al.: Interactive Atlas of Dermoscopy. EDRA Medical Publishing & New Media, Milan (2002)
19. Menzies, S.W., Crotty, K.A., Ingwar, C., McCarthy, W.H.: An atlas of surface microscopy of pigmented skin lesions. Dermoscopy. McGraw-Hill, Australia (2003)
20. ImageNet. http://www.image-net.org. Accessed 07 Jan 2021
21. Milczarski, P., Beczkowski, M., Borowski, N.: Blue-White Veil classification of dermoscopy images using convolutional neural networks and invariant dataset augmentation. In: Barolli, L., Woungang, I., Enokido, T. (eds.) AINA 2021. LNNS, vol. 226, pp. 421–432. Springer, Cham (2021). https://doi.org/10.1007/978-3-030-75075-6_34
22. Milczarski, P., Wąs, Ł: Blue-White Veil classification in dermoscopy images of the skin lesions using convolutional neural networks. In: Rutkowski, L., Scherer, R., Korytkowski, M., Pedrycz, W., Tadeusiewicz, R., Zurada, J.M. (eds.) ICAISC 2020. LNCS (LNAI), vol. 12415, pp. 636–645. Springer, Cham (2020). https://doi.org/10.1007/978-3-030-61401-0_59
23. Milczarski, P., Stawska, Z., Was, L., Wiak, S., Kot., M.: New dermatological asymmetry measure of skin lesions. Int. Journal of Neural Networks and Advanced Applications, Prague, pp. 32–38 (2017)

Contextual Image Classification Through Fine-Tuned Graph Neural Networks

Walacy S. Campos[✉][iD], Luis G. Souza[iD], Priscila T. M. Saito[iD], and Pedro H. Bugatti[iD]

Department of Computing, Federal University of Technology - Parana, 1640 Alberto Carazzai Ave, Cornelio Procopio, Parana, Brazil
{walacycampos,souza.1998}@alunos.utfpr.edu.br, {psaito,pbugatti}@utfpr.edu.br

Abstract. Nowadays, computer vision techniques have become popular in several domains (e.g., agriculture, industry, medicine, and others). Their success derives from the advances in computational resources and the large volume of complex data (i.e., images). These factors led to an increase in the use of convolutional neural networks. However, such deep learning architectures do not appropriately explore the relationships between the data (e.g., images) and their respective structure. To better gather and encodes these affinity connections into a deep neural network, we can use the so-called graph neural networks (GNNs). These graph-based networks also present drawbacks. The high number of relationships in a graph can be considered a bottleneck regarding the available resources and scalability. Hence, to mitigate this issue, we propose to use GNNs automatically tuned, defining their well-suited connections according to a given image context, which improves their efficiency and efficacy. We performed experiments considering different types of state-of-the-art deep features aggregated with the GNNs. The results demonstrate that our proposed method can achieve equal accuracy (statistically) to GNNs with complete and random connections. Moreover, we decreased the number of edges to a great extent (up to 96%), testifying to our method's effectiveness.

Keywords: Deep learning · Convolutional neural network · Graph neural network · Graph pruning · Computer vision

1 Introduction

Computer vision has been used widely in different contexts like agriculture, medicine, industrial, etc. Part of this success is due to the massive volume of complex data generated daily by these domains. Moreover, with advances in computational resources (e.g., GPUs), it was possible to process (e.g., classify) all these data in an efficient way [2].

A powerful computer vision technique that was boosted by the recent advances was the so-called Convolutional Neural Network (CNN). Although CNNs present good results, they are incapable of gathering and harnessing the relationship between the data. These bonds between the data can considerably improve the effectiveness of the

© Springer Nature Switzerland AG 2021
L. Rutkowski et al. (Eds.): ICAISC 2021, LNAI 12855, pp. 15–24, 2021.
https://doi.org/10.1007/978-3-030-87897-9_2

classification process. Then, to reach this important property, it is possible to use GNNs [21] aggregated with CNNs.

As well-known, a graph structure can be highly affected by factors like the number of nodes and edges. This behavior leads to a significant drawback when using GNNs to image classification. The higher the number of edges, the higher the memory footprint and the cost to generate a suitable learning model. The cost of training a GNN is proportional to the number of edges. Moreover, we cannot generate a random or complete graph regarding image classification since it neglects semantic connections and impacts the accuracy of the model. A typical approach to solve this problem and build an appropriate graph is the brute force policy (e.g., grid-search). The problem with this strategy is a high computational cost. Some works tried to consider a priori graph (e.g., knowledge graph) to specific domains to solve this issue [4, 5, 14] but, unfortunately, in dynamic and real scenarios is almost impossible to obtain such a priori structure.

Thus, to mitigate these drawbacks, in this paper, we propose to use GNNs automatically tuned, defining their well-suited connections according to a given image context, improving their efficiency and efficacy. To do that, we based our approach according to the similarity between the graph nodes (i.e., images). Despite its simplicity, our approach achieved notable results. It significantly reduced memory footprint when using GNNs aggregated with CNNs and obtained good accuracy compared with random and complete graphs. Also, we achieve better representativeness of the semantic between an image and its objects to define the global classification context. It is worth mentioning that the literature works [6, 10, 13] tried to apply similar approaches, however disregarding GNNs aggregated with CNNs to define the global image context.

In summary, our main contributions are: i) we proposed an approach capable of tuning and reaching a well-suited GNN structure to improve the effectiveness of image classification context; ii) our approach also diminishes the computational cost of GNNs aggregated with CNNs; iii) through our tuning policy we provide greater structural semantics to the learning model; iv) since our tuning process is based on similarity, our approach can be straightforwardly extended using several literature practices that use the same concept (e.g., different distances, indexes, pruning policies, among others).

2   Background

In [18] the authors presented the seminal work regarding GNNs. They discussed issues about the traditional neural networks handling information between their features. Their work has guided many other GNNs approaches [7, 8, 16]. Indeed, GNNs were motivated by CNNs due to their ability to extract spatial features in many scales and build expressive representations [22]. In [3] the authors introduced convolutional filter usage in GNNs, generalizing the concept of CNNs applied to Euclidean spaces to non-Euclidean ones. Equation 1 formally defines the convolution operation on graphs.

$$g_\theta \star \approx \sum_{k=0}^{K} \theta' T_k(L_{sym})s \tag{1}$$

where $s \in R^n$ is a graph signal, $g\theta$ is a spectral filter, the symbol $\star$ is the convolutional operator, T_k refers to the Chebyshev polynomials, $\theta' \in R^K$ is a vector of Chebyshev coefficients, and L_{sym} represents the normalized Laplacian graph $L_{sym} := D^{-\frac{1}{2}}LD^{-\frac{1}{2}}$, considering the value for Laplacian Graph as $L := D - A$, $A = [a_{ij}]$ as the adjacency matrix non-negative and $D = diag(d1, d2, ..., dn)$ as the degree matrix of A where $d_i = \sum j a_{ij}$ is the degree of vertex i [13].

In [12] it was proposed a graph convolutional network (GCN) as defined by Eq. 2. The authors simplified the model limiting nearest neighbors value in the convolution to $k = 1$, and approximating the largest eigenvalue (λ_{max}) of L_{sym} by 2.

$$g_\theta \star s = \theta(I + D^{-\frac{1}{2}}AD^{-\frac{1}{2}})s, \tag{2}$$

where θ is considered the remained Chebyshev coefficient. The next process consists in applying a normalization trick to the convolution matrix (see Eq. 3):

$$I + D^{-\frac{1}{2}}AD^{-\frac{1}{2}} \rightarrow \widetilde{D}^{-\frac{1}{2}}\widetilde{A}\widetilde{D}^{-\frac{1}{2}}, \tag{3}$$

where $\widetilde{A} = A + I$ and $\widetilde{D} = \sum j\widetilde{A}_{ij}$.

Finally, the GCN itself can be described in Eq. 4.

$$H^{(l+1)} = \sigma(\widetilde{D}^{-\frac{1}{2}}\widetilde{A}\widetilde{D}^{-\frac{1}{2}}H^{(l)}\Theta^{(l)}) \tag{4}$$

where $H^{(l)}$ is the matrix of activations for the l-th layer, while $H^{(0)} = X$, $\Theta^{(l)} \in R^{a \times f}$ is considered as the trainable weight matrix in the layer l, and σ the activation function (i.e., $ReLU(.) = max(0, .)$) [13].

Although very promising, the work of [12] presents some issues regarding the memory footprint, and it is applied to text classification (i.e., the graph structure is intrinsically defined). Moreover, it assumes that the edges have equal relevance to generate the learning model. Hence, it neglects that in real scenarios, each connection can show a different contribution. Besides, in [13] it is described that GCNs lose their representation power when several convolutional layers are added to the architecture, and the performance complexity is prohibitive.

We believe that our approach can cope with these issues since we consider different relevant levels to edges according to the image context. Besides, through this relevance, we can remove useless edges (i.e., not contribute to gather and harness the semantic relationship between an image and its objects). In [6] the authors proposed a technique based on sharing network information to define the importance of a given edge in a tree structure regarding traditional data. However, to the best of our knowledge, our work is the first to consider a relevance mechanism in GNNs aggregated with CNNs to define the global context of images (complex data).

3   Proposed Approach

To define the well-suited connections for the GNNs according to a given image context, we present the proposed approach pipeline, as illustrated in Fig. 1. In the first step, we obtain the objects (bounding boxes) from the images of the dataset. After that, in Step 2, features of the bounding boxes are extracted through pre-trained CNNs (e.g., ImageNet transfer learning). In Step 3 the bounding boxes belonging to the same image become nodes in a complete subgraph.

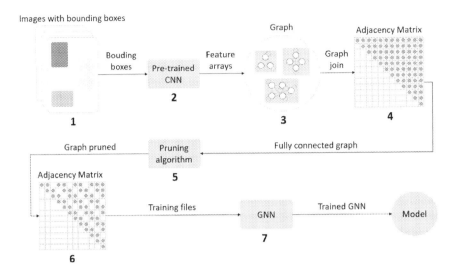

Fig. 1. Pipeline of the proposed approach.

Next, for each image is built a subgraph $G(V, E)$, where V represents the bounding boxes of the images and E their relations. Each bounding box from a given image is connected with their respective siblings (i.e., remaining bounding boxes from the same image). Then, in Step 4, an affinity matrix is generated to describe the entire graph (composed of several subgraphs). It is easy to see that the entire graph comprises several subgraphs, and the number of subgraphs is equal to the number of images in the dataset.

In Step 5, we automatically fine-tune the generated graph, defining their well-suited connections according to a given image context. To do so, we compute the relevance of each vertex according to the mean distance regarding its incident vertices. It is worth mentioning that we assign for each edge a relevance based on the dissimilarity between its nodes. The vertex's relevance is used as a threshold (τ) to suppress useless incident edges. Equation 5 formally defines the threshold setting. Figure 2 illustrates an example of the complete graph (left graph), and it is respectively fine-tuned one (right), obtained after applying our proposed approach suppressing edges according to the automatic threshold.

$$\tau = \frac{\sum_{k=0}^{l-1} w(n'_k, n''_k)}{l} \tag{5}$$

where l represents the number of edges, k is an iterator, $r(v'_k, v''_k)$ is a function to obtain the relevance of the edge connecting vertices v'_k and v''_k.

After, in Step 6, our approach creates the affinity matrix of the fine-tuned graph. Finally, in Step 7, this matrix is the input to train a GNN, generating a learning model.

In this paper, we create an instance of our proposed approach to calculate the edge's relevance through the well-known Euclidean distance. However, our approach can be straightforwardly extended to different kinds of distance functions, as well as tuning

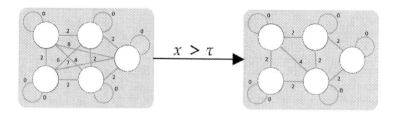

Fig. 2. Illustration of the tuning process in a fully connected graph

mechanisms. Algorithms 1 and 2 detail our proposed approach and the tuning mechanism considered in the present paper.

An important property of our proposed approach is that we do not need to know the bounding boxes labels to predict the global context of the image. As aforementioned in Sect. 2, an a priori knowledge graph is costly and tiresome to obtain. Hence, a complete graph is easier to build. However, it also demands a high computational cost and can link incoherent objects, confusing the learning model.

Algorithm 1: Proposed approach

---

input   : set of images S, a given CNN C
output   : learning model M
auxiliaries: sets of extracted features Z_i, training sets Z', testing sets Z'', a vanilla GNN
$\quad\quad\quad\quad$ G, a fine-tuned GNN G^Ω

for *each* $S_j, j = 1, 2, ..., |S|$ **do**
$\quad$ $B \leftarrow$ boundingBoxes(S_k);
$\quad$ **for** *each* $B_i, i = 1, 2, ... |B|$ **do**
$\quad\quad$ $Z \leftarrow Z \cup$ deepFeatures(B_i, C);

Split Z;
$Z' \leftarrow$ random training samples from Z;
$Z'' \leftarrow$ random testing samples from $Z \setminus Z'$;
$G \leftarrow$ completeGraph(Z', Z'');
$G^\Omega \leftarrow$ fineTunedGraph(G.graph);

---

4 Experiments

This section discusses the scenarios of the experiments, describes the image dataset used in the experiments, and presents the discussion about the obtained results comparing our proposed approach against complete and random graphs, respectively.

Algorithm 2: Tuning mechanism

---

input   : an undirected graph G
output   : fine-tuned undirected graph G^{Ω}
auxiliaries: $\mathcal{V}$: set of vertices from the input graph, $\mathcal{A}$: list of edges incident to vertex v, τ: threshold to suppress irrelevant edges

for *each* $v \in V$ **do**
    $A \leftarrow getIncidentEdges(v)$
    $\tau \leftarrow calculateMeanDistance(A)$
    foreach $a \in A$ **do**
        if $edgeRelevance(a) > \tau$ **then**
            $removeEdge(a)$

4.1 Scenarios

To obtain the best hyperparameters to the GCN we executed a grid-search strategy [1]. To do so, we defined the number of hidden layers (16, 64, 256), epochs (2000), learning rate (0.001, 0.005, 0.01, 0.05), and dropout (0.3, 0.5, 0.8, 0.9). These possibilities of hyperparameters resulted in 384 experiments.

We combined the GNN with different CNN architectures. Then, we used Efficient-NetB7 [20], InceptionV3 [19], ResNet50 [9] and VGG19 [17]. Finally, to corroborate the effectiveness of our approach we compared it against a fully connected and random graphs. As optimizer we used ADAM [11].

To perform the experiments, we used a computer with Intel Core i7 from the 6th generation with 8 cores, 16 threads, and 3.40 GHz; 32 GB of RAM and an Nvidia GeForce RTX 2080Ti GPU, with 4352 CUDA cores. The experiments were executed in GPU mode.

4.2 Dataset Description

For the experiments, the MIT67 [15] dataset was chosen because it provides the requirements for the development of this work, as the: images, bounding boxes of each image, and global classes (image classes).

The MIT67 is a dataset to solve indoor scene recognition, including 67 different classes. However, because some classes have few examples, annotation errors, missing data, or images without bounding boxes, data cleaning was required, which resulted in the exclusion of the following classes: *auditorium, bowling, elevator, jewelry shop, locker room, hospital room, restaurant kitchen, subway, laboratory wet, movie theater, museum, nursery, operating room, waiting room*. Thus, obtaining 53 classes, 2607 images, and 50.8 68 bounding boxes, that were divided into the training (80% of the data) and test (20%) sets in a random and stratified way.

Table 1 shows the distribution of images/objects by class and bounding boxes by image. It is possible to note that the dataset is unbalanced, having a high number of samples for some classes, such as: "kitchen" (308 images, 7511 bounding boxes), "bedroom" (350, 5112); and low numbers for others like "cloister" (16, 159) and "winecellar" (16, 222).

Table 1. MIT67 dataset distribution.

Analysis	Mean	Std	Min	Median	Max
Images per class	49.2	68.1	16	19	350
Objects per class	959.8	1311.7	159	475	7511
Objects per image	18.8	12.7	1	17	95

4.3   Results

To perform the experiments, we trained GCNs aggregated with the four CNNs architectures cited in Sect. 4.1. Tables 2 and 3 show the obtained results regarding the fully connected graph against our approach, respectively. We calculated metrics considering the efficacy, such as accuracy, precision, recall, and F1, to analyze the results. We also show the total number of edges generated and the dimensionality of the feature vectors used to represent the nodes.

Table 2. Results for top 1 performance for fully connected graph on MIT67 dataset.

	Fully connected graph on MIT67 dataset					
Architecture	Accuracy	Precision	Recall	F1	Edges (10^5)	Dimension
EfficientNetB7	69.90	58.68	58.30	56.33	13.95	2560
InceptionV3	63.47	55.67	56.10	52.61	13.95	2048
ResNet50	64.89	56.57	52.69	52.22	13.95	2048
VGG19	57.81	47.90	44.80	42.98	13.95	512

Analyzing Tables 2 and 3 we can see that our approach statistically ties with the vanilla graph (fully connected graph). Moreover, our approach decreased to a great extent the number of edges of the graph.

For instance, the fully connected graph and our approach presented 13.95×10^5 and 7.13×10^5, respectively, regarding the ResNet50. Thus, our approach reduced the memory footprint up to 96%. We observed this same behavior when analyzing the other CNNs aggregated with our GNN. According to the results, our approach with EfficientNetB7, InceptionV3, and VGG19 achieved reductions of up to 73%, 83%, and 51%, respectively.

To better visualize all the considered metrics, in Figs. 3 (a) and (b), we generated the radar plots with the obtained results regarding the fully connected graph and our approach, respectively. It is clear to note that our approach accomplished a considerable edge reduction while maintaining efficacy.

We also performed experiments considering random connections. To create the random edges and obtain a fair comparison, we used the same number of edges obtained by our approach. It is important because using a higher or lower number of random edges, when compared with ours, could lead to a false degeneration of the graph or a false improvement (higher computational cost). Thus, our approach can also answer which is the best number of edges to reach a suitable trade-off between efficacy and efficiency.

Table 3. Results for top 1 performance considering our approach on MIT67 dataset.

| Architecture | Our approach on MIT67 dataset | | | | | |
	Accuracy	Precision	Recall	F1	Edges (10^5)	Dimension
EfficientNetB7	67.58	58.18	57.37	57.37	8.08	2560
InceptionV3	59.10	50.94	48.07	47.49	7.62	2048
ResNet50	58.21	46.93	42.67	42.50	7.13	2048
VGG19	53.33	45.17	42.06	40.91	9.26	512

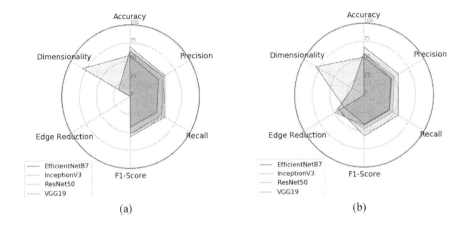

Fig. 3. Radar plot using different CNN architectures; (a) Fully connected graph; (b) Pruned graph.

The experiment regarding the random edges achieved an accuracy of up to 60% with InceptionV3. The same behavior, where our approach presented the best results, was observed regarding the other CNNs. This testifies that our approach effectively defines the edges' relevance, capturing the semantic relationship between the objects of an image to define its global context.

Thus, considering the obtained results, we can argue that our proposed approach was capable of automatically tunning the GNN graph structure aggregated with a given CNN, defining well-suited connections according to a given context, improving the entire process.

5   Conclusions

In this paper, we proposed an approach capable of automatically fine-tuning a given GNN. To do so, it defines the well-suited connections according to a given image context, improving the effectiveness of the entire process. Our approach was based on the similarity between the graph nodes (i.e., images).

Despite its simplicity, it reached considerable results. It not only successfully reduced the memory footprint when using GNNs aggregated with CNNs, but also maintained good accuracies when compared with random and complete graphs. This testifies that our approach achieved better representativeness of the semantic between an image and its objects to define the global classification context. Our results showed that the fine-tuned GNN reached up to 96% regarding the memory footprint while maintaining the accuracy.

For future works, we intend to explore other image datasets. We also aim to extend our approach regarding different distances and detection mechanisms for edge relevance.

References

1. Andonie, R.: Hyperparameter optimization in learning systems. J. Membr. Comput. 1–13 (2019). https://doi.org/10.1007/s41965-019-00023-0
2. Bronstein, M.M., Bruna, J., Lecun, Y., Szlam, A., Vandergheynst, P.: Geometric deep learning: going beyond euclidean data. IEEE Sig. Process. Mag. **34**(4), 18–42 (2017)
3. Bruna, J., Zaremba, W., Szlam, A., LeCun, Y.: Spectral networks and deep locally connected networks on graphs. In: Proceedings of the International Conference on Learning Representations, pp. 1–14 (2014)
4. Chen, X., Gupta, A.: Spatial memory for context reasoning in object detection. CoRR abs/1704.04224 (2017)
5. Chen, X., Li, L., Fei-Fei, L., Gupta, A.: Iterative visual reasoning beyond convolutions. CoRR abs/1803.11189 (2018)
6. Chung, H., Huang, H., Chen, H.: Predicting future participants of information propagation trees. In: Proceedings of the IEEE/WIC/ACM International Conference on Web Intelligence, pp. 321–325 (2019)
7. Gallicchio, C., Micheli, A.: Graph echo state networks. In: Proceedings of the International Joint Conference on Neural Networks, pp. 1–8 (2010)
8. Gori, M., Monfardini, G., Scarselli, F.: A new model for learning in graph domains. In: Proceedings of the IEEE International Joint Conference on Neural Networks, vol. 2, pp. 729–734 (2005)
9. He, K., Zhang, X., Ren, S., Sun, J.: Deep residual learning for image recognition. CoRR abs/1512.03385 (2015)
10. Hong, L., Zou, L., Lian, X., Yu, P.S.: Subgraph matching with set similarity in a large graph database. IEEE Trans. Knowl. Data Eng. **27**(9), 2507–2521 (2015)
11. Kingma, D.P., Ba, J.: Adam: a method for stochastic optimization. arXiv preprint arXiv:1412.6980 (2014)
12. Kipf, T.N., Welling, M.: Semi-supervised classification with graph convolutional networks. In: Proceedings of the International Conference on Learning Representations, pp. 1–14 (2019)
13. Li, Q., Han, Z., Wu, X.: Deeper insights into graph convolutional networks for semi-supervised learning. CoRR abs/1801.07606 (2018)
14. Marino, K., Salakhutdinov, R., Gupta, A.: The more you know: Using knowledge graphs for image classification. In: Proceedings of the IEEE Conference on Computer Vision and Pattern Recognition, pp. 20–28 (2017)
15. Quattoni, A., Torralba, A.: Recognizing indoor scenes. In: Proceedings of the IEEE Conference on Computer Vision and Pattern Recognition, pp. 413–420 (2009)
16. Scarselli, F., Gori, M., Tsoi, A.C., Hagenbuchner, M., Monfardini, G.: The graph neural network model. IEEE Trans. Neural Netw. **20**(1), 61–80 (2009)
17. Simonyan, K., Zisserman, A.: Very deep convolutional networks for large-scale image recognition. arXiv:1409.1556 (2014)

18. Sperduti, A., Starita, A.: Supervised neural networks for the classification of structures. IEEE Trans. Neural Netw. **8**(3), 714–735 (1997)
19. Szegedy, C., Vanhoucke, V., Ioffe, S., Shlens, J., Wojna, Z.: Rethinking the inception architecture for computer vision. CoRR abs/1512.00567 (2015)
20. Tan, M., Le, Q.V.: Efficientnet: rethinking model scaling for convolutional neural networks. CoRR abs/1905.11946 (2019)
21. Wu, Z., Pan, S., Chen, F., Long, G., Zhang, C., Yu, P.S.: A comprehensive survey on graph neural networks. CoRR abs/1901.00596 (2019)
22. Zhou, J., Cui, G., Zhang, Z., Yang, C., Liu, Z., Sun, M.: Graph neural networks: a review of methods and applications. CoRR abs/1812.08434 (2018)

Architecture Monitoring and Reliability Estimation Based on DIP Technology

Faisal Mehmood Shah[1] , Zohaib Mehmood Shah[2] , Sarmad Maqsood[3] ,
Robertas Damasevicius[3(✉)] , Muhammad Ali Shahzad[4] ,
Michał Wieczorek[5] , and Marcin Woźniak[5]

[1] Pakistan Space & Upper Atmosphere Research Commission, Karachi, Pakistan
[2] Research Center for Modeling and Simulation,
National University of Sciences and Technology, Islamabad, Pakistan
[3] Department of Software Engineering, Kaunas University of Technology,
Kaunas, Lithuania
{sarmad.maqsood,robertas.damasevicius}@ktu.lt
[4] School of Nuclear Science and Engineering, North China Electric Power University,
Beijing, China
[5] Faculty of Applied Mathematics, Silesian University of Technology, Gliwice, Poland
marcin.wozniak@polsl.pl

Abstract. All civil infrastructure units demand regular inspection to avoid functional and structural damages. Periodic examinations are in accordance with classification society's standards which contain both non-destructive tests and visual surveys, to search structural damage, reliability, cracks, thickness measurement, and Water dripping generally documented by manually or with measurements tape. But, it is very hard to search cracks by visually monitoring much larger structures. Hence, the advent of crack detecting and monitoring systems has been a major issue. In this proposed study a crack detection algorithm in reference to digital image processing technology is suggested. Obtaining information of a surface crack by using an image pre-processing pipeline and to estimate the failure is proposed which helps and detects the structure cracks and body health information. It provides the identification system more portable and integrated, estimates the crack more precisely and reduction in expenditure as well. The proposed algorithm accuracy is 93.8% as compared to the traditional and recent work.

Keywords: Crack detection · Image smoothing · Morphological processing · Structural inspections.

1 Introduction

Civil infrastructure components need regular inspection every two years to ensure public safety. The durability of structure is a significant element not only in construction also in the maintenance aspect. The consequence of safety

© Springer Nature Switzerland AG 2021
L. Rutkowski et al. (Eds.): ICAISC 2021, LNAI 12855, pp. 25–35, 2021.
https://doi.org/10.1007/978-3-030-87897-9_3

is risen as the construction of super long span bridges, asymmetric and high rise buildings are populations. Therefore, construction safety control systems are being developed these days. Traditionally, structure assessment is carried out by long ladders, vehicles, or ropes for access, the methods associated are dangerous, expensive and challenging shown in Fig. 1 [1].

Fig. 1. Traditional method of inspection.

The need of the moment is to establish an efficient infrastructure assessment management system which can contribute to construction authorities, allowing them to make best decisions about maintenance. By establishing an efficient infrastructure management system, there is a possibility to secure stability, minimize the inspector's role, reduce maintenance cost and inspection time. Furthermore we can analyze the condition and structural health by processing and acquiring the data [2]. Compared with traditional inspection approaches today, visual inspection is a new primary way of structure assessment to solve this problem. Because, scientific attention has emerged in the use of unmanned aerial vehicles (UAVs) means to deliver cheaper, faster, smoother data acquisition and safer, in addition digital record compared to personal visual assessment. In this way a surveyor performs inspection by camera based monitoring or fixes sensors. Majority of the constructions/infrastructures are made from concrete [3]. In these constructions, the one way to judge structural health is to investigate cracks on the external surface and the body of the structure [4]. There are various cracks caused among which are fractures, transverse cracks, massive cracks and longitudinal cracks. If cracks are not timely looked after and repaired, they will become increased seriously on snowy and rainy days in case of bridges they will greatly affect vehicles safety [5]. So the health of the structure can be directly and easily determined by monitoring the surface crack, the crack evaluation must take place on a regular basis to ensure safety and durability within its lifespan [10]. At present several studies are conducted about automated crack detection methods. In [11] an unsupervised method for crack detection on the road surface is performed by using Otsu thresholding. Using image processing technique is mentioned in [13]. Two stage algorithm based on CNN and edge detection is in [14,15] and other models of deep learning [8,22]. Many methods involve image feature extraction and indexing [6,7]. For shape extraction and classification often are also used fuzzy rules [9]. Computer vision based an improved I-UNet Convolutional Networks to extract different scale of features and carry out multi-scale feature fusion in [12,16] and crack detection in Pyramid are [17]. A deep learning approach, test of concrete surface and analysis of crack on bridge structure is performed in [18]. The need of the hour is to establish an effective, skillful and practical infrastructure assessment management system which can help classification societies for monitoring structure work and also enable them for better decision-making with respect to maintenance planning. To solve this problem, this paper objective is to develop an image processing algorithm pipeline that Illustrates workflow for data quality determination, data acquisition and has a major impact on the current approaches. To address these issues, the aim of this analysis is to explore the application of a remote sensing technique by using image processing algorithms to help and detect the structure cracks and body health information. The method of identifying cracks and system process is shown Fig. 2. Section 2 describes the complete methodology of a proposed approach.

2   Automatic Detection Algorithm

This section presents our proposed algorithm which comprises of five phases that include grayscale image closing, image enhancement, crack regions rapid extraction, morphological noise removal, and calculation of cracks areas as depicted in Fig. 1. These phases are detailed in the following subsections.

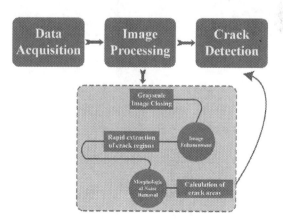

Fig. 2. Image processing flow chart.

2.1   Grayscale Image Closing

Image closing technique belongs to a morphological operations family, generally, these measures are implemented to binary images (i.e. which and black); but, grayscale equivalents could be used. Applying these measures need a structure element that can be considered as a stencil which is slid within the image for specific effects to occur.

The resulting consequences rely on the operating executing, in addition to type and size of the structuring element. Image closing maintains the background areas that are similar in shape that the structure elements are using, or that could entirely restrain the structuring element themselves. Other portions that do not match the criteria cannot be further processing. A complete history about the effect of image closing is shown in [19]. The developed algorithm for detection of the crack uses line shaped constructing elements, and excludes the area from an image that is not line-shaped. This operation is rarely seen in the grayscale. Overall five different line constructing elements of magnitude 3 pixels oriented 0, 45, 90, −45 and 60° were used. Grayscale closing approach effect of the original picture is shown in fig. In Fig the closing image appears to be blurred, and the small circular dots in Fig. 3(a) are not so visible like in Fig. 3(b). This practice minimizes the noise in the picture and provides far better detection capabilities.

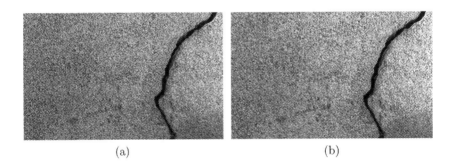

(a)   (b)

Fig. 3. (a) Original image (b) After applying grayscale image closing.

2.2   Image Enhancement

During the image closing process or image generation, transformation or transmission. The quality of the picture devalued and degraded because of many factors like illumination, so image enhancement is needed. In order to process the pixel points whose orbital picture is distributed and highly concentrated equally on every period to increase the contrast of orbital picture, the topology is used to histogram equalization because less level of gray loss to process him.

Image contrast can be viewed from the distribution of the histogram for gray picture. After that histogram equalization action the pixels are evenly distributed in the entire gray scale limit, hence demonstrating that the processing picture has strong contrast, also non-track surface and track surface portion both obtain a good distinction.

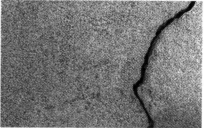

Fig. 4. Image enhancement.

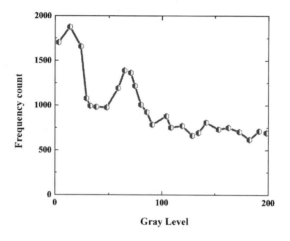

Fig. 5. Enhanced gray histogram.

2.3 Crack Regions Rapid Extraction

Before the picture understanding and recognition, the picture of the crack image is segmented, towards the goal is to examine it separated from the complex background, and the picture feature to the research target is selected. Picture segmentation approach plays a crucial role for picture understanding and recognition.

In this study, shadow region and crack coefficient are segmented from the picture through coefficient of variation and resemblance among pixels are calculated to additionally distinguish the shadow region and crack region. The resemblance calculation results are using adaptively adjusted, mean scale smoothing to determine the background picture model. The original picture is eminent with the background picture. Finally, fusion of an iterative and custom approach applies to calculate the threshold and reach the crack extraction region.

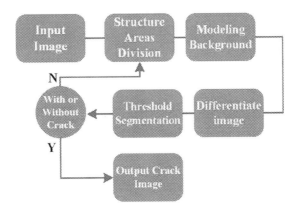

Fig. 6. Block diagram model of a crack extraction workflow.

Whenever necessary to compare the two sets of data for the degree of dispersion, if the two sets of data measurement scales differ too much, or different data size, it is not appropriate to use standard deviation directly for comparison. In that case, the measurement scale must be eliminated, the dimension influence and variation or coefficient can make it, and it is the ratio among standard deviations of average of original date to the original data. The coefficient variation e.g. standard deviation, range and variance, are absolute numbers that can reflect the data dispersion degree. The data size affects as well as the variable values of the degree of dispersion, but also the intermediate-level of the variable values. Extracted crack picture is shown in Fig. 7.

2.4 Morphological Noise Removal

In algorithm two ways to remove the blobs. Firstly, image blobs area can normalize into range 0–1 by utilizing their area properties. Secondly, identifying noise blobs in accordance with predefined blob area, 0.25 threshold value. As in, any blob area below this threshold is eliminated out of the picture. This threshold arbitrary value determined assumption that blob related with concrete cracks would be a wider area with noisy blobs. The outcome from the image processing detection algorithm is shown in Fig. 8(a). The results show by utilizing the arbitrary threshold 0.25, a small portion of crack is eliminated; but, this does not affect the outcomes from the following classification crack algorithm procedure. Figure 8(b) demonstrated the overlaid crack mask on the primary image.

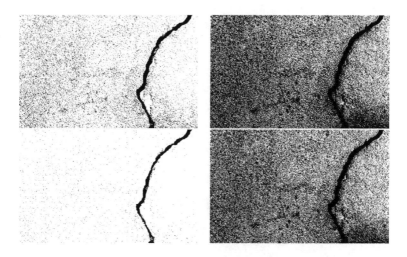

Fig. 7. Crack extraction images.

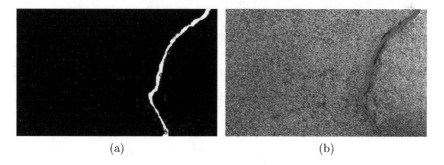

(a)   (b)

Fig. 8. (a) Crack detection final mask picture (b) Overlaid on original picture.

2.5   Crack Areas Calculation

Area is a sort of appropriate measurement regarding the whole dimension of an object. Within the segmented picture, the region of the goal means the amount of pixels concerning the goal's edge that are unassociated about the gray lever of each point however the size from that goal. Area only considers the goal's weight and mass. Scan the total region from goal R and calculate precisely pixels whose "1" is gray lever.

The most effective way of calculating the area is to count the number of pixels within space of the border (containing border itself), its description is Eq. (1).

$$A = \sum_{X=1}^{n} \sum_{Y=1}^{m} F(X, Y). \tag{1}$$

The rate of crack detection and the accuracy detection are determined in Eq. (2) and Eq. (3), respectively.

$$C_R = \frac{I_{CR}}{I_{OR}} \times 100\%, \qquad (2)$$

$$C_A = \frac{I_{CR}}{I_C} \times 100\%, \qquad (3)$$

where I_{CR} represents the length of detected cracks and I_C represents the length of all detected objects. It can be acquired from the automatic detecting results. I_{OR} denotes the original cracks length, which is calculated by on-the-spot inspection.

2.6   Crack Perimeter Calculation

Perimeter is the entire from its outer boundary or edge which encircle all of the pixels in the picture. In the normal calculations: defined by the area of edge, in particular the sum of the edge points.

2.7   Crack Circularity Calculation

A is the area, P parameter. Once an object is in a circle, C= 1 in case of a lathy area, C>1, the roundness could be considered for measurement to estimate how rough or complicated the object is. If the measured roundness is to a specified extent, we ensure it is a crack. In Fig. 9 Crack picture after binarization, data is shown.

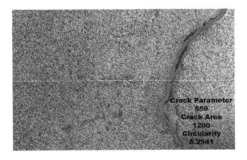

Fig. 9. Indentify result.

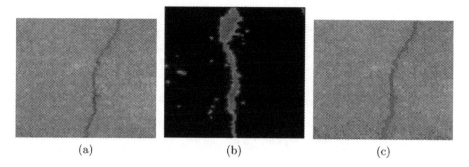

(a)   (b)   (c)

Fig. 10. (a) Original image from [20] (b) Algorithm results of [20] (c) Our proposed algorithm result.

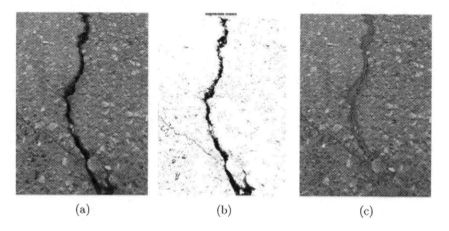

(a)   (b)   (c)

Fig. 11. (a) Original image from [21] (b) Algorithm results of [21] (c) Our proposed algorithm result.

3 Preliminary Experimental Result Comparison

The experimental setup is performed on Apple MacBook Pro 2020, 16 GB RAM and M1 processor in Python environment. To check the robustness of the proposed algorithm we come to results with [20,21]. The dataset used in this experiment is taken from [20,21]. The results clearly show the effectiveness of the algorithm. A comparative analysis with their approach and our suggested method are shown in Figs. 10 and 11. Figure 12 illustrate the average accuracy and efficiency of a proposed algorithm with comparison [20,21] which shows the practicality of our work.

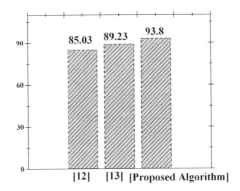

Fig. 12. Average accuracy and efficiency (%).

4 Conclusions

With the purpose of to estimate the safety of a concrete structure, a methodology to discover and identify the cracks has been suggested. In this study we implement a crack detection algorithm for structure assessment is performed by using an image preprocessing pipeline. It has been made possible to detect and visualize the structure crack smoothly by using this approach. The experiment results show that the algorithm identifies and classifies cracks in comprehensive manners and the proposed study results performance are better compared to conventional methods.

References

1. Chen, S., Laefer, D.F., Mangina, E., Zolanvari, S.I., Byrne, J.: UAV bridge inspection through evaluated 3D reconstructions. J. Bridge Eng. **24**(4), 05019001 (2019)
2. Guédé, F.: Risk-based structural integrity management for offshore jacket platforms. Mar. Struct. **63**, 444–461 (2019)
3. Lv, Y., et al.: Quality control of the continuous hot pressing process of medium density fiberboard using fuzzy failure mode and effects analysis. Appl. Sci. **10**(13), 4627 (2020)
4. Urbonas, A., Raudonis, V., Maskeliūnas, R., Damaševičius, R.: Automated identification of wood veneer surface defects using faster region-based convolutional neural network with data augmentation and transfer learning. Appl. Sci. **9**(22), 4898 (2019)
5. Capizzi, G., Lo Sciuto, G., Woźniak, M., Damaševicius, R.: A clustering based system for automated oil spill detection by satellite remote sensing. In: Rutkowski, L., Korytkowski, M., Scherer, R., Tadeusiewicz, R., Zadeh, L.A., Zurada, J.M. (eds.) ICAISC 2016. LNCS (LNAI), vol. 9693, pp. 613–623. Springer, Cham (2016). https://doi.org/10.1007/978-3-319-39384-1_54
6. Grycuk, R., Wojciechowski, A., Wei, W., Siwocha, A.: Detecting visual objects by edge crawling. J. Artif. Intell. Soft Comput. Res. **10**(3), 223–237 (2020)

7. Grycuk, R., Najgebauer, P., Kordos, M., Scherer, M.M., Marchlewska, A.: Fast image index for database management engines. J. Artif. Intell. Soft Comput. Res. **10**(2), 113–123 (2020)

8. Guo, L., Woźniak, M.: An image super-resolution reconstruction method with single frame character based on wavelet neural network in internet of things. Mobile Netw. Appl. **26**, 1–14 (2020)

9. Korytkowski, M., Scherer, R., Szajerman, D., Połap, D., Woźniak, M.: Efficient visual classification by fuzzy rules. In: 2020 IEEE International Conference on Fuzzy Systems (FUZZ-IEEE), July 2020

10. Ma, Z., Liu, S.: A review of 3D reconstruction techniques in civil engineering and their applications. Adv. Eng. Inform. **37**, 163–174 (2018)

11. Mubashshira, S., Azam, M.M., Ahsan, S.M.M.: An unsupervised approach for road surface crack detection. In: 2020 IEEE Region 10 Symposium (TENSYMP), pp. 1596–1599 (2020)

12. Połap, D., Woźniak, M. Bacteria shape classification by the use of region covariance and convolutional neural network. In: 2019 International Joint Conference on Neural Networks (IJCNN), July 2019

13. Shifani, S.A., Thulasiram, P., Narendran, K., Sanjay, D.R.: A study of methods using image processing technique in crack detection. In: 2020 2nd International Conference on Innovative Mechanisms for Industry Applications (ICIMIA), pp. 578–582 (2020)

14. Wang, G., Liu, Y., Xiang, J.: A two-stage algorithm of railway sleeper crack detection based on edge detection and CNN. In: 2020 Asia-Pacific International Symposium on Advanced Reliability and Maintenance Modeling (APARM), pp. 1–5 (2020)

15. Kumar, B., Ghosh, S.: Detection of concrete cracks using dual-channel deep convolutional network. In: 2020 11th International Conference on Computing, Communication and Networking Technologies (ICCCNT), pp. 1–7 (2020)

16. Wang, L., Ye, Y.: Computer vision-based Road Crack Detection Using an Improved I-UNet convolutional networks. In: 2020 Chinese Control and Decision Conference (CCDC), pp. 539–543 (2020)

17. Yang, F.: Feature pyramid and hierarchical boosting network for pavement crack detection. IEEE Trans. Intell. Transp. Syst. **21**(4), 1525–1535 (2020)

18. XingQi, G., Quan, L., MeiLing, Z., HuiFeng, J.: Analysis and test of concrete surface crack of railway bridge based on deep learning. In: 2020 IEEE 5th Information Technology and Mechatronics Engineering Conference (ITOEC), pp. 437–442 (2020)

19. Sundararajan, D.: Edge detection. In: Digital Image Processing, pp. 257–280. Springer, Singapore (2017). https://doi.org/10.1007/978-981-10-6113-4_9

20. Yuhan, Z., Juan, Q., Zhiling, G., Kuncheng, J., Shiyuan, C.: Detection of road surface crack based on PYNQ. In: 2020 IEEE International Conference on Mechatronics and Automation (ICMA), Beijing, China, pp. 1150–1154 (2020)

21. Ahmad, A.R., Osman, M.K., Ahmad, K.A., Anuar, M.A., Yusof, N.A.M.: Image segmentation for pavement crack detection system. In: 2020 10th IEEE International Conference on Control System, Computing and Engineering (ICCSCE), Penang, Malaysia, pp. 153–157 (2020)

22. Woźniak, M., Wieczorek, M., Siłka, J., Połap, D.: Body pose prediction based on motion sensor data and recurrent neural network. IEEE Trans. Ind. Inform. **17**(3), 2101–2111 (2020)

An Efficient Technique for Filtering of 3D Cluttered Surfaces

Piyush Joshi[1,2(✉)], Alireza Rastegarpanah[1,2], and Rustam Stolkin[1,2]

[1] The Faraday Institution, Quad One, Harwell Science and Innovation Campus, Didcot, UK

[2] School of Metallurgy and Materials, University of Birmingham, Birmingham B15 2TT, UK

{a.rastegarpanah,r.stolkin}@bham.ac.uk

Abstract. 3D Object recognition is a rapidly growing research field in the area of computer vision. The presence of cluttered surfaces is the main obstacle to object recognition. In this paper, we propose to present a technique for automatic filtering of cluttered surfaces. First, the technique clusters a point cloud and then based on three features that are size, distance and spatial information, cluttered surfaces are separated. To the best of our knowledge, this is the first technique that can remove any cluttered surface (plane or irregular) from a point cloud. We have experimented on two complex RGBD datasets containing heavily cluttered surfaces and using the proposed metric, we measure the remaining cluttered surfaces after filtering of a point cloud. The proposed clutter filtering has removed 87.60% and 89.68% cluttered surfaces for Challenge and Willow datasets respectively.

Keywords: Cluttered surface · Cluttered surface filtering · 3D object recognition.

1 Introduction

Nowadays, 3D object recognition has become an active research topic due to its advantages over 2D recognition and the availability of low-cost 3D cameras e.g., Microsoft Kinect, Intel RealSense etc. [2,8,10,12] . These 3D cameras are lightweight, small and can provide high-resolution 3D data in real-time. The role of a 3D recognition technique is to find an object in a scene acquired by a 3D camera. The computational cost of object recognition technique mainly depends on size of a scene [5]. Generally, a scene acquired by a 3D camera consists of different cluttered surfaces other than objects. For example, a scene may contain a table (plane surface) and background (irregular surfaces) along with objects

This research was conducted as part of the project called "Reuse and Recycling of Lithium-Ion Batteries" (RELIB). This work was supported by the Faraday Institution [grant number FIRG005].

P. Joshi and A. Rastegarpanah are identified as joint lead authors of this work

© Springer Nature Switzerland AG 2021
L. Rutkowski et al. (Eds.): ICAISC 2021, LNAI 12855, pp. 36–43, 2021.
https://doi.org/10.1007/978-3-030-87897-9_4

such as bottles, cups, boxes etc. These surfaces increase the search time of a recognition technique. Moreover, they may resemble with surfaces of an object which can lead to false recognition. Therefore, filtering of cluttered surfaces before recognition can improve the efficiency and reduce the computation cost of searching [7,9].

Cluttered surfaces are divided into two categories of plane and irregular surfaces. A widely used plane surface filtering technique is random sample consensus (RANSAC) [6]. This method estimates parameters of a mathematical model from data. It first randomly selects the smallest number of elements (3 points for the plane detection) that can form the mathematical model from the input data, and use these elements to calculate the parameters of the corresponding model. Further, all points are tested with this estimated model and points that are greater than a threshold are considered as outliers, otherwise, they are regarded as inliers. After defined iterations, a shape that consists of the largest number of inliers is considered as a final output. A computational efficient version of RANSAC is proposed in [13]. A 3D Hough transform based plane extraction technique is proposed in [3]. This technique is mainly based on voting rule and estimate shape parameters using collected votes. A randomized Hough transform [4] is an efficient version of Hough transform that is effective for plane detection in point clouds. There is one effort that has been made in [5] towards automatic filtering of plane surfaces. This technique first transforms normal vectors into Gaussian space. Further, density-based clustering is utilized to cluster the Gaussian sphere and major parallel planes are removed from the point cloud.

The above-discussed techniques have performed well in the removal of plane surfaces. However, along with plane surfaces, irregular surfaces are also part of a scene. An example of a scene is presented in Fig. 1 that consists of objects, plane and irregular surfaces. More than plane surfaces, irregular surfaces can be an obstacle to recognition techniques as they cause false recognition. Therefore, there is a need to remove these surfaces along with plane surfaces. In this paper, we have made the first effort to remove all cluttered surfaces.

2   Cluttered Surface Filtering

A framework of the proposed filtering of cluttered surfaces is presented in Fig. 2. The technique first removes plane surfaces using RANSAC algorithm and clusters a point cloud. Then, it computes a bounding box for each cluster. Further, features based on size, distance and spatial information of each bounding box are computed to classify irregular cluttered and non-cluttered surfaces.

2.1   Plane Surface Filtering

For filtering of plane surfaces, we have utilized RANSAC algorithm [6]. This algorithm fits a model to hypothetical inliers (a random subset of the original data). Further, the fitted model is used to test all data using threshold T_f. Points

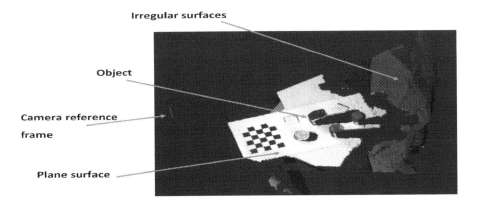

Fig. 1. A representation of plane and irregular surfaces in a point cloud

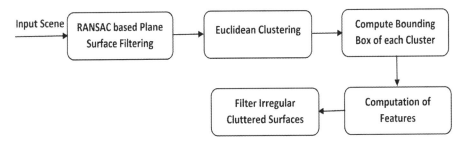

Fig. 2. A framework of cluttered surfaces filtering

that are fit the estimated model are considered as a part of the consensus set. This procedure is repeated a fixed number of times say I_{max}. Figure 3 shows a filtered point cloud by RANSAC algorithm. It is observed that plane surfaces are detected and removed well by RANSAC. However, we can also see that a major part of cluttered surfaces (i.e. irregular surfaces) are still present in the scene along with objects. Therefore, filtering of irregular cluttered surfaces with plane surfaces is an important task.

2.2 Euclidean Clustering

We divide a point cloud into different regions for filtering of irregular cluttered surfaces. The proposed technique utilizes a Euclidean Cluster algorithm proposed in [11] for clustering of a point cloud. For clustering, this technique has considered neighbors of a point that have a radius less than a threshold d_{th}. After computation of clusters, we have generated a bounding box of each cluster for estimation of features such as size, distance and spatial information. The smallest bounding box for a point set is the box with the smallest measure (volume, area or hypervolume) within which all the points lie. In this paper, we have

(a) A scene before filtering   (b) A scene after filtering

Fig. 3. An example of RANSAC plane surface filtering

utilized the axis-aligned bounding box (AABB) method to generate bounding boxes of each cluster [12].

2.3   Computation of Features

We have utilized size, distance and spatial information of each bounding box to remove irregular cluttered surfaces. These features are computed as follows.

Feature Based on Size. Generally, the size of an object is quite different from other cluttered surfaces. For example, size of a bottle is smaller than the size of a table or background surfaces (Fig. 1). We have considered the difference in size to separate an object from irregular cluttered surfaces. In order to estimate a size feature (S), we have computed the volume of a bounding box. A normalized size feature S_i for i-th bounding box is given as.

$$S_i = \frac{(H_i \times W_i \times L_i) - (H_{min} \times W_{min} \times L_{min})}{(H_{max} \times W_{max} \times L_{max}) - (H_{min} \times W_{min} \times L_{min})} \tag{1}$$

where H, W and L are height, width and length of a bounding box respectively.

Feature Based on Spatial Information. It is observed that a non-cluttered surface consists of fewer neighboring surfaces than irregular cluttered surfaces. For example, an object has limited neighboring clusters, however, an irregular background contains many surfaces close to each other. Therefore, we have utilized spatial information (I) as a number of neighbors around each cluster within a radius r_i. Figure 4 shows a concept of spatial information of clutter (irregular surfaces) and non-clutter (object). In this figure, a red box indicates a bounding box for a cluster. We can clearly see that a bounding box of an irregular surface (i.e. background) has many neighboring bounding boxes while a bounding box of an object has a very less number of neighboring boxes. Therefore, an irregular cluttered surface has high spatial information while a surface of an object has

Irregular surfaces: High spatial information

Object: Low spatial information

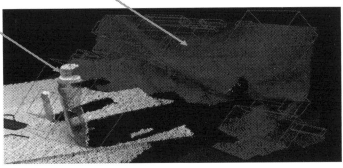

Fig. 4. A representation of spatial information

low spatial information. Spatial information of each cluster is normalized using min-max normalization.

Feature Based on Distance. Usually, an object is found to be near and a cluttered background is far from the origin of a camera frame in a scene. We measure the distance (D) between the center of each bounding box and the origin of a camera frame to separate irregular cluttered and non-cluttered surfaces. A normalized distance D feature is given as.

$$D_i = \frac{d(O, C_i) - d(O, C_{min})}{d(O, C_{max}) - d(O, C_{min})} \tag{2}$$

where O is origin of a camera reference frame, C_i is a center of i-th bounding box, C_{max} and C_{min} are the maximum and minimum distances between the origin and a bounding box and d defines the Euclidean distance.

2.4   Classification of Cluttered and Non-cluttered Surfaces

We used three normalized features (size S, spatial information I and distance D) of every bounding box to remove irregular cluttered surfaces. An object has a small S feature, a low I feature and a small D feature in comparison to other surfaces in a scene. The following steps are required for all clusters to remove cluttered surfaces.

1. First, normalized features based on size and spatial information of each cluster are combined i.e. $SI = (S + I)/2$.
2. A cluster that has SI feature less than a threshold T_1 is further checked for distance feature. Otherwise, the cluster is labeled as a cluttered surface and removed.

Table 1. Description of datasets.

Datasets	Model/Scene	Cluttered surfaces	Data type
Challenge [1]	35/176	High	RGBD
Willow [1]	35/353	Very high	RGBD

(a) A scene of Challenge dataset   (b) A scene of Willow dataset

Fig. 5. Scenes of considered datasets.

3. Finally, a cluster (that satisfied constraint in the second step) has a distance less than a threshold T_2 is classified as a non-cluttered surface i.e. an object surface. Otherwise, the cluster is considered as a cluttered surface and then removed from a scene.

3   Experimental Evaluation

The performance of the proposed technique is evaluated on two heavily cluttered datasets: Challenge and Willow [1]. The detail of these datasets is shown in Table 1. Challenge and Willow datasets include RGBD data of typical household objects with cluttered surfaces (plane and irregular surfaces). Scenes of considered datasets are shown in Fig. 5. These datasets provide pose (ground truth) of every object present in the scene. Objects present in a scene are partial due to occlusions or one-sided camera view and it is observed by experiment that on average, one-third of points of objects are present in the scenes of considered datasets. We present a metric for the measurement of remaining clutter after filtering of cluttered surfaces. The remaining clutter $R_{clutter}$ is given as follows.

$$R_{clutter} = 100 \times \frac{R_{af} - T_O/3}{T_{bf} - T_O/3} \tag{3}$$

where R_{af} represents remaining points after filtering, T_{bf} indicates total points present before filtering and T_O represents total points of all objects present in a scene.

Table 2. Comparison based on $R_{clutter}$

Techniques	Proposed	Plane removal [5]
$R_{clutter}$ (%):Challenge	12.40	69.10
$R_{clutter}$ (%):Willow	10.32	76.50

 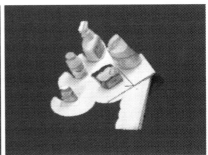

(a) A scene before filtering   (b) A scene after filtering

Fig. 6. Results of proposed technique

In order to filter plane surfaces using RANSAC, we set T_f to 0.01 and maximum iteration I_{max} to 100. For clustering, radius d_{th} is set to 0.008 and a radius r_i used in computation of spatial information is set to 0.5. Both thresholds T_1 and T_2 are set to 0.60. All the above parameters have set by experimentation on 10 scenes of Challenge datasets. These parameters have kept unchanged for the remaining scenes of Challenge dataset and the whole Willow dataset.

To the best of our knowledge, there is no technique available in the literature that can filter both plane and irregular surfaces. For comparison, we have included a technique that has automatically removed plane surfaces. Table 3 shows remaining clutter $R_{clutter}$ of proposed and a technique proposed in [5].

From the table, it is observed that the proposed technique has filtered 87.60% ($R_{clutter} = 12.40\%$) and 89.68% ($R_{clutter} = 10.32\%$) clutter for Challenge and Willow datasets respectively and a technique proposed in [5] has removed only 30.90% ($R_{clutter} = 69.10\%$) and 23.50% ($R_{clutter} = 76.50\%$) clutter for Challenge and Willow datasets respectively. The better performance of the proposed technique is due to the filtering both plane and irregular surfaces while the compared technique has removed only plane surfaces. Figure 6 shows a final filtered scene using the proposed technique. It is clearly observed from the figure that most of the cluttered surfaces (plane and irregular) are removed by the proposed technique.

4   Conclusion

In this paper, we have proposed a cluttered surface filtering technique that can remove both plane and irregular surfaces from a point cloud. The proposed technique has considered size, distance and spatial information of clusters to separate cluttered and non-cluttered surfaces. The technique has evaluated on two heavily cluttered datasets: Challenge and Willow. Our experimental results demonstrate that the proposed technique has removed 87.60% and 89.68% clutter for Challenge and Willow datasets respectively.

References

1. Aldoma, A., Fäulhammer, T., Vincze, M.: Automation of ground truth annotation for multi-view RGB-D object instance recognition datasets. In: Proceedings of the IEEE/RSJ International Conference on Intelligent Robots and Systems (IROS 2014), pp. 5016–5023 (2014)
2. Aldoma, A., Tombari, F., Di Stefano, L., Vincze, M.: A global hypotheses verification method for 3D object recognition. In: Proceedings of European Conference on Computer Vision, pp. 511–524 (2012)
3. Ballard, D.: Generalizing the hough transform to detect arbitrary shapes. Patt. Recogn. **13**(2), 111–122 (1981)
4. Borrmann, D., Elseberg, J., Lingemann, K., Nüchter, A.: The 3D hough transform for plane detection in point clouds: a review and a new accumulator design. 3D Research **2**, 1–13 (2011)
5. Czerniawski, T., Nahangi, M., Walbridge, S., Haas, C.: Automated removal of planar clutter from 3D point clouds for improving industrial object recognition. In: Proceedings of the 33rd International Symposium on Automation and Robotics in Construction (ISARC), pp. 357–365 (2016)
6. Fischler, M.A., Bolles, R.C.: Random sample consensus: a paradigm for model fitting with applications to image analysis and automated cartography. Commun. ACM **24**(6), 381–395 (1981)
7. Joshi, P., Rastegarpanah, A., Stolkin, R.: Are current 3D descriptors ready for real-time object recognition? In: Proceedings of 8th IEEE International Conference on Control, Mechatronics and Automation (ICCMA 2020), pp. 217–221 (2020)
8. Joshi, P., Rastegarpanah, A., Stolkin, R.: A survey on training free 3D texture-less object recognition techniques. In: Proceedings of IEEE International Conference on Digital Image Computing: Techniques and Applications (DICTA), pp. 1–3 (2020)
9. Joshi, P., Rastegarpanah, A., Stolkin, R.: A training free technique for 3D object recognition using the concept of vibration, energy and frequency. Comput. Graph. **95**, 92–105 (2021)
10. Kaiser, M., Xu, X., Kwolek, B., Sural, S., Rigoll, G.: Towards using covariance matrix pyramids as salient point descriptors in 3D point clouds. Neurocomputing **120**, 101–112 (2013)
11. Rabbani, T., Van Den Heuvel, F., Vosselmann, G.: Segmentation of point clouds using smoothness constraint. Int. Arch. Photogrammetry, Remote Sens. spat. Inf. Sci. **36**(5), 248–253 (2006)
12. Rusu, R.B., Cousins, S.: 3D is here: point cloud library (PCL). In: Proceedings of IEEE International Conference on Robotics and Automation, pp. 1–4 (2011)
13. Schnabel, R., Wahl, R., Klein, R.: Efficient ransac for point-cloud shape detection. Comput. Graph. Forum **26**(2), 214–226 (2007)

A Computer Vision Based Approach for Driver Distraction Recognition Using Deep Learning and Genetic Algorithm Based Ensemble

Ashlesha Kumar[1], Kuldip Singh Sangwan[2], and Dhiraj[3(✉)]

[1] Department of Computer Science and Information Systems (CSIS),
Birla Institute of Technology and Science (BITS) Pilani, Pilani Campus, Pilani, India
f20180760@pilani.bits-pilani.ac.in
[2] Department of Mechanical Engineering, Birla Institute of Technology
and Science (BITS) Pilani, Pilani Campus, Pilani, India
kss@pilani.bits-pilani.ac.in
[3] Central Electronics Engineering Research Institute (CSIR-CEERI), Pilani, India
dhiraj@ceeri.res.in

Abstract. As the proportion of road accidents increases each year, driver distraction continues to be an important risk component in road traffic injuries and deaths. The distractions caused by increasing use of mobile phones and other wireless devices pose a potential risk to road safety. Our current study aims to aid the already existing techniques in driver posture recognition by improving the performance in the driver distraction classification problem. We present an approach using a genetic algorithm-based ensemble of six independent deep neural architectures, namely, AlexNet, VGG-16, EfficientNet B0, Vanilla CNN, Modified DenseNet and InceptionV3 + BiLSTM. We test it on two comprehensive datasets, the AUC Distracted Driver Dataset, on which our technique achieves an accuracy of 96.37%, surpassing the previously obtained 95.98%, and on the State Farm Driver Distraction Dataset, on which we attain an accuracy of 99.75%. The 6-Model Ensemble gave an inference time of 0.024 s as measured on our machine with Ubuntu 20.04(64-bit) and GPU as GeForce GTX 1080.

Keywords: Distraction · Ensemble · Genetic algorithm · Deep learning

1 Introduction

Road accidents are increasingly becoming one of the leading causes of fatalities, affecting both the developed and developing countries alike. The Global Status Report on Road Safety [1] revealed that such crashes and accidents claim roughly a million lives on a yearly basis. The count of people enduring disabling injuries as a result is even higher (ranging from approximately 20 to 50M). Most of these road traffic accidents are caused by drivers being distracted due to various factors such as conversations with co passengers or use of mobile devices. This leads to

© Springer Nature Switzerland AG 2021
L. Rutkowski et al. (Eds.): ICAISC 2021, LNAI 12855, pp. 44–56, 2021.
https://doi.org/10.1007/978-3-030-87897-9_5

delayed response time and thus increases chances of accidents. Hence it is imperative to develop an accurate system which is capable of real-time detection of drivers
distracted by various factors in their environment and so an alarm can be raised
to alert them in time. Distracted driving, as defined by the NHTSA [2], is "any
activity that diverts attention from driving". The Center for Disease Control and
Prevention (CDC) puts forward a more inclusive definition by bifurcating distractions into those caused visual, manual and cognitive sources. Cognitive distractions divert an individual's mind away from the situation at present, manual distractions directly involve a person taking his hand off the steering wheel whereas
visual distractions cause a person to take their eyes off the road.

Our research is focused on the manual category, targeting distractions of the
form of talking or texting on phone, adjusting radio, fixing hair and makeup,
eating or drinking and reaching behind to pick up stuff, sample images of which
are depicted in Fig. 1 and Fig. 2. We propose a technique involving a genetically
weighted ensemble of six end -to- end deep learning architectures consisting of
convolutional neural networks and their combinations with recurrent neural networks which surpasses the state-of-the-art accuracies obtained on two of the most
popular driver distraction datasets. We use AlexNet, VGG-16, InceptionV3 +
BiLSTM, EfficientNet B0, Vanilla CNN and a modified hierarchical variant of
DenseNet-201 as our individual branches in the ensemble. Our experiments are
divided into two case studies, first on the AUC Distracted Driver (V1) Dataset
results of which we compare with another study on GA weighted ensemble conducted by [3], and second on the State Farm Driver Distraction Dataset, where
we evaluate our studies in comparison to three previous studies done on the same
dataset.

The paper is arranged as follows: Sect. 2 talks about the related literatures
available in the field of distracted driving detection, while Sect. 3 explains the
structure and composition of the two datasets used. Section 4 discusses our proposed approach in length, explaining the details of various branches used to form
the ensemble and the genetic algorithm adopted to construct the same. Section 5
presents the evaluation results obtained from our experiments and a comparative analysis with respect to the previous studies conducted on the datasets.
Section 6 presents the conclusion of our paper along with the scope of future
research based on it.

2 Related Works

Advances in technologies endorsed by fields such as Machine Learning and Deep
learning have permitted researchers in the last two decades to come up with a
myriad of distraction detection techniques, the earliest ones being dominated
by simple classifiers such as SVM and Decision Trees. Research in the field of
distracted driver detection can broadly be classified into categories of traditional
machine learning approaches and modern deep learning solutions. Zhao et al. [4]
proposed another inclusive dataset- the South East University (SEU) Driving
Posture dataset, containing images of drivers taken in a side-view fashion, covering a broader range of activities such as talking on the cell- phone, driving

safely, operating the lever and eating foodstuff. They adopted an approach of extracting features by making use of contourlet transform, assessing the performance using four classifiers, out of which the Random Forests classifier showed the best performance, achieving an accuracy of 90.63%. Subsequently, Yan et al. [5] presented an approach to identify driving postures by making use of deep convolutional neural networks, attaining an accuracy of 99.78% on the SEU dataset. Eraqi et al. [3] proposed a novel dataset, the AUC Distracted Driver dataset, comprising 10 classes of distractions.

Fig. 1. Ten classes of the AUC Distracted Driver (V1) Dataset

Fig. 2. Ten classes of the State Farm Driver Distraction Dataset

The inputs to the individual models in the ensemble were not same, with one of the networks being supplied with only raw images, another with skin-segmented images, while the remaining three networks were provided with images of hands, face and both hands-and-face respectively. Further progress in the field witnessed researchers and scientists combining both CNNs and RNNs to achieve better performance and accuracy in the detection of driver distraction. In 2019, Munif Alotaibi and Bandar Alotaibi [6] introduced a combination of three elaborate deep learning architectures. Their approach was tested on State Farm and AUC Datasets, yielding an excellent accuracy of 99.3% on a fifty percent train/test split of the former one. Another benchmark was set by the Vanilla CNN architecture proposed by Jamsheed et al. [7] in 2020, giving an overall accuracy of 97% on the State Farm Dataset.

3   Dataset Information

3.1   American University of Cairo (AUC) Distracted Driver (V1) Dataset

The AUC Distracted Driver (V1) dataset [8] was the first publicly available dataset for distracted driver detection. Sample frames of the dataset are shown in Fig. 1. It consists of participants from seven countries consisting of both males (22) and females (9). Videos were shot in 4 different cars: Proton Gen2 (26), Mitsubishi Lancer (2), Nissan Sunny (2), and KIA Carens (1) [8]. The dataset comprises of 12977 frames divided into 10 classes, as depicted by Table 1.

Table 1. Summary of the AUC Distracted Driver (V1) Dataset

Class	Activity	No. of frames
0	Safe Driving	2764
1	Phone Right	975
2	Phone Left	1020
3	Text Right	1480
4	Text Left	917
5	Adjusting Radio	915
6	Drinking	1209
7	Hair or Makeup	901
8	Reaching Behind	869
9	Talking to Passenger	1927
Total		12977

3.2   State Farm Driver Distraction Dataset

To assess the strength and performance of our approach in a more generalized fashion, we also used another recent dataset as part of our second case study on driver distraction classification, the State Farm Driver Distraction dataset. Sample frames from the dataset are depicted in Fig. 2. This dataset was released as part of State Farm's Distracted Driver Detection competition [9] organized on Kaggle in 2016.

Table 2. Summary of the State Farm Dataset

Class	Activity	No. of frames
0	Safe Driving	2489
1	Text Right	2267
2	Talk Right	2317
3	Text Left	2346
4	Talk Left	2326
5	Adjusting Radio	2312
6	Drinking	2325
7	Reaching Behind	2002
8	Hair or Makeup	1911
9	Talking to Passenger	2129
Total		22424

Similar to the AUC dataset, it consists of images belonging to 10 classes, with postures ranging from safe driving to distractions such as texting on the phone while driving (using left or right hands), talking on the phone (using left or right hands), conversing with a co-passenger, operating the radio, reaching behind, adjusting hair or makeup, and drinking. The dataset thus consists of 22,424 images, distributed among the ten classes as shown in Table 2.

4　Proposed Methodology

Our approach takes advantage of a genetically weighted ensemble of convolutional neural networks, all of which were trained on raw images of the driver distraction datasets. The images were resized to a fixed size of 224×224.

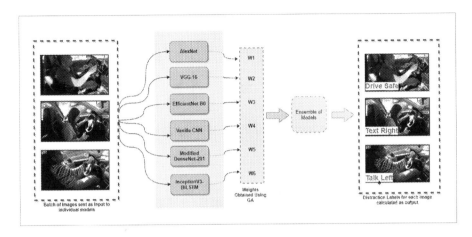

Fig. 3. Summary of our proposed approach, employing a genetically weighted ensemble of net-works. The batch containing input images is fed to the networks, and the output produces labels for each of the individual images

For our purpose, we conducted an exploratory analysis in which several branches were trained on each of the datasets separately, each branch having a different net-work architecture. We then calculate the weighted sum of the outputs of all classifiers employing a genetic algorithm, yielding the final class probability vector. The characteristics of each of the six branches are described in the following sections. The system overview is shown in Fig. 3.

4.1　Independent Classifier Branches

We trained several neural network architectures separately, adopting some of the finest ones previously achieving state-of-the-art performance on various computer vision tasks.

AlexNet. The AlexNet model was proposed in 2012 by Krizhevsky et al. [10] This model clinched the first place in the 2012 ImageNet competition with a top-5 error rate of 15.3%. It has five convolutional layers, some of which are followed by max-pooling layers, and three fully-connected layers. ReLU activation function is applied to the output feature maps of every convolutional layer. We trained our AlexNet models from scratch without any modifications to the original architecture, except the addition of batch normalization before all ReLU activations and the final SoftMax layer. We used an RMSProp optimizer with a learning rate of 10^{-4} for training, and ran the model for 50 epochs with a batch size of 32 images.

VGG-16. VGG-16 was introduced in 2014 by Simonyan and Zisserman [11]. It is deeper than AlexNet, and consists of 13 convolutional layers followed by three fully connected layers. The input to the VGG ConvNets is a fixed-size 224×224 RGB image. The image is passed through a stack of convolutional layers, where filters are used with a very small receptive field: 3×3. The stack of convolutional layers is followed by three fully-Connected (FC) layers: the first two have 4096 channels each. The final layer is the soft-max layer.

For our purpose, we make use of a VGG-16 model pretrained on the ImageNet dataset as an integrated feature extractor, and the layers of the pre-trained model were frozen during training. An RMSProp optimizer with a learning rate of 10^{-4} was used, keeping the number of epochs at 50 during the training process along with a batch size of 64 images.

EfficientNet B0. Tan and Le [12] proposed a novel model scaling method in 2019 that uses a simple yet highly effective compound coefficient to scale up CNNs in a more structured manner. They developed a new baseline network, EfficientNet B0, by leveraging a multi-objective neural architecture search using the AutoML MNAS framework, that optimizes both accuracy and efficiency (FLOPS). As our third branch, we make use of the Efficient Net B0 model provided by Keras, pretrained on ImageNet Dataset. The pre-trained model is integrated into a new model, where two dense layers and a softmax activation are added to it, and the layers of the pre-trained model are trained along with the new model. For the training process, we used an RMSProp optimizer, keeping the number of epochs as 30 and the batch size as 64 images.

Vanilla CNN. Vanilla CNN architecture was proposed by Jamsheed et al. in 2020 [7], giving an overall accuracy of 97% on the State Farm Dataset. The model is constructed with a total of 3 convolutional layers, a flatten layer and 3 dense layers. Activation function for all the convolutional layers and dense layers is ReLU, while the last layer em-ploys a softmax activation. Table 3 depicts the summary of the Vanilla CNN Model. This forms the fourth branch of our ensemble.

Modified DenseNet-201 Hierarchical Model. Huang et al. [13] proposed an upgrade of the ResNet model, the DenseNet, which subsequently clinched the best paper award in CVPR2017. Verma et al. [14] proposed modified variants of DenseNet-201 employing hierarchical structures for posture detection on the Yoga-82 Dataset. Their main motive was to utilize hierarchy structure in the proposed dataset. One of the variants which we make use of consists of hierar-chical connections built upon the DenseBlock 2 and DenseBlock 3 for three class

Table 3. Vanilla CNN architecture

Layer	Type	Output shape
1	Convolutional layer	(None, 224, 224, 60)
2	Max pooling layer	(None, 112, 112, 60)
3	Convolutional layer	(None, 112, 112, 90)
4	Max pooling layer	(None, 56, 56, 90)
5	Convolutional layer	(None, 56, 56, 200)
6	Max pooling layer	(None, 28, 28, 200)
7	Flatten layer	(None, 156800)
8	Dense layer	(None, 512)
9	Dense layer	(None, 128)
10	Dense layer	(None, 10)

levels. Level 1 originally consists of six classes, and level 2 is composed of 20 classes. The third level contains 82 classes, and the corresponding classification branch forms the main branch of the network.

InceptionV3 + BiLSTM. Inception architecture was first introduced in 2014 by Szegedy et al. [15]. It utilizes numerous kernel sizes in every convolutional layer to harness the power of varied kernel sizes, while at the same time prevents overfitting by avoiding deeper architectures. LSTMs are an extension of recurrent neural networks, capable of learning long-term dependencies. LSTMs make use of the mechanism of cell states and its various gates to promote information flows, using which they can selectively choose to forget or remember things. The deep-bidirectional LSTMs [16,17] are an extension of the described LSTM models in which two LSTMs are applied to the input data. A technique using pre-trained InceptionV3 CNNs [18] integrated with Bidirectional Long Short-Term Memory layers gave good results on the AUC Distracted Driver Dataset, outperforming other state-of-the-art CNN's and RNN's in terms of average loss and F1-score. We adopt this architecture as our sixth branch for the ensemble.

For our purpose, we employ this variant with slight modifications, since the datasets used in our case studies are not explicitly hierarchical in nature. We set the class levels as 512, 128 and 10 respectively, and use the third variant proposed in [14] as our fifth branch for the main ensemble. A batch size of 64 images was used, and the network was trained for 30 epochs using an RMSProp optimizer.

4.2   Genetic Algorithm(GA) Based Ensemble

Our technique uses a classifier ensemble based on the idea that ensembles can serve as strong classifiers or a more accurate mechanism for prediction than the individual weak classifiers they consist of. Suppose that there are N such individual classifiers, $C_1, C_2...C_N$. Solving a prediction problem involving m classes

$M_1, M_2...M_m$. Let us assume that depending on the classification problem, the features used and the training set, classifier C_1 is more efficient in classifying M_1, C_2 is more accurate on another class, say $M_2 or M_3$ and so on. If we form an ensemble of classifiers, $C_1, C_2...C_N$ we cannot assign equal weights to all of them and hence a method has to be employed to determine the weight of vote given to each model. We adopt the genetic algorithm [19], a search heuristic that automatically and effectively finds the proper weights of all the eligible models. If the individual probability vectors are designated by $V_1...V_N$, produced as the output of the last softmax layer of each model, then in a weighted voting system, the final predictions are calculated in Eq. (1) as:

$$V_{final} = \frac{\sum_{i=1}^{N} w_i * V_i}{\sum_{i=1}^{N} w_i} \quad (1)$$

where each chromosome consists of N genes, $w_1, w_2 \ldots w_N$. We adopt the Mean Squared Error (MSE) as our objective function and create an initial population of 48 individuals. We run the algorithm for 30 generations. The chromosome with the highest fitness score is selected in the end.

5   Experiments and Results

As a first step, we constructed a GA weighted ensemble of all the six branches taken together. We conducted our experiments on the AUC Driver Distraction Dataset as our first case study using all the six models as an ensemble, and then on the State Farm Dataset as our second case study to further assess the strength of our approach by comparing it with the existing state-of-the-art obtained for the same. The training plots for all the individual branches depicting the training accuracy and training loss as a function of the number of epochs are shown in Fig. 4.

5.1   Results on AUC Driver Distraction (V1) Dataset

The results of all the individual branches trained on the AUC Dataset are shown in Table 4.

Table 4. Results obtained on Individual Branches for AUC Dataset

Model	Accuracy
AlexNet	96.24%
InceptionV3-BiLSTM	95.28%
VGG-16	95.3%
Modified DenseNet-201	94.42%
Vanilla CNN	95.76%
EfficientNet B0	95.3%

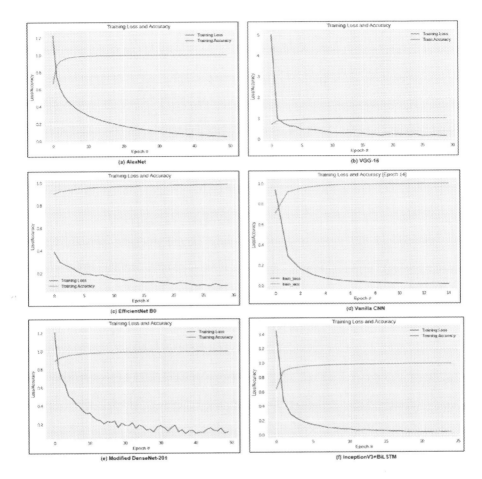

Fig. 4. Training curves of each of the individual architectural branches

		Predicted Label									
		C0	C1	C2	C3	C4	C5	C6	C7	C8	C9
	C0	**0.9740**	0.004	0.0	0.0	0.0019	0.0029	0.0069	0.0019	0.0039	0.0039
	C1	0.0168	**0.9739**	0.0018	0.00746	0.0	0.0	0.0	0.0	0.0	0.0
	C2	0.0	0.0509	**0.9431**	0.00299	0.0	0.0	0.0029	0.0	0.0	0.0
	C3	0.0028	0.0083	0.0055	**0.9695**	0.0083	0.0027	0.0027	0.0	0.0	0.0
True	C4	0.0029	0.0028	0.0	0.034	**0.9541**	0.0057	0.0	0.0	0.0	0.0
Labels	C5	0.0419	0.0	0.0	0.0	0.0089	**0.9401**	0.0089	0.0	0.0	0.0
	C6	0.0095	0.0	0.0	0.0	0.0	0.0071	**0.9762**	0.0023	0.0	0.0047
	C7	0.038	0.0	0.0	0.0	0.0	0.0	0.0058	**0.95321**	0.0	0.0029
	C8	0.0232	0.0	0.0	0.0	0.0	0.0	0.0038	0.0038	**0.9612**	0.0077
	C9	0.0272	0.0	0.0	0.0	0.0	0.0	0.0090	0.0	0.0015	**0.9621**

Fig. 5. Confusion matrix for the GA ensemble formed using the six branches stated earlier

6-Model Ensemble. Our 6-Model GA-Weighted Ensemble gave an overall accuracy of 96.37%, which was greater than the accuracies of all the individual branches used. This particular combination also showed marked improvement in the classification of the classes Safe Driving (2.06%), Phone Right (0.76%), Text Right (0.80%) and Reaching Behind (3.36%) involve the use of mobile phones by the same hand.

The confusion matrix shown in Fig. 5 reveals that there is a confusion between the classes Phone Right and Text Right and between the classes Phone Left and Text Left. A possible reason could be because both actions involve the use of mobile phones by the same hand (right or left respectively). Another observation drawn was that "Hair and Makeup" was majorly confused for the "Talking to Passenger" positions, due to the striking similarity in the two positions, as depicted in Fig. 6.

Fig. 6. Comparison of "Hair or Makeup" and "Talking to Passenger" postures. The similarity in the positions cause the driver to bend towards the other side in both cases with movement of the right hand, hence causing the confusion in the decision making process of the network

5.2   Results on State Farm Driver Distraction Dataset

We tested the ensemble on the State Farm Dataset and compared the results with three other studies conducted on the same dataset, [6,7] and [20] using similar train/test partitions employed by the original authors, as depicted in Table 5.

Munif Alotaibi and Bandar Alotaibi [6] experimented with three different train/test partitions, achieving the highest accuracy of 99.3% on the fifty percent train/test split. Our 6-Model ensemble exceeded the benchmark accuracy on almost all of the partitions, except one where it gave a similar performance. The comparison for all the references is represented by Table 5.

Vanilla CNN [7] architecture was originally tested on the State Farm Dataset, and since our ensemble employed it as one of the branches, we used similar train/dev/test splits to compare our results with it as well. Our proposed method gave an accuracy of 99.57% as compared to 97.66% previously attained by [7]. This revealed that using a combination of classifiers gave better results as compared to using only a single model.

K. R. Dhakate and R. Dash [19] made use of a stacking ensemble technique on a different split of the State Farm Dataset, on which our technique achieves an accuracy gain of 2.75%.

Table 5. Results obtained on Different Train/Test Partitions of the State Farm Dataset and its performance comparison with other published results

Reference	Data split			Accuracy		
	Train data %	Val data %	Test data %	Obtained by authors	Our ensemble	Accuracy gain
[6]	10%		90%	96.23	97.16%	0.93%
	30%	–	70%	98.92	98.923 %	0%
	50%		50%	99.3	99.42%	0.12%
[20]	64% (13K Images taken)	18%	18%	97%	99.75%	2.75%
[7]	80% (Total 17939 taken)	–	20%	97.66%	99.57%	1.91%

6   Conclusion

As the use of mobile phones and technological devices increases at an aggressive pace, potential risk to road safety due to the distractions caused by these is also in-creasing exponentially. We present an approach that incorporates the strength of a number of advanced deep learning architectures to accurately predict whether a driver is distracted or driving safely. We make use of the nature inspired "genetic algorithm" to efficiently create a weighted ensemble. Our approach has been tested on two comprehensive datasets, and shows excellent results in terms of accuracy based performance gain on both of them with respect to already published results in literature. As a future work, the GA ensemble can be deployed on an embedded device and its performance can be evaluated in real-time. Research also needs to be done on how to effectively reduce the confusion between some of the classes (i.e., "Hair and Makeup" and "Talking to Passenger", or "Phone Right and Text Right") showing similar trends in most of the experiments. The network combination can also be used for generating labels for the unlabeled images in State Farm Dataset for further detailed studies.

References

1. World Health Organization: Management of Substance Abuse Unit. Global Status Report on Alcohol and Health, 2014. World Health Organization, Geneva (2014)
2. Pickrell, T.M., Li, H.R., KC, S.: Traffic safety facts (2016). https://www.nhtsa. gov/risky-driving/distracted-driving

3. Eraqi, H.M., Abouelnaga, Y., Saad, M.H., Moustafa, M.N.: Driver distraction identification with an ensemble of convolutional neural networks. J. Adv. Transp. Mach.
   Learn. Transp. (MLT) (2019)
4. Zhao, C.H., Zhang, B.L., He, J., Lian, J.: Recognition of driving postures by contourlet transform and random forests. IET Intell. Transp. Syst. **6**(2), 161–168
   (2012)
5. Yan, C., Coenen, F., Zhang, B.: Driving posture recognition by convolutional neural
   networks. IET Comput. Vis. **10**(2), 103–14 (2016). https://doi.org/10.1049/iet-cvi.
   2015.0175
6. Alotaibi, M., Alotaibi, B.: Distracted driver classification using deep learning. Sig.
   Image Video Process. **14**(3), 617–624 (2019). https://doi.org/10.1007/s11760-019-
   01589-z
7. Abdul Jamsheed, V., Janet, B., Reddy, U.S.: Real time detection of driver distraction using CNN. In: 2020 Third International Conference on Smart Systems and
   Inventive Technology (ICSSIT), Tirunelveli, India, pp. 185–191 (2020). https://
   doi.org/10.1109/ICSSIT48917.2020.9214233
8. Abouelnaga, Y., Eraqi, H., Moustafa, M.: Real-time distracted driver posture
   classification. In: Neural Information Processing Systems (NIPS 2018), Workshop on Machine Learning for Intelligent Transportation Systems, December 2018.
   arXiv:1706.09498
9. State farm distracted driver detection. https://www.kaggle.com/c/state-farm-
   distracted-driver-detection
10. Krizhevsky, A., Sutskever, I., Hinton, G.E.: ImageNet classification with deep con-
    volutional neural networks. Commun. ACM **60**(6), 84–90 (2017). https://doi.org/
    10.1145/3065386
11. Simonyan, K., Zisserman, A.: Very deep convolutional networks for large-scale
    image recognition. In: Proceedings of the International Conference on Learning
    Representations (2015)
12. Tan, M., Le, Q.: EfficientNet: rethinking model scaling for convolutional neural net-
    works. In: Proceedings of the 36th International Conference on Machine Learning,
    pp. 6105–6114 (2019)
13. Huang, G., Liu, Z., Van Der Maaten, L., Weinberger, K.Q.: Densely connected
    convolutional networks. In: 2017 IEEE Conference on Computer Vision and Pat-
    tern Recognition (CVPR), Honolulu, HI, pp. 2261–2269 (2017). https://doi.org/
    10.1109/CVPR.2017.243
14. Verma, M., Kumawat, S., Nakashima, Y., Raman, S.: Yoga-82: a new dataset for
    fine-grained classification of human poses, April 2020
15. Szegedy, C., Vanhoucke, V., Iofe, S., Shlens, J., Wojna, Z.: Rethinking the inception
    architecture for computer vision. In: Proceedings of the 2016 IEEE Conference on
    Computer Vision and Pattern Recognition, CVPR 2016, pp. 2818–2826, July 2016
16. Schuster, M., Paliwal, K.K.: Bidirectional recurrent neural networks. IEEE Trans.
    Sig. Process. **45**(11), 2673–2681 (1997)
17. Siami-Namini, S., Tavakoli, N., Namin, A.S.: A Comparative analysis of
    fore-casting financial time series using ARIMA, LSTM, and BiLSTM.
    arxiv: cs.LG/1911.09512 (2019)
18. Mafeni Mase, J., Chapman, P., Figueredo, G.P., Torres Torres, M.: Benchmarking
    deep learning models for driver distraction detection. In: Nicosia, G., et al. (eds.)
    LOD 2020. LNCS, vol. 12566, pp. 103–117. Springer, Cham (2020). https://doi.
    org/10.1007/978-3-030-64580-9_9

19. Holland, J.H.: Genetic algorithms. Sci. Am. JSTOR **267**(1), 66–73 (1992). www.jstor.org/stable/24939139
20. Dhakate, K.R., Dash, R.: Distracted driver detection using stacking ensemble. In: 2020 IEEE International Students' Conference on Electrical, Electronics and Computer Science (SCEECS), Bhopal, India, pp. 1–5 (2020). https://doi.org/10.1109/SCEECS48394.2020.184

Multimodal Image Fusion Method Based on Multiscale Image Matting

Sarmad Maqsood[1] , Robertas Damasevicius[1(✉)] , Jakub Siłka[2] ,
and Marcin Woźniak[2]

[1] Department of Software Engineering, Kaunas University of Technology,
Kaunas, Lithuania
{sarmad.maqsood,robertas.damasevicius}@ktu.lt
[2] Faculty of Applied Mathematics, Silesian University of Technology, Gliwice, Poland
marcin.wozniak@polsl.pl

Abstract. Multimodal image fusion combines the complementary information of multimodality images into a single image that preserves the information of all the source images. This paper proposes a multimodal image fusion method situated on image enhancement, edge detection, multiscale sliding window, and image matting to obtain the detailed region information of the input images. In the proposed system, firstly the multimodality input images are rectified via a contrast enhancement method through which the intensity distribution is refined for clear vision. The spatial gradient edge detection method is utilized for separating the edge information from the contrast-enhanced images. These edges are then used by a multiscale sliding window method to provide global and local activity level maps. These activity maps further generate trimap and decision maps. Finally, by employing the improved decision maps and fusion rule the fused image is acquired.

Keywords: Multimodal medical image · Image fusion · Multiscaling · Image matting

1 Introduction

Multimodality image fusion combines important information from various imaging modalities to provide an improved fused image. The obtained fused image is highly informative and is very dedicated to further processing for disease diagnosis and remedy [1,2]. Multimodal image fusion techniques have an extensive variety of applications for their improved and precise illustration of informative outcomes [3,5,6]. Diversified medical modalities i.e., CT and MRI that have allowed radiologists to analyze the important body patterns and parts positioned in the internal human body for reports generating and clinical analysis [7–9]. Therefore, multimodal images of various modalities are needed to be merged to produce a single image that can allow functional information. To obtain the fused image with significant details has drawn researchers attention to multimodal fusion of medical images [4,10].

© Springer Nature Switzerland AG 2021
L. Rutkowski et al. (Eds.): ICAISC 2021, LNAI 12855, pp. 57–68, 2021.
https://doi.org/10.1007/978-3-030-87897-9_6

Image fusion normally has two core divisions, one is the spatial domain the other is the transform domain [11]. The spatial domain technique produces the fused image by taking pixels/sections/blocks of images deprived of alteration [12]. Transform domain techniques merge the complementary transform coefficients and apply an inverse transformation to form the image fusion. Many research present also machine learning models [13].

The multiscale transformation fusion method has great attention in medical image fusion. In transform domain-based technique, variational adaptive PDE [14], contourlet transform [15], discrete wavelet transform [16], non-subsampled contourlet transform [17], curvelet transform [18] and sparse representation [11] methods have been employed in multimodal image fusion. In the current intervals, numerous image fusion approaches have been presented i.e., multiscale and optimization-based image fusion. Multiscale approaches give detailed images representation, these techniques well preserve the details of the edge [2]. Multiscale transformation and sparse representation methods have obtained a significant interest in the transform domain and work positively in the analysis of medical images [19]. However, these methods exhibit several limitations, i.e., undesirable side effects such as reduced contrast and high spatial distortion, the appearance of artifacts in the fused images.

This paper proposes a multimodal image fusion approach using contrast enhancement and edge detection method to extract the edges from the input images. A multiscale sliding window technique is employed for identifying the global and local intensity alterations to produce the initial activity level maps. These various activity level maps then create a trimap. An improved image matting method is employed for creating the final decision maps. Finally, by using the enrich decision maps and fusion rule the fused image is formed.

The remaining paper is structured as follows. Section 2 explains the detailed method of the proposed framework. Section 3 presents the fusion metrics. The experiment results and discussion are presented in Sect. 4 and finally conclusion is given in Sect. 5.

2   The Proposed Multiscaling Image Matting (MSIM) Technique

Let, I_i is the source image with dimensions of $R \times S$, where $r = 1, 2, 3, \ldots, R$, $s = 1, 2, 3, \ldots, S$ and $i \in [1, 2]$ shows the CT and MR images, respectively. The proposed MSIM fusion method is detailed in Algorithm 1.

Algorithm 1. Proposed multimodal image fusion framework

Input Image: $I_i, i \in [1, 2]$.
Output Image: Fused Image I_F.
begin

Improve contrast, $I_i \xrightarrow{\text{contrast enhancement}} Í_i$.

Compute edges, B_p and $B_q \xrightarrow{\text{edge maps}} B_i$.

for k = 9, l = 9 **do**

    Compute activity maps, w, $B_i \xrightarrow{9 \times 9} Ǵ_i$.

    Compute smooth activity maps, $Ǵ_i \xrightarrow{\text{sum filter}} G_i$.

    Compute score maps, $G_i \rightarrow \xi_1$.

end for

for k = 27, l = 27 **do**

    Compute activity maps, w, $B_i \xrightarrow{27 \times 27} Ḫ_i$.

    Compute smooth activity maps, $Ḫ_i \xrightarrow{\text{sum filter}} H_i$.

    Compute score maps, $H_i \rightarrow \xi_2$.

end for

Compute focus maps, $\xi_{1,2} \xrightarrow{\text{AND}} D_1$ and D_2.

Generate trimap, $D_{1,2} \xrightarrow{\text{trimap}} T$.

Create alpha matte, T, $I_i \xrightarrow{\text{matting}} \alpha$.

Obtain fused image, $I_F = \alpha \times I_i$.

end

2.1 Preprocessing

Contrast enhancement is the most extensive method to improve the images having low contrast. No-reference image quality assessment (NR-IQA) [20], is employed to refine the contrast and conserve the mean brightness of the input images I_i.

$$Í_i \xleftarrow{\text{NR-IQA [20]}} I_i. \tag{1}$$

I_i is employed to the input images to get the contrast enhanced images $Í_i$. Image enhancement improved the edges of both source images. Figure 1 illustrates the improvement in edge information. Figure 1 (a, b) shows the source images of CT and MR images, respectively, and their map edges are displayed in Fig. 1 (c, d). Contrast enhancements of CT and MR images are illustrated in (Fig. 1 (e, f)) and their corresponding gradient maps are displayed in Fig. 1 (g, h). It is worth mentioning that after the contrast enhancement approach, there is a prominent improvement in the edges of the source images.

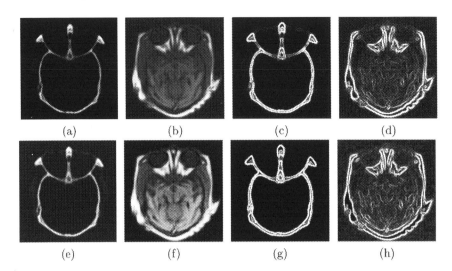

Fig. 1. The "CT and MRI" input images for edge detection after image enhancement. (a, b) Input images, (c, d) edges of (a, b) achieved by sobel operator [21], (e, f) contrast enhancement using No-reference image quality assessment (NR-IQA) [20], (g, h) edges of (e, f) attained by sobel operator.

2.2   Feature Extraction

The Sobel Operator [21] is used to achieve the edges of an image by calculating an approximation of an image gradient. At each location the result is either the norm of this vector or corresponding gradient vector.

Firstly, the gradient is obtained in X direction by convolving the image with the first kernel from left to right given in Eq. (2) as,

$$B_p = \begin{bmatrix} 1 & 0 & -1 \\ 2 & 0 & -2 \\ 1 & 0 & -1 \end{bmatrix}. \tag{2}$$

Similarly, the image is convoluted with the first kernel from top to bottom and the gradient is obtained in Y direction given in Eq. (3) as,

$$B_q = \begin{bmatrix} 1 & 2 & 1 \\ 0 & 0 & 0 \\ -1 & -2 & -1 \end{bmatrix}. \tag{3}$$

At this point the vectors of the gradient of image are obtained. The magnitude of each vector B_i is calculated by using Eq. (4) to obtain the edges.

$$B_i = \sqrt{(B_p)^2 + (B_q)^2}. \tag{4}$$

2.3   Multiscaling

A multiscale sliding window method is employed to obtain various salient features from activity maps B_i. The windows which are created for this experiment are 9×9 and 27×27. Firstly, by applying the spatial domain filters the activity maps are classified into a set of 9×9 components, as given Eq. (5) and Eq. (6):

$$\acute{G}_i(r, s) = \sum_{a=-k}^{k} \sum_{b=-l}^{l} w(a, b) B_i(x + a, y + b). \tag{5}$$

$$G_i(u, v) = \sum_{(r,s) \epsilon \Omega} \acute{G}_i(r, s). \tag{6}$$

Each set activity is preserved in the map scores form. Moreover, the intensity values calculated in each block ($G_1(u, v)$ and $G_2(u, v)$) is deliberated to improve the score maps ξ_a^1 and ξ_b^1 by compared with one another as in Eq. (7) and Eq. (8).

$$\xi_a^1(r, s) = \begin{cases} 1, & \text{if } G_1(u, v) > G_2(u, v) \\ 0, & \text{Otherwise} \end{cases} \tag{7}$$

$$\xi_b^1(r, s) = 1 - \xi_a^1(r, s). \tag{8}$$

Similarly, the activity map for 27×27 block of pixels is given as in Eq. (9) and Eq. (10).

$$\acute{H}_i(r, s) = \sum_{a=-(k \times 3)}^{k \times 3} \sum_{b=-(l \times 3)}^{l \times 3} w(a, b) B_i(x + a, y + b). \tag{9}$$

$$H_i(u, v) = \sum_{(r,s) \ \epsilon \ \Omega} \acute{H}_i(r, s). \tag{10}$$

Each block size activity is also preserved in the map scores form. The intensity levels calculated in each 27×27 block ($H_1(u, v)$ and $H_2(u, v)$) is deliberated to improve the score maps ξ_a^2 and ξ_b^2 by compared with one another as in Eq. (11) and Eq. (12).

$$\xi_a^2(r, s) = \begin{cases} 1, & \text{if } H_1(u, v) > H_2(u, v) \\ 0, & \text{Otherwise} \end{cases} \tag{11}$$

$$\xi_b^2(r, s) = 1 - \xi_a^2(r, s). \tag{12}$$

The different sliding windows maps provide different features information which refine the maps quality. The information from the input images at various scales are selected by multiscale sliding windows technique and provide disparate information for image fusion i.e., small window size extracts local characteristics and large window size focuses on global intensity variations of an image. The multiscale information of all maps (ξ_a^1 and ξ_a^2) and (ξ_b^1 and ξ_b^2) are fused together to form a single focus map, conveying the features of both scales as given in Eq. (13).

$$D_i(r, s) \xleftarrow{\text{AND}} \xi_b^i(r, s), \xi_b^i(r, s). \tag{13}$$

After that the trimap is generated. It roughly divides the input images into following regions i.e., definite focused, definite defocused and the unknown regions. The trimap T is processed by using D_1, D_2 as in Eq. (14).

$$T \xleftarrow{\text{Trimap}} D_i. \tag{14}$$

The trimap (T), source images (I_1, I_2) are combined to construct an alpha matte (α) by using the closed-form matting technique [22], as in Eq. (15).

$$\alpha \xleftarrow{\text{Alpha Matte}} T, I_i. \tag{15}$$

where α is a value between $[0, 1]$, which means that these pixels are mixed by the focused and defocused pixels. $\alpha(r, s) = 1$ or 0 means that the point (r, s) of the source image A_i is in the focus or defocus, respectively. Finally, the fused image is formed by calculating the weighted sum of the source image with alpha matte performed as a weight map as in Eq. (16).

$$I_F(r, s) = \sum_{n=1:i} \alpha(r, s) \times I_i(r, s). \tag{16}$$

3   Objective Evaluation Metrics

To assess the superiority and the effectiveness of the proposed MSIM system with others image fusion algorithms, five most commonly metrics are used for quantitative analysis, i.e., Mutual Information (MI) [23], Entropy (EN) [24], Feature Mutual Information (FMI) [25], Spatial Structural Similarity (SSS) [26], and Visual Information Fidelity (VIF) [27]. For these metrics, the bold value exhibits the higher result. These metrics are defined as follows.

3.1   Mutual Information (MI)

MI [23] calculates the mutual information between two discrete variables and is defined as,

$$MI = \sum_{x=1}^{n}\sum_{y=1}^{n} H_{ij}(x,y)\log_2 \frac{H_{ij}(x,y)}{H_i(x)H_j(y)}, \qquad (17)$$

where $H_{ij}(x,y)$ shows the combined probability density distribution of the grayscale image in i and j. $H_i(x)$ and $H_j(y)$ shows the probability density distribution of the grayscale image in i and j, respectively. It calculates the sum of common information among source images and the fused image. Highest MI value reveals that the fused image has more information than the source images.

3.2   Entropy (EN)

The EN [24] is expressed as,

$$EN(x) = -\sum_{x=0}^{N-1} H_i(x)\log_2 H_i(x), \qquad (18)$$

where N is the number of gray level, which is fixed to 256 in this test, and $H_i(x)$ is the normalized histogram of fused image i.

3.3   Feature Mutual Information (FMI)

FMI [25] is determined as,

$$FMI_y^{i,j} = \frac{1}{N}\sum_{x=1}^{N} \frac{I_x(y,i)}{S_x(y)S_x(i)} + \frac{I_x(y,j)}{S_x(y)+S_x(j)}, \qquad (19)$$

where N shows sliding windows number, $S_x(y)$ is the entropy of nth window in an image y, $I_x(y,i)$ is the regional common information between n^{th} window of image y and i. Similarly, $I_x(y,j)$ is the regional mutual information between the nth window of image y and j. $FMI_y^{i,j}$ computes the source images edge information. Greater value of $FMI_y^{i,j}$ shows the better fused image quality.

3.4   Spatial Structural Similarity (SSS) $Q^{AB/F}$

$Q^{AB/F}$ SSS [26] is presented by Xydeas and Petrovic. This metric decides the quantity of transmitted information of edges from input images into the fused image. $Q^{AB/F}$ for the two source images can be stated as follow:

$$Q^{AB/F} = \frac{\sum_{l=1}^{m}\sum_{w=1}^{n}(Q^{AB}(p,q)W^A(p,q) + Q^{BF}(p,q)W^B(p,q))}{\sum_{l=1}^{m}\sum_{w=1}^{n}(W^A(p,q) + W^B(p,q))}, \qquad (20)$$

where $Q^{AB/F}(p,q)$ denotes the information moves from source image A into the fused image F for the pixel location (p,q) and $W^B(p,q))$ is the weight for a pixel location (p,q). The pixel with higher gradient value influences more to the $Q^{AB/F}$ than the lower gradient value. Thus $W^A(p,q) = [Grad(x,y)]^T$. Where T is a constant.

3.5   Visual Information Fidelity (VIF)

VIF [27] is developed on human visual system. VIF evaluates the performance of fusion between the source image and the fused image by calculating the data common among them. In most possibility, a standard source image is hard to achieve. To estimate the fusion performance in this experiment an adapted model of VIF, is determined between the source images and the fused result by averaging the values of VIF.

4   Results and Discussion

4.1   Experimental Setup

In this section, the proposed image fusion approach is compared both subjectively and objectively with three image fusion methods, i.e., Guided Filtering based fusion (GFF) [28], Laplacian Pyramid (LP) [29] and Convolutional Neural Network (CNN) [30]. All the aforementioned approaches are implemented based upon the codes available by the authors. The medical image fusion datasets are acquired from [31] and the dimensions of the source images standardized as 256 × 256 pixels. The experiments were performed on a laptop Intel(R) Core i7 2.59 GHz processor with 16 GB RAM using MATLAB R2020b. The proposed MSIM method is analyzed by performing both qualitative and quantitative evaluation processes.

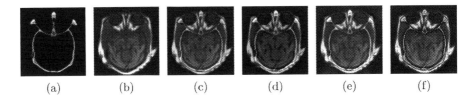

(a)   (b)   (c)   (d)   (e)   (f)

Fig. 2. Source images "Med-1" and the fused images acquired by different fusion algorithms. (a, b) Source images. (c)–(f) Fused images acquired using GFF [28], LP [29], CNN [30] and proposed method respectively.

4.2   Fusion Results

Three sets of multimodal images are used in this experiment. Figures 2(a, b), 3(a, b), 4(a, b) displayed the source images and the fused results achieved by GFF, LP, CNN, and the proposed method are displayed in Figs. 2(c)–(f), 3(c)–(f), 4(c)–(f) respectively. The source CT image provides the information about bone structures and hard tissues and MRI source image provides soft tissue information. Figures 2(c), 3(c), 4(c) shows the results obtained by GFF, which is unable to fully preserved the information in the fused image because of over enhances the structural features and fail to identify salient brain structures. Figures 2(d), 3(d), 4(d) displayed the results acquired by LP. This approach also

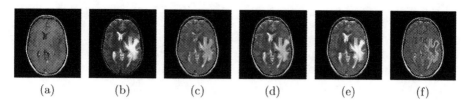

Fig. 3. Source images "Med-2" and the fused images acquired by different fusion algorithms. (a, b) Source images. (c)–(f) Fused images acquired using GFF [28], LP [29], CNN [30] and proposed method respectively.

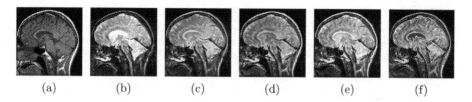

Fig. 4. Source images "Med-3" and the fused images acquired by different fusion algorithms. (a, b) Source images. (c)–(f) Fused images acquired using GFF [28], LP [29], CNN [30] and proposed method respectively.

Table 1. The quantitative evaluation of different fusion algorithms. Bold values show the highest result.

Images	Fusion methods	MI [23]	EN [24]	FMI [25]	SSS [26]	VIF [27]
Med-1	GFF [28]	3.4313	6.7971	0.9032	0.7849	0.4864
	LP [29]	2.5508	6.2724	0.7412	0.6321	0.4141
	CNN [30]	3.5248	6.7541	0.7712	0.7992	0.8991
	Proposed MSIM	**3.8554**	**6.9324**	**0.9539**	**0.8013**	**0.9342**
Med-2	GFF [28]	3.8595	5.8459	0.8596	0.5919	0.4295
	LP [29]	3.5908	5.6692	0.8568	0.6571	0.4352
	CNN [30]	4.2014	7.8421	0.7458	0.6969	0.8015
	Proposed MSIM	**4.6387**	**8.0138**	**0.8759**	**0.7132**	**0.8583**
Med-3	GFF [28]	3.4514	4.4081	0.9047	0.6470	0.4961
	LP [29]	3.4733	4.6547	0.7690	0.6391	0.9255
	CNN [30]	4.2540	5.1748	0.8421	0.7441	0.9408
	Proposed MSIM	**4.5982**	**5.7483**	**0.9641**	**0.7949**	**0.9822**

failed to detect the structural details information in the fused image and suffers from undesirable artifacts. Figures 2(e), 3(e), 4(e) shows the results obtained by CNN, which provides slightly better results than the remaining methods, but still shows the lack of sharpness and distorted regions. The results of the proposed method are displayed in Figs. 2(f), 3(f), 4(f) which provides high contrast, clear information and fully preserved the details in the fused image. Table 1 illustrates that the proposed MSIM method also exhibits superior performance quantitatively than the other algorithms. The top value of evaluation metrics shows in the bold highlighted.

5 Conclusions

In this paper, a constructive multimodality image fusion approach is proposed using the Sobel operator and sliding window method to detect the salient pixels of each input image. The input images are firstly pre-processed using the no reference image quality assessment (NR-IQA) contrast enhancement method and their gradients are computed using the Sobel operator. Afterward, the multiscale sliding window method is proposed for the perfect creation of trimap. Then an image matting method is employed to get the accurate region for the decision maps, and the fused images. More importantly, the closed-form matting method is adept to construct full use of the spatial connections between the nearby pixels for weight estimation. The proposed method achieved the highest mutual information, entropy, feature mutual information, spatial structural similarity, and visual information fidelity of 4.638, 8.013, 0.964, 0.801, and 0.982 respectively. Experimental results demonstrate that the proposed image fusion approach well preserves the complementary information of various source images and achieves superior fusion performance both subjectively and objectively when compared with other image fusion algorithms.

References

1. Muzammil, S.R., Maqsood, S., Haider, S., Damaševičius, R.: CSID: a novel multimodal image fusion algorithm for enhanced clinical diagnosis. Diagnostics **10**(11), 904 (2020)
2. Maqsood, S., Javed, U., Riaz, M.M., Muzammil, M., Muhammad, F., Kim, S.: Multiscale image matting based multi-focus image fusion technique. Electronics **9**(2), 472 (2020)
3. Grycuk, R., Wojciechowski, A., Wei, W., Siwocha, A.: Detecting visual objects by edge crawling. J. Artif. Intell. Soft Comput. Res. **10**(3), 223–237 (2020)
4. Grycuk, R., Najgebauer, P., Kordos, M., Scherer, M.M., Marchlewska, A.: Fast image index for database management engines. J. Artif. Intell. Soft Comput. Res. **10**(2), 113–123 (2020)
5. Woźniak, M., Wieczorek, M., Siłka, J., Połap, D.: Body pose prediction based on motion sensor data and recurrent neural network. IEEE Trans. Ind. Inform. **17**(3), 2101–2111 (2020)
6. Juočas, L., Raudonis, V., Maskeliūnas, R., Damaševičius, R., Woźniak, M.: Multi-focusing algorithm for microscopy imagery in assembly line using low-cost camera. Int. J. Adv. Manufact. Technol. **102**(9), 3217–3227 (2019). https://doi.org/10.1007/s00170-019-03407-9

7. Guo, Z., Li, X., Huang, H., Guo, N., Li, Q.: Deep learning-based image segmentation on multimodal medical imaging. IEEE Trans. Radiat. Plasma Med. Sci. **3**(2), 162–169 (2019)
8. Ke, Q., Zhang, J., Wei, W., Damaševičius, R., Wozniak, M.: Adaptive Independent Subspace Analysis (AISA) of Brain Magnetic Resonance Imaging (MRI) data. IEEE Access **7**(1), 12252–12261 (2019)
9. Khan, M.A., et al.: Multimodal brain tumor classification using deep learning and robust feature selection: a machine learning application for radiologists. Diagnostics **10**(8), 1–19 (2020)
10. Manchanda, M., Sharma, R.: An improved multimodal medical image fusion algorithm based on fuzzy transform. J. Vis. Commun. Image Represent. **51**(2), 76–94 (2018)
11. Maqsood, S., Javed, U.: Biomedical signal processing and control multi-modal medical image fusion based on two-scale image decomposition and sparse representation. Biomed. Sig. Process. Control **57**, 101810 (2020)
12. Li, H., Qiu, H., Yu, Z., Li, B.: Multifocus image fusion via fixed window technique of multiscale images and non-local means filtering. Sig. Process. **138**, 71–85 (2017)
13. Woźniak, M., Siłka, J., Wieczorek, M.: Deep neural network correlation learning mechanism for CT brain tumor detection. Neural Comput. Appl., 1–16 (2021). https://doi.org/10.1007/s00521-021-05841-x
14. Wei, W., Zhou, B., Połap, D., Woźniak, M.: A regional adaptive variational PDE model for computed tomography image reconstruction. Pattern Recogn. **92**, 64–81 (2019)
15. Yang, S., Wang, M., Jiao, L., Wu, R., Wang, Z.: Image fusion based on a new contourlet packet. Inf. Fusion **11**(2), 78–84 (2010)
16. Yang, Y.: A novel DWT based multi-focus image fusion method. Procedia Eng. **24**(1), 177–181 (2011)
17. Li, H., Qiu, H., Yu, Z., Zhang, Y.: Infrared and visible image fusion scheme based on NSCT and low-level visual features. Infrared Phys. Technol. **76**, 174–184 (2016)
18. Nencini, F., Garzelli, A., Baronti, S., Alparone, L.: Remote sensing image fusion using the curvelet transform. Inf. Fusion **8**(2), 143–156 (2007)
19. Yang, B., Li, S.: Visual attention guided image fusion with sparse representation. Optik (Stuttg) **125**(17), 4881–4888 (2014)
20. Yan, J., Li, J., Fu, X.: No-reference quality assessment of contrast-distorted images using contrast enhancement. arXiv preprint arXiv:1904.08879 (2019)
21. Gao, W., Zhang, X., Yang, L., Liu, H.: An improved Sobel edge detection. In: Proceedings of the 3rd International Conference on Computer Science and Information Technology, vol. 9, no. 11, pp. 67–71 (2010)
22. Levin, A., Lischinski, D., Weiss, Y.: A closed-form solution to natural image matting. IEEE Trans. Pattern Anal. Mach. Intell. **30**(2), 228–242 (2007)
23. Hossny, M., Nahavandi, S., Vreighton, D.: Comments on information measure for performance of image fusion. Electron. Lett. **44**(18), 1066–1067 (2008)
24. Liu, Y., Liu, S., Wang, Z.: A general framework for image fusion based on multiscale transform and sparse representation. Inf. Fusion **24**, 147–164 (2015)
25. Haghighat, M.B.A., Aghagolzadeh, A., Seyedarabi, H.: A non-reference image fusion metric based on mutual information of image features. Comput. Electr. Eng. **37**(5), 744–756 (2011)
26. Petrović, V.S., Xydeas, C.S.: Sensor noise effects on signal-level image fusion performance. Inf. Fusion **4**(3), 167–183 (2003)
27. Han, Y., Cai, Y., Cao, Y., Xu, X.: A new image fusion performance metric based on visual information fidelity. Inf. Fusion **14**(2), 127–135 (2013)

28. Li, S., Kang, X., Hu, J.: Image fusion with guided filtering. IEEE Trans. Image Process. **22**, 2864–2875 (2013)
29. Du, J., Li, W., Xiao, B.: Union Laplacian pyramid with multiple features for medical image fusion. Neurocomputing **194**, 326–339 (2016)
30. Liu, Y., Chen, X., Cheng, J., Peng, H.: A medical image fusion method based on convolutional neural networks. In: Proceedings of the 2017 20th International Conference on Information Fusion (Fusion), pp. 10–13 (2017)
31. Zhu, Z., Chai, Y., Yin, H., Li, Y., Liu, Z.: A novel dictionary learning approach for multi-modality medical image fusion. Neurocomputing **214**, 471–482 (2016)

Targeting the Most Important Words Across the Entire Corpus in NLP Adversarial Attacks

Reza Marzban$^{(\boxtimes)}$ ⓘ, Johnson Thomas ⓘ, and Christopher Crick ⓘ

Computer Science Department, Oklahoma State University, Stillwater, OK, USA
`reza.marzban@okstate.edu`

Abstract. In recent years, deep learning has revolutionized many tasks, from machine vision to natural language processing. Deep neural networks have reached extremely high accuracy levels in many fields. However, they still encounter many challenges. In particular, the models are not explainable or easy to trust, especially in life and death scenarios. They may reach correct predictions through inappropriate reasoning and have biases or other limitations. In addition, they are vulnerable to adversarial attacks. An attacker can subtly manipulate data and affect a model's prediction. In this paper, we demonstrate a brand new adversarial attack method in textual data. We use activation maximization to create an importance rating for each unique word in the corpus and attack the most important words in each sentence. The rating is global to the whole corpus and not to each specific data point. This method performs equal or better when compared to previous attack methods, and its running time is around 39 times faster than previous models.

Keywords: Adversarial attacks · Natural language processing · Deep learning · LSTM · Transformers · CNN

1 Introduction

Machine learning with deep learning models has achieved impressive advances, empowered by the era of big data and the proliferation of cheap computing power. Such models have reached or even surpassed human expert performance in a number of tasks (e.g. machine vision and natural language processing). However, despite their accuracy and performance, many other shortcomings and limitations must be improved. The most important limitation of deep learning models is that they are like black boxes and are not interpretable. As a result, we may not trust them in life and death situations. In addition, we have seen that they are extremely vulnerable to adversarial attacks. If an attacker creates adversarial examples by adding a little well-chosen noise to an input, although humans still classify them appropriately, the models can be deceived and classify them incorrectly, potentially leading to catastrophic results.

Adversarial attacks come in two flavors: white box and black box attacks. In a white box attack, the attacker has access to the information, details and

© Springer Nature Switzerland AG 2021
L. Rutkowski et al. (Eds.): ICAISC 2021, LNAI 12855, pp. 69–80, 2021.
https://doi.org/10.1007/978-3-030-87897-9_7

weights of the attacked model. Conversely, in the black box context, the attacker does not have any internal model information. Obviously, in real life scenarios, black box attacks are more realistic. There have been many works on adversarial attacks and creating defense mechanisms against them. Most have concentrated on machine vision and image processing, as it is much easier to visualize and compare images in a 2-dimensional format, whereas in the Natural Language Processing (NLP) field, there has significantly less work, as it is much more challenging to visualize and compare textual data. NLP is the field of allowing machines to communicate in humans' languages, and many tasks, from sentiment analysis to natural language generation, fall under its umbrella.

Adversarial attacks in NLP consist of two phases: choosing the words to attack and choosing the technique of manipulation or perturbation. There are many different word-level perturbation techniques in NLP (e.g. replace, delete, add, swap); these may seem different to humans but all of them result in the same behavior in NLP models. They will create a noisy word that is not in the model dictionary, and the model will assign it an 'unknown' label. Most existing adversarial attacks in NLP choose their targeted words to attack one input at a time, where each attack is applied to the most important words for a particular input instance. While such a technique can be successful, it requires access to the texts to be attacked in advance, in order to train the attacker model. They cannot attack brand new texts even when they are from the same source and domain.

In this paper, we have created a brand new black box adversarial attack technique that uses and trains a Convolutional Neural Network (CNN) on a portion of a dataset. It then analyses the CNN filters to establish a global importance rate for all words in the corpus using activation maximization. These importance rates apply across the model and reflect the overall model logic, rather than being local to a single data point. These importance rates can be used to attack even examples that have never been seen before by targeting the most important words. Our technique has equal or better performance in comparison to previous techniques in various tasks, and it is much faster as it does not need to analyze each and every new data point to develop an optimal attack.

We tested our adversarial attack on two benchmarks: a sentiment analysis (binary classification) and a tagging dataset with 20 possible classes. The attacks on these datasets were evaluated on three different models to see if the attack technique is generalizable to multiple architectures. We used Long Short Term Memory (LSTM), Convolutional Neural Networks (CNN), and attention-based Transformers. Performance was equal to previous techniques in the sentiment analysis task and much better in the tagging task. When we are attacking 40 words per text, our model reduces the test accuracy 6% more than previous techniques on average. In addition, in both tasks, our technique is much faster than the previous versions. Our technique, applied to a forty-thousand-element dataset, takes 164 s, while the baseline comparison model takes 6,323 s, a 39-fold improvement.

2   Related Works

NLP is a subfield of Artificial Intelligence (AI) that enables us to interact with computers in human languages [4,7,8]. To apply standard deep learning architectures to NLP, textual data must be transformed into an amenable numerical format. There are multiple ways of doing this, such as a one-hot encoder or n-gram representation, but by far the most common approach is to use the Word2Vec [17] algorithm. This creates a custom-length vector that represents the semantic meaning of a particular word, and attempts to reflect the relationship of words with each other. Word2Vec mappings can be trained from scratch, but it is also common to use one of the available pre-trained word representations like Glove [18].

After preprocessing texts, and transforming them into numerical vectors, the data can be fed into a deep learning model. There are various famous model architectures that work well on NLP tasks. The first natural choice is LSTM [9] which is an advanced version of a Recurrent Neural Network (RNN). It is designed with time-series data in mind, and it has an internal memory. According to its gate weights, it will decide what to remember and what to forget. Another famous deep learning architecture is the CNN. We know that CNNs have revolutionized image processing tasks, but CNNs can also be applied to textual data [10–12,21]. Yin [22] compared RNNs and CNNs on various NLP tasks and studied the performance of each. Activation Maximization (AM) is a method that can be applied on CNN models; some research has focused on creating an importance rate for each unique word in a corpus using AM on CNNs by analyzing the convolution filter weights [15]. We utilize this technique to create an adversarial attack method.

In 2017, a new generation of NLP models appeared, starting with Vaswani's first Transformer attention-based architecture [19]. Instead of remembering an entire text, it assigns an attention weight to each token, which allows it to process much longer texts. The attention technique enabled the creation of much more advanced transformer-based models like BERT [5], RoBERTa [13], and GPT-3 [3].

The various deep learning models have contributed to rapid improvement in NLP task performance, but they also have created challenges. To start with, deep learning models are not intuitive or explainable. They also have some limitations; for instance, most NLP models can accept only limited sequence lengths. Some authors have tried to overcome this limitation [16]. In addition, they can easily be manipulated by adversarial examples. Adversarial attacks have recently been a major research focus in machine vision [1,2], as it is very easy for humans to recognize patterns in two dimensions. However, very few have worked on NLP adversarial attacks. The same adversarial attacks techniques can be applied on textual data as well [20,23]. Gao [6] created a technique called WordBug that analyses each and every input to find the most important words in that text, then targeting them in an attack. They have used a temporal score extracted from a bidirectional LSTM to detect important words in each text.

In this paper, our contribution offers a similar technique that is much faster, and its performance is either equal or better than WordBug. We evaluate our algorithm on two datasets with different levels of difficulty: an IMDb sentiment analysis dataset [14] and a Stack Overflow dataset with 20 possible tags or classes.

3 Technical Description

3.1 Benchmark Datasets

In order to test our approach, we chose and used two datasets with different tasks. The first one is the IMDb movie review dataset,[1] which is a binary classifcation task for sentiment analysis. The second one is the Stack Overflow dataset,[2] in which each question is tagged with one of 20 possible tags. Obviously the first task is easier for models as it contains only two classes.

Our preprocessing step was very straightforward. We converted everything to lower case and dropped all stop-words and hapax legomena (words that appear only once in an entire corpus). We also dropped all special characters and numerical values. After preprocessing, the Stack Overflow dataset contained 40,000 rows and around 28,000 unique words. The IMDb dataset contained 42,928 rows and around 23,000 unique words. We split each of them into training and test sets.

3.2 Choosing and Targeting Words to Attack

The attack sequence begins by targeting particular words, hopefully selected to have the greatest possible effect on the model's performance and accuracy. The mechanism for choosing words to attack is the core comparison between techniques.

WordBug. Gao's technique [6] created a bidirectional LSTM and trained the model on each sentence to be attacked, identifying the most important words in each sentence in turn. The approach used three scores: Temporal, Temporal Tail and combined (which is the mean of the two previous scores). In order to create the Temporal score and Temporal Tail score of the ith word in a sentence, they used Eqs. 1 and 2, in which n is the number of tokens or words in each data point, $x[1:n]$ are the n words in each data points and $F()$ is a function that maps input words to the probability of belonging to the actual class using the bidirectional LSTM. They observed that their combined score outperformed other approaches.

$$TemporalScore(x_i) = F(x_1, x_2, ..., x_{i-1}, x_i) - F(x_1, x_2, ..., x_{i-1}) \qquad (1)$$

---

[1] https://ai.stanford.edu/~amaas/data/sentiment/.
[2] https://console.cloud.google.com/marketplace/product/stack-exchange/stack-overflow.

$$TemporalTailScore(x_i) = F(x_i, x_{i+1}, x_{i+2}, ..., x_n) - F(x_{i+1}, x_{i+2}, ..., x_n) \quad (2)$$

The advantage of this technique is that each text is studied and produces the most important words local to that specific text. However, it has two big disadvantages: it needs to be trained on each and every sentence to be attacked, and it cannot be applied to other sentences even from the same context. It is also very slow, as it needs to run the LSTM $m * n$ times in which m is the number of rows and n is the number of tokens in each one.

Activation Maximization. This method [15] uses a 1-dimensional CNN, trains on a subset of data, and uses the CNN layer filters to find the importance rate of all unique words in the corpus based on Eq. 3.

$$importance = \left\{ \sum_{f=1}^{F} \sum_{s=1}^{S} \sum_{i=1}^{I} |w_i * Filter_{f*s*i}| \, | w \in Corpus, Filter \right\} \quad (3)$$

In Eq. 3, F is the number of filters in the CNN layer, S is the size of the filters, and I is the embedding length. w is a word embedding vector with a length of I. Corpus is a matrix of our entire word embedding of size $m * I$, in which m is the count of unique words in our corpus dictionary. Filter is a 3-D tensor of size $F * S * I$. This equation calculates the sum of activations of all filters caused by a single word.

The advantage of this method is that it provides a general importance rate applicable to all sentences that come from the same source and distribution. As a result, we can use its insight on new never-before-seen sentences. The other advantage is its speed, as it needs only to train the CNN model once, for a couple of epochs, and then its filters can be used (via Eq. 3 to create the importance rate. The downside of this method is that the globally important words may or may not be the best option for each sentence.

Choosing Words in Each Sentence. After using one of the above models, we target the top t words in each sentence that have the highest importance rate and attack them. In our experiments we used different following values for t: 0, 10, 20, 40. When the $t = 0$, it means we are not attacking at all; this is a baseline performance for comparison with our attacks. We tested the attacks on three different models to determine whether these techniques are applicable on all models. We used an LSTM, a CNN and a Transformer model to see how the different attacks behave on each.

3.3 Word Perturbation

After selecting the words to attack, a word manipulation or perturbation method must also be chosen. There are four main word attack methods in NLP: replace, delete, add, swap. To human observers, these might have different effects, but all of these methods have the same result, to create a noisy word that is not in the NLP model dictionary. Such words end up with an 'unknown' label assigned by the model. We used swap in this paper; we randomly chose two adjacent middle character in each targeted word and swapped them. Examples are presented in Table 1. Note that, for purposes of simplifying the presentation, a very short subset of texts is presented in this table. In the original data, the texts are much longer, averaging around 200 words or tokens in each.

Table 1. Word manipulation and perturbation (with $t = 2$)

Text before the perturbation	Text after the perturbation (with $t = 2$)
The movie was extremely boring	The moive was extremely broing
How can i import csv file in python	How can i improt csv file in pyhton

3.4 Comparison of Methods and Evaluation

In order to compare WordBug vs our Activation Maximization method, we tested both on three brand new models created from scratch on our two benchmark datasets. We used a CNN, an LSTM and a Transformer model, all of which were built with basic, straightforward architecture. We trained all of these models on the training set. As our attack method is a black-box adversarial attack, we attacked only the test set (since the attack does not have access to the models' information and details).

In the WordBug method, each text in the test set was analyzed and attacked in turn, but in our Activation Maximization method, we fitted our importance rate on a subset of texts and used it for attacking the test set as a whole. We measured the run-time of both techniques while they were learning which words to attack and compared them. In the next step we evaluated our attacks on the three models we had trained, to observe the attack performance. In both methods, we used swap as the word perturbation, and used 0, 10, 20, 40 for our t.

4   Experimental Results

In the experiments, t is set to four different values: $0, 10, 20, 40$. We also included $t = 0$ which shows the baseline model without any inserted attack, so the effectiveness of each adversarial attack method can be seen. We tested our three models on two datasets (IMDb and Stack Overflow) with 3 different techniques of attack: WordBug, Activation Maximization and random. In the random case, t words are chosen at random to attack. In interpreting the attack performance, a lower test accuracy means a more effective attack.

4.1   Evaluation on IMDb

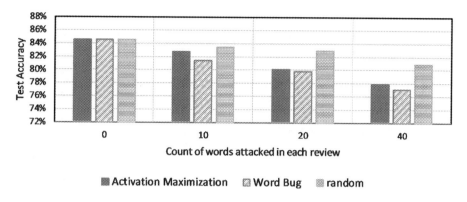

Fig. 1. Evaluation of attack methods in CNN on IMDb dataset

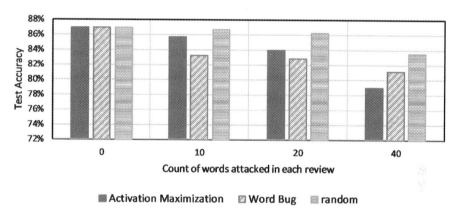

Fig. 2. Evaluation of attack methods in LSTM on IMDb dataset

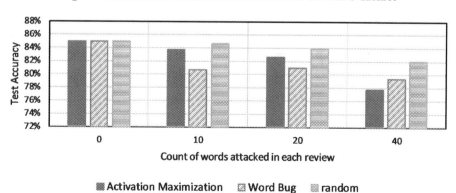

Fig. 3. Evaluation of attack methods in transformer on IMDb dataset

Figures 1, 2, and 3 show that, for the IMDb dataset in most cases, the WordBug technique works slightly better than Activation Maximization but not significantly. On average, the WordBug attacks reduce the models' accuracies less than 1% more than the AM attacks.

Our results when using WordBug to attack the IMDb dataset differs from Gao's original results [6]. Ours is a precise implementation of the WordBug attack, so the only potential reasons for this variation might be different preprocessing steps or different LSTM architectural hyperparameters. However, for the purposes of comparing WordBug and Activation Maximization, this difference is immaterial. The same preprocessing steps and model hyperparameters are used to compare the approaches fairly.

4.2   Evaluation on Stack Overflow

The Stack Overflow dataset task is much more complex than the naive Sentiment Analysis task in IMDb. The task requires classifying data into 20 possible classes or tags, and the texts are much longer. As a result, Figs. 4, 5, and 6 show that the Activation Maximization method is performing significantly better than WordBug. On average, the Activation Maximization attack leads to a 4.6% lower performance than WordBug's.

4.3   Transfer Learning from IMDb to Stack Overflow

After comparing the performance of WordBug and Activation Maximization, we show that in the AM method, we can train the attack on a subset of data and then use it to attack newer never-seen-before data from the same distribution and context. This performs equally well or better than WordBug, which has to study every single input to be able to attack it.

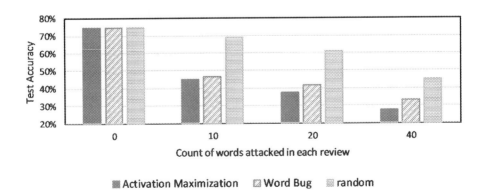

Fig. 4. Evaluation of attack methods in CNN on Stack Overflow dataset

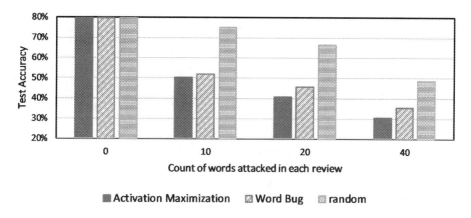

Fig. 5. Evaluation of attack methods in LSTM on Stack Overflow dataset

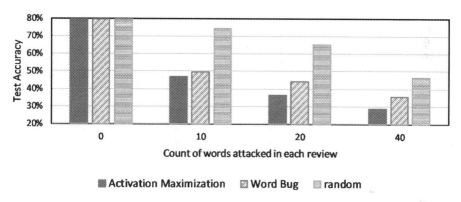

Fig. 6. Evaluation of attack methods in transformer on Stack Overflow dataset

We also wanted to show whether the same transfer learning logic can apply more generally, on new data but from a different context. In order to check this, we trained our Activation Maximization attacker on the IMDb dataset and used it on Stack Overflow. Figure 7 shows that the experiment does not support this, as the performance is even worse than a random attack.

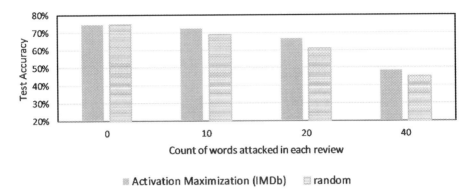

Fig. 7. Evaluation of using IMDb importance rate extracted from IMDb dataset and used to attack Stack Overflow dataset

4.4   Run Time Comparison

In addition to their effect on a model's accuracy, another important factor in adversarial attacks is the speed. We applied the attacks on corpora with different sizes varying from 20 to 40,000 data elements and measured the time needed for each method to train and choose the most important words to target.

Figure 8 shows that Activation Maximization is much faster than WordBug. This is to be expected, as WordBug must analyze each word in each data element one by one, while the Activation Maximization approach just trains a CNN model for a couple of epochs and uses the resultant filters to identify globally important words. With a corpus size of 40,000, Activation Maximization is 39 times faster than WordBug.

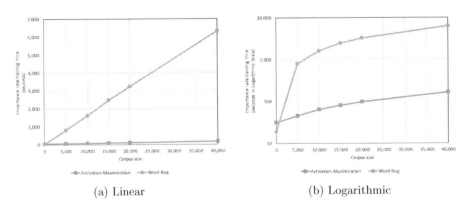

(a) Linear   (b) Logarithmic

Fig. 8. Runtime comparison of WordBug and Activation Maximization

4.5   Results Analysis

We have demonstrated that the Activation Maximization approach is not only much faster than WordBug, but that it is equally good or better in its attack performance (depending on the task). In addition, Activation Maximization can be used to attack new input without any need for further training, whereas WordBug must be trained on each sentence prior to a successful attack. However, we observed that this ability of Activation Maximization only applies to data from the same context and distribution.

Both of these models are black-box attacks and do not have access to the details of the attacked model. In WordBug, the extracted information about most important words is local to each specific input, whereas in Activation Maximization, the attacker finds the most important words to target based on a global evaluation of a large corpus.

5   Conclusion and Future Work

Our new Activation Maximization adversarial attack has many benefits: it is significantly faster, its performance is equal to or better than WordBug, and it can be used on new inputs (from the same context) without any further training. Our attack method is a black box attack which means it does not need access to the structure or weights of an attacked model. Developing these kinds of effective and efficient attacks will enable us to evaluate new models to find their blind spots.

Our future work involves upgrading the attack so that it can be used on other contexts as well through transfer learning. As a result, we may be able to train an attacker without reference to a specific kind of data, and still effectively attack many different kinds of data targets.

References

1. Akhtar, N., Mian, A.: Threat of adversarial attacks on deep learning in computer vision: a survey. IEEE Access **6**, 14410–14430 (2018)
2. Brendel, W., Rauber, J., Bethge, M.: Decision-based adversarial attacks: reliable attacks against black-box machine learning models. arXiv preprint arXiv:1712.04248 (2017)
3. Brown, T.B., et al.: Language models are few-shot learners. arXiv preprint arXiv:2005.14165 (2020)
4. Collobert, R., Weston, J., Bottou, L., Karlen, M., Kavukcuoglu, K., Kuksa, P.: Natural language processing (almost) from scratch. J. Mach. Learn. Res. **12**(Aug), 2493–2537 (2011)
5. Devlin, J., Chang, M.W., Lee, K., Toutanova, K.: BERT: pre-training of deep bidirectional transformers for language understanding. arXiv preprint arXiv:1810.04805 (2018)
6. Gao, J., Lanchantin, J., Soffa, M.L., Qi, Y.: Black-box generation of adversarial text sequences to evade deep learning classifiers. In: 2018 IEEE Security and Privacy Workshops (SPW), pp. 50–56. IEEE (2018)

7. Goldberg, Y.: A primer on neural network models for natural language processing. J. Artif. Intell. Res. **57**, 345–420 (2016)

8. Hirschberg, J., Manning, C.D.: Advances in natural language processing. Science **349**(6245), 261–266 (2015)

9. Hochreiter, S., Schmidhuber, J.: Long short-term memory. Neural Comput. **9**(8), 1735–1780 (1997)

10. Kalchbrenner, N., Grefenstette, E., Blunsom, P.: A convolutional neural network for modelling sentences. arXiv preprint arXiv:1404.2188 (2014)

11. Kim, Y.: Convolutional neural networks for sentence classification. arXiv preprint arXiv:1408.5882 (2014)

12. Le, H.T., Cerisara, C., Denis, A.: Do convolutional networks need to be deep for text classification? In: Workshops at the Thirty-Second AAAI Conference on Artificial Intelligence (2018)

13. Liu, Y., et al: RoBERTa: a robustly optimized BERT pretraining approach. arXiv preprint arXiv:1907.11692 (2019)

14. Maas, A.L., Daly, R.E., Pham, P.T., Huang, D., Ng, A.Y., Potts, C.: Learning word vectors for sentiment analysis. In: Proceedings of the 49th Annual Meeting of the Association for Computational Linguistics: Human Language Technologies, vol. 1, pp. 142–150. Association for Computational Linguistics (2011)

15. Marzban, R., Crick., C.: Interpreting convolutional networks trained on textual data. In: Proceedings of the 10th International Conference on Pattern Recognition Applications and Methods - Volume 1: ICPRAM, pp. 196–203. INSTICC, SciTePress (2021). https://doi.org/10.5220/0010205901960203

16. Marzban, R., Crick., C.: Lifting sequence length limitations of NLP models using autoencoders. In: Proceedings of the 10th International Conference on Pattern Recognition Applications and Methods - Volume 1: ICPRAM, pp. 228–235. INSTICC, SciTePress (2021). https://doi.org/10.5220/0010239502280235

17. Mikolov, T., Sutskever, I., Chen, K., Corrado, G.S., Dean, J.: Distributed representations of words and phrases and their compositionality. In: Advances in Neural Information Processing Systems, pp. 3111–3119 (2013)

18. Pennington, J., Socher, R., Manning, C.D.: GloVe: global vectors for word representation. In: Proceedings of the 2014 Conference on Empirical Methods in Natural Language Processing (EMNLP), pp. 1532–1543 (2014)

19. Vaswani, A., et al.: Attention is all you need. In: Advances in Neural Information Processing Systems, pp. 5998–6008 (2017)

20. Wallace, E., Feng, S., Kandpal, N., Gardner, M., Singh, S.: Universal adversarial triggers for attacking and analyzing NLP. arXiv preprint arXiv:1908.07125 (2019)

21. Wood-Doughty, Z., Andrews, N., Dredze, M.: Convolutions are all you need (for classifying character sequences). In: Proceedings of the 2018 EMNLP Workshop W-NUT: The 4th Workshop on Noisy User-Generated Text, pp. 208–213 (2018)

22. Yin, W., Kann, K., Yu, M., Schütze, H.: Comparative study of CNN and RNN for natural language processing. arXiv preprint arXiv:1702.01923 (2017)

23. Zhang, W.E., Sheng, Q.Z., Alhazmi, A., Li, C.: Adversarial attacks on deep-learning models in natural language processing: a survey. ACM Trans. Intell. Syst. Technol.gy (TIST) **11**(3), 1–41 (2020)

RGB-D Odometry for Autonomous Lawn Mowing

Marcin Ochman[1,2](✉) ⓘ, Magda Skoczeń[1,2] ⓘ, Damian Krata[1], Marcin Panek[1],
Krystian Spyra[1], and Andrzej Pawłowski[1] ⓘ

[1] Unitem, Wroclaw, Poland
{marcin.ochman,magda.skoczen,damian.krata,marcin.panek,
krystian.spyrakrystian.spyra,andrzej.pawlowski}@unitem.pl
[2] Faculty of Electronics, Wroclaw University of Science and Technology,
Wroclaw, Poland

Abstract. Localization for outdoor mobile robots is crucial to accomplish complex tasks in difficult environments. One of the examples is an autonomous mower operating in various lawns placed in parks, airports, home gardens and many more. To ensure all navigation algorithms' requirements are met, first accurate estimation of current position and orientation needs to be found. Scientists proposed many approaches using encoders, RADARs, LIDARs or vision/depth cameras. However, this is the first attempt to investigate odometry performance for autonomous lawn mowing using RGB-D cameras. The contribution is twofold. First, several odometry algorithms in autonomous mower environments were examined in terms of localization accuracy and execution time. Secondly, a new dataset was collected containing sequences from a city park and home lawn. The dataset contains aligned color and depth images. This study aimed to extend knowledge about RGB-D odometry and analyze how RGB-D cameras may be used in agricultural robots, where the environment is often an open space without many feature points or distinctive objects used in the odometry algorithms.

Keywords: Autonomous mower · Visual odometry · RGB-D camera · RGB-D odometry · Agricultural robot · Mobile robot · Localization

1 Introduction

One of the major problems for unmanned ground vehicles and autonomous robots is accurate localization. This task is especially challenging in unknown outdoor environments where robots meet with the presence of uneven and rough terrain as well as changing weather and lighting conditions. Furthermore, the localization task is part of the navigation pipeline in real-time applications such

This work was supported by The National Centre for Research and Development [grant number POIR.01.01.01-00-1069/18].

© Springer Nature Switzerland AG 2021
L. Rutkowski et al. (Eds.): ICAISC 2021, LNAI 12855, pp. 81–90, 2021.
https://doi.org/10.1007/978-3-030-87897-9_8

as self-driving cars. The recent development of accurate, robust and fast RGB-D sensors, such as Intel RealSense [8] enhances the use of vision-based motion estimation methods which is visual odometry.

The term visual odometry(VO) refers to the process of estimating the motion of the robot-based only on the visual input from the cameras attached to the platform [11,13].

The rest of the paper is organized as follows. Section 2 presents the existing solutions for outdoor mobile robots. Section 2.1 describes the algorithms used in our study. In Sect. 3 we formulate the research problem tackled in this paper. Sections 3.1 and 4 contain the dataset description and the results discussion, respectively. Finally, some conclusions are drawn in Sect. 5.

2 Related Work

Over the years, researchers proposed multiple approaches to localization systems using a variety of sensors. The most commonly used methods for navigation in the outdoor environment utilize signals from Global Navigation Satellite System (GNSS) [7,12] providing a global position of a robot. However, satellite signals suffer from signal blockage and multi-reflection for many robot's working areas, such as robots working in agriculture, forestry or mining industry, which makes the measurements inaccurate and requires the usage of alternative solutions [17]. One of them is called dead reckoning also known as relative positioning, which is a method that evaluates the position of the mobile robot based on the measurements of its internal sensors so it does not depend on external information about the environment. It incorporates the signal from encoders attached to the wheels (wheel odometry) and from the Inertial Measurement Unit (IMU) [4]. The next is a group of range-based localization techniques, which utilize the measurements of sonar or laser rangefinder sensors. These methods include algorithms such as landmark-based localization [16] or scan matching algorithm [9]. Visual odometry is another commonly used technique. It utilizes the camera's signal to estimate the robot's motion. Three main types of algorithms can be distinguished: feature-based, dense and learning-based methods. Initial work on visual odometry focused on sparse, feature-based methods [6]. First, this technique extracts a sparse set of feature points from a sequence of images and matches them. Then the camera pose is estimated by minimizing the reprojection errors between feature pairs. With the emergence of the consumer RGB-D cameras, direct methods were developed [10,15]. A more extensive description of the dense approach can be found in Sect. 2.1. In recent years, there has been growing interest in the learning-based methods due to advances of deep learning algorithms [18]. Usually in real-life applications, the robot pose is calculated by fusing the multiple types of data using methods such as extended Kalman filter (EKF) [2].

For a detailed review of localization systems for robots in agriculture, we refer the readers to [1] and the references therein.

2.1 Algorithms Overview

The algorithms used in this study belong to a group of dense methods, which means that they estimate the robot pose directly from the whole image. Compared with the feature-based methods, this approach allows to reduce the execution time of algorithms by omitting the feature extraction and matching step. Moreover, usage of the whole image instead of chosen points allows to more efficiently exploit data acquired by the sensors. The algorithms are implemented in the OpenCV library [3].

The **RgbdOdometry** algorithm [15] estimates the robot poses directly and densely from the RGB-D images. This method finds a rigid body transformation which represents the camera motion by minimizing the photometrical error between two consecutive images captured by camera. This method is based on the photo-consistency assumption. That implies that for good performance, it requires a static environment and small camera displacements.

The **Iterative Closest Point** (ICP) algorithm [10] utilizes only the depth maps. The camera pose estimation process comprises two iteratively repeating steps. The first step aims to find the correspondence between the predicted model of the surface and currently acquired depth data using a projective data association algorithm. Then, the second step uses the correspondence to estimate the transformation (pose estimation) which aligns the predicted model of the surface with the depth by minimizing the point-plane distance. The new pose estimation is then used in the first step. The process repeats until convergence.

The **FastICPOdometry** algorithm is an accelerated implementation of the ICP algorithm. Some of the improvements are parallel computations, interpolation of the points and the normals or the use of the universal intrinsics.

The **RgbdICPOdometry** algorithm combines results obtained with *RgbdOdometry* and *ICPOdometry* by minimizing the sum of their energy functions.

3 Problem Description

Autonomous mower may operate in various and demanding environments, especially for vision-based systems. It seems to be a challenging area for odometry algorithms due to changing lightning conditions, dynamic scenes or dust occurance. Many studies have only focused on indoor applications [14] or outdoor applications where environment is very diversified i.e. autonomous car on a street [5]. The aim of the research is to verify whether RGB-D cameras and RGB-D odometry algorithms may be used outdoors, especially in natural environments of a mower and measure both localization errors and execution time. The platform used in the study is shown in Fig. 1a. The mower is equipped with several sensors including RGB-D cameras, encoders, LIDARS, GNSS and safety bumpers. All computations are executed on onboard computer powered by Robot Operating System. Its task is to accomplish plan generated by navigation module based on given contour. Navigation algorithms is out of the scope of the article thus it is not going to be discussed. However, accurate localization is obligatory to fulfill mowing planner requirements. The platform is a 4-wheeled synchro drive system. As a result, the orientation of the mower is constant. Consequently, this study is focused mainly on a translational vector.

(a)   (b)

Fig. 1. Autonomous Mower: a) platform used in the research, b) Intel RealSense D435i mounted on autonomous mower

3.1   Dataset Overview

One of the main contribution of this research was dataset containing various sequences from Intel RealSense D435i (Fig. 1b). Based on sensors mounted on autonomous mower (Fig. 1a) such as encoders, IMU and GNSS, reference path for each sequence was measured. The dataset was divided into two categories depending on the place where images and ground-truth were recorded i.e. city park and home lawn. Dataset contains 18 sequences of different duration. Each sequence consists of depth and RGB images stored in PNG format. Depth data is stored as a 16-bit single-channel image, while RGB as an 8-bit three-channel image. In addition, every sequence has a CSV file containing ground truth represented by translational vector and orientation quaternion. Figure 2 presents examples of RGB images and theirs equivalent of depth image from home lawn and city park. In both groups, one may find numerous trajectory shapes. Moreover, in the city park sequences there are dynamic objects e.g. pedestrians which make dataset more demanding and diversified for evaluation purposes.

(a)   (b)

Fig. 2. Examples of RGB and depth images with applied colormap from prepared dataset. There are two different environments: a) home lawn, b) city park.

3.2   Error Metrics

To compare algorithms accurately, we defined and measured three different error types. Positional error (PE_n) is a distance between the n-th reference point x_n^{ref} and n-th point x_n of estimated trajectory (Eq. 1). A major disadvantage of the positional error is an accumulation of former errors. As a result, big error in one algorithm step may result in higher values of latter PE.

$$\text{PE}_n = ||x_n - x_n^{\text{ref}}|| \tag{1}$$

To avoid described drawback, relative positional error (RPE_n) is introduced (Eq. 2). In RPE_n the translations $t_n = x_n - x_{n-1}$ and $t_n^{\text{ref}} = x_n^{\text{ref}} - x_{n-1}^{\text{ref}}$ between consecutive points of estimated and reference trajectory is computed. Finally, the Euclidean norm of subtracted vectors is evaluated.

$$\text{RPE}_n = ||t_n - t_n^{\text{ref}}|| = ||x_n + x_{n-1}^{\text{ref}} - x_n^{\text{ref}} - x_{n-1}|| \tag{2}$$

Additionally to RPE_n, relative orientation error (ROE_n) is calculated (Eq. 3) which is angle between vectors t_n and t_n^{ref}.

$$\text{ROE}_n = \arccos \frac{t_n \cdot t_n^{\text{ref}}}{||t_n||||t_n^{\text{ref}}||} \tag{3}$$

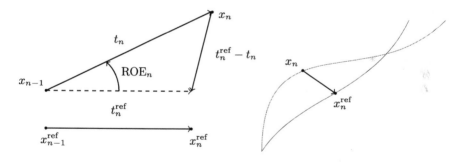

Fig. 3. Visualization of RPE_n, ROE_n and PE_n.

4   Research and Results

In our tests, we used four algorithms described in Sect. 2.1. For every sequence of the prepared dataset we compared mean and standard deviation of three error metrics according to the Eq. 4, where E is one of defined errors $\{\text{PE}, \text{RPE}, \text{OE}\}$ and N is total number of sequence frames. Execution times were also taken into considerations.

$$\bar{E} = \frac{1}{N} \sum_{n=1}^{N} E_n, \quad \sigma_E = \sqrt{\frac{1}{N} \sum_{n=1}^{N} (E_n - \bar{E})^2} \tag{4}$$

Results are presented in Table 1. The best localization accuracy is provided by *RgbdICPOdometry* algorithm. However, best performance is coupled with higher computation requirements so algorithm execution takes longer. It is caused by its complexity. *RgbdICPOdometry* combines outcomes of two independent algorithms which are *ICPOdometry* and *RgbdOdometry*. Nonetheless, 50 ms is still good result comparing to our demand which is 15 fps. Tests were carried out on PC with 16 GB RAM and Intel Core i7 7700 CPU. Experiment proves also accumulative characteristics of PE.

Figure 4 presents PE in time. Figures 5a and 5b show that *RgbdICPOdometry* is the most similar to the reference trajectory in both plots. By contrast, *FastICPOdometry* and *RgbdOdometry* give poor results. Their estimated trajectories in numerous sequences are completely different to the reference ones, so they are not suitable for our autonomous mower. These observations are in line with Table 1 data. The analysis of the sequences' RGB images shows a significant disparity of color values for consecutive images due to lighting condition changes and camera post-processing. This makes photo-consistency-based approaches such as *RgbdOdometry* not appropriate for outdoor applications.

In spite of the fact that *RgbdICPOdometry* handles quite well, its accuracy is still not acceptable. In fact, none of the examined algorithms are satisfactory.

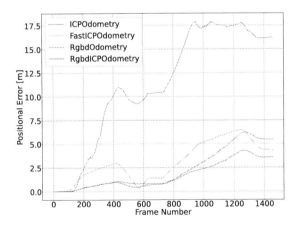

Fig. 4. Positional error accumulation in time for home lawn sequence 2.

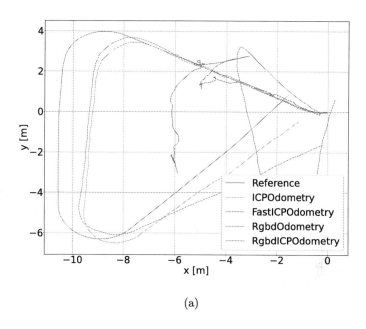

(a)

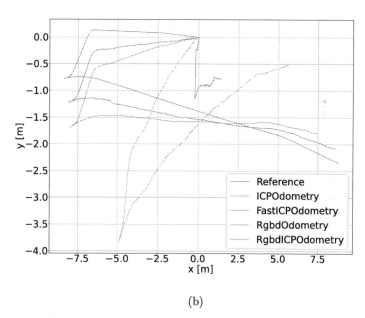

(b)

Fig. 5. Comparison of reference trajectory and estimated trajectories using four algorithms: ICPOdometry, FastICPOdometry, RgbdICPOdometry and RgbdOdometry. a) home lawn 7, b) city park 6

Table 1. Means and standard deviation of algorithm's translational, rotational error and execution time for each sequence of the prepared dataset categorized on various environments.

Dataset	No.	ICPOdometry								FastICPOdometry							
		PE [m]	σ_{PE} [m]	RPE [cm]	σ_{RPE} [cm]	ROE [°]	σ_{ROE} [°]	t [ms]	σ_t [ms]	PE [m]	σ_{PE} [m]	RPE [cm]	σ_{RPE} [cm]	ROE [°]	σ_{ROE} [°]	t [ms]	σ_t [ms]
Backyard	0	2.91	1.57	0.61	0.97	25.62	48.22	26.15	11.05	3.34	1.73	0.46	0.69	30.90	47.73	17.10	3.28
	1	1.72	0.86	0.98	0.76	22.92	31.92	35.63	2.05	6.14	5.51	2.22	1.97	73.45	59.25	19.95	1.55
	2	2.21	2.11	1.34	0.90	28.73	31.52	35.73	1.80	3.10	2.00	1.20	0.94	26.28	33.77	20.10	0.91
	3	0.90	0.60	1.34	0.96	26.11	28.07	35.68	1.97	2.42	2.05	1.82	1.03	37.06	33.56	19.81	0.79
	4	0.76	0.67	1.23	0.81	17.78	21.14	35.47	2.09	2.48	2.02	1.37	0.98	18.17	21.56	19.86	0.84
	5	1.14	0.49	1.54	2.40	15.09	17.32	36.17	2.25	4.34	2.11	2.28	2.13	29.38	19.58	21.77	0.73
	6	0.25	0.12	1.14	0.93	11.48	13.80	35.18	2.36	0.73	0.32	1.58	1.03	20.72	15.83	22.26	0.52
	7	0.91	0.40	1.10	0.75	13.85	15.45	35.46	2.14	4.86	2.55	1.61	0.79	22.95	16.65	19.30	0.70
	8	0.52	0.30	1.30	0.85	17.93	18.76	35.39	1.84	1.78	1.04	1.67	0.83	25.13	19.39	19.25	0.45
City park	0	1.41	0.33	0.49	1.02	23.65	37.36	35.53	2.04	3.90	1.01	0.69	1.30	35.51	45.40	22.81	0.58
	1	1.70	2.11	1.02	1.55	44.23	43.18	35.60	2.16	2.60	2.33	1.16	1.66	56.01	48.41	22.98	0.59
	2	2.90	1.71	2.21	2.23	43.56	34.07	35.70	3.98	3.94	2.18	2.72	2.17	68.70	37.84	22.82	0.51
	3	0.01	0.01	0.16	0.11	12.11	30.57	35.87	4.16	0.11	0.06	0.17	0.07	13.51	33.93	23.37	0.63
	4	3.45	1.69	1.30	1.04	38.04	29.85	35.70	2.26	7.28	3.46	1.74	1.35	71.66	36.21	23.80	0.48
	5	1.56	0.96	1.22	0.84	37.70	30.20	35.96	2.84	3.77	2.04	1.69	1.09	77.40	36.04	24.27	0.52
	6	2.12	1.30	1.48	0.94	30.57	20.34	35.45	2.37	4.17	2.37	2.08	1.16	56.35	32.84	23.58	0.61
	7	4.93	3.36	1.24	0.81	38.32	27.76	36.29	2.28	9.93	5.68	1.67	0.90	66.49	37.14	23.53	0.42
	8	10.11	4.74	4.83	4.22	51.80	45.14	36.46	1.88	12.97	6.55	4.12	3.68	33.57	27.98	21.90	0.94
Average		2.20	1.30	1.36	1.23	27.75	29.15	35.19	2.86	4.33	2.50	1.68	1.32	42.40	33.59	21.58	0.84

Dataset	No.	RgbdOdometry								RgbdICPOdometry							
		PE [m]	σ_{PE} [m]	RPE [cm]	σ_{RPE} [cm]	ROE [°]	σ_{ROE} [°]	t [ms]	σ_t [ms]	PE [m]	σ_{PE} [m]	RPE [cm]	σ_{RPE} [cm]	ROE [°]	σ_{ROE} [°]	t [ms]	σ_t [ms]
Backyard	0	1.02	0.52	0.35	0.81	22.09	40.84	8.74	3.15	2.37	1.32	0.51	0.84	24.94	46.51	32.51	14.77
	1	5.15	3.35	1.67	1.88	51.96	58.74	12.17	1.31	1.48	0.70	0.70	0.59	19.00	30.24	45.23	1.90
	2	11.25	5.96	3.82	2.66	97.34	60.97	12.33	1.16	1.70	1.34	1.02	0.82	21.89	26.47	46.17	1.22
	3	5.76	3.90	3.48	1.82	63.33	52.85	10.52	2.76	0.80	0.47	1.11	0.91	19.62	21.45	45.94	1.46
	4	2.92	2.27	2.08	1.58	50.66	56.37	11.09	0.42	0.60	0.52	0.91	0.69	13.22	18.15	45.08	1.27
	5	4.25	2.32	3.67	3.11	68.08	53.86	12.82	1.31	0.91	0.51	1.25	2.48	11.22	15.62	45.50	1.75
	6	1.59	1.37	2.25	2.34	40.33	50.08	14.52	1.23	0.24	0.13	0.83	0.89	8.61	13.03	46.91	1.84
	7	4.57	2.77	3.72	2.27	77.23	56.52	12.35	2.38	0.99	0.61	1.01	0.72	14.12	14.56	45.22	1.98
	8	2.05	1.29	2.95	1.96	73.37	59.18	11.70	1.21	0.45	0.27	1.08	0.78	14.05	17.05	45.96	1.44
City park	0	0.70	0.11	0.51	0.95	20.95	39.70	19.49	0.71	0.48	0.09	0.48	0.93	19.01	35.52	51.98	1.50
	1	0.66	1.07	0.90	1.42	30.09	40.03	20.48	2.44	1.14	1.03	0.71	1.14	26.38	35.75	52.87	1.64
	2	0.60	0.53	1.67	2.13	26.16	38.29	20.59	2.78	1.12	0.63	1.54	1.68	22.34	30.01	53.30	2.95
	3	0.01	0.00	0.25	0.12	13.13	33.05	18.49	0.16	0.01	0.01	0.24	0.14	12.58	31.81	51.13	3.10
	4	0.50	0.14	0.97	1.20	16.35	28.41	20.07	1.31	1.09	0.40	0.90	0.94	18.56	28.06	52.73	2.12
	5	0.17	0.06	0.88	0.84	17.17	31.32	19.81	1.29	0.35	0.18	0.88	0.78	21.54	36.78	52.94	2.28
	6	0.30	0.12	1.00	0.96	7.49	14.52	18.73	1.44	0.69	0.34	1.02	0.90	9.23	12.61	51.88	1.84
	7	1.19	0.77	0.82	0.69	14.58	25.54	21.54	3.20	1.57	1.11	0.86	0.67	17.40	23.91	54.19	3.46
	8	11.93	8.30	5.46	4.83	54.84	52.31	21.42	4.08	7.51	4.47	4.74	4.22	49.75	43.89	55.79	4.08
Average		3.03	1.94	2.03	1.75	41.40	44.03	15.94	1.80	1.31	0.79	1.10	1.12	19.08	26.75	48.63	2.81

4.1   Errors Handling

Odometry algorithms used in the study may fail during computations. It means that for some source and destination frames they cannot estimate camera movement. In our dataset failures also occur. For that reason, some failure strategy needs to be chosen. When an algorithm cannot find a solution the next frame is taken and the current position remains the same causing errors increases. Our strategy is relatively simple but sufficient. However, other alternatives are also possible. In opposite to the described method, previous source frame may be taken. More complex methods may be also introduced including trajectory interpolation. Errors handling is crucial part of the system and it is interesting future research topic.

5 Conclusions

In this paper, we examined the performance of the dense visual odometry algorithms in outdoor environment conditions. Furthermore, we have collected a new dataset for testing VO algorithms.

The results of this study indicate that *RgbdICPOdometry* outperforms other algorithms investigated in the paper. However, we found out that none of the examined algorithms fulfill our requirements.

Further research will investigate the influence of the different types of failure handling strategies, as explained in Sect. 4.1, on the visual odometry algorithms performance. We also plan to expand our dataset with various outdoor scenes typical for an autonomous mower. We believe that more advanced algorithms for Visual Simultaneous Localization and Mapping (VSLAM) may give better outcomes. Our future work will concentrate on VSLAM methods. We hope that loop-closure may boost localization accuracy, especially for typical mowing trajectory.

References

1. Aguiar, A.S., dos Santos, F.N., Cunha, J.B., Sobreira, H., Sousa, A.J.: Localization and mapping for robots in agriculture and forestry: a survey. Robotics **9**(4), 97 (2020). https://doi.org/10.3390/robotics9040097
2. Bishop, G., Welch, G., et al.: An introduction to the Kalman filter. Proc. SIG-GRAPH, Course **8**(27599–23175), 41 (2001)
3. Bradski, G.: The openCV library. Dr. Dobb's J. Softw. Tools **25**(11), 120–123 (2000)
4. Cho, B.S., sung Moon, W., Seo, W.J., Baek, K.R.: A dead reckoning localization system for mobile robots using inertial sensors and wheel revolution encoding. J. Mech. Sci. Technol. **25**(11), 2907–2917 (2011). https://doi.org/10.1007/s12206-011-0805-1, http://www.springerlink.com/content/1738-494x
5. Geiger, A., Lenz, P., Urtasun, R.: Are we ready for autonomous driving? the KITTI vision benchmark suite. In: 2012 IEEE Conference on Computer Vision and Pattern Recognition, IEEE June 2012. https://doi.org/10.1109/cvpr.2012.6248074
6. Howard, A.: Real-time stereo visual odometry for autonomous ground vehicles. In: 2008 IEEE/RSJ International Conference on Intelligent Robots and Systems, pp. 3946–3952 (2008). https://doi.org/10.1109/IROS.2008.4651147
7. Jurišica, L., Duchoň, F., Kaštan, D., Babinec, A.: High precision GNSS guidance for field mobile robots. Int. J. Adv. Rob. Syst. **9**(5), 169 (2012). https://doi.org/10.5772/52554
8. Keselman, L., Woodfill, J.I., Grunnet-Jepsen, A., Bhowmik, A.: Intel(R) realsense(TM) stereoscopic depth cameras. In: 2017 IEEE Conference on Computer Vision and Pattern Recognition Workshops (CVPRW), pp. 1267–1276 (2017). https://doi.org/10.1109/CVPRW.2017.167

9. Lingemann, K., Nüchter, A., Hertzberg, J., Surmann, H.: High-speed laser localization for mobile robots. Robo. Auton. Syst. **51**(4), 275–296 (2005) https://doi.org/10.1016/j.robot.2005.02.004, http://www.sciencedirect.com/science/article/pii/S0921889005000254

10. Newcombe, R.A., et al.: Kinectfusion: real-time dense surface mapping and tracking. In: 2011 10th IEEE International Symposium on Mixed and Augmented Reality, pp. 127–136 (2011). https://doi.org/10.1109/ISMAR.2011.6092378

11. Nister, D., Naroditsky, O., Bergen, J.: Visual odometry. In: Proceedings of the 2004 IEEE Computer Society Conference on Computer Vision and Pattern Recognition, CVPR 2004, vol. 1, pp. I-I (2004). https://doi.org/10.1109/CVPR.2004.1315094

12. Nizette, B., Tridgell, A., Yu, C.: Low-cost differential GPS for field robotics. In: 2014 IEEE/ASME International Conference on Advanced Intelligent Mechatronics, pp. 1521–1526 (2014). https://doi.org/10.1109/AIM.2014.6878299

13. Scaramuzza, D., Fraundorfer, F.: Visual odometry [tutorial]. IEEE Rob. Autom. Mag. **18**(4), 80–92 (2011). https://doi.org/10.1109/MRA.2011.943233

14. Silva, Bruno M. F.., Gonçalves, Luiz M. G..: Visual odometry and mapping for indoor environments using RGB-D cameras. In: Osório, Fernando S.., Wolf, Denis Fernando, Castelo Branco, Kalinka, Grassi, Valdir, Becker, Marcelo, Romero, Roseli A. Francelin. (eds.) LARS/Robocontrol/SBR -2014. CCIS, vol. 507, pp. 16–31. Springer, Heidelberg (2015). https://doi.org/10.1007/978-3-662-48134-9_2

15. Steinbrücker, F., Sturm, J., Cremers, D.: Real-time visual odometry from dense RGB-D images. In: 2011 IEEE International Conference on Computer Vision Workshops (ICCV Workshops), pp. 719–722 (2011). https://doi.org/10.1109/ICCVW.2011.6130321

16. Strack, Andreas, Ferrein, Alexander, Lakemeyer, Gerhard: Laser-based localization with sparse landmarks. In: Bredenfeld, Ansgar, Jacoff, Adam, Noda, Itsuki, Takahashi, Yasutake (eds.) RoboCup 2005. LNCS (LNAI), vol. 4020, pp. 569–576. Springer, Heidelberg (2006). https://doi.org/10.1007/11780519_55

17. Szrek, J., Trybała, P., Góralczyk, M., Michalak, A., Ziętek, B., Zimroz, R.: Accuracy evaluation of selected mobile inspection robot localization techniques in a GNSS-denied environment. Sensors **21**(1), 141 (2020). https://doi.org/10.3390/s21010141

18. Wang, S., Clark, R., Wen, H., Trigoni, N.: Deepvo: towards end-to-end visual odometry with deep recurrent convolutional neural networks. In: 2017 IEEE International Conference on Robotics and Automation (ICRA), pp. 2043–2050 (2017). https://doi.org/10.1109/ICRA.2017.7989236

Using PMI to Rank and Filter Edges in Graphs of Words

Marcela Ribeiro de Oliveira$^{(\boxtimes)}$ and Eduardo J. Spinosa

Federal University of Paraná, Curitiba, Brazil
{mroliveira,spinosa}@inf.ufpr.br

Abstract. Text classification is a classic problem in Natural Language Processing. An essential task in text classification is the construction of the representation, which must provide relevant information to the classifier. One of the most effective representation models uses graphs to represent documents. This paper presents an approach that uses this representation model but with weighted graphs. We propose to use a popular word association measure, the Pointwise Mutual Information (PMI), to calculate the weights of graph edges. These weights then serve as a guide to identify and remove edges between words with low association levels. Then, using node2vec, we extract the features of each graph and use a text convolutional neural network for classification. We conducted experiments in order to compare different kinds of graph modeling in terms of classification score and the proportion of edges that were removed. The results obtained indicate that this approach makes it possible to reduce the number of edges in the graphs maintaining classification performance.

Keywords: Graph of words · Word association · Edge filtering

1 Introduction

Text classification is a widely explored problem in the Natural Language Processing (NLP) field. It can be applied in several tasks like document organization, spam filtering and news filtering. In general, text classification has the following steps: data processing, text processing, feature extraction, training and evaluation of the model. In particular, feature extraction transforms the data into a representation that the algorithms can process. Among the existing forms of text representation, such as bag-of-words and n-grams, one of the most effective is based on graphs, which is the focus of this work.

Graphs can capture important information in text, such as term co-occurrence and term relationships, which are not considered by the bag-of-words model [16]. One possible transformation of a text to a graph, known as graph of words, considers words as nodes and their co-occurrence inside a sliding window of fixed size as edges. A graph representation makes it possible to use a node embedding algorithm, that maps each graph node to a feature vector and uses these feature vectors for classification.

In this paper we present an approach to remove edges from text graphs without causing loss in text classification performance. For this, we propose the use of Pointwise

© Springer Nature Switzerland AG 2021
L. Rutkowski et al. (Eds.): ICAISC 2021, LNAI 12855, pp. 91–100, 2021.
https://doi.org/10.1007/978-3-030-87897-9_9

Mutual Information (PMI), an existing co-occurrence measure based on word association, to calculate graphs' weights. Then, ranking the edges by their weight we can remove the edges containing the lower weights.

By identifying and removing redundant graph edges we aim to generate a representation that, despite containing fewer edges than graphs generated by co-occurrence alone, does not lead to a loss in the classification score. In this way, it is possible to represent the same document by a graph using fewer edges.

Our approach also allows to define the percentage of edges to be removed, which we call cut percentage p. Thus, through experiments with several values for the cutting percentage p, it is possible to define how many graph edges can be removed without causing significant loss in the representation.

We performed experiments in four text classification datasets that encompass different text classification tasks. The results obtained indicate that this approach makes it possible to reduce the number of edges in the graphs maintaining classification performance.

The rest of this paper is organized as follows. Section 2 gives an overview of the related work using graphs to represent texts. Section 3 provides details of our proposed approach. Section 4 presents the experiments methodology, including implementation details, the datasets employed and the evaluation process. Section 5 presents the results and their discussion and Sect. 6 draws conclusions and describes future work.

2   Related Work

Graphs have been used to represent text in several tasks, including emotion recognition [6], sentiment classification [8, 19], text generation [12] and question answering [5].

In the text classification problem specifically, graphs are used to represent documents even when deep neural network models are not used, which is the case of [15] and [17]. In Rousseau et al. [15], each text is represented using the graph of words model, so they treated the problem as a graph classification problem. In this approach, frequent subgraphs were extracted and used as features for a linear SVM classifier. Skianis et al. [17] also proposed graphs that represent collections and labels. A collection graph is the union of all graphs that belong to a collection, and a label graph represents a class in which all the words of documents belonging to the respective class are the nodes and their co-occurrence are the edges.

Recently several works have been conducted using deep learning and graphs to text classification [9, 11, 14, 18]. The one that is closest to ours was proposed by Bijari et al. [1]. As the before-mentioned works, each sentence is represented by a graph. In this research, node2vec is applied to the sentence graphs to obtain the feature representation in an unsupervised manner. Then, a slight variant of the ConvNet architecture of Kim [10] and Collobert et al. [3] is used for sentiment classification.

These works motivated ours to investigate the graph construction and evaluate the impact of removing edges in the graph representation learning process in order to obtain a more concise representation.

3   Proposed Approach

The proposed approach considers a classification problem where each document is represented by a graph. In the experiments, we compare unweighted graphs with all co-occurrence edges with weighted graphs where a predefined percentage of edges were removed. We used node2vec to extract the representation of the graphs and then a convolutional neural network to perform the classification.

Figure 1 presents an overview of our approach composed of the following steps: text pre-processing, graph modeling, representation learning and classification. Next, we describe each of these components.

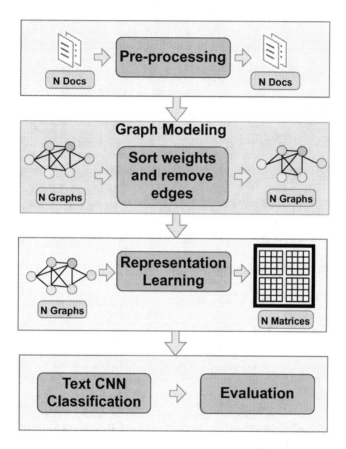

Fig. 1. Proposed approach overview. First, we pre-process the documents and model each of them as a graph. Then we sort the edges by weight and remove a certain percentage of them. Finally, we generate the embedding matrices and classify the documents with the Text CNN.

3.1   Text Pre-processing

In this step, transformations are applied in the documents, so they can be adequately represented and processed by an algorithm. In our approach, we have adopted the following text processing techniques: lowercase conversion, tokenization, non-alphabetic characters and stop words removal and stemming.

3.2   Graph Modeling

The graphs in our approach follow the graph of words model [15]: each document is represented by a graph, the nodes are the unique terms of the document, and the edges represent the co-occurrences of the terms within a sliding window of fixed size. This graph is also known as the co-occurrence graph.

In our work, we generated two representation models to be compared, one with unweighted and another with weighted graphs; the first one is a co-occurrence graph only, the other one has edge weights calculated by a co-occurrence word association measure. This kind of measure estimates the association between two words by comparing the word pair's corpus-level bigram frequency to some function of the unigram frequencies of the individual words [4].

A popular word association measure is the Pointwise Mutual Information (PMI) [2], calculated by Eq. 1:

$$\mathbf{PMI} = log \frac{f(x,y)}{f(x) * f(y)/W} \tag{1}$$

W represents the number tokens in the corpus; $f(x)$, $f(y)$ are the unigram frequencies of x and y words respectively in the corpus; and $f(x, y)$ is the (x, y) word pair frequency with amplitude restriction (frequency of co-occurrence of the word pair within a fixed size window).

Based on PMI, we generate two weighted graphs: **Local PMI** and **Global PMI**, that differ in how the weight is computed. Local PMI considers a single document in the computation, so the same word pair can have different values depending on the document that generated the graph. And Global PMI considers the set of all documents in the computation so that the same word pair will have the same values for all graphs.

After the computation of the weights, for each graph we sort the edges by their weight and remove a certain percentage p of the edges starting from the ones with lower values, where p is a parameter of the algorithm. If that process results in the disconnection of a word (vertex) from the graph, the vertex is removed and not considered by the representation learning algorithm. This implies that the graph that represents the text may have fewer words than the original text.

We expect that this method for edge removal guided by a word association measure leads us to find a more concise representation maintaining quality. In other words, if there is no loss in the representation quality using the proposed method to remove edges, it means that we can represent a text by a graph using less edges than the ones captured by the co-occurrence window alone. And also that the measure that we propose to apply is useful in ranking the edges by relevance to the representation and finding the unnecessary ones.

3.3   Graph Representation Learning

We used a node embedding algorithm to extract the graph's representation. The idea of this type of algorithm is to automatically learn to encode the graph structure into low-dimensional embeddings. So, every graph node is represented by an embedding.

We chose node2vec [7], a semi-supervised graph representation learning algorithm that works through random walks starting from each vertex of the graph. Two node2vec characteristics are: a flexible notion of the neighborhood of a vertex and a random biased walk, since it is guided by parameters, in which several neighborhoods of the vertex are explored [7].

Since node2vec gives one embedding per graph node, an embedding matrix $E_{n \times d}$ represents an entire graph, where n is the number of nodes, and d is the embedding dimension. Moreover, since each document is represented by a graph, we have one embedding matrix for each document.

3.4 Text CNN for Classification

For the classification step, we use a convolutional neural network (CNN), based on the Text CNN proposed by Kim [10], which receives an embedding matrix where each line is a word embedding from the document. The main difference compared to Kim's approach is that, instead of having an embedding matrix with word embeddings, our approach will have node embeddings.

To ensure that all embedding matrices have the same size before submitting them as an input to the CNN, we calculated a threshold value to ensure that the matrix would fit 75% of the documents in the dataset. Texts larger than this value were truncated, and the smaller ones were zero-padded.

4 Experiment Setup

This section provides details about how we conducted the experiments, including implementation, datasets and evaluation methodology.

The experiments were performed in a server running Linux Mint version 19.5 with 16 cores of Intel(R) Xeon(R) E5620 processors and 70 GB of RAM.

4.1 Implementation Details

Since in the graph representation learning step we used node2vec, we had to set the values for some parameters. Table 1 presents these parameters, a brief description of each one of them and the value used for all the experiments.

Table 1. node2vec parameters, their descriptions and the values used in the experiments

Parameter	Description	Value
Dimensions	Embeddings dimension	100
walk_length	Number of nodes in each walk	2
num_walks	Number of walks per graph node	10
p	Return parameter	2
q	In-out parameter	0.7
Workers	Number of workers for parallel execution	4

For the classification step, the Text CNN had six convolutional layers, three with kernel size 2 and filter sizes 32, 64 and 128, and three with kernel size 3 and filter sizes 32, 64 and 128. The feature maps generated by all six convolutional blocks were globally max-pooled and fed into a fully connected layer with 256 neurons. We also included a dropout layer after each convolutional layer and one before the fully connected layer to avoid overfitting. The fully connected layer output was fed into a softmax layer for classification. We trained the Text CNN for 20 epochs.

4.2 Datasets

We evaluate our approach across two different text classification tasks: sentiment analysis and topic classification. For this, we chose the following datasets, which are widely used in text classification literature:

- **Polarity v2.0:** a sentiment analysis dataset composed by 2000 movie-reviews, where 1000 are positive and 1000 are negative.
- **WebKB:** a topic classification dataset containing web pages collected from computer science departments. The original dataset contains 7 classes, but we used a subset with 4 classes (student, faculty, course, project), composed of 7287 documents.
- **R8:** a topic classification dataset containing news documents labeled into 8 categories. This is a subset of the Reuters-21578 dataset [13].
- **20-Newsgroups (20NG):** a topic classification dataset which contains approximately 20000 newsgroup posts, partitioned (nearly) evenly across 20 different topics. We removed the duplicated news documents, resulting in 18846 documents.

Table 2 summarizes the characteristics of each dataset.

Table 2. Dataset characteristics

Dataset	Documents	Classes
Polarity	2000	2
WebKB	7287	4
R8	7674	8
20 Newsgroups	18846	20

4.3 Evaluation

To preserve the percentage of samples of each class, we used stratified cross-validation. Also, to ensure that all the folds had the same samples in all experiments, we fixed the seed in fold generation. We split each dataset in ten folds. For each fold, 90% of the samples were used to train, and 10% of the samples were used to test.

To evaluate the classification performance we used F1-score. As we are using 10-fold cross-validation, we have ten F1-scores, one for each fold, so we report the mean of these ten F1-scores in the results to give a single estimation.

To evaluate if there is a significant difference between the classification score in the graphs constructed using our approach to remove edges compared to the unweighted graphs without edge removal, we performed a statistical test. We chose the Wilcoxon signed-rank test, a non-parametric test, where the null hypothesis is that two related paired samples come from the same distribution. We rejected the null hypothesis when the p-value was less than 0.05.

5   Results and Discussion

We compare the following representations: Unweighted graphs ($\overline{W}$) against Local PMI weighted graphs, and Unweighted graphs ($\overline{W}$) against Global PMI weighted graphs. For each representation, we generate graphs with window sizes of 4, 12 and 20. And for each combination of graph type and window size we evaluate three values for the cut percentage p: 5%, 10% and 20%.

Table 3 presents the mean F1 classification scores over 10 folds (cross-validation) for each dataset and parameter setting. Then, we apply the Wilcoxon test to verify if there is statistical difference between the baseline representation ($\overline{W}$) and the proposed approach with edge removal (either using Local or Global PMI). Only the classification scores in bold are statistically different: either better (marked with "+") or worse (marked with "−") than the baseline. Our main interest is to maintain the classification score even with a smaller number of edges, indicating that these removed edges were not necessary to maintain the quality of the representation.

Table 3. 10-fold mean F1-score for each dataset, window size and cut percentage using unweighted graphs ($\overline{W}$), weighted graphs with Local PMI and weighted graphs with Global PMI.

	Graph	Cut p	Polarity Window size			WebKB Window size			R8 Window size			20 NG Window size		
			4	12	20	4	12	20	4	12	20	4	12	20
Baseline	$\overline{W}$	0	78.98	78.28	76.42	89.88	89.58	90.15	81.42	82.82	81.09	83.19	82.73	83.61
Proposed approach	Local PMI	5	77.73	78.49	78.71	90.48	89.10	89.87	81.47	81.83	**85.40⁺**	83.05	82.64	**82.63⁻**
		10	78.83	78.76	**80.04⁺**	89.80	89.48	89.90	80.82	80.27	82.44	83.07	82.68	**82.73⁻**
		20	79.93	79.01	77.84	89.14	89.34	89.31	79.29	81.74	81.83	**81.24⁻**	**82.01⁻**	**82.56⁻**
	Global PMI	5	78.63	76.93	78.63	89.68	88.92	88.63	79.65	80.10	**85.22⁺**	83.06	82.51	83.17
		10	79.21	77.22	76.68	89.67	88.56	**88.11⁻**	80.57	80.39	82.57	83.46	**83.66⁺**	83.00
		20	79.19	76.09	79.48	89.71	88.69	89.37	80.45	81.21	82.02	83.51	83.07	83.09

In general, we can observe that the results were stable for most of the datasets, except for the 20 Newsgroups, where there were many statistically worse results using Local PMI specifically. The WebKB dataset also obtained one statistically worse result, but this time using Global PMI to calculate edge weights. Polarity and R8 datasets were those who have obtained more stability, presenting only positive statistically significant variations.

Observing the window size parameter, we can conclude that larger windows do not improve the classification score. And smaller windows sizes, like 4 and 12, present more stability in terms of the classification score.

The weight measure Global PMI performs better than the Local PMI, and it is possible to observe that when we use Global PMI with window size 4, even removing up to 20% of the edges, there is no statistically significant loss in the F1-score.

In this way, we can conclude that it is possible to calculate the edges' weights from the co-occurrence graph using PMI and sort these weights in order to remove a percentage of them from the graph, without causing a loss in the representation quality. Also, the experiments pointed that generating graphs with window size 4, and Global PMI to calculate the edges' weights, it is possible to use graphs with 20% less edges and still obtain a statistically equivalent classification score.

6   Conclusion and Future Work

In this paper, we presented an approach to rank and filter edges from graphs of words. Based on a popular word association measure, the Pointwise Mutual Information (PMI), we propose two ways to calculate the edge weights: the Local and Global PMI measures. In this context, our approach first ranks graph edges by their weights and then removes the smaller ones from the graph until it reaches a fixed percentage p.

The results showed that both measures, Local PMI and Global PMI, help to improve the representation in the sense of identifying edges that can be removed from the graph without causing a loss in the classification score. Moreover, the Global PMI measure generated more stable results.

Future work includes: exploring other word association measures to calculate the graph-of-words' edge weights; evaluating more cut thresholds in order to get the best value to remove edges without causing a statistically significant loss of representation quality; and evaluating and comparing graph representation learning time and memory usage for the graphs constructed using our approach and the unweighted graphs.

Acknowledgements. We thank the financial support from the Brazilian National Council for Scientific and Technological Development (CNPq). We also would like to thank Alana AI for its financial support.

References

1. Bijari, K., Zare, H., Kebriaei, E., Veisi, H.: Leveraging deep graph-based text representation for sentiment polarity applications. Expert Syst. Appl. **144**, 113090 (2020)
2. Church, K.W., Hanks, P.: Word association norms, mutual information, and lexicography. Comput. Linguist. **16**(1), 22–29 (1990)
3. Collobert, R., Weston, J., Bottou, L., Karlen, M., Kavukcuoglu, K., Kuksa, P.: Natural language processing (almost) from scratch. J. Mach. Learn. Res. **12**(1), 2493–2537 (2011)
4. Damani, O.: Improving pointwise mutual information (PMI) by incorporating significant co-occurrence. In: Proceedings of the Seventeenth Conference on Computational Natural Language Learning, Sofia, Bulgaria, pp. 20–28. Association for Computational Linguistics, August 2013. https://www.aclweb.org/anthology/W13-3503

5. De Cao, N., Aziz, W., Titov, I.: Question answering by reasoning across documents with graph convolutional networks. In: Proceedings of the 2019 Conference of the North American Chapter of the Association for Computational Linguistics: Human Language Technologies, Volume 1 (Long and Short Papers), Minneapolis, Minnesota, pp. 2306–2317. Association for Computational Linguistics, June 2019. https://doi.org/10.18653/v1/N19-1240, https://www.aclweb.org/anthology/N19-1240

6. Ghosal, D., Majumder, N., Poria, S., Chhaya, N., Gelbukh, A.: DialogueGCN: a graph convolutional neural network for emotion recognition in conversation. In: Proceedings of the 2019 Conference on Empirical Methods in Natural Language Processing and the 9th International Joint Conference on Natural Language Processing (EMNLP-IJCNLP), Hong Kong, China, pp. 154–164. Association for Computational Linguistics, November 2019. https://doi.org/10.18653/v1/D19-1015, https://www.aclweb.org/anthology/D19-1015

7. Grover, A., Leskovec, J.: node2vec: scalable feature learning for networks. In: Proceedings of the 22nd ACM SIGKDD International Conference on Knowledge Discovery and Data Mining, pp. 855–864. ACM (2016)

8. Huang, B., Carley, K.: Syntax-aware aspect level sentiment classification with graph attention networks. In: Proceedings of the 2019 Conference on Empirical Methods in Natural Language Processing and the 9th International Joint Conference on Natural Language Processing (EMNLP-IJCNLP), Hong Kong, China, pp. 5469–5477. Association for Computational Linguistics, November 2019. https://doi.org/10.18653/v1/D19-1549, https://www.aclweb.org/anthology/D19-1549

9. Huang, L., Ma, D., Li, S., Zhang, X., Wang, H.: Text level graph neural network for text classification. In: Proceedings of the 2019 Conference on Empirical Methods in Natural Language Processing and the 9th International Joint Conference on Natural Language Processing (EMNLP-IJCNLP), Hong Kong, China, pp. 3444–3450. Association for Computational Linguistics, November 2019. https://doi.org/10.18653/v1/D19-1345, https://www.aclweb.org/anthology/D19-1345

10. Kim, Y.: Convolutional neural networks for sentence classification. In: Proceedings of the 2014 Conference on Empirical Methods in Natural Language Processing (EMNLP), Doha, Qatar, pp. 1746–1751. Association for Computational Linguistics, October 2014. https://doi.org/10.3115/v1/D14-1181, https://www.aclweb.org/anthology/D14-1181

11. Kipf, T.N., Welling, M.: Semi-supervised classification with graph convolutional networks. In: International Conference on Learning Representations (ICLR) (2017)

12. Koncel-Kedziorski, R., Bekal, D., Luan, Y., Lapata, M., Hajishirzi, H.: Text generation from knowledge graphs with graph transformers. In: Proceedings of the 2019 Conference of the North American Chapter of the Association for Computational Linguistics: Human Language Technologies, Volume 1 (Long and Short Papers), Minneapolis, Minnesota, pp. 2284–2293. Association for Computational Linguistics, June 2019. https://doi.org/10.18653/v1/N19-1238, https://www.aclweb.org/anthology/N19-1238

13. Lewis, D.D.: An evaluation of phrasal and clustered representations on a text categorization task. In: Proceedings of the 15th Annual International ACM SIGIR Conference on Research and Development in Information Retrieval, pp. 37–50 (1992)

14. Linmei, H., Yang, T., Shi, C., Ji, H., Li, X.: Heterogeneous graph attention networks for semi-supervised short text classification. In: Proceedings of the 2019 Conference on Empirical Methods in Natural Language Processing and the 9th International Joint Conference on Natural Language Processing (EMNLP-IJCNLP), Hong Kong, China, pp. 4821–4830. Association for Computational Linguistics, November 2019. https://doi.org/10.18653/v1/D19-1488, https://www.aclweb.org/anthology/D19-1488

15. Rousseau, F., Kiagias, E., Vazirgiannis, M.: Text categorization as a graph classification problem. In: Proceedings of the 53rd Annual Meeting of the Association for Computational Linguistics and the 7th International Joint Conference on Natural Language Processing (Volume 1: Long Papers), Beijing, China, pp. 1702–1712. Association for Computational Linguistics, July 2015. https://doi.org/10.3115/v1/P15-1164, https://www.aclweb.org/anthology/P15-1164

16. Shanavas, N., Wang, H., Lin, Z., Hawe, G.: Structure-based supervised term weighting and regularization for text classification. In: Métais, E., Meziane, F., Vadera, S., Sugumaran, V., Saraee, M. (eds.) NLDB 2019. LNCS, vol. 11608, pp. 105–117. Springer, Cham (2019). https://doi.org/10.1007/978-3-030-23281-8_9

17. Skianis, K., Malliaros, F., Vazirgiannis, M.: Fusing document, collection and label graph-based representations with word embeddings for text classification. In: Proceedings of the Twelfth Workshop on Graph-Based Methods for Natural Language Processing (TextGraphs-12), New Orleans, Louisiana, USA, pp. 49–58. Association for Computational Linguistics, June 2018. https://doi.org/10.18653/v1/W18-1707, https://www.aclweb.org/anthology/W18-1707

18. Yao, L., Mao, C., Luo, Y.: Graph convolutional networks for text classification. In: Proceedings of the AAAI Conference on Artificial Intelligence, vol. 33, pp. 7370–7377 (2019)

19. Zhang, C., Li, Q., Song, D.: Aspect-based sentiment classification with aspect-specific graph convolutional networks. In: Proceedings of the 2019 Conference on Empirical Methods in Natural Language Processing and the 9th International Joint Conference on Natural Language Processing (EMNLP-IJCNLP), Hong Kong, China, pp. 4560–4570. Association for Computational Linguistics, November 2019. https://doi.org/10.18653/v1/D19-1464, https://www.aclweb.org/anthology/D19-1464

Development and Research of Quantum Models for Image Conversion

Samoylov Alexey[iD], Gushanskyi Sergey, and Potapov Viktor[(✉)][iD]

South Federal University, Taganrog, Russia
{asamoylov,smgushanskyi,vpotapov}@sfedu.ru

Abstract. Quantum imaging is a new trend that is showing promising results as a powerful addition to the arsenal of imaging techniques. Per-pixel representation of an image using classical information requires a huge amount of computational resources. Hence, exploring techniques for representing images in a different information paradigm is important. This paper describes the variety of options for representing images in quantum information. Image processing is a well-established area of computer science with many applications in today's world such as face recognition, image analysis, image segmentation and noise reduction using a wide range of techniques. A promising first step was the exponentially efficient implementation of the Fourier transform in quantum computers versus the FFT in classical computers. In addition, images encoded in quantum information can obey unique quantum properties such as superposition or entanglement. The laws of quantum mechanics can reduce the required resources for some tasks by many orders of magnitude if the image data is encoded in the quantum state of a suitable physical system. The aim of this work is to develop and study quantum models of image transformation.

Keywords: Quantum algorithm · Classes of algorithms complexity · Polynomial time · Confusion · Model of a quantum computer · Quantum parallelism · Qubit

1 Introduction

Quantum image processing focuses on the use of quantum computing [1] and quantum information processing to create and work with quantum images. Due to some of the amazing properties inherent in quantum computing, notably entanglement and parallelism, quantum information processing technologies are expected to offer capabilities and characteristics that are unmatched in their traditional equivalents. These improvements can be associated with computational speed, guaranteed security, minimal storage requirements, etc. However, modern image processing methods require expensive computational resources, storage and image processing. There are many computational techniques, such as Fast Fourier Transform (FFT), which provide acceleration for image processing. Quantum computing, on the other hand, defines a probabilistic approach for representing classical information using methods from quantum theory. The idea of quantum information was first introduced in 1980 by Paul Benioff. In the same year,

© Springer Nature Switzerland AG 2021
L. Rutkowski et al. (Eds.): ICAISC 2021, LNAI 12855, pp. 101–112, 2021.
https://doi.org/10.1007/978-3-030-87897-9_10

Yuri Manin proposed a model of a quantum computer in his book "Computable and Uncomputable". In 1982, the field of quantum information theory was formalized and described by Richard Feynman, who made it popular with his paper on modeling physics in computers. Devid Deutsch took this field forward with the development of the quantum Turing machine. Thanks to these advances, quantum computing [2] has shown many promises for the future of computers. This new class of problem computability has been grouped into a computational class called bounded error quantum polynomial time (BEQP).

The basic idea behind quantum computing is the qubit [3], the quantum analogue of the classical computer bit. A classic bit is only capable of storing a specific value (0 or 1). A qubit can store both 0 and 1 in an undefined state called a superposition:

$$\alpha|0> +\beta|1>, \tag{1}$$

where α and β are normalized probability amplitudes.

Formulation of the problem. The aim of the study is to build a modular specialized system for new information technologies and algorithms, analyze their performance and computational complexity. Realization of the theoretical foundations for the creation of software systems for new information technologies of a quantum orientation. It also consists in the development of theoretical, algorithmic and practical foundations for building a modular system for the interaction of information processes and algorithms with an open architecture.

2   Mathematical Apparatus

The logic and mathematical apparatus of quantum computing devices differs significantly from the logic and mathematics of classical computing technology due to the nature and specific properties of quantum particles (qubits), which allow such calculations to be carried out. The quantum computer operates, as described above, with quantum bits, which have two basic states $|0\rangle = \begin{pmatrix} 1 \\ 0 \end{pmatrix}$ and $|1\rangle = \begin{pmatrix} 0 \\ 1 \end{pmatrix}$. The qubit state $|0\rangle$ corresponds to the spin up state of an electron in the atom [4], the spin down state corresponds to $|1\rangle$, and the mixed state (superposition of states) corresponds to the intermediate spin position. Spin is the proper angular momentum of elementary particles, but the spin is not associated with motion in space, it is the eigenvalue of a quantum particle, which cannot be explained from the standpoint of classical mechanics. In the general case, the state of a qubit is described by a wave function (or a state vector):

$$|\psi\rangle = \alpha_0|0\rangle + \alpha_1|1\rangle \tag{2}$$

where $\alpha_0, \alpha_1 \in C$, α_0 and α_1 are the complex amplitudes of the reading probability $|0\rangle$ or $|1\rangle$. The fact is that when measuring quantum bits in a standard basis with the probability $|\alpha_0|^2$, $|0\rangle$ will be obtained, and with the probability $|\alpha_1|^2 - |1\rangle$. In this case, the qubit will go into a state of quantum zero or one. The effect of a quantum mixed state exists only until the moment the qubit is measured, that is, when reading, we bring the qubit into one of the basic states, and which one depends on the probability obtained in the calculations. Another visual option for representing the qubit state vector is the so-called Bloch sphere [5]. It is a mapping of the complex plane of psi-function values onto a sphere, in fact, on it you can visually see "spin up", "spin down". However, this is the space of values of the qubit state, while the spin pseudovector itself cannot be negative. Analyzing Fig. 1, one can notice that the multiplication of the state of the qubit $|\psi\rangle$ by $e^{i\varphi}$ does not change the probability of reading zero and one.

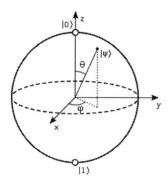

Fig. 1. Bloch sphere.

The state of a system of two quantum bits is described with respect to four basic states (in a classical computer, four states can be encoded with two bits): $|00\rangle, |01\rangle, |10\rangle, |11\rangle$.

$$|\psi\rangle = \alpha_0|00\rangle + \alpha_1|01\rangle + \alpha_2|10\rangle + \alpha_3|11\rangle, \tag{3}$$

$$|\alpha_0|^2 + |\alpha_1|^2 + |\alpha_2|^2 + |\alpha_3|^2 = 1 \tag{4}$$

That is, the multi-qubit system in formula (4) describes all possible states (combinations of zeros and ones), the number of combinations is found by the formula:

$$N = 2^q \tag{5}$$

where N is the number of basis states that a given quantum system of qubits describes, and q is the number of qubits in the system. At the beginning of the analysis, it was indicated that the states of zero and one are encoded as $|0\rangle = \begin{pmatrix} 1 \\ 0 \end{pmatrix}$ and $|1\rangle = \begin{pmatrix} 0 \\ 1 \end{pmatrix}$ state function $|\psi\rangle = \alpha_0|0\rangle + \alpha_1|1\rangle$, but if we expand the product $\alpha_0|0\rangle$, then we get the following:

$$\alpha_0|0\rangle = \alpha_0 \begin{pmatrix} 1 \\ 0 \end{pmatrix} = \begin{pmatrix} \alpha_0 \\ 0 \end{pmatrix} \tag{6}$$

those, the first line contains the amplitude of the probability of reading zero. And for $\alpha_1|1\rangle$ we get that the second line contains the amplitude of the probability of reading one:

$$\alpha_1|1\rangle = \alpha_1 \begin{pmatrix} 0 \\ 1 \end{pmatrix} = \begin{pmatrix} 0 \\ \alpha_1 \end{pmatrix}. \tag{7}$$

Then it becomes clear how the function $|\psi\rangle = \alpha_0|0\rangle + \alpha_1|1\rangle$ contains the probabilities of all states:

$$|\psi\rangle = \alpha_0|0\rangle + \alpha_1|1\rangle = \begin{pmatrix} \alpha_0 \\ 0 \end{pmatrix} + \begin{pmatrix} 0 \\ \alpha_1 \end{pmatrix} = \begin{pmatrix} \alpha_0 \\ \alpha_1 \end{pmatrix}. \tag{8}$$

The first line contains the amplitude of the probability to get 0 when reading, and the second to get the probability when reading 1. If we consider a system consisting of two qubits, then its state vector is described as follows:

$$|\psi\rangle = \alpha_0|00\rangle + \alpha_1|01\rangle + \alpha_2|10\rangle + \alpha_3|11\rangle$$

$$= \alpha_0 \begin{pmatrix} 1 \\ 0 \\ 0 \\ 0 \end{pmatrix} + \alpha_1 \begin{pmatrix} 0 \\ 1 \\ 0 \\ 0 \end{pmatrix} + \alpha_2 \begin{pmatrix} 0 \\ 0 \\ 1 \\ 0 \end{pmatrix} + \alpha_3 \begin{pmatrix} 0 \\ 0 \\ 0 \\ 1 \end{pmatrix}$$

$$= \begin{pmatrix} \alpha_0 \\ 0 \\ 0 \\ 0 \end{pmatrix} + \begin{pmatrix} 0 \\ \alpha_1 \\ 0 \\ 0 \end{pmatrix} + \begin{pmatrix} 0 \\ 0 \\ \alpha_2 \\ 0 \end{pmatrix} + \begin{pmatrix} 0 \\ 0 \\ 0 \\ \alpha_3 \end{pmatrix} = \begin{pmatrix} \alpha_0 \\ \alpha_1 \\ \alpha_2 \\ \alpha_3 \end{pmatrix} \tag{9}$$

3   Quantum Image Processing Theory

Consider the case of using quantum information by encoding 8 bits of binary information into 3 qubits. The key feature of a qubit is the ability to store an undefined value as a superposition [6] of "0" and "1". Therefore, any operation will inherently receive 0 or 1, as shown in Table 1.

Table 1. Possible states of a 3-qubit register.

Qubits	φ1	φ2	φ3	φ4	φ5	φ6	φ7	φ8
Ψ1	0	0	0	0	1	1	1	1
Ψ2	0	0	1	1	0	0	1	1
Ψ3	0	1	0	1	0	1	0	1

A three-bit quantum register ($\Psi = \psi1\psi2\psi3$) can store 8 bits of information in superposition, where each configuration is assigned a probability amplitude as shown below:

$$\varphi = \alpha_0|000> + \alpha_1|001> + \ldots + \alpha_{2^n-1}|111 \tag{10}$$

The probability amplitudes $(\alpha_1, \alpha_2, \ldots)$ represent the probability that the quantum register will be in this configuration. It is also very important to note that quantum logic gates [7] have some unique advantages that classical systems cannot take advantage of. For example, the 2-qubit CNOT gate is reversible. This means that if this gate is applied twice, the qubit will return to its original state. To represent quantum images, you can use this gate to build a circuit for preparing a quantum image. The CNOT gate can be used as a conditional switch, since it only flips the target bit if the input is 1. It is also possible to attach any other unitary operation to the conditional gate instead of the NOT gate to implement the conditional operation. Technically innovative efforts in the field of quantum pattern and object recognition, with subsequent research related to them, can be divided into three main groups:

- Quantum image processing. These applications are aimed at improving the tasks and applications of digital or classical image processing.
- Optical quantum visualization;
- Quantum image processing based on classical methods.

A review of the quantum representation of an image was published in [8]. In addition, the recently published book Quantum Image Processing [9] provides a comprehensive introduction to quantum image processing, which focuses on extending traditional image processing problems to quantum computing platforms [10]. It briefly presents the available representations of quantum images and their operations, discusses possible applications of quantum images and their implementation, and discusses open questions and future development trends.

3.1 Quantum Image Manipulation

Much of the effort has been focused on developing algorithms [11] for manipulating position and color information encoded using a quantum image representation and its many variations. For example, fast geometric transformations were initially proposed, including (two-point) swap, flip, orthogonal rotations, and constrained geometric transformations, to constrain these operations to a specific area of the image. Recently, a rather interesting idea of quantum images was implemented based on a new extended quantum representation to match the position of each image element in the input image with a new position in the output image and upscale the quantum image to resize the quantum image. While based on the representation of a quantum image, the general form of color transforms was first proposed using single-bit gates such as gates X, Z and H.

Using basic quantum elements and the aforementioned operations, researchers have contributed to the extraction of quantum characteristics of an image, quantum image segmentation, quantum image morphology, quantum image comparison, quantum image filtering, quantum image classification, quantum image stabilization.

3.2 Quantum Image Transformation

By coding and processing image information in quantum mechanical systems [12], the basis of quantum image processing is presented, where a pure quantum state encodes image information: to encode pixel values in probability amplitudes and pixel positions in basic computational states. Given an image $F = (F_{i,j})_{M*L}$, where $F_{i,j}$ represents the pixel value at position (i, j) with $i = 1,...,M$ and $j = 1,..., L$, a vector f with ML elements can be formed by placing the first elements of M in the vector f will be the first column of F, the next M elements will be the second column, and so on.

A large class of image operations are linear, such as unitary transformations, convolutions, and linear filtering. In quantum computing, a linear transformation can be represented as $|g> = U|f>$ with the input image state $|f>$ and the output image state $|g>$. A unitary transformation can be implemented as a unitary evolution. Some basic and commonly used image transformations (for example, the Fourier, Hadamard and Haar wavelet transforms) can be expressed as $G = PFQ$ with the resulting image G and the row (column) transformation matrix P (Q). The corresponding unitary operator U can be written as $U = Q^T \cdot P$. Several commonly used 2D image transformations such as Haar wavelet, Fourier transform and Hadamard transform have been experimentally demonstrated on a quantum computer with exponential acceleration compared to their classical counterparts. In addition, a new highly efficient quantum algorithm has been proposed and experimentally implemented for detecting the boundary between different regions of the image: it requires only one one-bit gate at the processing stage, regardless of the image size.

4   Implementing a Quantum Computing Simulation Environment

In Fig. 2 shows the developed model of a quantum computing device and the result of executing the quantum algorithm for transforming a set of pixels. This software development is an offline desktop computing system with an open architecture, but in the future, it is planned to be broadcast online.

The measurement process consists in playing a random number from the interval $[0,1]$. When the value falls into the interval $[0,|c_1|^2]$, the output value of the measuring process is the basic state of the bra of vector 0, otherwise (interval $[|c_1|^2,1]$) - the quantum state of the bra of vector 1.

In the latest quantum computer technology, only a limited number of qubits can be well controlled and controlled. Whereas, on the other hand, the real problem of optical recognition of symbols in the quantum support vector method requires tens of qubits, which cannot be implemented at present. Thus, we will restrict the problem to the case with minimal costs, in which only two options ("6" or "9") are in the list and only two properties (horizontal or vertical position, determined later) constitute the problem. This makes it possible to demonstrate this quantum artificial intelligence algorithm based on the 4-qubit moment of a quantum processor core at room temperature. The developed algorithm uses quantum principles, such as superposition of quantum states of a computing system, entanglement of quantum states, and transformation of a classical image into a quantum state by encoding the color palette of a pixel set within complex amplitude quantum states.

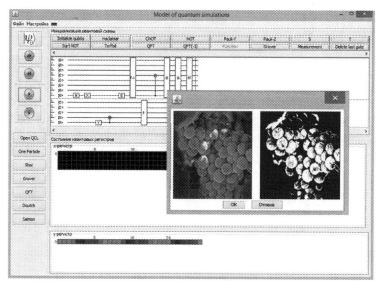

Fig. 2. Model of a quantum computing device.

5 Comparison of Image Transformation Models and Their Classification

Technically, these pioneering efforts, with the subsequent research associated with them, can be divided into three main groups, of which (1) and (2) are considered outside the scope of this review:

(1) Quantum digital image processing: these applications are designed to take advantage of some of the properties responsible for the efficiency of quantum computing algorithms to improve some well-known digital or classical image processing problems and applications

(2) Optics-based quantum imaging: These applications focus on the development of new methods of optical imaging and parallel processing of information at the quantum level by exploiting the quantum nature of light and the intrinsic parallelism of optical signals.

(3) Classical style QIP: These applications take inspiration from the expectation that quantum computing hardware will soon be physically realized and hence such research aims to expand classical imaging problems and applications to the quantum computing framework.

Venegaz-Andraza and Bose introduced the representation of images on quantum computers by proposing a lattice of qubits method where each pixel was represented in its quantum state and then a quantum matrix was created with them. However, this is just a quantum analogue of the classical image and there is no additional advantage in quantum form. However, the next huge advance is the creation of a flexible quantum imaging model (FQI) for multiple intensity levels. The GPKI representation is expressed mathematically as follows:

$$|I(\theta) >= \frac{1}{2^n} \sum_{i=0}^{2^{2n}-1} (\sin(\theta_i)|0 > +\cos(\theta_i)|1 >)|i > \tag{11}$$

where θ corresponds to the intensity of the i-th pixel. Since the intensity values are encoded in the amplitudes of the quantum state, it is relatively easy to apply various transforms, such as the quantum Fourier transform, since they are applied directly to the amplitude of the image. Another approach describes the representation of the pixel values of the image at baseline states instead of amplitudes as shown below:

$$|I> = \frac{1}{2^n} \sum_{X=0}^{2^{2n}-1} \sum_{Y=0}^{2^{2n}-1} |f(X, Y)\rangle|XY > \tag{12}$$

where $f(X, Y)$ refers to the intensity of the pixels in (X, Y). These methods are pretty comprehensive, but they have their own drawbacks. The first major drawback is that they require a square image $(2^N * 2^N)$ along with the need for qubits to encode positions along with pixel intensities. Comparison of the existing and the proposed quantum model of images can be seen in Table 2.

Table 2. Comparison of quantum imaging models.

Model	Flexible presentation of quantum images	Representation of image pixel values in basic states	Two-dimensional quantum states and normalization amplitude
Form	$2^N * 2^N$	$2^N * 2^N$	$2^N * 2^M$
Number of qubits	$2N + 1$	$2N + 1$	$M + N$
Computational complexity	$O(2^{4m})$	$O(2^{2m})$	Pure state

Most quantum imaging models can be prepared with Hadamard gates and controlled operator flip. For $2^N * 2^N$, the first step is to initialize n qubits when $|0 >^{\otimes 2n +1}$. We then apply a Hadamard gate to each qubit to place them in superposition.

$$H\left(|0 >^{\otimes 2n +1}\right) = \frac{1}{2^n}|0 > \otimes \sum_{i=0}^{2^{2n}-1} |i > \tag{13}$$

Next, we apply the guided rotation operators to each base state resulting in a shared state:

$$R\left|H >= \left(\prod_{i=0}^{2^{2n}-1} R_i\right)\right|H \geq |I(\theta) > \tag{14}$$

Each of the R_i rotations correspond to a specific pixel rotation. Therefore, we encode the intensities by rotating these quantum states. The implementation scheme can be achieved by combining Hadamard, CNOT and phase shifting gates. We will introduce FPQI because of its simple and efficient function, which allows us to preserve the intensity of the image as the amplitude of the probability of a quantum state. Gates of the same type are also used in the quantum Fourier transform scheme. We implement FPQI both on a classical computer and on an IBM quantum simulator (Fig. 3).

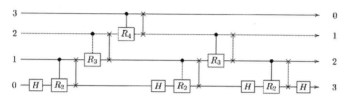

Fig. 3. Quantum Fourier transform scheme.

Modeling quantum systems in a classical computer is computationally expensive. This is because we need to explicitly define the superposition (in other words, store the superposition in memory). In a real quantum computer, superposition is the property of the system and therefore does not need to be stored. For example, consider encoding N bits of information into $\log_2 N$ bits, a superposition of all possible combinations of 0 and 1 can be stored in a quantum register as opposed to a quantum register in which we can only store one of all possible states. A classical computer would have to store all possible states as a list (taking up extra space), unlike a quantum computer. This also means that we do not access these quantum bits as in the classical system. The n bits of information in these n qubits can be accessed only through measurement. The state we want to achieve is shown in Eq. 3. To build the circuit, use a 2×2 image as shown below:

$$\begin{bmatrix} \theta_1 & \theta_2 \\ \theta_3 & \theta_4 \end{bmatrix} \tag{15}$$

Such coding is possible by using a controlled rotation gate to rotate individual base states based on intensity. Let's test this algorithm on a 2×2 image. The IBM simulator allows a small number of qubits to store an image larger than 2×2. Hence, we present the mechanics of a circuit encoding a 2×2 image in an IBM quantum register. We will use the server side of the simulator for the IBM Q 5 Tenerife. Let's use a simple 2×2 image:

$$I = \begin{bmatrix} 5 & 1 \\ 2 & 3 \end{bmatrix} \tag{16}$$

Note that these intensities will be normalized. Therefore, we will not be able to accurately reconstruct the image. The image probability distribution is shown in Fig. 4 (Table 3).

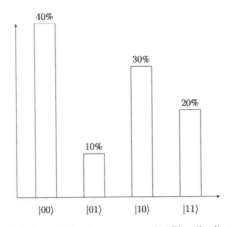

Fig. 4. The resulting image as a probability distribution.

Table 3. Open source quantum computing libraries.

Library	Computing environment
QETLAB	MATLAB
QISKit	Python
Strawberry fields	Python
IBM Q experience	Web-interface
Quantum dev kit	Q#
OpenFermion	Python

Since manipulation of intensities simply means controlled rotation, we can use the circuit in Fig. 1 to define specific gates. For example, negating an image would simply

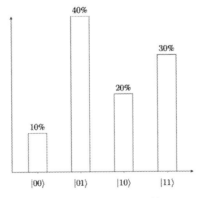

Fig. 5. The resulting inverted image.

mean rotating the amplitude by π. Operations on a quantum image can be done by rotating quantum bits around a Bloch sphere. For example, to invert a quantum image, we can simply rotate the cube-bits around the Bloch sphere 180° and get Fig. 5.

6   Conclusions

Quantum image processing relies on the application of quantum computing and quantum information processing to create and manipulate quantum images. The field of quantum recognition of patterns and objects with subsequent research related to them can be divided into three main groups:

- Quantum image processing;
- Optical quantum visualization;
- Quantum image processing based on classical methods.

The article considers the theory of quantum image processing and quantum image manipulation (fast geometric transformations, including (two-point) swap, flip, orthogonal rotations and limited geometric transformations). In this work, the study of the field of quantum image processing was carried out and the FPQI model was tested. Models (developed in the work, etc.) were investigated by explicitly modeling quantum behavior on a large quantum system. A basic scheme for encoding 2×2 images has been created to get a clearer understanding of the mechanics of quantum image models.

Acknowledgments. The research was funded by the Russian Foundation for Basic Research according to the project № 19-07-01082.

References

1. Guzik, V., Gushanskiy, S., Polenov, M., Potapov, V.: The computational structure of the quantum computer simulator and its performance evaluation. In: Silhavy, Radek (ed.) CSOC2018 2018. AISC, vol. 763, pp. 198–207. Springer, Cham (2019). https://doi.org/10.1007/978-3-319-91186-1_21
2. Samoylov, A., Gushanskiy, S., Polenov, M., Potapov, V.: The quantum computer model structure and estimation of the quantum algorithms complexity. In: Silhavy, R., Silhavy, P., Prokopova, Z. (eds.) CoMeSySo 2018. AISC, vol. 859, pp. 307–315. Springer, Cham (2019). https://doi.org/10.1007/978-3-030-00211-4_27
3. Beach, G., Lomont, C., Cohen, C.:Quantum image processing (QuIP). Proceedings of the 32nd Applied Imagery Pattern Recognition Workshop, pp. 39–40 (2003). https://doi.org/10.1109/AIPR.2003.1284246. ISBN 0-7695-2029-4
4. Yan, F., Iliyasu, A.M., Venegas-Andraca, S.E.: A survey of quantum image representations. Quant. Inf. Process. **15**(1), 1–35 (2015). https://doi.org/10.1007/s11128-015-1195-6
5. Boneh, D., Zhandry, M.: Quantum-secure message authentication codes. In: Johansson, T., Nguyen, P.Q. (eds.) EUROCRYPT 2013. LNCS, vol. 7881, pp. 592–608. Springer, Heidelberg (2013). https://doi.org/10.1007/978-3-642-38348-9_35

6. Potapov, V., Gushanskiy, S., Polenov, M.: Optimization of models of quantum computers using low-level quantum schemes and variability of cores and nodes. In: Silhavy, R. (ed.) CSOC 2019. AISC, vol. 986, pp. 264–273. Springer, Cham (2019). https://doi.org/10.1007/978-3-030-19813-8_27

7. Kleppner, D., Kolenkow, R.: An Introduction to Mechanics, 2nd edn, 49 p. Cambridge University Press, Cambridge (2014)

8. Venegas-Andraca, S.E.:. Quantum Image Processing. Springer, Singapore (2020). ISBN 978-9813293304

9. Sukachev, D.D., Sipahigil, A., Lukin, M.D.: Silicon-vacancy spin qubit in diamond: a quantum memory exceeding 10 ms with single-shot state readout. Phys. Rev. Lett. **199** (2017)

10. Lukin, M.D.: Probing many-body dynamics on a 51-atom quantum simulator. Nature **551** (2013)

11. Potapov, V., Gushansky, S., Guzik, V., Polenov, M.: Architecture and software implementation of a quantum computer model. In: Silhavy, R., Senkerik, R., Oplatkova, Z.K., Silhavy, P., Prokopova, Z. (eds.) Software Engineering Perspectives and Application in Intelligent Systems. AISC, vol. 465, pp. 59–68. Springer, Cham (2016). https://doi.org/10.1007/978-3-319-33622-0_6

12. Raedt, K.D., et al.: Massively parallel quantum computer simulator. Comput. Phys. Commun. **176**, 121–136 (2007)

Selecting the Optimal Configuration of Swarm Algorithms for an NLP Task

George Tambouratzis[(✉)] [iD]

ILSP/Athena Research Centre, 6 Artemidos Street, 15125 Maroussi, Greece
giorg_t@athenarc.gr

Abstract. The present article involves the generation of phrase boundaries in unconstrained free text. Particle Swarm Optimisation (PSO) is applied to determine the optimal values for a set of parameters, based on a limited amount of training data. The starting point is a detailed analysis of generated solutions, which leads to a reformulation of the phrasing task. Based on this reformulation, the optimal swarm configuration is investigated, including the interconnection between swarm particles as well as velocity initialisation. Another aspect studied is whether a homogeneous or a heterogeneous swarm provides a better optimisation, established by comparing three PSO variants. The experimental results are further analysed with statistical tests to determine their significance.

Keywords: Natural language parsing · Syntactically-based phrasing · Particle Swarm Optimization (PSO) · Adaptive PSO (AdPSO) · Predator Prey Optimization (PPO)

1 Introduction

The present article studies the creation of a phrasing model that segments unconstrained sentences of text into syntactically-valid phrases. Such a model forms a key component for a variety of applications, in the areas of natural language processing (NLP) and understanding.

The phrasing model assumes attractive and repulsive forces between groups of consecutive words in the sentence. These forces are each multiplied accordingly by weights which depend on the relative locations of these words. A PSO-type metaheuristic [1] is then utilized to optimize these weights, and has been shown [2] to successfully extrapolate an effective phrasing model from limited training data.

This task is revisited here to determine why in some cases the phrasing results have been sub-optimal. Specific improvements are investigated, drawing

The authors acknowledge support of this work by the project "DRASSI" (MIS5002437) which is implemented under the Action "Reinforcement of the Research and Innovation Infrastructure", funded by the Operational Programme "Competitiveness, Entrepreneurship and Innovation" (NSRF2014-2020) and co-financed by Greece and the European Commission (European Regional Development Fund).

© Springer Nature Switzerland AG 2021
L. Rutkowski et al. (Eds.): ICAISC 2021, LNAI 12855, pp. 113–125, 2021.
https://doi.org/10.1007/978-3-030-87897-9_11

inspiration from [3], which underlines the importance of precise formulation in optimizing the solution. Here, the reformulation is aimed to distinguish more precisely between positive and negative training patterns.

At the same time, a rethinking of specific swarm aspects is performed, to determine whether performance can be improved over previous results. In this effort, findings have been incorporated from recently published work. One particular aspect examined is the use of different communication topologies between the swarm particles. It has been reported [4] that the use of a local instead of a global neighborhood of particles to propagate best solutions leads to an improved optimization performance, where improvements may be more prominent for extended runs. On the other hand, [5] reports no clear superiority of either neighbourhood, the best choice being task-dependent.

Swarms studied previously for the phrasing task [2] were homogeneous, with all particles sharing the same position and velocity update rules, as specified in [6]. Silva et al. [7] propose Predator Prey Optimization (PPO) which adds a new type of swarm particle, namely a predator. This special-role particle is designed to track the swarm particle that has determined the best solution, and chase it away from this so-far optimal solution, to prevent the swarm from stagnating. A refined heterogeneous PPO swarm has also been proposed [8] that beyond standard particles comprises both a predator and two types of scouts that perform a local search around the best solution found so far.

The velocity initialization approach in PSO is claimed to have a substantial effect in the optimization process [9]. To verify if this applies to the given task, experiments are run using four metrics to evaluate the phrasing accuracy.

2   The Phrasing Task

2.1   Review of Relevant Work

The phrasing model studied here uses the concept of attraction and repulsion forces [2] and is termed ARG (Attraction-Repulsion phrase Generator). ARG is general-purpose and was used in machine translation [10]. To use ARG with lesser-resourced languages, only a limited amount of training data is assumed.

The concept of attraction and repulsion replicates natural systems where opposing forces contribute to reach a stable solution. The work reported here aims to improve the phrasing quality, by taking into account recently proposed research directions. More specifically, it aims to: (1) Investigate alternative task formulations, to determine if the final solution determined by the swarm can be improved. (2) Identify whether within a pool of PSO algorithms, one consistently outperforms the others, and if the increased complexity of certain algorithms leads to better performance. (3) Regarding the initial velocity, are better results obtained by adopting a zero-magnitude initial velocity instead of a random initialization? (4) Which is the most effective interconnection approach for propagating the best solutions within the swarm?

2.2 Description of the Attraction–Repulsion Phrasing Model

To achieve effective generalization from smaller training sets, the sequences of words are replaced by their grammatical tags. Therefore, the example sentence "the child holds the dog" is transformed into "article noun verb article noun".

Splitting a sentence into phrases involves deciding where to place phrase boundaries, considering all the points between consecutive words. ARG assumes attractive and repulsive forces (denoted as $Fatt(i)$ and $Frep(i)$ respectively) between consecutive words. To implement the attraction-repulsion model, a set of window-type templates sampling consecutive words is defined. The frequencies of these templates are calculated for all instantiations of words within the training data. For instance, assuming a window of 2 words around the kth segmentation point, if pattern $[W_k; W_{k+1}]$ occurs 80% of times within the same phrase, and only 20% of times in different phrases, this represents a strong attraction. This is the case for the first and second words of the example, namely "the" and "child". On the other hand, a frequent placement of words W_k and W_{k+1} in different phrases equates to a strong repulsion (for example the second and third words of the aforementioned sentence, i.e. "child" and "holds").

Four templates are chosen for ARG, with window sizes of between one and four words [2]. In the ARG phrasing model, all these forces are combined to generate the potential, Pot, of the specific point being a phrase boundary. In (1), the first term corresponds to attractive forces, the second to repulsive forces and the third to a threshold. When using 4 windows, there are 9 ARG parameters.

$$Pot(i) = \sum_{j=1}^{m}(a_j * Fatt_j(i)) - \sum_{j=1}^{m}(b_j * Frep_j(i)) - thr \qquad (1)$$

The decision on whether to include a phrase boundary at a given point depends on the potential of this point. For a negative potential, a phrase boundary is placed. If the potential is positive, then no boundary is placed.

3 Essentials of the Swarm Algorithm

3.1 Description of Particle Swarms Studied

Three particle swarm variants are studied comparatively here. The first is sPSO, which implements the standard PSO algorithm [1]. In sPSO two variables define a particle, namely its velocity $v_l(t)$ and its location $x_l(t)$ in the pattern space. The next position of particle i depends on its current location and velocity. At each step, the particle's velocity is revised based on its proximity to the best solution $y_{i,j}(t)$ established by this particle, and the best solution $\widehat{y_{(i,j)}}(t)$ by all particles within the neighbourhood of particle i, as defined by (3):

$$x_{i,j}(t+1) = x_{i,j}(t) + v_{i,j}(t) \qquad (2)$$

$$v_{i,j}(t+1) = \omega \cdot v_{i,j}(t) + c_1 \cdot r_{i,j}(t) \cdot [y_{i,j}(t) - x_{i,j}(t)] + c_2 * r_{i,j}(t) \cdot [\widehat{y_{i,j}(t)} - y_{i,j}(t)] \qquad (3)$$

The main hyperparameters affecting PSO operation are ω, c_1 and c_2, termed as the inertia, cognitive and social coefficients. The basic PSO model is expanded in sPSO by adding an elitism component and a local search element. The optimal values for main PSO hyperparameters have been researched systematically [11], where 14 suggested sets of coefficients were compared. Several different sets have been applied to the ARG task [12], leading to the most effective set of values (ω = 0.7298, $c_1 = c_2 = 1.49618$), this set being originally suggested in [13].

The second swarm variant is AdPSO (Adaptive PSO, [14]), proposed as a substantial enhancement of the PSO algorithm, which switches between four distinct states during optimization, namely exploration, exploitation, convergence and jumping out. AdPSO search hyperparameters are modified internally, and there is no need to pre-define their values.

The third PSO variant is PPO [7], which is heterogeneous and supplements the standard PSO particles with scout and predator particles. More specifically, the predator particle chases swarm particles located near the best solution away from this position, to search for new optima, while the two scout particles search for new solutions at different locations of the search space. We have re-implemented the method of [8], with two modifications. We initialize the predator at the very middle of the pattern space to prevent any initial bias, and set the initial predator velocity to zero across all dimensions. PPO hyperparameters share the values of sPSO, in preference to those of [8].

3.2   Swarm Configurations

In all experiments presented here, 9 fully independent weights are used, being optimized using the three PSO variants. Each swarm comprises 20 particles, that evolve for 1,000 epochs. Fully-informed swarms had been used earlier [12], where all particles are aware of the best solution found so far by any particle, as this was claimed to give the best results [15]. Recent research has shown [5] that the choice of global or local connectivity between particles should depend on the characteristics of the specific task. Consequently both connectivities are studied here. The PPO swarm, applied here for the first time to the phrasing task, contains 20 standard swarm particles as well as one predator and two scout particles. Thus its complexity is increased over sPSO and AdPSO.

For each swarm configuration, 100 randomly initialized runs have been performed, where initialization affects the initial location and velocity of each particle. These populations are evaluated with respect to the phrasing accuracy in comparison to a gold standard phrasing. When reporting results, the lowest error, the highest error and the average error over the 100 runs are quoted.

4   Evaluating the Phrasing Accuracy

To evaluate phrasing results, two languages are used, English and Greek. Segmentation accuracy is calculated as the fraction of discovered segmentation points which coincide with the grammatically correct segmentation between phrases, over a set of test sentences. The test set for each language comprises 200 sentences, with circa 4,000 words and 1,500 phrases. An objective metric determines the segmentation accuracy, based on tp (true positives), tn (true negatives), fp (false positives) and fn (false negatives). F-measure focuses mainly on the correct identification of positive patterns (here, absence of erroneous boundaries). Three additional metrics are used to comprehensively evaluate the swarms [12]:

- The complement of F-measure (termed FCompMeas) emphasizes accurate recognition of true negatives (phrase boundaries) rather than true positives.
- Youden's J coefficient weighs equally false positives and false negatives.
- Cohen's k coefficient incorporates the possibility of an agreement occurring by chance.

5 Analysing the Models Generated During Phrasing

The 9 weights of Eq. (1) take real values in the range from 0 to 1. Earlier results [12] showed that many PSO-derived weights had a value coinciding with one of the extreme values of the available optimization range. The optimized weight values are summarized in the first part of Table 1 over 100 randomly-initialized runs. This particular example corresponds to the results obtained using the F-Measure metric, with randomly-initialized velocities and a fully-connected sPSO swarm, but is typical of phrasing outcomes for all configurations.

Table 1. PSO weight values with original scheme and NF refinement.

	Original formulation		NF formulation	
	Weights set to 1	Weights set to 0	Weights set to 0	Weights set to 1
English	200 (22.2%)	139 (15.4%)	181 (20.1%)	155 (17.2)%
Greek	187 (20.8%)	165 (18.3%)	163 (20.0%)	122 (13.6%)

The left half of Table 1 reports the standard formulation in which the percentage of weights having an extreme value (either 0 or 1) exceeds 35%. When a weight is set by the PSO algorithm to 1 it means that this weight has the maximum possible contribution to the calculation of potentials. Conversely, a weight set to 0 means that notwithstanding the frequency of the pattern, it is not used in deciding whether to add a phrase boundary or not. It is counter-intuitive to have zero values for many weights, meaning that frequency-related information is consistently ignored. A detailed analysis of the phrasing process has shown that one likely reason for the abundance of zero-values is the actual formulation used. True positives and true negatives are defined as follows:

$$\textbf{set boundary } if \ \sum_{j=1}^{m}(a_j * Fatt_j(i)) - \sum_{j=1}^{m}(b_j * Frep_j(i)) \geq thr \qquad (4)$$

$$\text{no boundary} \;\; if \;\; \sum_{j=1}^{m}(a_j * Fatt_j(i)) - \sum_{j=1}^{m}(b_j * Frep_j(i)) < thr \qquad (5)$$

Though Eq. (4) is appropriate for most cases, a likely problem arises when a boundary decision is made on-the-fence, i.e. when the potential is equal to zero. Based on (4), a potential of zero is equivalent to a true positive. In the extreme case that all parameter weights become zero, (4) will always produce a true-positive (without taking into account the frequencies of patterns) and a correct phrasing boundary is assumed. If many model weights converge to 0, the phrasing produced by (4) and (5) is judged as perfect, and the ARG system is rewarded for setting weights to 0 values, even though this results in ignoring frequency information. This is potentially further aggravated for metrics such as F-Measure, that are less sensitive to the false positives. To avoid such events, on-the-fence cases need to be avoided, and (4) and (5) are replaced by:

$$\text{set boundary} \;\; if \;\; \sum_{j=1}^{m}(a_j \cdot Fatt_j(i) - \sum_{j=1}^{m}(b_j \cdot Frep_j(i) > thr \qquad (6)$$

$$\text{no boundary} \;\; if \;\; \sum_{j=1}^{m}(a_j \cdot Fatt_j(i) - \sum_{j=1}^{m}(b_j \cdot Frep_j(i) \leq thr \qquad (7)$$

This modification of Eqs. (4) and (5) is termed as the NF (non-fence) variant. The results obtained for the NF variant are shown in the right-hand part of Table 1, indicating a reduction in the number of weights with extreme values. Experiments show that for both languages the NF mechanism forms larger, more meaningful phrases from which to extract the necessary patterns. The number of single-word phrases (frequently trivial) is substantially reduced by NF. For example, for Greek the average phrase length rises by 25%, from 2.0 to 2.5.

6   Experiments on Swarm Parameters

6.1   Introductory Material

Experiments have involved varying two factors, namely velocity initialisation and connectivity of the particles in the swarm. Regarding the first factor, the effort has previously focused on comparing the best solution found by the swarm when the velocity is initialized to non-zero random values [12]. The alternative [9] is to set the velocity of all particles to zero at epoch 0 and then allow it to grow depending on the particle's location in relation to the local/global minima.

Regarding topology, the results of [4] have reopened the issue of whether a fully-informed topology (Gbest, also termed global topology, cf. [5]) is better or not than a local topology (Lbest, also known as ring topology [5]). Topology affects the speed of propagation of the best solution found across a swarm. Though [15] has claimed that a fully-interconnected swarm is best, [4] has recently reported that a localised exchange of information is best.

Table 2. Errors (expressed as percentages) when phrasing English using FMeasure

Swarm	veloc = 0 & Gbest			veloc = 0 & Lbest			veloc $\neq$ 0 & Gbest			veloc $\neq$ 0 & Lbest		
	avg	min	max	avg	min	max	avg	min	max	avg	min	max
sPSO	5.87	5.71	6.61	5.86	5.74	6.00	5.87	5.71	6.63	5.86	5.72	5.98
AdPSO	6.24	5.69	6.74	6.15	5.82	6.60	6.25	5.69	6.63	6.14	5.69	6.58
PPO	6.75	5.71	7.78	6.39	5.68	7.29	6.37	5.7	7.34	6.18	5.7	6.83

Table 3. Error (expressed as percentages) when phrasing Greek using FMeasure

Swarm	veloc = 0 & Gbest			veloc = 0 & Lbest			veloc $\neq$ 0 & Gbest			veloc $\neq$ 0 & Lbest		
	avg	min	max	avg	min	max	avg	min	max	avg	min	max
sPSO	4.02	3.86	4.24	4.02	3.87	4.25	4.05	3.89	4.38	4.02	3.86	4.18
AdPSO	4.06	3.84	4.45	4.07	3.9	4.33	4.08	3.87	4.57	4.06	3.9	4.25
PPO	4.52	3.92	4.83	4.42	4.00	4.71	4.3	3.87	4.72	4.16	3.87	4.55

In our experiments, the three swarm types (sPSO, AdPSO and PPO) are compared. For each swarm (i) a global topology is used, as well as (ii) a local topology where all particles form a one-dimensional structure and each particle only accesses information from its two immediate neighbours. The results obtained for English and Greek are depicted in Tables 2 and 3 respectively. Due to space restrictions, only F-Measure results are reported. Having 2 velocity initializations, 2 topologies and 4 metrics, leads to 16 configurations. The convergence characteristics (minimum and average number of epochs for swarm convergence, as well as standard deviation) are depicted in Tables 4 and 5.

Table 4. Convergence characteristics when phrasing English texts using FMeasure

Swarm	veloc = 0 & Gbest			veloc = 0 & Lbest			veloc $\neq$ 0 & Gbest			veloc $\neq$ 0 & Lbest		
	avg	min	stdev	avg	min	stdev	avg	min	stdev	avg	min	stdev
sPSO	637.6	181	252.2	742.8	214	190	683.8	113	248.7	754.3	213	200.5
AdPSO	652.9	170	243.2	856.7	373	133	663	189	238.9	856.1	413	116.2
PPO	504.1	22	319.5	607.6	75	283.1	440.5	18	309.1	605.9	73	280.3

6.2 Comparing the Swarm Topologies

The first question is how Lbest compares to Gbest. This is determined by studying all 12 configurations (three swarm types and four optimization metrics). Based on Table 6, when the particle velocity is initialized to 0, the minimum error over the 100 runs is lower for Gbest than for Lbest in most cases. The superiority of the Gbest approach is more marked for non-zero initial velocities.

On the other hand, Lbest achieves consistently a lower average error than Gbest. Thus, the Lbest topology is better suited to cases where few independent runs can be performed. This observation applies to both languages tested.

Table 5. Convergence characteristics when phrasing Greek texts using FMeasure

Swarm	veloc = 0 & Gbest			veloc = 0 & Lbest			veloc ≠ 0 & Gbest			veloc ≠ 0 & Lbest		
	avg	min	stdev	avg	min	stdev	avg	min	stdev	avg	min	stdev
sPSO	552.7	73	260.1	709.8	84	247.6	608.1	53	267.6	704.8	132	222.5
AdPSO	638.7	39	249.2	783.1	227	175.2	630	109	252.5	761.1	214	178.3
PPO	357.3	6	331.8	488.9	22	290	380.1	12	322.2	586.6	37	247.9

Table 6. Comparison of Lbest and Gbest topologies for swarm models

	Error type	English texts		Greek texts	
		min error	avg. error	min error	avg. error
velocity = 0	Gbest < Lbest	6	0	8	3
	Gbest > Lbest	5	12	3	8
	Gbest = Lbest	1	0	1	1
velocity ≠ 0	Gbest < Lbest	8	0	7	1
	Gbest > Lbest	0	12	3	9
	Gbest = Lbest	4	0	2	2

6.3   Comparing the Particle Swarm Variants

A further question is which algorithm achieves a lower error. In this case there are 16 different configurations, (four metrics multiplied by two velocity initializations and by two topologies). For each configuration, a 3-way comparison between the swarm types is performed. The results are summarized in Tables 7 and 8. To cater for ties between two (or three) swarm variants, fractional scores are awarded (for instance if two swarm types score the same error, each is awarded half a point).

Table 7. Comparison of the swarm algorithms depending on velocity initialisation.

	Error type	English texts		Greek texts	
		Min error	Avg. error	Min error	Avg. error
velocity = 0	sPSO > others	2	7	4	8
	AdPSO > others	2	1	3	0
	PPO > others	4	0	1	0
velocity ≠ 0	sPSO > others	3.3	7	1	8
	AdPSO > others	2.83	1	2	0
	PPO > others	1.83	0	5	0

According to Table 7, when particle velocity is initialized to zero, the average error over 100 runs is minimized by using the sPSO variant, for almost all cases.

Table 8. Comparison of the swarm algorithms as a factor of the neighbourhood used.

	Error type	English texts		Greek texts	
		Min error	Avg. error	Min error	Avg. error
Gbest	sPSO > others	3.33	8	1	8
	AdPSO > others	2.33	0	4	0
	PPO > others	2.33	0	3	0
Lbest	sPSO > others	2	6	4	8
	AdPSO > others	2.50	2	1	0
	PPO > others	3.50	0	3	0

The same applies if the velocity is initialized to a non-zero value. In only one case per language, does AdPSO give a lower error than sPSO, whilst PPO never produces the lowest average error of all swarm types.

Turning to the minimum error, no swarm variant is clearly superior. When velocity is initialized to zero (cf. Table 7), for English PPO gives the lowest minimum error for 50% of the configurations. However, for Greek, it is sPSO that most frequently achieves the lowest minimum error over the 100 runs.

Table 8 compares the three algorithms in terms of the topology used. Considering the average error, sPSO is the swarm type that achieves the best optimization in the vast majority of cases, irrespective of topology. In only 2 configurations when using LBest, does AdPSO improve over sPSO, whilst PPO never achieves the lowest average error.

Concerning the minimum error, all three swarm types have a similar amount of configurations for which they achieve the lowest minimum value. On the whole, PPO achieves slightly better results than the other swarm variants, in particular for the Lbest neighbourhood topology, even though all swarm types achieve the lowest minimum error for some combinations. However, PPO is not consistently superior to the two other swarm types, in a similar manner to sPSO, which dominated other swarm types regarding the average error.

6.4   Determining the Statistical Significance of Results

The emergence of sPSO coupled with Lbest as a more effective swarm configuration has opened the possibility for even larger improvements. In agreement to the results of [4], sPSO with Lbest is less susceptible to stagnation and succeeds in discovering new better solutions. To determine the significance of these results, statistical paired sample t-tests have been carried out. These show that with respect to the average error, the performance of sPSO is significantly better than that of AdPSO and of PPO at a 0.05 level, for both languages.

Turning to the minimum error over 100 runs, for the English language, both PPO and sPSO generate significantly lower errors than AdPSO, at a 0.05 level. The errors of PPO and sPSO are statistically equivalent to each other.

In the case of the Greek language, the difference in minimum error achieved between the three swarm variants is not significant at a 0.05 level.

6.5   Convergence of Homogeneous vs. Heterogeneous Swarms

The comparison of homogeneous swarms (sPSO and AdPSO) and heterogeneous swarms (PPO) has shown that the smallest error corresponds to a homogeneous rather than a heterogeneous swarm. Conversely, PPO has an advantage in the speed of convergence to a solution. In Tables 4 and 5 the convergence speed of each swarm configuration is depicted, when phrasing English and Greek texts respectively. Typically, when phrasing English using a Gbest topology, PPO settles in ca. 20% fewer epochs than both sPSO and AdPSO. Adopting an Lbest topology, convergence is expectedly slower, but again the heterogeneous PPO swarm settles much more quickly to a solution than the homogeneous swarms (sPSO and AdPSO). A very similar situation applies when phrasing Greek texts.

The downside to the fast convergence of PPO is that the actual solution achieved is not as good as for homogeneous swarms. An example of the convergence is depicted in Fig. 1, for a typical run, when using the F-Measure metric, a Gbest topology and the particle velocity is initialised to zero. PPO initially converges to a lower error much faster than other swarm algorithms, but then fails to improve further. The homogeneous swarms evolve over more epochs and reach lower errors.

Studying in more detail the PPO dynamics, the predator chased away one of the standard swarm particles in 7.3% of cases. Scouts of type 1 were successful in finding a better solution in 34.1% of the epochs, while scouts of type 2 led to better solution only in 3.3% of cases, indicating that type-2 scouts are not essential in PPO, at least for the phrasing task.

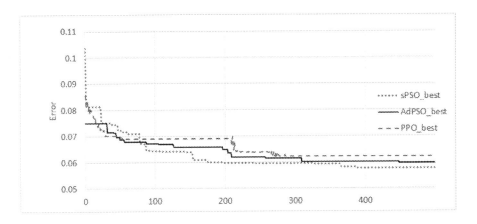

Fig. 1. Convergence trajectory when using F-Measure and zero initial velocity

7   Theoretical Insights on Swarm Complexity

Ever since PSO was introduced, researchers have striven to develop new variants able to determine ever better optimisation results. A recent trend has been to model more elaborate search behaviours as displayed by groups of natural organisms, most frequently higher-order mammals. One question that arises is if it is better to introduce new algorithms and to what extent these more specialised algorithms are effective over diverse tasks. A number of researchers have instead persevered with studying in depth the basic choices of the main PSO algorithms (including [4] and [5]), to determine if there exists a set of guidelines to achieve the best possible performance for a wide variety of tasks. The research described here follows the latter approach. Based on the experimental results, neither AdPSO nor PPO systematically achieve a better performance than the basic PSO. The increased complexity of advanced swarm variants does not confer any substantial advantage in comparison to the standard PSO algorithm for the given task.

Of particular interest has been the heterogeneous nature of PPO with the inclusion of specialized particles to improve exploitation and exploration. The use of a predator particle is an intriguing, intuitively promising concept, though it turns out not to be particularly effective for the given task. Here, different artefacts could be at play. The way a predator chases particles away from the current best solution by only displacing along one dimension may not be sufficient. Our findings warrant studying in more detail the dynamics of the predator behavior.

The cause of PPO and AdPSO failing to surpass the standard PSO may be their increased complexity. PSO is a relatively simple concept that works effectively for a wide range of problems and requires limited parameter tuning (and the parameters have been widely studied). The fact that PPO and AdPSO fail to improve accuracy may well be due to the difficulty of setting up the additional internal parameters, leading to a case of simpler-is-better.

8   Conclusions

This article has been spurred by the intention to improve the accuracy of a text phrasing model, whose parameters are optimised via particle swarm algorithms. Initially a refinement of the problem formulation is proposed, that results in syntactically more meaningful, longer phrases.

A second research direction has been to determine whether a global or local topology is more effective in the given task. The global topology gives slightly lower minimum errors over 100 runs, but Lbest results in lower average errors. In this respect, the conclusions of [5] are confirmed by our experiments for the phrasing task. In addition, regarding the velocity initialization, setting the initial particle velocities to zero results in a slightly better optimization behavior by the swarm and a lower final error. A similar observation applies to the initial velocity of predator particles.

Heterogeneous swarms have been investigated in comparison to homogeneous ones, and have been found to possess a faster convergence. However, in the latter phases of evolution, heterogeneous swarms do not match the lower errors achieved by homogeneous ones.

A direction for future research involves improving heterogeneous swarms. The introduction of specialist-role particles (especially the predators) substantially reduces the number of epochs required for convergence of the swarm. Though this is not critical for the current application, it can be of higher importance in other applications. Scout particles have been much less effective in discovering new solutions, and do not justify their use. The aim is to study and evaluate more advanced fine-tuned variations of the predator concept in the near future.

References

1. Kennedy, J., Eberhart R.C.: Particle Swarm Optimisation. In: Proceedings of the IEEE International Conference on Neural Networks, Perth, Australia, pp. 1942–1947 (1995)
2. Tambouratzis, G.: Applying PSO to natural language processing tasks: optimizing the identification of syntactic phrases. In: Proceedings of CEC-2016 Conference, Vancouver, Canada, pp. 1831–1838 (2016)
3. Ashlock, D., McGuinness, C., Ashlock, W.: Representation in evolutionary computation. In: Liu, J., Alippi, C., Bouchon-Meunier, B., Greenwood, G.W., Abbass, H.A. (eds.) WCCI 2012. LNCS, vol. 7311, pp. 77–97. Springer, Heidelberg (2012). https://doi.org/10.1007/978-3-642-30687-7_5
4. Blackwell, T., Kennedy, J.: Impact of communication topology in particle swarm optimization. IEEE Trans. Evol. Comput. **23**(4), 689–702 (2019)
5. Engelbrecht A.P., Cleghorn, C.W.: Recent advances in particle swarm optimization analysis and understanding. In: Proceedings of the 2020 GECCO Conference Companion, pp. 747–774 (2020)
6. Engelbrecht, A.P.: Heterogeneous particle swarm optimization. In: Dorigo, M., et al. (eds.) ANTS 2010. LNCS, vol. 6234, pp. 191–202. Springer, Heidelberg (2010). https://doi.org/10.1007/978-3-642-15461-4_17
7. Silva, A., Neves, A., Costa, E.: An empirical comparison of particle swarm and predator prey optimisation. In: O'Neill, M., Sutcliffe, R.F.E., Ryan, C., Eaton, M., Griffith, N.J.L. (eds.) AICS 2002. LNCS (LNAI), vol. 2464, pp. 103–110. Springer, Heidelberg (2002). https://doi.org/10.1007/3-540-45750-X_13
8. Silva, A., Neves, A., Gonçalves, T.: An heterogeneous particle swarm optimizer with predator and scout particles. In: Kamel, M., Karray, F., Hagras, H. (eds.) AIS 2012. LNCS (LNAI), pp. 200–208. Springer, Heidelberg (2012). https://doi.org/10.1007/978-3-642-31368-4_24
9. Engelbrecht, A.P.: Particle swarm optimization: velocity initialization. In: IEEE CEC Congress, Brisbane, Australia (2012)
10. Tambouratzis, G., Vassiliou M., Sofianopoulos, S.: Machine Translation with Minimal Reliance on Parallel Resources. Springer, Cham (2017). https://doi.org/10.1007/978-3-319-63107-3
11. Harrison, K.R., Engelbrecht, A.P., Ombuki-Berman, B.M.: An adaptive particle swarm optimization algorithm based on optimal parameter regions. In: Proceedings of the IEEE SSCI-2017, pp. 1–8 (2017)

12. Tambouratzis G.: PSO optimal parameters and fitness functions in an NLP task. In: Proceedings of IEEE CEC-2019, Wellington, New Zealand, pp. 611–618 (2019)
13. Eberhart R.C., Shi Y.: Comparing inertia weights and constriction factors in particle swarm optimization. In: Proceedings of the Congress on Evolutionary Computation, La Jolla, CA, USA, vol. 1, pp. 84–88 (2000)
14. Zhan, Z.-H., Zhang, J., Li, Y., Chung, H.S.-H.: Adaptive particle swarm optimisation. IEEE Trans. Syst. Man Cybern. Part B: Cybern. **39**(6), 1362–1381 (2009)
15. Kennedy, J., Mendes, R.: Neighborhood topologies in fully informed and best-of-neighborhood particle swarms. IEEE Trans. Syst. Man Cybern. Part C: Appl. Rev. **36**(4), 515–519 (1996)

Active Learning Strategies and Convolutional Neural Networks for Mammogram Classification

João Marcelo Tozato$^{(\boxtimes)}$ [ID], Pedro Henrique Bugatti[ID],
and Priscila Tiemi Maeda Saito[ID]

Department of Computing, Federal University of Technology - Parana,
1640 Alberto Carazzai Ave, Cornelio Procopio, Parana, Brazil
tozato@alunos.utfpr.edu.br, {pbugatti,psaito}@utfpr.edu.br

Abstract. Deep learning has been used successfully in a variety of applications due to the large data availability and the growth in computing power. However, some domains present a shortage of both samples and labels, for instance, the medical area. In this work, we propose machine learning approaches that include traditional supervised classifiers and active learning methods for the breast lesion domain, in order to aid breast cancer diagnosis. We propose the introduction of active learning strategies in this process, to sort out the most informative samples in the dataset. The active learning process reduces the burden of the dataset annotation, while also improving the robustness of our models. Hence, we achieved considerable gains with fewer labeled training images, minimizing the specialist's annotation effort. The validation of our proposed methodology is done on a public breast lesion-related dataset and our results show considerable accuracy gains over the traditional supervised learning approach and reductions of up to 68% in the labeled training sets.

Keywords: Machine learning · Active learning · Pattern recognition.

1 Introduction

The usage of deep learning for classification or segmentation tasks has become extremely efficient in a variety of application domains, such as medical problems. For instance, in hemorrhage detection in CT scans [6], segmentation of MRI scan images of the brain [1], breast cancer detection [9], amongst other applications.

According to the American Institute for Cancer Research [4], breast cancer is the most common type and the fifth major cause of death in women worldwide. Hence, applying intelligent systems in this scenario can assist in the decision process of the professional in charge of the diagnosis. Furthermore, the early diagnosis of this disease can reduce its mortality rate.

Despite some efforts found in the literature [10,14,18], most of these consider learning approaches that require a substantial amount of labeled data and do not take into account some mandatory restrictions of medical applications (e.g. mostly related to computational time and resources).

© Springer Nature Switzerland AG 2021
L. Rutkowski et al. (Eds.): ICAISC 2021, LNAI 12855, pp. 126–134, 2021.
https://doi.org/10.1007/978-3-030-87897-9_12

In addition, there are other challenges related to data acquisition and labeling processes. The gathering of mass-related breast lesion images is not trivial due to the maintenance of patients' privacy, availability of data by hospitals, among others. Moreover, generally, there is the requirement of sample labeling by one or more specialists (e.g. considering radiologists with different levels of experience) to ensure that the correct labels are assigned. It also impacts and contributes to the lower availability of labeled samples. The data labeling process requires time and effort from the specialist and is highly susceptible to errors.

Therefore, this paper addresses the study, development and validation of active learning strategies, in order to compare them to the traditional supervised learning approach. Active learning strategies have been widely used and successful in several other application domains. Such strategies allow for obtaining a reduced set of the most informative samples to the learning process of pattern classifiers. More effective and efficient classifiers can be obtained, achieving higher accuracies faster and minimizing the effort of the expert in the labeling process.

2   Background

The active learning approach considers the usage of the classifier in the selection of the most informative samples from a designated dataset. This method is advantageous on tasks that require hundreds or even thousands of labeled data (such as images), mainly by reducing the burden of annotating the whole dataset, which demands plenty of time and effort from a specialist in a given domain to execute the labeling of these samples [3].

It is an iterative process that makes use of a selection strategy to gradually obtain a fixed number of samples and incorporate them in the training dataset of the learning algorithm. In such way, by using the active learning approach, it is possible to create robust classification models with far less labeled instances and hence reducing the cost of the data labeling process.

At each iteration of the active learning process, a fixed number of samples (for our methodology twice the number of existing classes) is selected by the selection strategy and then gradually incorporated into the training set. New instances of the classifier are obtained and evaluated in the test set. Each active learning method used in this work to select the most informative samples is related to the uncertainty criterion [16].

There are different active learning strategies in the literature, one of them is based on Entropy (EN) [17], which can be understood as the degree of uncertainty of a variable, prioritizing samples that have a greater value for this measure. It calculates this according to the Eq. 1, where y is the probability of a given label for a sample x.

$$EN(x) = -\sum p_i(y|x) \log p_i(y|x) \qquad (1)$$

In the technique called Least Confidence (LC) [11], the model selects the sample that presents a lower confidence for the most probable class. Equation 2 shows the inner working of this technique, where y' is the highest probability given by the model for a sample x. Thus, a lower value for the probability of the most probable class leads to a higher chance of this sample being selected, due to the low confidence assigned to it.

$$LC(x) = 1 - p(y'|x) \tag{2}$$

Equation 3 shows the Margin Sampling (MS) [15] technique, which takes into account not only the most likely label, as in the previous strategy, but it is based on the smallest difference between the first and second most likely labels for sample selection, where y' and y'' represent the highest probabilities for a sample x.

$$MS(x) = p(y''|x) - p(y'|x) \tag{3}$$

3   Proposed Methodology

Initially, according to the first step of our pipeline (Fig. 1) we obtained and organized our dataset as described in Subsect. 4.1. Our methodology consists in two key approaches (traditional supervised learning and active learning, respectively) as shown in Steps 3 and 4 of the pipeline. Both approaches depend on the extraction of deep features through the use of CNNs (according to Step 2), which are acquired by removing the classification layers of a CNN model and getting the output of a given layer. In the present work we consider the last layer before the fully connected layers for this process.

For the feature extraction process we apply the Transfer Learning strategy, which allowed the initialization of our network's weights based on the weights of another neural network that was already trained on the ImageNet dataset [5]. The parameters of the old network are reused for the inference process of this new network, therefore reducing the computational cost to train neural networks from scratch.

We have also applied normalization to our input data. It is a technique that aims to adjust the mean and standard deviation of the input data values of a given neural network on a common scale, such as close to zero and one, respectively. It becomes especially important when using pre-trained neural networks, due to the fact that the model only knows how to work with data of the type that it has seen before. If the inputs of the new network using these parameters, do not share these normalization statistics, the results will not be as expected.

The fourth step of our methodology consisted in using active learning strategies alongside with traditional classifiers. Active learning strategies allow the selection of the most informative samples for the learning process. Therefore, we can reduce the amount of annotated images required for classification tasks while achieving significant or equivalent results when compared to the supervised approach, which requires a completely annotated training set.

We performed comparisons between the active learning strategies and the random sample selection at each iteration. In addition, we also compared the

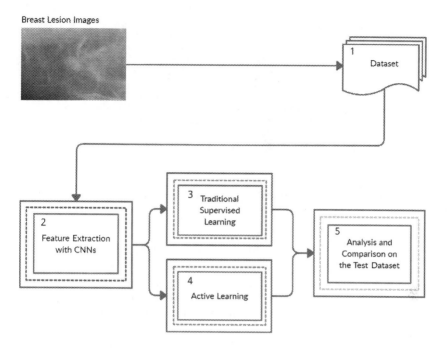

Fig. 1. Pipeline of the proposed methodology.

traditional supervised learning approach (which require a fully annotated training set) and the active learning strategy.

4   Experiments

4.1   Dataset

We used in this work the public dataset called MAMMOSET [12], which contains images regarding to three types of lesions: mass, calcification and normal (with no kind of lesion). For this work we have considered exclusively the subset of mass-related lesions, which are divided into malignant or benign. Figure 2 shows samples of the two distinct classes of the dataset.

The subset contains 1381 images in total and are distributed in the following manner: the training and test set are composed of 568 and 67 images for the malignant class and 671 and 75 images for the benign class, respectively.

4.2   Scenarios

Our training set is divided into 10 mutually exclusive stratified splits so that the percentage of each class samples is preserved. Each split has its own validation set, which is used to check the performance and model's biases during its training

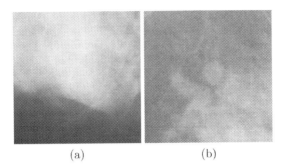

(a)   (b)

Fig. 2. Examples of images from the MAMMOSET dataset: (a) malignant sample, (b) benign sample.

in that split. At the end of the training of a given split, we start the inference process in a fixed test set that is the same throughout the process for each split. The traditional supervised learning and active learning experiments were conducted using the deep features extracted from the CNN architectures (DenseNet-121, DenseNet161, EfficientNetB3, EfficientNetB4 ResNet34 and ResNet50) in conjunction with traditional classifiers such as k-Nearest Neighbors (k-NN) [13], Naive Bayes (NB) [7], Random Forest (RF) [2] and Support Vector Machines (SVM) [8]. The images of the dataset were resized to 224 x 224 pixels before being used as input to the CNNs. For every network considered in the feature extraction process, we have not updated its weights, these architectures were just used as fixed feature extractors. Table 1 shows the dimensionality of each feature map of the CNNs used in the feature extraction process. The hyper-parameters chosen for the classifiers are the standard as provided in their literature.

Table 1. Description of the deep feature extractors with respect to their feature map dimensionality.

Extractor	Feature dimensionality
DenseNet121	1024
DenseNet161	1664
EfficientNetB3	1536
EfficientNetB4	1792
ResNet34	512
ResNet50	2048

5   Results and Discussion

Initially, we show the results obtained by the traditional learning approach (Fig. 3), considering the deep features extracted from each of the CNN architectures and the supervised classifiers (k-NN, NB, RF and SVM, respectively). We

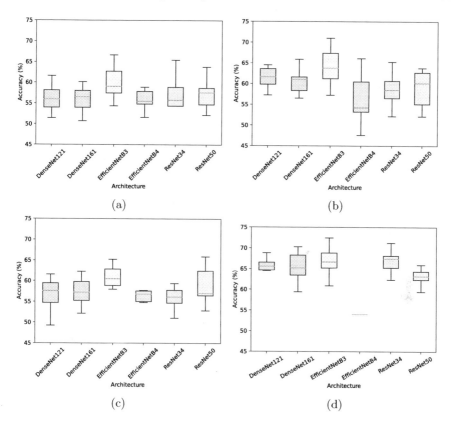

Fig. 3. Mean accuracies obtained by the traditional learning approach considering the deep features extracted from each architecture (DenseNet121, DenseNet161, Efficient-NetB3, EfficientNetB4, ResNet34 and ResNet50) and the supervised classifiers: (a) k-NN, (b) NB, (c) RF and (d) SVM.

can notice that, in general, the deep features obtained from the EfficientNetB3 architecture presented higher accuracy values in relation to the other architectures for all classifiers. The EfficientNetB3 architecture achieved the highest accuracy (up to 66.98 ± 3.43) with the SVM classifier. It is important to note that these experiments requires all samples from the dataset labeled.

Then, we performed experiments considering the active learning approach with the features extracted from the EfficientNetB3 architecture, in order to reduce the need to annotate all samples in the dataset. The reason for choosing this particular architecture is mainly due to its consistency and overall high accuracies presented in the supervised learning experiment. Figure 4 presents the mean accuracies obtained by the selection strategies (EN, LC, MS and RANDOM) along the iterations of the learning process, considering each of the supervised classifiers (k-NN, NB, RF and SVM, respectively).

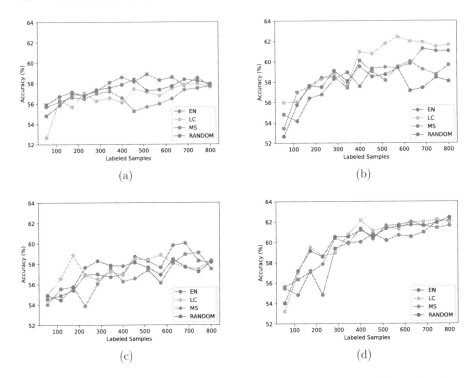

Fig. 4. Mean accuracies obtained by the active learning approach considering the selection strategies (EN, LC, MS and RANDOM), features extracted from the Efficient-NetB3 architecture and the supervised classifiers: (a) k-NN, (b) NB, (c) RF and (d) SVM.

It is possible to note that the active learning approach achieves better results with a reduced labeled training dataset, when comparing to the traditional supervised learning approach, which requires the dataset to be completely labeled. The active learning strategies allow a significant reduction (up to 44%, 68%, 61%, 60%) in the labeled training set required to reach accuracies equivalent to those obtained by the traditional supervised approach, considering the k-NN, NB, RF and SVM classifiers, respectively.

Table 2 shows the computational times for each combination of active learning strategy and classifier. We can verify that, as a more complex classifier, the SVM model presented classification times much higher than the others. Moreover, the random strategy (RANDOM) achieved smaller selection times, since it does not have a specific criterion for the selection of the samples that will integrate the training dataset.

Table 2. Total selection and classification times in seconds obtained by the active learning approach considering the selection strategies (EN, LC, MS and RANDOM), features extracted from the EfficientNetB3 architecture and the supervised classifiers (k-NN, NB, RF and SVM).

Strategy	Classifier	Selection time	Classification time
EN	k-NN	1.002e+00	1.477e-01
LC		8.151e-01	1.101e-01
MS		8.132e-01	1.103e-01
RANDOM		1.619e-02	4.839e-02
EN	NB	9.541e-02	3.975e-02
LC		5.920e-02	2.184e-02
MS		6.288e-02	2.509e-02
RANDOM		2.500e-02	1.145e-02
EN	RF	2.273e-02	1.181e-01
LC		1.303e-02	8.439e-02
MS		1.565e-02	9.341e-02
RANDOM		2.785e-02	9.621e-02
EN	SVM	6.496e-01	3.309e+00
LC		5.404e-01	2.652e+00
MS		7.282e-01	3.657e+00
RANDOM		1.288e-02	2.666e+00

6   Conclusion

In the present work, we conduct extensive experiments following the proposed methodology in order to assist in the diagnosis of breast cancer lesions. We compared the results obtained from two main different approaches to this image classification task: supervised learning and active learning strategies alongside with traditional classifiers. Regarding the active learning approach, we were able to verify that, in contrast to the supervised approach, which requires the whole labeled dataset for its learning process, the selection strategies provide a way to create representative training sets and achieve high accuracies for this particular classification task.

We also explored the significance of using active learning strategies with CNNs on image classification tasks, exhibiting that it is possible to reach expressive results with a reduced labeled dataset, specially when comparing to traditional supervised learning. The results obtained in the carried out experiments show the benefits of the usage of active learning strategies in the development process of a classifier, since in this medical context is not common that all the available images are annotated. Then, by selecting the most informative samples for the model's learning there is a reduction in both time and effort required for the labeling process of a dataset by a specialist.

References

1. Akkus, Z., Galimzianova, A., Hoogi, A., Rubin, D.L., Erickson, B.J.: Deep learning for brain MRI segmentation: state of the art and future directions. J. Digit. Imag. **30**(4), 449–459 (2017)
2. Breiman, L.: Random forests. Mach. Learn. **45**(1), 5–32 (2001). https://doi.org/10.1023/A:1010933404324
3. Bressan, R.S., Bugatti, P.H., Saito, P.T.: Breast cancer diagnosis through active learning in content-based image retrieval. Neurocomput. **357**, 1–10 (2019)
4. American Institute of Cancer Research: Breast cancer: how diet, nutrition and physical activity affect breast cancer risk. https://www.wcrf.org/dietandcancer/breast-cancer
5. Deng, J., Dong, W., Socher, R., Li, L.J., Li, K., Fei-Fei, L.: Imagenet: a large-scale hierarchical image database. In: 2009 IEEE conference on computer vision and pattern recognition, pp. 248–255. IEEE (2009)
6. Grewal, M., Srivastava, M.M., Kumar, P., Varadarajan, S.: Radnet: radiologist level accuracy using deep learning for hemorrhage detection in CT scans. In: 2018 IEEE 15th International Symposium on Biomedical Imaging (ISBI 2018), pp. 281–284. IEEE (2018)
7. Hand, D.J., Yu, K.: Idiot's bayes—not so stupid after all? Int. Stat. Rev. **69**(3), 385–398 (2001)
8. He, Z., Xia, K., Niu, W., Aslam, N., Hou, J.: Semisupervised SVM based on cuckoo search algorithm and its application. Math. Prob. Eng. **2018**, 1–13 (2018). https://doi.org/10.1155/2018/8243764
9. Huynh, B.Q., Li, H., Giger, M.L.: Digital mammographic tumor classification using transfer learning from deep convolutional neural networks. J. Med. Imag. **3**(3), 034501 (2016)
10. Kooi, T., et al.: Large scale deep learning for computer aided detection of mammographic lesions. Med. Imag. Anal. **35**, 303–312 (2017)
11. Lewis, D.D., Catlett, J.: Heterogeneous uncertainty sampling for supervised learning. In: Machine Learning Proceedings 1994, pp. 148–156. Elsevier, Amsterdam (1994)
12. Oliveira, P., de Carvalho Scabora, L., Cazzolato, M., Bedo, M., Traina, A., Jr., C.: Mammoset: An enhanced dataset of mammograms. In: Proceedings of the satellite events - Brazilian Symposium on Databases, pp. 256–266 (2017)
13. Rani, P., Vashishtha, J.: An appraise of KNN to the perfection. Int. J. Comput. Appl. **170**(2), 13–17 (2017). https://doi.org/10.5120/ijca2017914696
14. Ribli, D., Horváth, A., Unger, Z., Pollner, P., Csabai, I.: Detecting and classifying lesions in mammograms with deep learning. Sci. Rep. **8**(1), 1–7 (2018)
15. Scheffer, T., Decomain, C., Wrobel, S.: Active hidden Markov models for information extraction. In: Hoffmann, F., Hand, D.J., Adams, N., Fisher, D., Guimaraes, G. (eds.) IDA 2001. LNCS, vol. 2189, pp. 309–318. Springer, Heidelberg (2001). https://doi.org/10.1007/3-540-44816-0_31
16. Settles, B.: Active learning literature survey. University of Wisconsin-Madison Department of Computer Sciences, Technical reports (2009)
17. Shannon, C.E.: A mathematical theory of communication. Bell Syst. Tech. J. **27**(3), 379–423 (1948)
18. Valério, L.M., Alves, D.H., Cruz, L.F., Bugatti, P.H., de Oliveira, C., Saito, P.T.: Deepmammo: deep transfer learning for lesion classification of mammographic images. In: 2019 IEEE 32nd International Symposium on Computer-Based Medical Systems (CBMS), pp. 447–452. IEEE (2019)

Data Mining

Exploiting Time Dynamics for One-Class and Open-Set Anomaly Detection

Lorenzo Brigato[1], Riccardo Sartea[2], Stefano Simonazzi[2], Alessandro Farinelli[2], Luca Iocchi[1], and Christian Napoli[1]([✉])

[1] Department of Computer, Automation and Management Engineering, La Sapienza University of Rome, Viale Ariosto 35, 06121 Roma, Italy
{brigato,iocchi,cnapoli}@diag.uniroma1.it
[2] Department of Computer Science, University of Verona, Strada le Grazie 15, 37134 Verona, Italy
{riccardo.sartea,alessandro.farinelli}@univr.it

Abstract. In this paper we describe and compare multiple one-class anomaly detection methods for Cyber-Physical Systems (CPS) that can be trained with data collected only during normal behaviors. We also consider the problem of detecting which group of sensors is most affected by the anomalous situation solving an open-set classification task. The proposed methods are domain independent and are based on a temporal analysis of data collected by the system. More specifically, we use different flavours of deep learning architectures, including recurrent neural networks (RNN), temporal convolutional networks (TCN), and autoencoders. Experimental results are conducted in three different scenarios with publicly available datasets: social robots, autonomous boats and water treatment plants (SWaT dataset). Quantitative results on these datasets show that our approach achieves comparable results with respect to state of the art approaches and promising results for open-set classification.

1 Introduction

Cyber-Physical Systems (CPS) such as robots, industrial machinery, autonomous vehicles, etc., are employed in an ever increasing variety of environments. In many cases these systems have a direct impact on society, e.g., power or water treatment plants. It is therefore crucial to ensure the correct and safe operation of CPS, as the consequence of unexpected failure or disruption could harm people and business. In this work we aim to detect behaviors of CPS that do not conform to the known normal operation, i.e., anomalies. Furthermore, since attacks to CPS cause unwanted behaviors that differ from the normal, detecting an anomaly allows also to detect the underlying attack. A key problem of anomaly detection is the lack of data regarding the anomalies. In fact, it is easy to profile the normal behavior of a CPS, however it is more complex to observe the CPS under a realistic anomalous behavior. Moreover, it is almost impossible to predict all the possible types of anomalies that may occur in complex CPS.

© Springer Nature Switzerland AG 2021
L. Rutkowski et al. (Eds.): ICAISC 2021, LNAI 12855, pp. 137–148, 2021.
https://doi.org/10.1007/978-3-030-87897-9_13

For this reason, the anomaly detection problem is often addressed as a one-class classification problem, i.e., the only available data for training belongs to the class referring to the normal behavior. The works in [1] proposes a multivariate anomaly detection method based on Generative Adversarial Networks (GAN), using Long Short-Term Memory Recurring Neural Networks (LSTM-RNN) for generator and discriminator models. After the training process, the discriminator is used to distinguish between normal and abnormal behavior. The DAGMM approach proposed in [2] makes use of autoencoders and Gaussian Mixture Model (GMM) that takes the low-dimensional input from the compression network and outputs mixture membership prediction for each sample. Anomaly detection for CPS, and specifically robots, has been recently addressed in [3], where authors propose to extract system logs from a set of internal variables of a robotic system, transform such data into images, and train different Autoencoder architectures to detect anomalies. In this work, we propose a new transformation of the input specifically designed for Convolutional Autoencoders that takes time into account. We show that such transformation increases the robustness of Convolutional Autoencoders, without changing the underlying model. In this work we also propose a regression approach to the open-set classification of anomalies that is capable of identifying which sensors of a CPS are most likely the cause of a detected anomaly. This information is very valuable as it allows for an interpretation of the anomaly and helps the analysis of its causes. In particular, the open-set refers to the different classes of possible anomalies that are not known at training time since the training process is one-class, i.e., only normal behavior. In the empirical evaluation we compare our solutions with several statistical methods and state of the art approaches for anomaly detection, considering techniques that use both explicit [4,5] and implicit [1,6–9] representations of the time dimension. We use three datasets for testing: autonomous water drone *Boat* [3], the *Pepper* social robot [3], and *SWaT* [10]. The first two were collected from robotic platforms[1], whereas the third one was collected from an industrial CPS of a water treatment plant. Moreover, we also show a use case for the *Pepper* social robot involved in a public demonstration subject to a simulated attack. This use case is employed as additional test set for the anomaly detection methods described in the paper. Crucially, in this test we run models trained on the mentioned dataset on a log that has been acquired in a very different situation, e.g., different environment, different means of interaction. The analysis confirms that the simulated attack can be detected by using our approach. Overall results of the experiments show that deep learning architectures are suitable to detect anomalies in one-class and open-set settings and that exploiting time dynamics generally improve such performance. In summary, the contributions of this paper are the following: *i)* we propose a new transformation of the input specifically designed for Convolutional Autoencoders that characterize the temporal evolution of system variables; *ii)* we propose a novel approach based on regression for the open-set classification of anomalies that is capable of identifying which sensors of a CPS are most likely the cause of a detected anomaly; *iii)* we con-

---

[1] https://sites.google.com/diag.uniroma1.it/robsec-data.

duct an extensive empirical evaluation of multiple techniques on real datasets of robotic platforms and industrial CSP. Moreover, we also present the results on a different instance of real usage scenario for the *Pepper* platform.

2   Problem Definition

Let us consider a general CPS $\mathcal{S}$ that can be described by a set of variables V characterizing its internal state over time. We denote $K = |V|$. The evolution over time of V characterizes the behavior of S, that can either be *normal* or fall into a class of *anomalies*, generally denoted as *abnormal*. Let r_t be a record, i.e., a tuple of values of V collected at time t, note that $|r_t| = K$. We define $\pi_{t,d} = \{r_{t-d}, \ldots, r_t\}$ a system log of a behavior containing the values of variables V recorded from time $(t-d)$ to time t. Let us consider a scenario in which the logs are taken when $\mathcal{S}$ is performing a normal behavior, we denote with D_r the set of collected normal behavior records, and D_t a set of unknown behaviors potentially containing both normal and abnormal records. The first problem we define aims at producing a binary classifier of normal and abnormal behaviors of system logs, trained only with normal samples. The practical importance of this problem is that it is not necessary to collect abnormal situations, which are often very difficult to acquire in a significant quantity and variety to train a reliable model.

Definition 1. One-class classification of system logs. *Given a dataset of system logs D_r, generate a model that is able to classify new instances $\pi_{t,d} \in D_t$.*

The second problem considers the use of the model computed as a solution to Problem Definition 1, to compute the subset of variables $H \in 2^V$ that better explains the anomaly, when this is detected. This problem is also of practical importance, as the information produced helps in directing the investigation of the causes towards the sensors that are most likely involved in the anomaly.

Definition 2. Open-set classification of system logs. *Given a dataset of system logs D_r, generate a model that is able to classify new instances $\pi_{t,d} \in D_t$ and, when an anomaly is detected, estimate the subset of variables $H \in 2^V$ that contribute the most to the anomaly.*

3   Log-to-Temporal-Image Anomaly Detection

We propose a new solution to the one-class classification problem based on transforming logs into images described in [3]. We focus our attention on the temporal aspect of the logs and aim to improve the performance of Convolutional Autoencoders that resulted to be the worst models in terms of performance [3]. A key element in our approach is the transformation of each record $r_t \in R$ into a squared image $I \in \mathcal{I}$ (denoting with $\mathcal{I}$ the set of images). To this end, we implemented a *log-to-image transformation function* $\sigma : R \to \mathcal{I}$ able to transform log records into images. $D_{r,I} = \{(\sigma(r_t), normal) \mid \forall\ r_t \in D_r\}$ computed from the system logs that will be used for generating the class model. More precisely, given dataset D_r we define a new dataset of labeled images $D_{r,I} = \{(\sigma(r_t), normal) \mid \forall\ r_t \in D_r\}$ computed from the system logs that will be used for generating the class model

3.1 Log-to-Temporal Image Transformation

Since logs represent temporal data, the temporal information includes important features that may be needed to detect anomalies. However, convolutions do not exploit such information since their filters work over spatial dimensions that do to take time into account. Moreover, features in data are expected to be dense. This condition is not fulfilled unless the input space is low dimensional. The standard min_max normalization implemented in [3] penalized Convolutional Autoencoders when tested on $Pepper$ dataset. Indeed, 1) each image does not consider the values of the previous record, 2) the vector is simply reshaped to generate an image (padding if needed) so, values are located without following any reasoning. To solve the mentioned issues we define a log-to-image transformation function $\sigma : R \rightarrow \mathcal{I}$ implemented as follows. Let $(r_k)_t$ be the k-th variable within a record at time t and let $(\rho_k)_t = \frac{\partial (r_k)_t}{\partial t}$ and $(\theta_k)_t = \frac{\partial^2 (r_k)_t}{\partial t^2}$ be respectively the first and second time derivatives. They characterize the temporal evolution of the logs variables. Variables that follow similar patterns through time should be closely located to each other. On the other hand, variables whose temporal profile greatly differ should be mapped far from each other. In this manner, convolution kernels can extract the information needed to characterize the normality of the current sample. The first step regards the transformation of the derivatives to polar coordinates. The first derivatives are interpreted as radii, in polar coordinates, that originates at the image center. Since the image edge has a defined dimension I_e, $(\rho_k)_t$ is remapped to the interval $[0, \frac{I_e}{2} - 1]$. The absolute value of the second derivatives $|(\theta_k)_t|$ is then interpreted as the polar coordinates' angle, and therefore normalized accordingly to belong to $[0, 2\pi]$. The image row $(i_k)_t$ and column $(j_k)_t$ to which a variable of the log record $(r_k)_t$ is mapped, are computed as:

$$(i_k)_t = \lfloor (\rho_k)_t \cos(\theta_k)_t + \tfrac{I_e}{2} \rfloor$$
$$(j_k)_t = \lfloor (\rho_k)_t \sin(\theta_k)_t + \tfrac{I_e}{2} \rfloor \tag{1}$$

Note that i and j are floored to integer values. Some variables could be mapped to the same pixel, for instance, constant variables. We will refer to this circumstance as pixel superposition. Generally, when such a superposition occurs it means that some variables are concurrently and consistently varying over time. In Eq. (2) we tackle this problem and the related formal complications. We could consider such covaring variables as correlated noise. In this latter fashion, it is useful to drop out such variables since they can become detrimental for correct classification. On the other hand, the produced image size matters for the quantization error. In fact, small outputs will determine a high probability of pixel superposition whether or not variables covariate. We let the transformation overwrite values over the same pixels. Experiments proved that such mapping

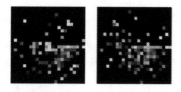

Fig. 1. Temporal image generated from a *normal* (left) and an *abnormal* (right) sample taken from *Pepper* dataset.

does not hurt performance. The image I is composed of three channels R, G and B. The red channel is filled with the normalized value of the log record $(\hat{r}_k)_t = \mathrm{minmax}((r_k)_t)$. The green channel refers to the semantic meaning of the variable while the blue channel is filled with zeros. In order to compose a mathematically consistent formalism, let define

$$\Lambda_t = \{((i_k)_t, (j_k)_t) \ \forall \ k \in [0, K] \cap \mathbb{N}\}$$
$$\bar{k}_{ijt} \triangleq \max\{k : (i_k)_t = i, (j_k)_t = j\} \qquad (2)$$
$$\chi_{ijt} \triangleq [(\hat{r}_{\bar{k}_{ijt}})_t, \tfrac{\bar{k}_{ijt}}{K}, 0]$$

It follows that for each pixel in position (i, j) at a time instant t is represented by an RGB vector computed as

$$\mathbf{I}_{ij}(t) \triangleq \begin{cases} \chi_{ijt} & (i, j) \in \Lambda_t \\ [0, 0, 0] & (i, j) \notin \Lambda_t \end{cases} \qquad (3)$$

Normalization bounds (minimum and maximum) are based on the training data distribution. At testing time, some values might be mapped outside the desired intervals. In this latter circumstance, these values are forced to the closest allowed value. An example of a couple of generated images from the *Pepper* dataset is shown in Fig. 1.

3.2 Class Modelling

Class modelling is performed through two different phases, as done in [3]. Since in this work we used images transformed following the previously described strategy, we will refer to this method as *(tConvEnc)*. The network is trained over the majority of the one-class dataset D_r to learn its latent representation. Then, the affiliation to the *normal* class follows a decision rule based on a threshold. Such threshold is computed through the loss function and estimated on a smaller sample that we will call $D_{thr} \subset D_r$. D_{thr} is not fed to the autoencoder. Given D_{thr} and the loss function $\mathcal{L}$ of a trained network, we compute $\mathbf{l} = \mathcal{L}(D_{thr})$, with $\mathbf{l}$ being a vector containing the loss values for each record $r_t \in D_{thr}$. We keep the approach used in [3] to compute the range of expected *normal* losses. We choose the upper and lower bounds as δ_u, $\delta_l = \mu(\mathbf{l}) \pm z \cdot \sigma(\mathbf{l})$, where μ is the mean and σ is the standard deviation of the values in $\mathbf{l}$. The value of z can be tuned to vary the interval. A testing sample is classified as *normal* if the value of its loss l_t is in between δ_u and δ_l.

3.3   Architecture of the Autoencoders

Pepper: The *tConvEnc* is composed by eight convolutional layers with kernel size of dimensions $(6, 6)$. The first three are followed by Max-pooling layers with stride dimensions equal to 2 whereas the central three layers with Up-sampling. The kernels for each convolutional layer are respectively $[100, 50, 30, 10, 30, 50, 100, 3]$. The batch size was set to 64. **Boat:** The *tConvEnc* is composed by six convolutional layers with kernel size of dimensions $(3, 3)$. The first two are followed by Max-pooling layers with stride dimensions equal to 2 whereas the central two layers with Up-sampling. The kernels for each convolutional layer are respectively $[32, 16, 8, 16, 32, 3]$. The batch size was equal to 64. **SWaT:** The *tConvEnc* is composed by six convolutional layers with kernel size of dimensions $(6, 6)$. The first two are followed by Max-pooling layers with stride dimensions equal to 2 whereas the central two layers with Up-sampling. The kernels for each convolutional layer are respectively $[32, 16, 8, 16, 32, 3]$. The batch size was equal to 1024. All input images are zero-padded. Each network was trained with Binary Cross-Entropy Loss and Adam optimizer (default learning rate). We fixed the number of epochs to 30. We used ReLUs for all hidden layers and Sigmoids for the output layers.

4   Regression

As an alternative approach we propose to use regression temporal models in order to address both problem definitions of Sect. 2. We used three different models: a Recurrent Neural Network (RNN), a dense neural network and a Temporal Convolutional Networks (TCN). The RNN has two layers and is composed of Gated Recurrent Units (GRU) gating mechanism with a hidden layer size of 64. The hidden layer is connected to the network output through a dropout layer. The dense neural network has a batch-norm layer, two hidden layers of dimension 64 units, both followed by dropout. The TCN uses 3 layers of size 30 and has kernel size of 3. All models take in input a contiguous time sequence of size (K, t), where t is the number of time intervals considered $(1 < t)$ and return an output of size K. We trained RNN with Backpropagation Through Time (BPTT). On the other hand, we created and shuffled all possible contiguous subsequences of length t for TCN and dense network.

The model, used for regression, takes in input a sequence $\pi_{t-1,d-1}$ and returns in output the predicted value of r_t, called $\widehat{r}_t$. To train the networks, Mean Squared Error (MSE) loss and Adam optimizer have been employed, using as labels the values in the log at time t, r_t. We fixed the dropout rate to 0.4, and the number of epochs has been tuned to maximize performance while avoiding over-fitting. The evaluation of a new sequence $\pi_{t,d} \in D_t$ follows these steps: if the MSE between the predicted values $\widehat{r}_t$ and the actual values $r_t \in \pi_{t,d}$ is above the threshold tr an attack is detected. The threshold is computed from the predictions $Pred_t$ on the training data D_r by selecting the p percentile (usually between 70 and 99.5).

$$Pred_t = \frac{\sum_{i=1}^{K}(r_{t,i} - \widehat{r}_{t,i})^2}{K} \tag{4}$$

$$tr = Perc_p(\{Pred_1, ..., Pred_{|D_r|}\}) \tag{5}$$

5 Open-Set Classification

To solve the problem in Definition 2, we consider a variation of the regression approach presented in Sect. 4 by changing how the results of the regression model are used (while the training does not change). In particular, instead of computing a single threshold tr as before, we compute a threshold t_i for each dimension $0 < i \leq K$ as follows

$$Pred_{t,i} = (r_{t,i} - \widehat{r}_{t,i})^2 \tag{6}$$

$$tr_i = Perc_p(\{Pred_{1,i}, ..., Pred_{|D_r|,i}\}) \tag{7}$$

Now, suppose we want to classify a new sequence $\pi_{t,d} \in D_t$. First, we retrieve the model prediction $\widehat{r}_t$ (just as in Sect. 4). Then, we compute the squared errors between every dimension $r_{t,i}$ and $\widehat{r}_{t,i}$ (as in Eq. 6) and divide the result by the respective threshold tr_i, in order to normalize between the thresholds of the dimensions. After that, dimensions whose normalized values are greater than 1 are ranked according to such values, and the first k dimensions determine a subset of variables, called H. The value k is domain dependent and can be set by using different criteria, such as the number of variables present in the dataset, by fixing a threshold on the ranking value, or by considering the number of variables typically involved in an anomaly (or attack). In our empirical evaluation we use this later criteria (see Sect. 6.2).

6 Experimental Results

6.1 Anomaly Detection Results

In the experimental results we follow the standard anomaly detection nomenclature: we consider abnormal samples as positives and normal samples as negatives. Therefore, true positive are abnormal recognized abnormal. Due to unbalanced data, typically used in anomaly detection, in addition to precision and recall, we consider F1-score. For all domains, we computed the results by varying the number of records t and the percentile value used to derive the threshold. Table 1 shows the values for *Pepper*, the *SWaT* behavior is similar, while for *Boat* most values are close to 1. We can see that all models have similar results, with slightly better results for the dense model, particularly in the recall. The precision remains high, exceeding 92% in some cases. As expected, we can observe that, as the percentile threshold value increases, the precision increases while reducing the recall and vice-versa. Instead, against expectations, we do not notice big differences by changing the number t of records considered. The best results, considering F1-score, were achieved by choosing as a percentile a value of 80 for

Table 1. One-class problem on *Pepper*. Results for the RNN, Dense and TCN. Results are given in terms of Precision (P), Recall (R), F1-score (F1).

	t	5			20		
	perc.	P	R	F1	P	R	F1
Dense	70	0.820	**0.970**	0.889	0.794	**0.971**	0.874
	80	0.852	0.949	**0.898**	0.852	0.947	0.897
	90	0.890	0.866	0.878	0.896	0.902	**0.899**
	95	**0.922**	0.784	0.847	0.921	0.750	0.827
	99	0.872	0.188	0.309	**0.970**	0.279	0.433
RNN	70	0.841	**0.794**	**0.817**	0.827	**0.797**	**0.812**
	80	0.863	0.708	0.778	0.900	0.711	0.794
	90	**0.894**	0.596	0.715	**0.923**	0.486	0.637
	95	0.884	0.388	0.539	0.920	0.394	0.552
	99	0.694	0.039	0.075	0.883	0.026	0.051
TCN	70	0.822	**0.726**	**0.771**	0.890	**0.729**	**0.802**
	80	0.841	0.607	0.705	0.888	0.614	0.726
	90	0.876	0.451	0.595	**0.929**	0.548	0.689
	95	**0.878**	0.370	0.521	0.917	0.373	0.530
	99	0.644	0.033	0.063	0.899	0.018	0.036

Pepper and *Boat* while 99.5 for *SWaT*. We suppose that this is affected primarily by two factors: (i) the different distribution of normal and attack data, dictated by the availability of data, (ii) *SWaT* has a regular periodic normal behavior, while for robots, their behaviors vary significantly over time. Table 2 presents the results obtained for the anomaly detection task. We compare our approach with several competitors: One-Class Support-Vector Machine (OC-SVM) [6], Isolation Forest (IF) [7], Kernel Density Estimator (KDE) [8,9], Generative Adversarial Networks-based Anomaly Detection (GAN-AD) [1], Principal Component Analysis (PCA), K-Nearest Neighbours (KNN), Feature Bagging (FB), Autoencoder (AE), Multivariate Anomaly Detection for time series data with Generative Adversarial Networks (MAD-GAN) [4] and Efficient Gan-based anomaly detection (E-GAN) [5]. In the *SWaT* dataset we observed that one sensor ($AIT201$), from about half the normal test data, has values 4 times higher than the maximum value of the training data, so we try to use only the first n components, obtained with PCA, as proposed in state of the art approaches [1,4,5]. We can see that if we use all the original data our regression approach is not very effective on *SWaT*. This is not surprising because of the abnormal sensor in the normal data that is recognized. If we use only the first n components instead the anomaly tends to fade and our approach improves performance, reaching, in the case of *SWaT*, the best approaches considered using GAN. It should be noted, however, that our method does not require the use of the principal components

Table 2. Comparison of approaches and domains. Results are given in terms of Precision (P), Recall (R) and F1-score (F1). The results are shown in groups (top-down: literature, past experiments, new experiments and new experiments with PCA).

	Boat			Pepper			Swat		
Method	P	R	F1	P	R	F1	P	R	F1
1SVM	0.609	1.000	0.757	0.982	0.366	0.533	0.149	0.907	0.256
LOF	0.918	0.071	0.132	0.885	0.323	0.473	0.122	0.996	0.217
EE	0.919	1.000	**0.958**	0.604	0.542	0.571	0.077	0.004	0.007
IF	0.538	0.446	0.487	0.823	0.512	**0.631**	0.256	0.791	0.387
GAN-AD [1]	–	–	–	–	–	–	0.933	0.636	0.750
PCA [4]	–	–	–	–	–	–	0.249	0.216	0.230
KNN [4]	–	–	–	–	–	–	0.078	0.078	0.080
FB [4]	–	–	–	–	–	–	0.102	0.102	0.100
AE [4]	–	–	–	–	–	–	0.726	0.526	0.610
EGAN [5]	–	–	–	–	–	–	0.406	0.677	0.510
MAD-GAN [4]	–	–	–	–	–	–	0.990	0.637	**0.770**
Enc [3]	–	–	**0.99**	–	–	0.90	0.973	0.612	0.751
DeepEnc [3]	–	–	**0.99**	–	–	**0.94**	0.985	0.608	0.752
ConvEnc [3]	–	–	**0.99**	–	–	0.77	0.992	0.618	**0.761**
tConvEnc	0.992	0.989	0.990	0.860	0.946	**0.901**	0.993	0.582	**0.734**
Dense	0.999	1.000	0.999	0.896	0.902	0.899	0.485	0.589	0.532
RNN	1.000	1.000	**1.000**	0.841	0.794	0.817	0.392	0.665	0.493
TCN	1.000	1.000	**1.000**	0.890	0.729	0.802	0.189	0.743	0.301
Dense (pc=8)	0.988	0.998	0.993	0.813	0.890	0.849	0.975	0.627	**0.763**
RNN (pc=8)	0.996	1.000	0.998	0.672	0.236	0.349	0.959	0.632	**0.762**
TCN (pc=8)	0.998	1.000	0.999	0.444	0.143	0.217	0.997	0.592	0.743

and indeed we have verified, with the other datasets, that using only the first n components, the performance drops. The F1-score values also decrease accordingly, from about 0.9 using all sensors: 0.89 with 100 components, 0.86 with 30 components, 0.80 with 10 components, 0.61 with 3 components and 0.24 with only one component. With $SWaT$, the performance increase because by reducing dimensions we are removing the abnormal behavior of sensor $AIT201$.

Considering the $tConvEnc$, we tried different image sizes, by changing the dimension of the edge I_e. We found out that large edges increase data sparsity and consequently reduce the performance of $tConvEnc$, while narrow edges increase the probability of pixel superposition. This is reasonable since convolutions prefer dense features. For the $Pepper$ dataset we generated images with edge of size $I_e = 24$, for the $Boat$ dataset $I_e = 16$ and for the $SWaT$ dataset $I_e = 12$. The hyperparameter z has been varied in the range $[0,6]$. The reported results are obtained choosing z that maximized the F1 score. It is clear that

Table 3. Open-set classification problem. Results are given in terms of Accuracy. Due to space constraints we only report a subset of all the attacks for $SWaT$

Boat	Dense	RNN	TCN	Pepper	Dense	RNN	TCN
Dos	0.004	0.692	**0.985**	Joint	0.445	0.533	0.494
DosPay	0.000	0.736	**0.963**	Led	0.462	0.674	**0.785**
GpsDown	0.000	**0.997**	0.000	Wheel	0.145	0.681	0.394
Stuck	0.454	0.550	**0.806**				
Avg	*0.115*	*0.744*	*0.688*	Avg	*0.351*	*0.630*	*0.558*
Swat	Dense	RNN	TCN		Dense	RNN	TCN
2	**0.933**	0.867	0.878	22	0.702	**0.904**	0.436
6	**0.875**	0.825	0.775	32	0.926	0.876	0.843
7	**0.931**	0.897	0.885	38	0.807	0.684	0.684
8	**0.974**	0.959	0.263	40	0.016	0.016	**0.869**
10	0.121	0.758	**1.000**	41	0.387	**0.834**	0.722
20	**0.925**	0.875	**0.888**	Avg	*0.290*	*0.317*	*0.298*

the new transformation increases the robustness of Convolutional Autoencoders. Indeed, *tConvEnc* reaches 0.901 F1-score while *ConvEnc* scored 0.77 on *Pepper*. For the *Boat* and *SWaT* datasets results are comparable with the normalized *ConvEnc*. We suppose that *ConvEnc* manages to maintain a good performance in these two datasets due to the small input dimensionality and the consequent small image size. The proposed temporal input transformation allows the *tConvEnc* to match the performance of the other autoencoders and state of the art methods. For instance, *tConvEnc* reaches an F1-score of 0.734 which is close the MAD-GAN F1-score of 0.770 on *SWaT*.

6.2   Open-Set Classification Results

In this set of experiments, we evaluate the performance of the system in providing classes of anomalies that are not seen at training time. In order to perform a quantitative analysis, we have used test sets with labels of specific attacks, corresponding to subsets of variables affected by such attacks. More specifically, for each kind of attack in the test set, e.g., an attack that takes control of the motors of the robots, we have specified the corresponding affected subset of variables $G \in 2^V$. We then compare all these subsets with the output of the open-set classification, i.e., the set H, by computing the Jaccard index J [11] between H and each subset G, and selecting the maximum. For these experiments the classification is evaluated with accuracy (since we are using the Jaccard index). However, the system has still been trained only with *normal* data, so test classes were never seen at training time. Table 3 reports the accuracy for each model with respect to each type of attack, whose names match those used in the documentation. From the *Boat* dataset, we can see that RNN correctly identifies each

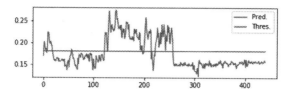

Fig. 2. Pepper prediction behavior under an attack: MSE value of Dense model (blue) vs. 99.7 percentile (red). (Color figure online)

attack with accuracy from 55% (Stuck) to 99% (GpsDown). The TCN model has higher accuracy for some kinds of attacks, but it is not able to identify the GpsDown attack. The dense model instead usually mis-classifies. Overall, the best model remains RNN. We also compared how the different types of attacks are predicted by the one-class classification (Problem Definition 1) and the open-set classification (Problem Definition 2). In particular, we observe that for some types of attacks, one approach is more effective than the other. For example, attack 28 on *SWaT* is detected very well as an anomaly (one-class), although the responsible sensors identified are wrong (open-set). Conversely, attack 32 is not detected (one-class), but sensors involved in this attack are correctly identified (open-set). This suggests that an approach with higher performance for anomaly detection may not necessarily be the most useful to detect the source of anomaly and hence to suggest viable interventions to address it. As an additional experiment on the open-set classification problem we tried to verify if our methodologies correctly identify the most involved sensors, before applying the classification with the Jaccard index. For each time interval in a specific attack, we have assigned an incremental score to the first 10 sensors involved, from 1 for the tenth to 10 for the first. In 21 types of attack out of 35 in *SWaT*, at least one of the sensors involved is detected. This result gives an idea of the operation at the basis of the classification and can already be used to understand the nature of the attack.

6.3   Pepper Use Case

In this final experiment, we show a use case for the *Pepper* social robot involved in a public demonstration subject to a simulated attack. In contrast to the other experiments, here we run the regression models trained on the previously mentioned *Pepper*, testing them on an additional log that has been acquired in a very different situation: a public space where several people interact with the robot with diverse goals and modalities. Figure 2 shows the outcome of the system during a portion of time where the *Pepper* robot was manually pushed back by a user. As shown in the figure, the system was able to detect the anomaly happening in the middle of the logged period, i.e., between time 120 to 250. Moreover, running the method for open-set classification we notice that the first sensor that appears in the ranking is the sensor that evaluates the stiffness of the front left wheel [*WheelFLStiff*] followed by some of the laser sensors positioned on the shovel. This confirms that the method is capable of providing useful indications on the possible source for the anomaly.

7   Conclusions

In this work, we have presented several one-class anomaly detection methods that can be trained using logs of robotic platforms and CPS. Moreover, we consider the problem of detecting variables or sensors that are most affected by the anomalous behavior addressing an open-set classification task. Quantitative comparison with state of the art methods on several datasets shows that our approach achieves comparable results for anomaly detection and promising results for open-set classification. Our work paves the way for several research directions. In particular, we believe that addressing the open-set classification problem is a key step to go beyond anomaly detection and move towards system analysis and diagnosis for robotic platforms. We believe that moving in this direction is crucial to widen the practical use of robotic technologies in real-world applications.

Acknowledgment. Sapienza University of Rome - funding for scientific research - year 2020.

References

1. Li, D., et al.: Anomaly detection with generative adversarial networks for multivariate time series. CoRR, vol. abs/1809.04758 (2018)
2. Zong, B., et al.: Deep autoencoding gaussian mixture model for unsupervised anomaly detection. In: International Conference on Learning Representations (2018)
3. Olivato, M., et al.: A comparative analysis on the use of autoencoders for robot security anomaly detection. In: Proceedings of IEEE/RSJ International Conference on Intelligent Robots and Systems (IROS) (2019)
4. Li, D., et al.: MAD-GAN: multivariate anomaly detection for time series data with generative adversarial networks. CoRR, vol. abs/1901.04997 (2019)
5. Zenati, H., et al.: Efficient gan-based anomaly detection. CoRR, vol. abs/1802.06222 (2018)
6. Schölkopf, B., et al.: Support vector method for novelty detection. In: Advances in Neural Information Processing Systems, pp. 582–588 (2000)
7. Liu, F.T., Ting, K.M., Zhou, Z.: Isolation forest. In: 2008 Eighth IEEE International Conference on Data Mining, pp. 413–422, December 2008
8. Parzen, E.: On estimation of a probability density function and mode. Ann. Math. Statist. **33**, 1065–1076 (1962)
9. Rosenblatt, M.: Remarks on some nonparametric estimates of a density function. Ann. Math. Statist. **27**, 832–837 (1956)
10. Goh, J., et al.: A dataset to support research in the design of secure water treatment systems. In: Critical Information Infrastructures Security, pp. 88–99. Springer International Publishing, Cham (2017)
11. Tan, P.-N., Steinbach, M., Kumar, V.: Introduction to Data Mining. Pearson Education, London (2006)

cgSpan: Pattern Mining in Conceptual Graphs

Adam Faci[1,2(✉)], Marie-Jeanne Lesot[1], and Claire Laudy[2]

[1] Sorbonne Université, CNRS, LIP6, F-75005 Paris, France
adam.faci@lip6.fr
[2] Thales, 1 Avenue Augustin Fresnel, 91767 Palaiseau, France

Abstract. Conceptual Graphs (CGs) are a graph-based knowledge representation formalism. In this paper we propose cgSpan a CG frequent pattern mining algorithm. It extends the DMGM-GSM algorithm that takes taxonomy-based labeled graphs as input; it includes three more kinds of knowledge of the CG formalism: (a) the fixed arity of relation nodes, handling graphs of neighborhoods centered on relations rather than graphs of nodes, (b) the signatures, avoiding patterns with concept types more general than the maximal types specified in signatures and (c) the inference rules, applying them during the pattern mining process. The experimental study highlights that cgSpan is a functional CG Frequent Pattern Mining algorithm and that including CGs specificities results in a faster algorithm with more expressive results and less redundancy with vocabulary.

Keywords: Conceptual graphs · Frequent pattern mining

1 Introduction

Conceptual Graphs (CGs) [3] represent knowledge as graphs containing concept nodes and relation nodes which refer to ontological knowledge in vocabulary. In their simplest form, they are similar to taxonomy-based labeled graphs (TLGs) i.e. labeled graphs with a partial order defined on the set of labels corresponding to an *is-a* hierarchy. While there are Frequent Pattern Mining (FPM) algorithms considering TLGs, an FPM algorithm taking CGs as input has not yet been proposed to the best of our knowledge. Yet mining patterns from sets of structures is a prevalent research subject [5,6,8,10]. Taking TLGs as input has been the work of a few propositions [2,7,9] and we consider them as a basis to design a CGs pattern mining algorithm.

We propose cgSpan based on these algorithms. We consider CGs as TLGs with more layers of information in order to reuse an existing algorithm of the state of the art. We propose to exploit three differences with the TLGs model; the relations fixed arity, the signatures and the inference rules. The first difference corresponds to the biparticity of CGs, the second one to the constraints on concept nodes labels connected to relation nodes and the third one to deduction mechanisms.

© Springer Nature Switzerland AG 2021
L. Rutkowski et al. (Eds.): ICAISC 2021, LNAI 12855, pp. 149–158, 2021.
https://doi.org/10.1007/978-3-030-87897-9_14

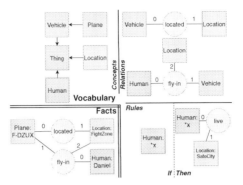

Fig. 1. Example of a CG with its vocabulary: CG in the bottom left part, a hierarchy on concepts in top left part, two signatures in the top right part and an inference rule in the bottom right part.

This paper has two goals. First it aims at defining a functional frequent pattern mining algorithm running on a CG database, taking the specificities and the additional knowledge in CGs into account, as compared to TLGs. Second it aims at showing that using such specificities results in a more efficient algorithm in memory space, speed and quality of output as compared to existing algorithms on TLGs.

Section 2 presents conceptual graphs as well as the current state of the art on taxonomy-based labeled graph pattern mining algorithms. Section 3 describes the proposed algorithm named cgSpan. Section 4 describes the experimental study and the considered quality criteria, as well as the obtained results. Section 5 concludes on this work and discusses some directions for future works

2   State of the Art

This section briefly recalls the definitions of conceptual graphs and summarizes some graph pattern mining algorithms.

2.1   Conceptual Graphs

Conceptual graphs [3] are a family of formalisms of knowledge representation, made of ontological and factual knowledge. They define sets of bipartite graphs, the CGs, representing the facts, all referring to ontological knowledge stored in the so-called vocabulary; an example is illustrated on Fig. 1 and commented below. Several levels of expressiveness have been defined, see [3] for an exhaustive review, the formal definitions of the notions used in the paper are recalled below.

The ontological part of a CG is a *vocabulary*, defined as a 5-tuple $V = (T_C, T_R, \sigma, I, \tau)$. T_C and T_R that respectively correspond to concept and relation types are two partially ordered disjoint finite sets, where ordering corresponds to generalisation. T_C is illustrated in the top left part of Fig. 1. It contains a greatest element $\top$, represented as "Thing" in Fig. 1. T_R is partitioned into subsets $T_R^1 \ldots T_R^k$, $1 \ldots k$ ($k \geq 1$) respectively, meaning that each relation type has an associated fixed arity. σ is a mapping associating a signature, illustrated on top right part of Fig. 1, to each relation. I is a set of individual markers. τ is a mapping from I to T_C.

The factual part are the CGs themselves. A CG is a 4-tuple $G = (C,R,E,label)$. G is a bipartite labeled multigraph as illustrated on the bottom left part of Fig. 1. C and R correspond to concept and relation nodes, where elements of C are pairs from $T_C \times I$ and elements of R are elements of T_C. E contains all the edges connecting elements of C and R and $label$ is a labelling function.

An extension to the previous basic CG setting consists in defining so-called inference rules, that define deduction principles to complete CG, either including additional nodes or updating their labels. As illustrated in bottom right part of Fig. 1, such a λ-rule is an ordered pair of λ-CG made of a hypothesis and a conclusion, where a λ-CG is a CG with defined connection nodes, i.e. nodes with generic marker with an associated variable used to match nodes from hypothesis to nodes from conclusion. Figure 1 illustrates the two connection nodes matching with the associated variable "$*x$". The application of a rule can be the extension of a CG, the conclusion then including the hypothesis, or the specialization of a pattern, the conclusion then being a copy of the hypothesis with more specific labels, but it can be both. Figure 1 illustrates the case of an extension rule. Formal definitions can be looked up to in [3].

We base our proposition on algorithms taking taxonomy-based labeled graphs (TLGs) [9] as input. A taxonomy is a set of labels with a partial order relation. For instance, the hierarchy of concepts in Fig. 1 is a taxonomy. The partial order relation is the generalisation, meaning that if A generalises B, if any instance of B is an instance of A. A TLG is a labeled graph whose labels are part of a taxonomy, less expressive than a CG. Figure 1 would represent a TLG if there were no distinction between concept and relation nodes, no relation fixed arity, no signatures, no inference rules and no individual markers.

2.2 Subgraph Mining

Frequent pattern mining (FPM) algorithms, e.g. summarized in [6], are central in the data mining community. They take a database as input and return a set of frequent patterns, by counting support, i.e. the number of occurrences, of each candidate, i.e. a potential frequent pattern, in the considered data. Patterns can be extracted from sets, also called transactions, sequences or structures for example. In the case of complex instances such as sequences or graphs, the support can be defined as the number of instances containing the candidate, not considering multiple occurrences within an instance as a weight. In this paper, we consider the classic choice where a graph supports a pattern if and only if there exists a homomorphism from the pattern to a subgraph of the considered graph.

One of the most used graph pattern mining algorithms is gSpan [10], that applies to labeled graphs and relies on encoding graphs as sequences, where a unique sequence corresponds to a unique graph, switching the context as well from subgraph mining to subsequence mining. It has been extended to the case of taxonomy-based labeled graphs [2,7] and then further enriched to add the possibility of mining directed TLGs with many taxonomies [9].

These TLGs pattern mining algorithms add the inclusion of hierarchies to find more patterns and gain in efficiency. Indeed, label generalization leads to a more lenient label comparison, as two different labels can be generalized to a similar one, increasing the number of returned patterns. Also, they first mine for structural patterns before considering labels by generalizing to the most general type, resulting in more patterns explored at once. They also tackle the problem of the massive amount of patterns returned in classic pattern mining approaches by pruning or not exploring irrelevant patterns. The pattern relevancy is determined by several measures such as over-generalization and statistical significance [9].

Our proposition is based on the DMGM-GSM algorithm (Directed Multi-Graph Miner - Generalized Subgraph Mining) [9]. The latter takes into account the taxonomy on labels in 3 steps: first for each node in a TLG, the label is replaced by the path to its upmost type. Then patterns are mined by considering only the top type in the newly encoded taxonomy paths labels. Finally, retrieved patterns are successively specialized until they are no more frequent, i.e. with a support count lower than the defined support threshold.

3   cgSpan: A CG Frequent Pattern Mining Algorithm

cgSpan is the first frequent pattern mining method running on Conceptual Graphs, to the best of our knowledge. Based on DMGM-GSM [9], it is conceived as taking consecutive enrichment steps from TLGs to CGs into account. We use the additional information in CGs to restrict the output and ensure both conformity of result and speed.

3.1   Overview

cgSpan is made of three modules, each one being dedicated to one specificity of CGs as opposed to TLGs: the rules fixed arity, the signatures and the inference rules. They can be combined to exploit simultaneously all characteristics of CGs, as illustrated in Sect. 4. They are described in turn in Sects. 3.2 to 3.4, this section describes their common points.

The essential part of cgSpan is a pre-processing step which translates each CG of the input database to TLG. The basic principle is similar to that of DMGM-GSM: labels are built by concatenating all concept types, from the top type to the node type, separating them with a '_'. For instance "Plane:F-DZUX" from Fig. 1 is replaced by "Thing_Vehicle_Plane_F-DZUX". Then this step is modified accordingly by each of the three modules to take into account CG specificities, especially by the neighborhood module (see Sect. 3.2).

Then the pattern mining is run with DMGM-GSM, only modified by the inference rule module described in Sect. 3.4 to increase efficiency.

Finally the post-processing translates the obtained frequent patterns back to the CG formalism. As described in Sect. 3.3, the signature module includes an additional pruning step.

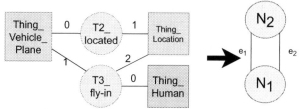

N_1 = T3(Thing,Thing,Thing)_fly-in(Human,Vehicle,Location)_fly-in(Human,Plane,Location)
N_2 = T2(Thing,Thing)_located(Vehicle,Location)_located(Plane,Location)
e_1 = Thing_Vehicle_Plane
e_2 = Thing_Location

Fig. 2. Translation from CG to TLG as sets of bricks. The taxonomy-path labels detailed below start from top types and specialize all node labels at the same time.

3.2 Exploiting Relation Arity and Neighborhood Nodes

The first specificity of CGs we propose to exploit is the arity of the relation nodes: any relation node is necessarily connected to a known number of concept nodes. In this regard, when operating on candidates, any relation node addition should immediately result in the addition of its connected nodes.

We therefore define an *elementary brick* as a node and its connected nodes of the input CGs, so as to avoid partial neighborhoods in returned patterns and to increase efficiency. Figure 2 illustrates how a CG can be encapsulated as a TLG of elementary bricks where "T3" is a relation type more general than "fly-in".

A brick is associated to a single TLG node. Its label is defined as follows: it is the concatenation of the taxonomy-enriched labels of the relation node and each of its associated concept nodes, in the order specified in CGs by the edge labels. Finally, edges are built between bricks that share a common concept node, as illustrated in the right part of Fig. 2. The edge labels are the taxonomy paths of the bricks common concept nodes. Figure 2 shows for instance the transformation from a brick in a CG with "A plane is flown by a human in a location" to N_1.

3.3 Exploiting Signatures

The second specificity of CGs considered in cgSpan is the signatures. They define for each relation type a restriction on concept types: for each relation node, they specify a maximal generic type for each connected concept node. We propose to exploit this information in the pre-processing and post-processing steps, respectively to restrict the label generalization of concept nodes and to prune patterns to avoid redundancy with signatures.

As an instance we consider in this section the relation "fly-in" with signature $(Human, Vehicle, Location)$ as illustrated in Fig. 3.

In the pre-processing step we propose to prune the taxonomy path, not building it up to the most general type, but up to the level indicated in the signature. For each relation node in a CG, concept nodes connected to this relation see

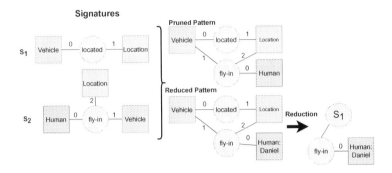

Fig. 3. Pattern pruning in post-processing step when considering signatures. The top right pattern is pruned as it is a mere aggregation of signatures. The bottom right pattern is reduced by replacing the redundant part to S_1 with a reference.

their labels compared to the corresponding signature. Each comparison is then followed by a cut of the taxonomy path down to the matching type in signature. For instance, "fly-in (Thing_Human, Thing_Vehicle_Plane, Thing_Location)" is replaced by "fly-in (Human, Vehicle_Plane, Location)" in Fig. 3.

In the post-processing step, two kinds of patterns are affected. First, we propose to remove patterns that can be reduced to a set of connected signatures. Second, we propose to modify patterns including parts that can be reduced to a set of connected signatures by replacing the latter by references to signatures.

For instance, "fly-in (Human,Vehicle,Location)" is ignored while "fly-in(Human, Vehicle_Plane, Location)" is not, since "Plane" is a specialization of "Vehicle" type specified in signature as illustrated in Fig. 3.

3.4   Exploiting Rules

The third specificity taken into account by cgSpan is positive rules, considering the ones including extension and the ones including specialization.

A positive rule including specialization specifies that when a CG includes the rule hypothesis, some nodes specified in rule conclusion can be specialized. During pre-processing, after the replacement by taxonomy paths in CGs, we propose to extend this path for each node matching such a rule to the type in conclusion.

On the contrary, a positive rule including extension enables the extension of CGs in database. During the pattern mining step, we propose to extend all patterns matching the rule hypothesis to the rule conclusion. In the process, the rule hypothesis and intermediate patterns are ignored. It can be applied more than once, thus extending a matching CG at each application, but we limit their use to one.

Figure 4 illustrates the application of the rule from Fig. 1. All intermediate patterns, from hypothesis to conclusion, are not explored when using such rules since the extension is in one step.

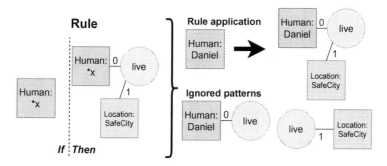

Fig. 4. Rule application example and examples of patterns not explored using this technique. The rule is only applied once on the top right part. Ignored patterns are potentially not explored if candidates start with a single concept node with type "Human".

4   Experimental Study

This section describes the experiments we conducted, presenting the considered synthetic data and quality criteria and discussing the obtained results.

4.1   Data Generation

To the best of our knowledge, no CG dataset of quality is available. Existing ones are either extremely simple or do not fully respect the CGs formal restrictions. One possibility would be to use a translation method such as T_{nat} [1] that makes it possible to obtain CG datasets from real datasets. However it usually results in datasets with mainly factual or mainly ontological knowledge. Moreover it does not allow to define the expected results in terms of frequent patterns to identify in the data, making it difficult to validate the proposed algorithm. Consequently we propose CGGen, a new algorithm to generate CG datasets from a set of constraints, either from a set of defined frequent patterns, used as components of generated CGs, or from frequency distributions that the generated CGs need to respect. CGGen thus enables the definition of expected results, as expected frequent patterns on one hand and frequency distribution on the other hand, that can be used to validate our algorithm.

We use two datasets in the experimental results, denoted D_1 nad D_2, consisting in a few hundreds of nodes. They consist of 1000 graphs of around 30 nodes each. The hierarchies contain 50 concept types and 20 relation types. D_2 CGs follow two distributions used for generation: one over their size and one over their labels.

D_1 enables the definition of expected results. First we expect the set of predefined CGs used as seed to generate these datasets to be present in the frequent patterns. Then we expect that some simple constraints verified by D_1 are also verified by the set of frequent patterns. Also we expect that the distribution on labels in D_2 is followed by the returned patterns.

4.2   Criteria

The comparison of the cgSpan returned patterns with the expected ones allows to make use of the classic precision and recall criteria, respectively defined as the proportion of expected patterns present in the returned ones, and reciprocally.

In addition, to assess the computational efficiency of the proposed cgSpan algorithm, the redundancy of patterns is assessed, defined as the proportion of pruned patterns w.r.t. all patterns, returned or pruned. The greater it is, the more redundancy has been removed, the better it is.

Finally, to assess efficiency, the run time of cgSpan is compared to that of DMGM-GSM.

4.3   Experimental Results

We process D_1 and D_2 translated to TLG with DMGM-GSM and process them with four variants of cgSpan, considering each module, as described in Sects. 3.2, 3.3 and 3.4, individually and full cgSpan that includes all three modules.

Table 1 gives the results obtained for D_1. All algorithms retrieve all expected patterns or combination of expected patterns so all versions attain the functional goal and obtain a recall equal to 100%. Regarding precision, all variants not taking elementary bricks into account return partial neighborhoods, reducing their performance for this criterion. The use of signatures prunes some of these incomplete patterns since most of them contain top concept nodes. We can spot that an increase in redundancy is correlated with an increase in precision. It can be interpreted as "Patterns that are pruned are mostly not expected patterns". Finally, the use of elementary bricks seems to increase the time efficiency significantly. Indeed, there is a gain of almost 20% in time while the gain with the use of rules is less than 10%. These observations have been confirmed by experiments with other generated databases with varying parameters. Tests on a huge database with millions of nodes are included in future works to observe how well this improvement keeps up with the increase in size.

Figure 5 presents the results with cgSpan on D_2, showing the number of occurrences of the expected and output patterns depending on their size. It shows the frequency distribution on the number of occurrences of patterns returned by cgSpan according to their size. The frequency distribution over the input dataset has been used to generate D_2, so we use both distributions to analyze results. Note that when counting patterns, only the maximal patterns, i.e. patterns not included in another pattern, contribute to this illustration: all counted patterns of size 3 are not present in the counted patterns of greater size. We can observe a global correlation between the input dataset and the frequent patterns. The fact that there is no frequent pattern of size 1 or 2 is explained with the elementary brick restriction that does not allow partial neighborhood while in D_2 these CGs correspond to isolated concept nodes.

Table 1. Comparison of cgSpan and its variants with D_1

Test	Rec. (%)	Prec. (%)	Red. (%)	T-Eff. (%)
DMGM-GSM	100	57	0	100
CgSpan with bricks	100	83	0	83
cgSpan with signatures	100	75	38	97
cgSpan with rules	100	63	20	93
Full cgSpan	100	85	46	79

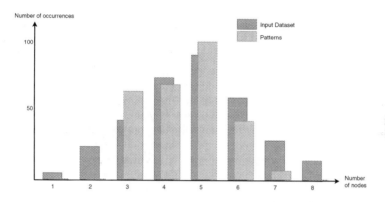

Fig. 5. Frequency distributions over D_2 and over the set of returned frequent patterns with cgSpan.

5   Conclusion and Future Works

This paper presents cgSpan, an algorithm to mine frequent patterns from a CG database. The particularities of CGs are exploited to produce a functional algorithm, and quality of result is increased w.r.t. defined criteria. We use the concept and relation difference and inference rules to speed up the process and we use signatures and other elements to prune less relevant patterns.

Future works will aim at extending the proposed cgSpan algorithm, considering even richer variants of CGs, in particular nested CGs [3]. Another direction will focus on the definition of quality criteria: it may be interesting to exploit statistical significance such as on based on Chung–Lu model [4] adapted to CGs to propose a functional statistical significance criterion for CGs mining algorithms such as cgSpan, used both to improve results and validation.

References

1. Baget, J.-F., Croitoru, M., Gutierrez, A., Leclère, M., Mugnier, M.-L.: Translations between RDF(S) and conceptual graphs. In: Croitoru, M., Ferré, S., Lukose, D. (eds.) ICCS-ConceptStruct 2010. LNCS (LNAI), vol. 6208, pp. 28–41. Springer, Heidelberg (2010). https://doi.org/10.1007/978-3-642-14197-3_7

2. Cakmak, A., Ozsoyoglu, G.: Taxonomy-superimposed graph mining. In: Proceedings of the 11th International Conferences on Extending Database Technology: Advances in Database Technology, pp. 217–228 (2008)
3. Chein, M., Mugnier, M.L.: A Graph-Based Approach to Knowledge Representation: Computational Foundations of Conceptual Graphs (Part. I). Springer (2008). https://doi.org/10.1007/978-1-84800-286-9
4. Chung, F., Lu, L.: The average distances in random graphs with given expected degrees. Proc. Nat. Acad. Sci. **99**(25), 15879–15882 (2002)
5. Elseidy, M., Abdelhamid, E., Skiadopoulos, S., Kalnis, P.: Grami: frequent subgraph and pattern mining in a single large graph. Proc. VLDB Endowment **7**(7), 517–528 (2014)
6. Han, J., Cheng, H., Xin, D., Yan, X.: Frequent pattern mining: current status and future directions. Data Min. Knowl. Discov. **15**(1), 55–86 (2007)
7. Inokuchi, A.: Mining generalized substructures from a set of labeled graphs. In: Proceedings of the 4th IEEE International Conferences on Data Mining (ICDM'04), pp. 415–418. IEEE (2004)
8. Iyer, A.P., Liu, Z., Jin, X., Venkataraman, S., Braverman, V., Stoica, I.: ASAP: fast, approximate graph pattern mining at scale. In: Proceedings of the 13th USENIX Symposium on Operating Systems Design and Implementation OSDI 18, pp. 745–761 (2018)
9. Petermann, A., Micale, G., Bergami, G., Pulvirenti, A., Rahm, E.: Mining and ranking of generalized multi-dimensional frequent subgraphs. In: Proceedings of the 12th International Conferences on Digital Information Management (ICDIM), pp. 236–245. IEEE (2017)
10. Yan, X., Han, J.: cgSpan: graph-based substructure pattern mining. In: Proceedings of IEEE International Conferences on Data Mining, ICDM'02, pp. 721–724. IEEE (2002)

Constrained Clustering Problems: New Optimization Algorithms

Hatem Ibn-Khedher[1]($\boxtimes$), Makhlouf Hadji[2], Mohamed Ibn Khedher[2], and Selma Khebbache[2]

[1] ALTRAN Labs, 78140 Velizy-Villacoublay, France
hatem.ibnkhedher@altran.com
[2] Institut de Recherche Technologique SystemX, 8 Avenue de la Vauve, 91120 Palaiseau, France
{makhlouf.hadji,mohamed.ibn-khedher,selma.khebbache}@irt-systemx.fr

Abstract. Constrained clustering problems are often considered in massive data clustering and analysis. They are used in modeling various issues in anomaly detection, classification, systems' misbehaviour, etc. In this paper, we focus on generalizing the K-Means clustering approach when involving linear constraints on the clusters' size. Indeed, to avoid local optimum clustering solutions which consists in empty clusters or clusters with few points, we propose linear integer programming approaches based on relaxation and rounding techniques to cope with scalability issues. We show the efficiency of the new proposed approach, and assess its performance using five data-sets from different domains.

Keywords: Constrained-clustering · K-Means · Combinatorial optimization

1 Introduction

In the era of big data, clustering approaches are more than necessary to cope with relevant issues and new challenges such as classification, anomaly detection, outliers' detection, etc. These issues are based on the analytics of big volumes of collected data using new cost efficient approaches that converge quickly or in negligible time. For instance, in the domain of cybersecurity, we need to investigate rapid solutions to analyze, in near real time, big volumes of data to detect any intrusion in the supervised system.

Existing classification techniques to detect abnormal situations are considering iterative processes that can be time consuming which is hampering in the case of real time decision making problems. One of the most relevant clustering algorithms is known as K-Means [9]. Classical K-Means approach is based on an iterative processing of two steps. The first step allows to identify the assignment of data-set points to fixed centroids or class centers, while the second step consists in moving the centroids according to the results of the first step. This iterative process is converging when the minimum value of the defined objective function, represented by the euclidean distance between data-set points and centroids, is reached or cannot be improved.

© Springer Nature Switzerland AG 2021
L. Rutkowski et al. (Eds.): ICAISC 2021, LNAI 12855, pp. 159–170, 2021.
https://doi.org/10.1007/978-3-030-87897-9_15

Nevertheless, classical K-Means approach encounters important drawbacks when it is applied to data sets with m data points in $n \geq 10$ dimensional real space $\mathbb{R}^n$. Indeed, in the case of the number of desired clusters (noted by k) is at least greater than 20 ($k \geq 20$), classical K-Means approach converges with one or more clusters which are either empty or contain only one data point [6]. In [5], one can find preliminary results on K-Means for two cases with $k = 50$ and $k = 100$, the average number of empty clusters are 4 and 12 respectively.

Moreover, empty clusters or clusters with few points may necessitate additional heuristics to process them. This will cause extra convergence-time to definitely eliminate empty clusters and then converge to a feasible clustering solution. In this paper, we focus on new mathematical optimization approaches based on linear programming techniques to improve the classical K-Means performance. In addition, we propose a mathematical formulation to considerably reduce the number of iterations which allows to rapidly converge to global optimum solutions for the different considered use-cases.

The rest of this paper is organized as follows: Sect. 2 addresses an overview on clustering approaches and their optimization processes while solving local or global optimization problems. Section 3 describes the constrained clustering problems and provides details on the proposed approach based on Integer Linear Programming (ILP). Section 4 is dedicated to a theoretical analysis discussing convergence conditions of the proposed algorithm. Performance evaluation using five real datasets in different domains are invoked in Sect. 5. We open a new research challenge addressing stochastic constrained clustering problems, in Sect. 6. Section 7 concludes the paper.

2   Related Work

Clustering can be considered as the most important unsupervised learning problem. It deals with finding a structure of an unlabeled data. A cluster is defined as a group of samples which are *similar* between them and are *"dissimilar"* to the samples belonging to other clusters. Clustering algorithms can be roughly regrouped into four categories: i) Partitioning ii) Hierarchical, iii) Density-based and iv) Constrained Clustering. Partitioning Clustering is the most popular technique.

Partitioning Clustering consists of fixing an initial number of k clusters, and iteratively reallocating samples among clusters to convergence. The number of clusters k is a pre-specified parameter by the user. The most popular algorithms are K-means and k-medoids.

Given n samples and an integer k, the k-means algorithm [13] consists in dividing the samples into k clusters by minimizing a cost function. Considering the distance from a sample to the average of the samples in its cluster; the cost function to be minimized is the sum of the squares of these distances. The principle of k-means is to assign each sample to the cluster whose center is nearest. The center is the average of all the samples in the cluster. The k-means algorithm is computationally difficult (NP-hard); however, efficient heuristic algorithms converge quickly to a local optimum. Its time complexity is $O(nkd)$, where d is the number of iterations. The major disadvantage of k-means is that clusters depend on the initialization and the used distance. The *a priori* selection of the k-parameter can be seen as a limitation or an advantage. In the case of bag of words calculation, this allows to set exactly the desired dictionary size. However, in some cases, it is preferable to avoid such a constraint. Moreover, as limitation, the K-means method is sensitive to outliers [11].

3   Clustering Algorithms: A Mathematical Formulation

Let $D = \{d^i\}_{i=1}^m$ be a data-set of m points in $\mathbb{R}^n$ and a number of k clusters (fixed). The K-Means clustering problem is defined as follows. Find cluster centers $C^1, C^2, ..., C^k$ in $\mathbb{R}^n$ such that the sum of the l_2-norm? distance squared between each point d^i and its nearest cluster center C^h is minimized. Specifically and mathematically formulated as follows:

$$\min_{C_1,...,C_k} \sum_{i=1}^m \min_{h=1,...,k} \left(\frac{1}{2}||d^i - C^h||_2^2\right) \tag{1}$$

Equation (1) is equivalent to the following assignment problem (Eq. 2) where we introduce selection or decision variables $x_{i,h}$.

$$\min_{C,x} \sum_{i=1}^m \sum_{h=1}^k x_{i,h} \times \left(\frac{1}{2}||d^i - C^h||_2^2\right) \tag{2}$$

subject to

$$\sum_{h=1}^k x_{i,h} = 1 \ \forall i = 1, .., m \tag{3}$$

$$x_{i,h} \in \{0,1\} \ \forall i = 1, ..., m \ ; \ \forall h = 1..., k \tag{4}$$

Constraints (3) are used to guarantee the assignment of all the data-set points to the different identified clusters. The selection binary variables $x_{i,h}$ indicate that the cluster h is selected to contain the data point i:

$$x_{i,h} = \begin{cases} 1 & \text{if the data point } d^i \text{ is closest to center } C^h \\ 0 & \text{Otherwise} \end{cases} \tag{5}$$

Hence, the objective function (2) under the constraints given by (3) and the integrity constraints (4) are defining a combinatorial optimization problem to assign data-set points to identified (k is known and constant) clusters. This formulation is providing the convex hull of incidence vectors represented by variables x of the clustering problem. The above defined problem is an Integer Linear Program (ILP), hence it is an NP-Hard instance, and necessitates efficient solutions to converge in negligible times. Before introducing our new approach, we propose a brief discussion of the classical K-Means algorithm.

Algorithm 1. Classical K-Means Algorithm

1: Given a database D of m points in $\mathbb{R}^n$ and cluster centers $C^{1.t}, C^{2.t}, ..., C^{k.t}$ at iteration t, compute $C^{1.t+}, C^{2.t+}, ..., C^{k.t+1}$ at iteration $t+1$ in the following 2 steps:

    1. **Cluster assignment:** For each data point $x^i \in D$:
       assign x^i to cluster $h(i)$ such that $C^{h(i).t}$ is closest to $x^{(i)}$ in the 2-norm.
    2. **Moving centroids (i.e., cluster update):** compute $C^{h,t+1}$ as the mean of all points assigned to cluster h.

2: stop when $C^{h,t} = C^{h,t+1}, h = 1, ..., k$

3.1 Classical K-Means Algorithm

The above optimization formulation (2) under constraints (3) and (4) is solved by the classical iterative K-Means algorithm. In each iteration, the considered problem is solved first for $x_{i,h}$ with fixed cluster centers C^h. Then, our problem is solved for C^h with fixed assignment variables $x_{i,h}$. It is important to notice that the calculated stationary point (using K-Means) satisfies the Karush-Kuhn-Tucker (KKT) conditions (see reference [14] for more details) for Problem (2) under assignment constraints and integrity constraints, which are necessary for optimality. In Algorithm 1, we describe the main stages of the classical K-Means clustering algorithm (an heuristic solution) that approximates the exact assignment problem described in Problem (2).

The solution provided using Algorithm 1 supposes initially that the number of desired clusters (i.e. k) is well known. Moreover, considering the iterative heuristic approach, the classical K-Means heuristic provides minimum local solutions in given convergence times, that will be discussed in next sections.

We need rapid approaches and exact solutions allowing to reduce the necessary convergence time to global optimum. Moreover, to avoid empty clusters, we investigate new hyper-plans to be added to the described mathematical formulation. To attend these objectives, we propose a new mathematical formulation to cope with the Constrained Clustering problem.

3.2 Constrained Clustering Algorithm

The classical K-Means algorithm uses an iterative refinement technique that converges to local optimum solutions and often provides empty clusters. More precisely, the naive algorithm assigns observations to a cluster with the nearest mean/centroid (i.e., a cluster with the least squared euclidean distance). Mathematically, it partitions the observations according to the Voronoi diagram generated by the mean. Then, it recalculates means/centroids for observations and often results with empty clusters. We propose, in the following, new constrained clustering approaches to eliminate empty clusters when converging to global optimum solutions for the considered problem.

Exact ILP Formulation. To totally eliminate empty clusters, we propose a new hyperplan, which indicates that each cluster h contains at least θ_h data points. This new valid inequality is defined as follows:

$$\sum_{i=1}^{m} x_{i,h} \geq \theta_h \ \forall h = 1, ..., k \tag{6}$$

where $\sum_{h=1}^{k} \theta_h \leq m$. This yields the following constrained K-Means problem:

$$\min_{C,x} \sum_{i=1}^{m} \sum_{h=1}^{k} x_{i,h} \times \left(\frac{1}{2} ||d^i - C^h||_2^2 \right) \tag{7}$$

subject to:

$$\sum_{h=1}^{k} x_{i,h} = 1 \ \forall i = 1, .., m \tag{8}$$

$$\sum_{i=1}^{m} x_{i,h} \geq \theta_h \ \forall h = 1, ..., k \tag{9}$$

$$x_{i,h} \in \{0,1\} \ \forall i = 1, ..., m \ ; \ \forall h = 1..., k \tag{10}$$

To optimally solve the mathematical formulation given by the objective function (7) under the linear constraints given by (8), (9) and (10), we need deep approaches (e.g. Branch and Bound techniques) to reach optimal solutions. To attend a global optimum, the above mentioned techniques can be time consuming. Hence, we need efficient solutions that can attend a global optimal solution in acceptable time. To cope with this problem, we propose the following approximation approach.

ILP Rounding Approach. To attend a global optimal solution for the constrained clustering problem, we propose an ILP rounding approach which consists in a linear relaxation of the integrity constraints (10). Hence, binary decision variables become continuous variables (see constraints (11)), and this allows to use linear programming methods (e.g. simplex approach) to efficiently solve the above problem.

$$0 \leq x_{i,h} \leq 1 \ \forall i = 1, ..., m \ ; \ \forall h = 1..., k \tag{11}$$

Using this relaxation, we can obtain global optimal solutions in few seconds even for large problem instances. In the case where the relaxation solution is not binary, we propose a simple rounding technique givens as follows: if $x \geq \beta$ then $x = 1$. Otherwise $(x < \beta)$ $x = 0$ where $0 \leq \beta \leq 1$ is a threshold fixed before simulations ($\beta = 0.8$ in our case). Using the rounding result, we simply need to check the feasibility (in terms of data-set points assignment) of the new solution. In the constrained clustering algorithm, the assignment step is important as it impacts the low complexity of this algorithm.

Minimum Cost Maximum Flow Assignment Approach. In [5], the authors proposed a graph-based approach to cope with the constrained clustering algorithm for large scales. This approach is based on a Minimum Cost Flow (MCF) technique. In the following, we propose a brief overview of this approach. Let N be the set of nodes. Each node $i \in N$ has associated with a value b_i indicating whether it is a supply node ($b_i > 0$), a demand node ($b_i < 0$), or a transshipment node ($b_i = 0$). The problem is feasible *if and only* the sum of the supplies equals the sum of the demands. Mathematically, this is equivalent to $\sum_{i \in N} b_i = 0$. Let A be the set of directed arcs. For each arc $(i, j) \in A$, the variable $y_{i,j}$ indicates amount of flow on the arc. Additionally, for each arc (i, j), the constant $c_{i,j}$ indicates the cost of shipping one unit flow on the arc. The mathematical formulation of MCF problem is provided as follows:

$$\min \sum_{(i,j) \in A} c_{i,j} y_{i,j} \tag{12}$$

subject to

$$\sum_j y_{i,j} - \sum_j y_{j,i} = b_i \; \forall i \in N \tag{13}$$

$$0 \leq y_{i,j} \leq u_{i,j} \; \forall (i,j) \in A \tag{14}$$

$$\sum_j y_{i,j} - \sum_j y_{j,i} = \begin{cases} 1 & \text{at } x^i \\ -\theta_h & \text{at } c^h \\ -m + \sum_{h=1}^{k} \theta_h & \text{at } a \end{cases} \tag{15}$$

4   Theoretical Analysis

In this section, we study the convergence guarantee and the computational complexity of the constrained clustering approaches (exact ILP and ILP rounding and relaxation).

Proposition 1. *(Convergence guarantee of ILP rounding approach)*
*ILP rounding and relaxation approach converges in a finite number of iterations at a cluster assignment that is **Globally** near-optimal.*

Proof. In the *cluster assignment* step of the ILP rounding algorithm, we guarantee the assignment of each points of the considered data set. This is ensured by the linear constraints given by (8). The relaxation of the Integer Linear Program approach consists in considering continuous decision variables in a compact and close polytope described by a finite number of hyperplans. Hence, and using a simple Simplex algorithm (see [14], for instance), one can reach a feasible solution. This solution can be a global optimum from time to time for the initial constrained clustering problem. In addition, convexity conditions for a linear optimization under linear constraints are also verified to reach at each iteration a feasible solution.

Proposition 2. *(Computational complexity of the ILP rounding approach)*
The proposed ILP rounding approach can be solved in polynomial time.

Proof. It is well known that solving ILP problem is time consuming as it is identified to be an NP-Complete problem. This complexity is due to using Integer decision variables making the scope of feasible solutions, for the considered constrained clustering problem, exponential and large to be investigated in negligible times, unless $\mathcal{P} = \mathcal{NP}$.

The ILP rounding techniques is based on the relaxation of these integer decision variables that become continuous values. This allows us to consider polynomial time approaches such as Simplex techniques, to polynomially attend optimal solutions in negligible times.

5   Performance Evaluation

We evaluate and assess the performance of our algorithm using public datasets to feed the proposed optimization models. Hence, we propose a short description of each used dataset before going through a deep investigation and application of our method.

5.1   Datasets Description

In this subsection, we describe the real traces used in the context of constrained clustering.

IRIS Trace. The dataset contains 3 classes of 50 instances each. Classes refer to iris plant types (Setosa, Versicolour, and Virginica) [4]. One class is linearly separable from the two others, which are not separable from each other.

Iris Dataset features are represented by sepal length/width and petal length/width.

5G-Uplink-Operations Trace. We use real traces from network operators.[1] This dataset is a time series of relevant key performance metrics used to evaluate optimization algorithms. We use the average reported Channel Quality Indicators (CQI), the average number of user connected to cellular network operator sites, and the physical Resource Blocks (PRB) at the uplink (UL) direction as key features [12]. The dataset is pre-processed using scaling methods in order to ease the clustering optimization. It is worth mentioning that we should not annotate the current dataset since the unsupervised nature of the K-Means algorithm.

---

[1] A dataset provided by Orange, a telecommunication service provider.

Call Detail Record (CDR) Trace. This benchmark is a Call Detail Record (CDR) traces collected from network operators [2, 10]. The mobile phone activity dataset is composed of one week of CDRs from the city of Milan and the Province of Trentino (Italy). When a user engages a telecommunication interaction, a Base Station (BS) is assigned by the operator and delivers the communication through the network. Then, a new CDR is created recording the time of the interaction and the BS which handled it. The dataset includes the following activities: Received and sent SMS, Incoming and outgoing call, and Internet activity. Network operators need clustering of end-users' profile for the sake of better subscribers management. Clustering is helpful also to minimize the deployment cost when a cluster head should be selected and managed.

COVID-19 Trace. Korea Centers for Disease Control & Prevention (KCDC) announced the information of COVID-19. Data Science for COVID-19 (DS4C) Project [1] makes a structured dataset based on the report materials of KCDC and local governments. The benchmark is a set of 4043 observations with raw data of COVID-19 patients. It includes: Patient moves/routes, Patient information, and Route types. This dataset is proposed to detect COVID-19 patients at an earlier stage. Intelligent clustering with extra constraints can help in detecting "COVID-19 Super Spreader" while minimizing the deployment cost (material, medicine, resuscitation bed, etc.) and efforts on each cluster.

Hydraulic System Maintenance Trace. This dataset is composed of 2205 observations (row data), for each row, we note 60 s of measurements and 17 sensors that record data with time [3]. For our simulation, we propose to select only sensors with a timestamp of 1 second. Hence, we resume the dataset to only 9 sensors. We need to realize efficient clusters containing points (several row data) indicating the following four maintenance faults: 1) Cooler conditions close to total failure 2) Valve conditions at least with severe lag 3) Severe internal pump leakage and 4) At least severely reduced pressure in hydraulic accumulator. Constrained clustering using this dataset is important. It consists in obtaining clusters with at least θ points before confirming the maintenance fault. This is often named by Diagnosis in the literature.

5.2 Clustering Optimization Evaluation

To deal with the multi-variate data and in order to use the proposed constrained clustering algorithm, we use Principle Component Analysis (PCA) to reduce data inputs from high dimensions to two dimensions. Then, we prepare our datasets by standardizing all the input features in order to ease the application of the proposed algorithms. Choosing the appropriate number of clusters in unlabel dataset is important in order to interpret and validate the consistency of data points. **Elbow method** [7] may be helpful in this stage to find the appropriate number of clusters. Indeed, this approach consists in representing the optimization function of K-Means according to the number of clusters k.

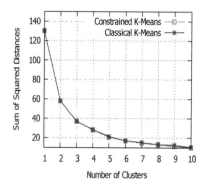

Fig. 1. Comparison between the constrained and the classical clustering approaches in terms of Sum of Squared Distances: case of IRIS traces

According to this method, the number of clusters is chosen when the elbow starts (i.e., for 3 clusters in Fig. 1). Therefore, from Fig. 1, we can deduce that the appropriate size of clusters (for Iris dataset) is $\theta_h = 3$. Hence we use this value in the next simulation and apply the same method (Elbow) to determine the best value of k for other datasets. Moreover, in Fig. 1, we show the efficiency of the proposed ILP approach which slightly improves (reduces) the sum of the squared distances compared to the K-Means state-of-the art algorithm.

To assess the efficiency and performance of the proposed optimization algorithms, we use the following three metrics:

- **Objective Functions Ratio**: it indicates the Sum of Squared Distances ratio between the constrained and the classical K-Means. It is provided by:

$$\frac{\min_{C,x} \sum_{i=1}^{m} \sum_{h=1}^{k} x_{i,h} \left(\frac{1}{2} \left|\left| d^i - C^h \right|\right|_2^2 \right)}{\min_{C^h,C^{h,i}} \frac{1}{m} \sum_{i=1}^{m} \left|\left| d^i - C^{h,i} \right|\right|_2^2} \quad (16)$$

where $C^{h,i}$ is the cluster centroid to which the data point d^i has been assigned using the classical K-Means.
- **Computational Time Ratio**: It indicates the execution time ratio between the constrained and the classical K-Means algorithms.
- **Average Number of Empty Clusters (or clusters with few points)**: it indicates the average number of clusters with less than θ data points.

Table 1 illustrates the weaknesses of the classical K-means approach when using the five aforementioned datasets. Indeed, the classical K-means algorithm converges with a large number of clusters with fewer data points or empty clusters in some cases. This result is hampering when we need clusters with at least θ points. The COVID-19 use case (see Table 1) depicts 19 clusters with less than 25 points represented by patients, in this use-case. Our approach is able to detect and distinguish important clusters compared to weak clusters. Hence, this can be usefully to concentrate and prioritize the aid and efforts on the new constrained clusters with at least θ patients.

Table 1. Average number of clusters with a number of data points less than $\theta_h = \theta$ using the classical K-Means

Datasets	θ_h				
	5	10	15	20	25
IRIS	0	3	5	7	9
5G-Uplink	7	7	7	7	7
CDR	1	1	1	2	3
Hydraulic	2	4	6	8	13
COVID-19	1	4	8	12	19

Table 2. Comparison through different datasets

Metrics	Dataset				
	IRIS ($\theta_h = 10$)	5G-UL ($\theta_h = 3$)	CDR ($\theta_h = 3$)	Hydraulic ($\theta_h = 5$)	COVID-19 ($\theta_h = 3$)
Objective functions ratio	0.99	1	1.25	1	1.17
Computational time ratio	1.5	1.57	1,31	2.44	2.27

Table 2 depicts the evaluation of the proposed constrained clustering approach (i.e., the ILP rounding algorithm) using objective function ratio and computational time ratio metrics. We run several simulations on the aforementioned 5 datasets to enlarge the grasp of our approach. These simulations illustrate the utility of the constrained clustering approach. Indeed, our approach is able to attend near-optimal solutions that are better then the classical K-Means solution, even in the case where new constraints on the size of clusters, are added (see cases where the objective function ratio is lower than 1). In the other cases (objective function ration is greater than 1), the obtained gap is negligible compared to 1, and this proves the efficiency of our approach to attend good solutions in the case of constrained clustering. The proposed constrained clustering approach does not require a significant computational time which justifies its efficiency in terms of convergence, in concrete use cases and applications.

In the next simulation, we highlight the impact of the "critical" or optimal value of the minimal clusters' size to be considered in a dataset clustering. We consider the Iris dataset (see the above description) and assess the objective function ratio for different θ_h values, as depicted in Fig. 2. One can see that the proposed constrained algorithm outperforms the classical K-Means approach especially for $\theta_h = 10$ and has a negligible gap for different values of θ_h with a maximum ratio of 1.12, in the worst case. This illustrates clearly the efficiency of our constrained clustering method that can be largely used in different use-cases that necessitate a minimum value of the cluster's size.

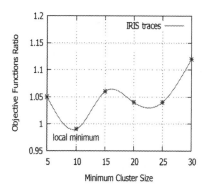

Fig. 2. Impact of θ_h parameter variation with fixed number of clusters ($K = 5$): case of IRIS trace

6   Open Research Challenge

In this paper, we considered the constrained clustering problem with already fixed and known number of clusters (i.e. k is a constant in our optimization). Moreover, for each cluster C^h, we desire a minimum number of points θ_h such as: $\theta_h \leq \sum_{i=1}^m x_{i,h} \ \forall h = 1, ..., k$. In this section, we propose to discuss an open research challenge when considering θ_h as a random variable according to a given probabilistic distribution. Hence, we would to minimize the number of existing points in each cluster, such that there will be no empty cluster?

Let's suppose that $\theta_h \sim N(\mu_{\theta_h}, \sigma_{\theta_h}^2)$ where μ_{θ_h} is the mean value of θ_h and $\sigma_{\theta_h}^2$ is its variance. We also suppose α as a prescribed probability and represents the reliability of our system. Then, we propose the following Stochastic Constrained Clustering Problem (SCCP), given as follows:

$$\min_{C,x} \ \mathbb{E}\left(\sum_{i=1}^m \sum_{h=1}^k x_{i,h} \times \theta_h \times \left(\frac{1}{2}||d^i - C^h||_2^2\right) \right) \tag{17}$$

Subject to:

$$\sum_{h=1}^k x_{i,h} = 1 \ \forall i = 1, .., m \tag{18}$$

$$\mathbb{P}\left(\sum_{i=1}^m x_i^h \geq \theta_h \right) \geq \alpha_h \ \forall h = 1, ..., k \tag{19}$$

$$x_{i,h} \in \{0,1\} \ \forall i = 1, ..., m \ ; \ \forall h = 1..., k \tag{20}$$

This stochastic program allows to find the optimal values of θ when considering constrained clustering constraints. Solving this stochastic program can be done using chance constrained approaches (see [8], for instance). These approaches will transform the initial stochastic program to non linear deterministic program that can be solved using Liapounov approaches (see [14]).

7 Conclusion and Future Work

In this paper, we considered a new optimization algorithm to cope with constrained clustering problems. These problems involve linear constraints on the size of each considered cluster in the same optimization. We proposed new approaches using rounding techniques to accelerate the data-set points' assignment of the classical K-Means formulation. Hence, we added new valid constraints to consider non-empty clusters and relaxed the integrity constraints of the exact proposed approach to guarantee the scalability of our solution. Our proposed algorithm outperforms clearly the classical K-Means approach, and go beyond when considering non-empty clusters.

For future work, we will consider stochastic optimization to determine the optimal values of the minimum number of points for each considered cluster (see SCCP formulation provided by equations and inequalities (17), (18), (19) and (20).

References

1. (2020). https://github.com/ThisIsIsaac/Data-Science-for-COVID-19
2. Call detail records (CDR) data set (2020). https://www.kaggle.com/marcodena/mobile-phone-activity
3. Condition monitoring of hydraulic systems data set (2020). https://www.kaggle.com/jjacostupa/condition-monitoring-of-hydraulic-systems
4. Iris data set (2020). https://archive.ics.uci.edu/ml/datasets/iris
5. Bradley, P.S., Bennett, K.P., Demiriz, A.: Constrained k-means clustering. Technical report, MSR-TR-2000-65, Microsoft Research (2000)
6. Bradley, P.S., Mangasarian, O.L.: K-plane clustering. J. Global Optim. **16**(1), 23–32 (2000)
7. Cui, M., et al.: Introduction to the k-means clustering algorithm based on the Elbow method. Acc. Auditing Finan. **1**(1), 5–8 (2020)
8. Geng, X., Xie, L.: Data-driven decision making with probabilistic guarantees. Optimization and Control (2019)
9. Hartigan, J.A.: Clustering Algorithms, 99th edn. Wiley, Hoboken (1975)
10. Italia, T.: Telecommunications - SMS. Call, Internet - MI (2015)
11. Jin, X., Han, J.: K-medoids clustering. In: Sammut, C., Webb, G.I. (eds.) Encyclopedia of Machine Learning, pp. 564–565. Springer, Boston (2010). https://doi.org/10.1007/978-0-387-30164-8
12. Khedher, H., Hoteit, S., Brown, P., Krishnaswamy, R., Diego, W., Veque, V.: Processing time evaluation and prediction in cloud-ran. In: ICC 2019–2019 IEEE International Conference on Communications (ICC), pp. 1–6 (2019)
13. MacQueen, J.B.: Some methods for classification and analysis of multivariate observations. In: Cam, L.M.L., Neyman, J. (eds.) Proceedings of the Fifth Berkeley Symposium on Mathematical Statistics and Probability, vol. 1, pp. 281–297 (1967)
14. Prékopa, A.: Contributions to the theory of stochastic programming. Math. Program. **4**(1), 202–221 (1973)

Robustness of Supervised Learning Based on Combined Centroids

Jan Kalina$^{(\boxtimes)}$ and Ctirad Matonoha

The Czech Academy of Sciences, Institute of Computer Science,
Pod Vodárenskou věží 2, 182 07 Prague 8, Czech Republic
{kalina,matonoha}@cs.cas.cz

Abstract. Recently, we proposed a novel sparse centroid-based supervised learning method, allowing to optimize a single centroid and its corresponding weights. The method is especially useful for localizing objects in images. Here, we extend the method to the task of joint localization of several objects in a 2D-image by means of combining several centroids. The novel approach, i.e. joint optimization of several centroids and a subsequent optimization of their weights, is illustrated on the task of localizing the mouth and both eyes in facial images. Because we are particularly interested in studying the robustness of the method to various modifications of the images, we evaluate the performance of the methods also over images artificially modified by additional noise, occlusion, changed illumination, or rotation. The novel centroid-based method is successful in the localization task, and the optimization turns out to ensure robustness with respect to the presence of noise or occlusion in the images. Moreover, combining the optimized centroids yields more robust results than a method using simple centroids with a highly robust correlation coefficient (with a high breakdown point).

Keywords: Machine learning · Sparsity · Regularization · Robust optimization · Outliers

1 Introduction

In a variety of tasks in image analysis, centroids are exploited as tools for the tasks of localization of objects or supervised segmentation. A centroid, also known as template or prototype, represents an ideal object, a typical form with ideal appearance. In [8], a review of more than 30 recent papers exploiting centroid-based methods is presented, with a focus on applications in biomedical or anthropological images. Centroid-based localization of objects in images is well known as template matching and remains popular, in spite of its simplicity, as a simple, powerful and comprehensible tool with a clear interpretation [2]. Theoretical results for centroid-based methods include available bounds on the classification performance of a (simplistic) centroid-based method of [6] or a Bayesian representation of deformable centroids proposed in [5]. The most common similarity measure between the centroid and the corresponding part of the image

© Springer Nature Switzerland AG 2021
L. Rutkowski et al. (Eds.): ICAISC 2021, LNAI 12855, pp. 171–182, 2021.
https://doi.org/10.1007/978-3-030-87897-9_16

is the Pearson product-moment correlation coefficient [11]. Centroids are often used as auxiliary tools within much more complex image analysis methods; the literature devoted to such applications is however often void of numerous crucial details, necessary for reproducing the results. In addition, centroids are often used on images after some suitable (e.g. Fourier or wavelet) transform, allowing to reduce computational costs.

While simple (naïve) centroids in the form of average objects have been commonly used, we proposed the first general optimized approach to centroid construction in [8]. The centroid-based supervised learning method exploits the weighted correlation coefficient, i.e. weighted extension of the Pearson's correlation coefficient, while the weights for the centroid are also optimized, ensuring many pixels to obtain zero weights (i.e. yielding a sparse solution). The optimization tasks require to solve a nonlinear loss function corresponding to a margin-like distance (inspired by [15]) evaluated for the worst pair over the database. In [8], the method was illustrated on the task of mouth localization in facial images. We compared the results of two optimization approaches:

– Linear (i.e. approximate) approach to solving the optimization task,
– Nonlinear optimization by means of the interior point method.

Linear approximation to the optimization yielded solutions very close to those of the nonlinear approach, where the latter was implemented in a highly optimized software package [10].

While combining several centroids for K objects ($K \geq 2$) is not uncommon in practical applications [18], we are not aware of a method for a joint optimization of several centroids. A different task, which is more elaborated in the literature, is a combination of several centroids for the same one object [14]. The methodological Sect. 2 of the current paper describes a joint localization of objects by means of centroids and as our main contribution extends the approach of [8] to localizing several objects in an image jointly. Section 3 investigates the performance of various centroid-based tools in the tasks of localizing the mouth and a joint localization of the mouth and both eyes in facial images; their performance is validated over a test dataset, possibly modified (contaminated) in various ways. Section 4 brings conclusions.

2   Methodology

Section 2.1 describes a centroid-based joint search of several objects in an image by means of given centroids. For this joint search of several objects in an image, the optimization of centroids of [8] will be extended in Sect. 2.2. An alternative to the approach of Sect. 2.1 based on a highly robust correlation coefficient is described in Sect. 2.3. The methods will be formulated in a general way, without using specific properties of faces.

2.1 Joint Localization of the Mouth and Eyes

Let us recall the weighted correlation coefficient $r_W(x, y; w)$ between two vectors $x = (x_1, \ldots, x_p)^T$ and $y = (y_1, \ldots, y_p)^T$ with non-negative weights $w = (w_1, \ldots, w_p)^T$ to be defined as

$$r_W(x, y; w) = \frac{\sum_{i=1}^{p} w_i(x_i - \bar{x}_w)(y_i - \bar{y}_w)}{\sqrt{\sum_{i=1}^{p}[w_i(x_i - \bar{x}_w)^2] \sum_{i=1}^{p}[w_i(y_i - \bar{y}_w)^2]}}, \tag{1}$$

where $\bar{x}_w = \sum_{j=1}^{p} w_j x_j = w^T x$ is the weighted mean of x, and $\bar{y}_w = w^T y$ is the weighted mean of y. In our task, the aim is to localize K objects in a single image by means of K given centroids $c_1, \ldots, c_K$ with one centroid for each object, where $c_k = (c_{k1}, \ldots, c_{kp_k})^T$ for $k = 1, \ldots, K$. Each of these objects c_k is a matrix. Throughout the whole study, all images (and also all such objects c_k) are considered after being converted to vectors.

Let us now have candidate areas $x_1, \ldots, x_K$ of the same sizes as the centroids $c_1, \ldots, c_K$ and let us have sets of weights $w_1, \ldots, w_K$ corresponding to the centroids. Such K-tuple (i.e. such set of K non-overlapping candidate areas), which maximizes

$$\sum_{k=1}^{K} r_W(x_k, c_k; w_k) \tag{2}$$

over all possible K-tuples in that particular image, will be classified as the true object (i.e. set of K true objects) in the given image.

In our experiments, we are primarily interested in localizing a triplet containing the mouth, the left eye, and the right eye in a facial image (i.e. with $K = 3$). The initial centroids and their construction together with the results will be presented later in Sect. 3.1. While each image contains one face (i.e. one mouth, one left eye and one right eye), we understand an area in the image to correspond to the mouth, if its midpoint is not more distant from the midpoint of the true mouth than 3 pixels (in the Euclidean distance). The same holds for each of the eyes.

2.2 Optimization for the Joint Localization

The optimization approach of [8] is limited to the task of localizing a single object by means of a single centroid. We will now propose a natural extension of this approach to optimizing centroids within a joint localization of K objects in an image. We need to carefully think about the notation needed for describing our approach.

We assume to have n images in the training dataset. We work with K centroids as explained in Sect. 2.1. This ordered K-tuple of centroids will be denoted as $\tilde{c}$, where the notation with a tilde will correspond to a K-tuple. Within the j-th image of the training dataset, where $j = 1, \ldots, n$, the set of all (admissible) positive examples is denoted as $\mathbb{E}_j^+$. Each positive example represents K subimages from the same image, i.e. K components (areas) from that image. We understand each of these subimages of a given positive example to be a positive example itself, i.e. each of the K subimages corresponds to the true object. To avoid confusion, any particular K-tuple $\tilde{x} \in \mathbb{E}_j^+$ has to have the same dimensions as $\tilde{c}$ (e.g. containing K rectangular matrices of the same sizes, transformed to vectors); we will now denote these as $x_1, \ldots, x_K$.

The set of all negative examples in the j-th image of the training dataset is denoted by $\mathbb{E}_j^-$. Here, we understand that none of the K subimages corresponds to the true object; in other words, each of the K components is a negative example. Also the negative examples have to have the same sizes as the centroids. A particular ordered K-tuple of $z_1, \ldots, z_K$ in $\mathbb{E}_j^-$ will be denoted as $\tilde{z}$. We work with weights $w_1, \ldots, w_K$ corresponding to the K centroids. We denote $w_k = (w_{k1}, \ldots, w_{kp_k})^T$ and assume $\mathbb{1}^T w_k = 1$ for $k = 1, \ldots, K$, where $\mathbb{1} = (1, \ldots, 1)^T$ is a vector of the corresponding length p_k. The ordered K-tuple of $w_1, \ldots, w_K$ is denoted by $\tilde{w}$. All the K-tuples, i.e. $\tilde{x}$, $\tilde{z}$ and $\tilde{w}$, have to have the same sizes as the K-tuple of centroids $\tilde{c}$.

For a particular choice of $\tilde{c}$ and $\tilde{w}$, we consider the loss function

$$f(\tilde{x}, \tilde{z}, \tilde{c}, \tilde{w}) = \sum_{l=1}^{K} r_W(x_l, c_l; w_l) - \sum_{l=1}^{K} r_W(z_l, c_l; w_l) \qquad (3)$$

and the margin-like measure

$$\min_{j=1,\ldots,n} \min_{\tilde{z} \in \mathbb{E}_j^-} \min_{\tilde{x} \in \mathbb{E}_j^+} f(\tilde{x}, \tilde{z}, \tilde{c}, \tilde{w}) \qquad (4)$$

corresponding to the worst positive example (i.e. the K-tuple with the worst separation from a negative example) over the training dataset.

It is beneficial (although not necessary) to consider the search of the best centroid only over a set of all admissible centroids denoted as $\mathbb{S}_c$. Using such approach, only K-tuples of centroids formulated under specific constraints are considered and each component c_k for $k = 1, \ldots, K$ must belong to a specified admissible region, which is given by (reasonable) conditions on the distances. We solve the optimization task (4) with the variable $\tilde{c}$ and with given fixed weights, i.e.

$$\arg\max_{\tilde{c} \in \mathbb{S}_c} \min_{j=1,\ldots,n} \min_{\tilde{z} \in \mathbb{E}_j^-} \min_{\tilde{x} \in \mathbb{E}_j^+} f(\tilde{x}, \tilde{z}, \tilde{c}, \tilde{w}). \qquad (5)$$

At the same time, $\tilde{z} \in \mathbb{E}_j^-$ will be understood as a K-tuple of images $z_1, \ldots, z_K$. Partial derivatives of the loss function have the form

$$\frac{\partial}{\partial c_{km}} f(\tilde{x}, \tilde{z}, \tilde{c}, \tilde{w}) = \frac{\partial}{\partial c_{km}} \sum_{l=1}^{K} (r_W(x_l, c_l; w_l) - r_W(z_l, c_l; w_l))$$

$$= \frac{\partial}{\partial c_{km}} r_W(x_k, c_k; w_k) - \frac{\partial}{\partial c_{km}} r_W(z_k, c_k; w_k) \qquad (6)$$

for $k = 1, \ldots, K$ and $m = 1, \ldots, p_k$; in this form, they are expressed as depending on a single centroid, so they can be expressed using the formulas explicitly given in [8].

In this paper, we use a linear approximation of f in all optimization computations. This linear approach contains several technical parameters controlling the speed of the computation; we use their values as in [8] and we use $\lambda = 0.005$ as the value of the (only) regularization parameter. After the centroid is optimized, the optimization of its weights according to

$$\arg \max \min_{\substack{j=1,\ldots,n}} \min_{\tilde{z} \in \mathbb{E}_j^-} \min_{\tilde{x} \in \mathbb{E}_j^+} f(\tilde{x}, \tilde{z}, c, w), \tag{7}$$

where the maximization is performed over $w_1 \in \mathbb{R}^{p_1}, \ldots, w_K \in \mathbb{R}^{p_K}$, is solved again exploiting a linear approximation. Partial derivatives

$$\frac{\partial}{\partial w_{km}} f(\tilde{x}, \tilde{z}, \tilde{c}, \tilde{w}) = \frac{\partial}{\partial w_{km}} \sum_{l=1}^{K} \left(r_W(x_l, c_l; w_l) - r_W(z_l, c_l; w_l) \right)$$

$$= \frac{\partial}{\partial w_{km}} r_W(x_k, c_k; w_k) - \frac{\partial}{\partial w_{km}} r_W(z_k, c_k; w_k) \tag{8}$$

for $k = 1, \ldots, K$ and $m = 1, \ldots, p_k$ may use the explicit formulas of [8].

Because a shifted mouth is still perceived as a true mouth, as explained in Sect. 2.1, we consider each image to contain more than a single positive example. This determines our $\mathbb{E}_j^+$ in every image. A negative example contains three regions of the given image, where the first does not correspond to the mouth, the second does not correspond to the left eye, and the third does not correspond to the right eye. The margin (4) is based on such 6-tuple (hexade), containing a positive example and a negative example, which has the smallest value of (2) over the whole training database. We optimize the triple of centroids for the mouth, left eye and right eye. The eyes are assumed to appear above the mouth; we assume the midpoint of each eye to be at least 20 pixels but not more than 55 pixels above the midpoint of the mouth. This defines our admissible region $\$_c$.

We do not use here the information that the left eye is (typically) similar to a mirror reflection of the right eye; this could be of course possible and also convenient from a computational point of view, but we prefer not to exploit this information to illustrate that the proposed method works reliably even without it.

2.3   An Approach Based on the Robust Correlation Coefficient

While the Pearson's correlation coefficient is well known to be highly vulnerable with respect to the presence of outlying values (outliers) in the data, its various robust versions have been available [13]. Therefore, we will also exploit the robust correlation coefficient r_{LWS} based on the least weighted squares (LWS) estimator in this paper. The LWS estimator (see [9,16]) represents a (possibly highly) robust estimator of parameters in the linear regression model with appealing properties. If suitable weights are chosen, the estimator combines high robustness (in terms of the breakdown point [4]) and high efficiency [3]. Formally, $r_{LWS}(x, y)$ is defined in Algorithm 1; it was proposed and investigated in [7].

Algorithm 1. $r_{LWS}(x,y)$: robust correlation coefficient based on the LWS regression estimator

---

Input: Vectors $x_1, \ldots, x_p$ and $y_1, \ldots, y_p$
Output: $r_{LWS}(x,y)$
   Compute the LWS estimator in the linear regression model

$$Y_i = \beta_0 + \beta_1 x_i + e_i, \quad i = 1, \ldots, p, \tag{9}$$

   with a given weight function

$$\psi(t) = \left(1 - \frac{t}{\tau}\right) \cdot \mathbb{1}[t < \tau], \quad t \in [0,1], \tag{10}$$

   where $\tau = 3/4$
   Denote the resulting weights assigned to individual observations as w^*
   $r_{LWS}(x,y) := r_W(x,y;w^*)$

If r_{LWS} is used as the correlation measure, then a joint localization of K objects in an image can be performed in an analogous way to (2), i.e. by finding the K-tuple with the largest value of the loss

$$\sum_{k=1}^{K} r_{LWS}(x_k, c_k), \tag{11}$$

while the LWS estimator finds a set of weights for each of the K individual fits. As the weights are assigned to individual values (in our case pixels) after a permutation, which is found automatically within the computation of the method, no optimization related to that of Sect. 2.2 is performed for the loss function (11).

3   Experiments over a Database of Facial Images

We use the database of 212 raw grey-scale 2D images of faces of healthy individuals of sizes 192×256 pixels, acquired at the Institute of Human Genetics, University of Duisburg-Essen. The database, obtained within research of genetic syndrome diagnostics based on facial images [1,2] under projects BO 1955/2-1 and WU 314/2-1 of the German Research Council (DFG), was analyzed in [8]. All images were taken under standardized conditions with one individual per image in a straight frontal view. Our two tasks

1. To localize the mouth, and
2. To localize jointly the mouth and both eyes

will be learned over the training set, which contains 124 randomly selected images, and applied over an independent validation set, which contains the remaining 88 images. In the whole study, we use our implementation of the procedure prepared in R software [12].

Fig. 1. Left: initial mouth centroid of size 26×56 pixels. Middle: optimal centroid with equal weights, where the left image was used as the initial centroid. Right: optimal weights for the optimal centroid, where equal initial weights were used.

Fig. 2. Left: initial centroid for the right eye of size 26×28 pixels. Middle: optimal centroid with equal weights, where the left image was used as the initial centroid. Right: optimal weights for the optimal centroid, where equal initial weights were used.

In addition to computing the results over the raw images (Sect. 3.1), we apply the methods to the validation set after being artificially modified in various realistic ways (Sect. 3.2). If optimization of centroids and/or their weights is used, it is always applied to raw (non-modified) images; in other words, the optimization procedures do not use the modified images at all.

All results are presented in the self-explaining Table 1. The classification accuracy is defined as the ratio of the properly localized objects divided by the total number of images. Particularly, in the task of joint localization of the mouth and both eyes, the definition of the classification accuracy considers the ratio of the properly localized three objects (i.e. the mouth and both eyes jointly have to be classified correctly) divided by the total number of images.

3.1 Initial Results

We start the study by constructing initial centroids. The mouth centroid (left image of Fig. 1) of size 26×56 pixels is constructed as the average of 7 true mouths of randomly selected individuals of the training dataset. The eye centroids (left

Fig. 3. Left: initial centroid for the left eye of size 26×28 pixels. Middle: optimal centroid with equal weights, where the left image was used as the initial centroid. Right: optimal weights for the optimal centroid, where equal initial weights were used.

Table 1. Classification accuracy evaluated in the tasks of localization of the mouth (left) and joint localization of the mouth and both eyes (right) over the test dataset. The initial centroids are described in Sect. 3.1; optimization of weights starts with equal initial weights. (A) Initial centroids with equal weights; (B) Optimal centroids with equal weights; (C) Optimal centroids with optimal weights; (D) Initial centroids with r_{LWS}.

Images	Localizing the mouth				Localizing the mouth and both eyes			
	(A)	(B)	(C)	(D)	(A)	(B)	(C)	(D)
Raw images	0.85	1.00	1.00	1.00	0.98	1.00	1.00	1.00
Noise, $\sigma = 0.01$	0.82	1.00	1.00	1.00	0.96	1.00	1.00	1.00
Noise, $\sigma = 0.02$	0.79	0.99	0.99	0.98	0.94	1.00	1.00	0.98
Noise, $\sigma = 0.03$	0.73	0.97	0.97	0.95	0.85	0.98	0.98	0.95
Occlusion, $k = 0.3$	0.83	1.00	1.00	1.00	0.97	1.00	1.00	1.00
Occlusion, $k = 0.6$	0.82	1.00	1.00	1.00	0.95	1.00	1.00	1.00
Occlusion, $k = 1.0$	0.79	0.97	0.98	0.98	0.90	1.00	1.00	1.00
Illumination,								
$\ell = 0.001$	0.83	0.97	0.96	0.96	0.97	1.00	1.00	1.00
$\ell = 0.003$	0.81	0.95	0.95	0.93	0.93	1.00	1.00	0.96
$\ell = 0.005$	0.72	0.85	0.86	0.82	0.88	0.97	0.97	0.92
Rotation (R1)	0.81	0.92	0.93	0.88	0.93	1.00	1.00	0.95
Rotation (R2)	0.56	0.83	0.85	0.69	0.90	1.00	1.00	0.94

images of Figs. 2 and 3) are obtained as a means of 10 randomly selected different eyes of the training dataset. Both eye centroids have the size 26×28 pixels. In all cases, we use equal weights as initial ones, i.e. $w_1 = \cdots = w_p = 1/p$, where p is now the number of pixels of the centroid. Using the initial mouth centroid with equal weights localizes the mouth correctly in 85 % of images of the test dataset, as presented in Table 1. Using the initial triplet of centroids localizes the mouth and both eyes jointly in 98 % of images of the test dataset, as presented again in Table 1. On the whole, we use four different approaches, where the first three exploit the loss (4).

(A) Initial centroids and initial weights,
(B) Optimal centroid and initial weights,
(C) Optimal centroid with optimal weights (i.e. the centroid is optimized first and the weights are optimized subsequently),
(D) Centroid-classification using initial centroids and r_{LWS} as the correlation measure (exploiting (11)).

Further, we perform the optimization of centroids for the joint localization of the mouth and both eyes, using the method of Sect. 2.2. Here, we use the naïve centroids shown in the left images of Figs. 1 to 3 as the initial centroids. The performance in the localization of the mouth and both eyes is presented again in Table 1. The optimal three centroids, obtained in the joint optimization, are shown in the middle images in Figs. 1 to 3. If the weights are further optimized for these optimal centroids (again in a joint optimization), the resulting optimal weights are shown in the right images of Figs. 1 to 3. The central parts of the eyes, shown as white, turn out to be the most relevant parts for the eye localization. On the other hand, about 30 % of pixels of the eye centroids obtain zero weights and do not contribute to the localization (black pixels in the right figures).

3.2   Modified Test Dataset

Noise. Pixel-independent noise generated from normal distribution $N(0, \sigma^2)$ with different values of σ is added to the grey intensities of every image of the test dataset. Without any surprise, the performance of the methods decreases with an increasing σ, as shown in Table 1. It is also natural that the joint localization performs better compared to localizing only the mouth. Still, it is interesting to realize that the optimization of centroids much improves the robustness to noise.

Occlusion. We study the effect of occlusion by modifying every image from the test dataset. We manually perform an artificial occlusion in the form of a local degeneration of every face by replacing the intensities in two square areas of 3×3 pixels in each face by a given constant k. The two square areas are placed to the image to have the same position with the location 12–14 pixels left and 12–14 pixels below the midpoint of the left pupil and 25–27 pixels left and 10–12 pixels below the midpoint of the mouth. The occlusion (even with a severe degeneration of the images) turns out to represent the smallest problem among all modifications considered in this section, especially if the optimized centroids are used. The performance of non-optimized methods slightly decreases with an increasing value of k.

Concerning the occlusion, the following interpretation is formulated based on a detailed analysis of results for individual images. As the optimality is performed under regularization, there are no pixels with an extreme influence. Still, the occlusion rapidly reduces the weighted correlation between the mouth and the centroid. However, centroid-based methods turn out to be very successful in localization tasks if the centroid is slightly shifted aside. Therefore, the mouth is still correctly localized when the centroid is shifted one or two pixels aside and the method is so robust to the presence of the occlusion. Thanks to the robustness to shift, the exact position of the occlusion is not so important and additional studies over occluded versions of the test dataset yield analogous results.

Illumination. To examine the effect of a different illumination, we add a column-dependent constant to every pixel of the image imitating a brighter illumination in the face. If we consider a pixel $[i, j]$ with intensity x_{ij} in an image with I rows and J columns, then the intensity in the modified image will be

$$x_{ij}^* = x_{ij} + \ell \cdot |j - j_0|, \quad i = 1, \ldots, I, \quad j = 1, \ldots, J, \tag{12}$$

where the midpoint of the mouth appears in the pixel $[i_0, j_0]$. The performance of the methods decreases with an increasing ℓ. We can say here that it is mainly the joint search that improves the performance over modified images, as its results are better compared to those in localizing only the mouth (even with optimized centroids).

Rotation. We perform two studies of rotation. In the study (R1), we consider each image of the validation set after rotations by $+5$, 0, and -5 degrees. Thus, we consider $3 \times 88 = 264$ test images. Such candidate areas are selected, for which the correlation values with the centroids are the largest over the set of all three rotated images of the same person. In the task of mouth localization, even the optimal centroids are not robust to a small rotation, while the results reveal the joint search to be very helpful here.

In the study (R2), rotating the set of centroids by angles of $0, 10, 20, \ldots, 350°$ is performed. Thus, each image is considered as a set of 36 images and the method classifies the objects in such out of the 36 versions of the image, where the loss (4) is the largest. Again, reliable results are obtained only for the joint localization, while the best results are obtained for joint localization with optimized centroids; its results work (and are practically invariant) under any rotation.

4   Conclusions

Centroids represent useful and in spite of their simplicity still popular tools in various image analysis (object localization or segmentation) tasks in biomedical or anthropological applications [17]. The optimization of centroids for the centroid-based localization (known as template matching) of [8] turned out to be comprehensible and at the same time powerful. This motivated us to propose a natural extension of the method of [8] to the task of a joint localization of several objects in an image in this paper. The new approach optimizes the centroids and subsequently their corresponding weights. The approach, which can be performed for localizing K objects jointly, is explained and illustrated on the task of localizing three objects in facial images, namely localizing the mouth and both eyes jointly. The optimization exploits no specific properties of faces.

Results over the study over a database of facial images reveal the optimization to be successful. The optimization improves the performance of the centroid-based method in the task of localizing the mouth and both eyes jointly. The optimal mouth centroid visually corresponds to a typical mouth appearance, and an analogous conclusion can be formulated for the eyes. If the weights are also optimized, the largest weights are assigned to central parts of the centroids, while their corners are trimmed away with zero weights. The optimization of weights ensures sparsity of the optimized centroids, although its effect on the localization performance is marginal; thus, the users may prefer to optimize only the centroids and to retain initial weights.

We are especially interested in the robustness of the novel method. Therefore, we present the results of the object localization in facial images, which are modified (contaminated) in several ways. The robustness is clearly better in the joint localization, compared to the mere localization of the mouth. The best results are obtained for optimal centroids and optimal weights. The results of such an approach even outperform those obtained with the highly robust r_{LWS}. For noise and occlusion, the optimization ensures robustness of the solution, while the situation is somewhat different for illumination and rotation. In the latter cases, it is the joint search that brings a stronger benefit for the robustness than the optimization. As future research topics, we plan to perform also theoretical investigations of the effects of noise or occlusion on centroid-based localization in images.

Acknowledgement. The research was supported by the grant GA19-05704S of the Czech Science Foundation and by the Ministry of Health of the Czech Republic, grant NU21-08-00432. The authors are thankful to Prof. Laurie Davies for inspiration and Prof. Michal Haindl for discussion. Patrik Janáček provided technical support.

The left image of Fig. 1 is reprinted from Biocybernetics and Biomedical Engineering, vol. 40, J. Kalina and C. Matonoha, "A sparse pair-preserving centroid-based supervised learning method for high-dimensional biomedical data or images", pp. 774–786, 2020, with permission from Elsevier.

References

1. Böhringer, S.: Syndrome identification based on 2D analysis software. Eur. J. Hum. Genet. **14**, 1082–1089 (2006)
2. Böhringer, S., de Jong, M.A.: Quantification of facial traits. Front. Genet. **10**, 397 (2019)
3. Čížek, P.: Semiparametrically weighted robust estimation of regression models. Comput. Stat. Data Anal. **55**, 774–788 (2011)
4. Davies, P.L., Gather, U.: Breakdown and groups. Ann. Statist. **33**, 977–1035 (2005)
5. Grenander, U.: General Pattern Theory. A Mathematical Study of Regular Structures. Oxford University Press, Oxford (1993)
6. Hall, P., Pham, T.: Optimal properties of centroid-based classifiers for very high-dimensional data. Ann. Statist. **38**, 1071–1093 (2010)
7. Kalina, J.: Implicitly weighted methods in robust image analysis. J. Math. Imaging Vis. **44**, 449–462 (2012)
8. Kalina, J., Matonoha, C.: A sparse pair-preserving centroid-based supervised learning method for high-dimensional biomedical data or images. Biocybern. Biomed. Eng. **40**, 774–786 (2020)
9. Kalina, J., Tichavský, J.: On robust estimation of error variance in (highly) robust regression. Meas. Sci. Rev. **20**, 6–14 (2020)
10. Lukšan, L., et al.: UFO 2017-interactive system for universal functional optimization. Technical report V-1252, ICS CAS, Prague (2017)
11. Mazurowski, M.A., Lo, J.Y., Harrawood, B.P., Tourassi, G.D.: Mutual information-based template matching scheme for detection of breast masses: from mammography to digital breast tomosynthesis. J. Biomed. Inform. **44**, 815–823 (2011)
12. R Core Team: A language and environment for statistical computing. R Foundation for Statistical Computing, Vienna (2016)
13. Shevlyakov, G.L., Oja, H.: Robust Correlation. Theory and Applications. Wiley, Chichester (2016)

14. Thomas, L.S.V., Gehrig, J.: Multi-template matching: a versatile tool for object-localization in microscopy images. BMC Bioinform. **21** (2020). Article 44
15. Vapnik, V.N.: The Nature of Statistical Learning Theory, 2nd edn. Springer, New York (2000). https://doi.org/10.1007/978-1-4757-2440-0
16. Víšek, J.Á.: Consistency of the least weighted squares under heteroscedasticity. Kybernetika **47**, 179–206 (2011)
17. Weber, W.B., Bookstein, F.L.: Virtual Anthropology. A Guide to a New Interdisciplinary Field. Springer, Wien (2011)
18. Wong, Z.H., Abdullah, K., Wong, C.J.: Template matching using multiple templates weighted normalised cross correlation. In: 2014 IEEE Symposium on Computer Applications & Industrial Electronics (ISCAIE), pp. 131–135 (2014)

Cognitive Consistency Models Applied to Data Clustering

Thales Vaz Maciel[1,2(✉)] and Leonardo Ramos Emmendorfer[2]

[1] Sul-rio-grandense Federal Institute for Education, Science and Technology,
Bagé 96418-400, Brazil
`thalesmaciel@ifsul.edu.br`
[2] Center for Computational Science, Federal University of Rio Grande,
Rio Grande 96203-900, Brazil
`leonardoemmendorfer@furg.br`

Abstract. Data clustering concerns the discovery of partitions in data such that items from the same groups are as similar to each other as possible, and items from different groups are as dissimilar to each other as possible. The literature presents vast diversity on data clustering approaches, including systems that model the behavior of social individuals from different species. This work proposes a clustering algorithm that is reasoned upon the social theory of cognitive dissonance and the psychological theory of balance. We investigate whether psychological balance aware decision-making capabilities would affect the data clustering task. Partial results revealed the superiority of the proposed approach according to 5 out of 9 clustering quality metrics, when compared to other clustering algorithms.

Keywords: Data clustering · Cognitive dissonance · Balance theory

1 Introduction

Data clustering is defined as the discovery of partitions in datasets, in the context of Machine Learning. The intuitive idea behind the notion of data clustering is the search for groups of objects which share some kind of similarity [7]. The resulting partitions (or clusters) contain elements that are as similar as possible to those that are in the same cluster and also as dissimilar as possible to those elements in other clusters. The task of clustering is also relevant due to the ability to reveal useful patterns from datasets. The topic has gained importance over recent years, mostly as a result of successful applications in a wide range of fields, such as biology, medicine, psychology, image processing, among others [11].

There is a large number of clustering algorithms available in the literature, which tend to specialize in specific characterizations of data [3,8]. The diversity in clustering strategies is very prominent. Centroid-based strategies, for instance, recurrently self-adapt until they meet stability criteria to the detection of data density decays that potentially translate into cluster borders.

© Springer Nature Switzerland AG 2021
L. Rutkowski et al. (Eds.): ICAISC 2021, LNAI 12855, pp. 183–191, 2021.
https://doi.org/10.1007/978-3-030-87897-9_17

Biologically-inspired models should also be mentioned, where the natural clustering behavior of individuals from certain species can be simulated computationally.

Clustering algorithms that are inspired by the social behavior of individuals from different species are described in the literature, including, for instance, the impact of conflicting cognitions on the individual's behavior. One can notice, however, a lack of the modeling of certain aspects of psychological processes that guide the behavior of individuals who compose society.

The theory of cognitive dissonance, proposed by Leon Festinger in 1957 [4], refers to a psychological phenomenon caused by situations where conflicting attitudes, beliefs or behaviors are present in the individual's psyche. According to the theory, such sort of situation causes mental discomfort, which leads the individual to alter one of the conflicting cognitions in order to reduce the discomfort and thus restore its psychological balance. Heider (1957) [6] presented the balance theory, which explains changes of attitude in individuals that seek psychological balance. According to the author, individuals exist in a constant need for maintenance of their own cognitive consistency, solving or avoiding conflicting values, beliefs, and situations that relate to the environment they are placed. Relations between two individuals and from those towards a general object, which may also be an equally cognitive-capable third individual, are represented in triads [6]. In a triad, the product of the three positive or negative relations represents the psychological state of the situation as a whole, either balanced or unbalanced.

Cartwright and Harary (1956) [1] proposed an extension of Heider's model that is based on graph theory, thus enabling the simultaneous analysis of multiple relations, whereas the original model considered strictly 3 entities. One of the authors' contributions to balance theory was the exact possibility of representing the relations between an unlimited set of entities. In that sense, Flament (1963) [5] presented a theorem stating that a signaled complete graph is balanced only if all possible triads, from the relations that compose the graph, are balanced. However, in a more profound application of graph theory to interpersonal relationships situation, the author revised their previous statement to be insufficient due to a consideration that different individuals may have different perceptions over the situations they are involved.

This work proposes a novel clustering algorithm which relies on theories of cognitive dissonance [4] and psychological balance [1,6] as the fundamental mechanisms responsible for group formation. The solution herein proposed works as follows: each data object is assigned to a single agent, which interacts with other agents similarly generated in order to iteratively achieve a stable group formation, where neighboring agents agree about group formation. Each agent senses its environment and reacts, at each iteration, aiming to reduce its own cognitive discomfort resulting from eventual disagreements with its neighbors.

The aim of this work is to investigate whether the adoption of human cognitive models related to cognitive consistency can be useful for data clustering, leading to increased quality of the discovered clusters. Specifically, we aim to define an original clustering algorithm which regulates the decision making processes related to the clustering task by using concepts from cognitive dissonance and balance theories.

This work is organized as follows: Sect. 2 introduces recent clustering algorithms that apply conflicting cognitions at some level. Section 3 presents the proposed method. In Sect. 4 the results from experiments are shown. Section 5 concludes the paper.

2   Related Work

This section presents a review of clustering algorithms that relate to cognitive consistency inherent concepts and that apply reasoning based on such concepts to the data clustering logic.

Cohen and Castro (2006) [2] presented a clustering algorithm based on the particle swarm optimization (PSO) algorithm, called Particle Swarm Clustering (PSC). As a clustering strategy, the authors modeled the human tendency of adapting their behavior according to the influence of the environment by minimizing the differences in opinions and ideas through time, but also considering the individual's past experiences as emergent behavior.

These ideas are closely related to the concepts of cognitive dissonance [4] and balance theory [6] as they propose that conflicting ideas may be reorganized by actions of an individual and also that these actions as performed in a search for psychological balance.

In recent years, there have been algorithms that share those basic premises and develop diversely from then on. Examples are the Ant Colony Optimization Clustering with Chaos (ACOCC) [12], Swarm Clustering Algorithm (SCA) [13] and Fully-Informed Particle Swarm Clustering (FIPS) [9].

Yang et al. (2018) [12] proposed an improvement to the ACO Clustering (ACOC) by including a chaotic function in the initialization and feromone update phases so that it produces disturbances in the ants' movements in the systems so that it leads them to find a globally optimized solution instead of a local one.

Zhu et al. (2018) [13] proposed an improvement to general PSC-based clustering algorithms, suggesting that existing related algorithms had been designed so that the swarm particles represented cluster centers. The contribution was that, instead, swarm particles would represent each data instance individually and that these particles would recurrently interact and movement towards more densely populated areas in the systems, therefore aggregating to for clusters.

Mansour and Ahmadi (2019) [9] proposed a PSO-based clustering algorithm with the hybrid application of different neighborhood topologies, namely total, ring, and distance-based. Authors considered that their contribution was in the use of the latest topology as it considers a certain number of closest particles in the clustering process and that neighborhoods change in each iteration.

Godois et al. (2018, 2020) [7] proposed a clustering algorithm that considers every data instance as an agent of a multiagent system. In the system, the agents have a multidimensional view circle, which is divided into sectors and, in each iteration, each agent would choose to move towards the center of the view circle sector that is most dense, in terms of population by other agents. This occurs recurrently in a way that the agents can change their assigned cluster in each iteration until no cluster changes are performed in a complete iteration. The difference between the first and the latest is that the first proposed that, after choosing a sector of their view circle towards which an agent would move in an iteration, it would assume the same cluster as the other agent from the chosen sector that is closest to it, while, in the latest, the assumed cluster would be the most representative from the chosen sector.

3   Cognitive Dissonance Clustering Algorithm

This section presents a novel algorithm, which is called Cognitive Dissonance Clustering (CDC). We have shown how cognitive dissonance can drive the clustering process.

The algorithm is based on the movement of agents which locally search for a better cognitive environment, where each agent represents a data point from the input dataset. An agent senses its environment, within a limited vision scope around its current coordinates. This resulting circumference is then divided into equally distributed sectors. Figure 1 illustrates the vision scope of an agent, which is split into 8 sectors.

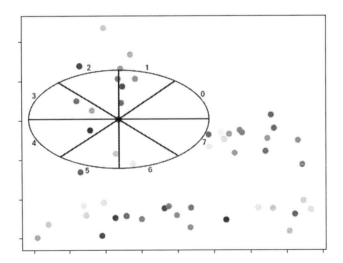

Fig. 1. The vision scope of an agent, which is split into 8 sectors labeled as 0,1, $\cdots$, 7.

Agents' decision making is parametrized. The parameters are, namely, the size of the step each agent takes in each iteration, the size of the radius of the agents' perception circle which limits the space in which each agent can perceive other agents, the number of sectors to which the agents' perception circles are divided (these are referred to as perception sectors) and the cognitive satisfaction threshold for the agents, which defines the psychological balance degree from which an interagent situation is classified as positive or negative. Figure 2 illustrates the algorithm proposed.

Algorithm 1 Cognitive Dissonance Clustering (CDC)

Input:
1. A dataset $\Delta = \{ \delta_1, \delta_2, \cdots, \delta_n \}$.
2. Vision radius Γ.
3. Step size λ.
4. Number of sectors ϕ around each agent.
Output:
5. A set of partitions P.

6. **for all** $\delta_i \in \Delta$ **do**
7.   Create an agent $\alpha_i \in A$.
8.   Set the coordinates of α_i as the same of data point δ_i.
9.   Set the cluster label of α_i as i.
10. **end for**
11. **while** the partitioning did not converge **do**
12. Ψ = set of positive and negative relations between agents in A
13.   **for all** $\alpha_i \in A$ **do**
14.   Divide the perception circle of α_i into ϕ sectors, $\Omega = \{ \omega_0, \omega_1, \cdots, \omega_{\phi-1} \}$.
15. $\omega^\star \leftarrow$ select the best $\omega_i \in \Omega$, which maximizes psychological comfort of $\alpha_i{}^\dagger$.
16.   Move α_i by λ towards the centroid of $\omega^\star$.
17.   Set the cluster label of α_i to the the prevalent cluster label in $\omega^{\star\dagger}$.
18.   **end for**
19. **end while**
20. **for all** $\delta_i \in \Delta$ **do**
21. $\delta_{i,\kappa} \leftarrow \alpha_{i,\kappa}$.
22. **end for**
23. †Ties are broken at random.

Fig. 2. Cognitive Dissonance Clustering (CDC).

At the beginning of each iteration, an interagent relation graph is built which links each agent to every other in the system. Initially, a positive relationship is established between each agent and its closest counterpart and also to all other agents that are at that same smallest distance to it. Figure 3 illustrates shows all positive relations at the first iteration for an illustrative dataset. The relation to the other agents is considered as negative, initially. At each iteration, the agent moves to the sector where it feels more comfortable, according to the psychological models proposed in [6] and [1]. The iterative process is described as follows.

Conflicting relationships between pairs of agents might appear. For instance, agent A might have a positive relationship with B but B has a negative relationship with A. In those cases, the positive relationship prevails. Reciprocal relations, for example, when an agent is the closest to another and vice-versa or

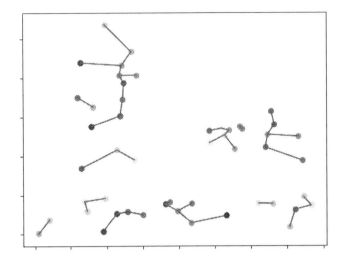

Fig. 3. Positive relations at the first iteration are represented as green links. (Color figure online)

an agent is not the closest to another and vice-versa, are just respectively defined as positive or negative. Each agent then verifies the psychological balance degree in each of its perception sectors at the moment.

In the process of verifying the balance degree of an agent's perception sector, only the interagent relations that are completely inside the sector are considered. From those, according to [1], relation triads are calculated as the product of the relations that compose the found relation triads. In order to discover the psychological state of the triads in each sector in which densities are at least 3, positive relations are labeled as +1 and negative relations are labeled as −1 [6]. Moreover, the product of the relations that form a triad corresponds to the psychological balance of the triad as a whole. This enables a triad to be evaluated as either positive or negative.

After defining the balance of every possible relation triad in an agent's perception sector, it is viable to calculate the balance of the sector as a whole. According to [5], a complete signaled graph, such as presented for verifications in the proposed algorithm, may be classified according to its balance degree. Balance degree calculation is applied in this context to define the psychological state for each sector of each circle of perception of each agent in the system. Later on, those sector states are summarized as either positive or negative, respectively, if the sector's balance degree reaches a defined cognitive satisfaction threshold, which is shared amongst every agent in the system.

Each agent then decides about the direction towards which it is to move at the current iteration. Agents decide to move towards the direction of the sector that leads to higher psychological comfort for itself. In the case of an agent detecting the same level of psychological comfort in more than one direction or when all the directions present the same neutral balance, the agent chooses to move towards the sector that presents the highest populational density at the moment. After performing its movement towards a sector, the agent assumes the cluster that represents the majority of the agents in that sector.

The algorithm runs iteratively until no agents change their cluster labels, or a threshold in the number of iterations is achieved. The final set of cluster labels represents the resulting partitioning from the algorithm.

4   Results and Discussion

This section describes the obtained results for the proposed clustering method, in terms of the quality of the resulting models for a specific dataset. Performance evaluation is assessed in comparison to other 3 data clustering algorithms, namely K-means, DL2, and AVC [7]. Except for K-means, which is applied here strictly as a comparative baseline, DL2 and AVC relate to the proposed algorithm in a more considerable level.

The performance metrics that were used to evaluate the clusters' quality in the discovered models where the adjusted Rand index (ARI), normalized mutual information (NMI), homogeneity, completeness, V-measure, and the coefficients of Fowles-Mallows (FM), Silhouette, Calinski-Harabasz (CH) and Davies-Bouldin (DB) according to definitions provided in [10].

The bidimensional dataset that was used in the tests was assembled specifically for the purpose of this comparison. It is composed of 50 instances, arranged over 3 classes that represent natural clusters with varying shapes, sizes, and densities. In total, each algorithm was executed 50 times using the dataset and the average values for each performance metric obtained from the discovered models are shown in Table 1.

Table 1. Average quality indicators for the partitionings discovered by K-means, DL2, AVC and CDC.

Average Performance Metric	K-means	DL2i	AVC	CDC
Adjusted Rand Index	0.481	0.788	0.845	0.932
Normalized Mutual Info	0.570	0.839	0.873	0.924
Homogeneity	0.546	0.930	0.928	0.943
Completeness	0.660	0.780	0.840	0.915
V-Measure	0.586	0.847	0.879	0.928
Fowles-Mallows	0.635	0.856	0.895	0.955
Silhouette	0.486	0.532	0.514	0.429
Calinski-Harabasz	55.966	68.794	62.557	45.154
Davies-Bouldin	0.737	0.565	0.606	0.663

As shown in Table 1, the CDC algorithm was superior on average to K-means, DL2, and AVC in 6 of the 9 cluster quality metrics that were applied in the cluster analysis. Namely, ARI, NMI, homogeneity, completeness, V-measure, and

FM were the clustering evaluation methods that better evaluated the models discovered by CDC and proposed that it is first-rated over the other 3 benchmarked algorithms. On the other hand, the coefficients of Silhouette, CH, and DB did not corroborate with the CDC superiority hypothesis, as it was only able to outperform K-means, according to those two measures.

From the results in Table 1, it is also possible to infer that CDC was superior to K-means, DL2, and AVC according to all internal evaluation metrics used. Even though competitive, CDC wasn't first-rated on average for the performance in the described tests according to the external cluster evaluation metrics adopted.

5   Conclusion

This paper revealed that cognitive consistency models, which are useful for understanding human behavior, can be adopted for the data clustering task. A novel algorithm was proposed, which mimics the response of actual human individuals under cognitive discomfort.

Initial results are promising, as shown by quantitative evaluation of the partitioning obtained. Further investigation should consider a wider range of study cases and datasets. Other types of cognitive models could also be adopted in the future.

References

1. Cartwright, D., Harary, F.: Structural balance: a generalization of Heider's theory. Psychol. Rev. **63**(5), 277 (1956)
2. Cohen, S.C., de Castro, L.N.: Data clustering with particle swarms. In: 2006 IEEE International Conference on Evolutionary Computation, pp. 1792–1798. IEEE (2006)
3. Ester, M., Kriegel, H.P., Sander, J., Xu, X.: Density-based algorithm for discovering clusters in large spatial databases with noise. In: KDD-96 Proceedings, pp. 226–231. AAAI Press (1996)
4. Festinger, L.: A Theory of Cognitive Dissonance. vol. 2, Stanford University Press, Redwood (1957)
5. Flament, C.: Applications of Graph Theory to Group Structure. Prentice-Hall, Hoboken (1963)
6. Fritz, H., et al.: The psychology of interpersonal relations (1958)
7. Godois, L.M., Adamatti, D.F., Emmendorfer, L.R.: A multi-agent-based algorithm for data clustering. Prog. Artif. Intell. **9**(4), 305–313 (2020). https://doi.org/10.1007/s13748-020-00213-3
8. MacQueen, J.B.: Some methods for classification and analysis of multivariate observations. In: 5th Berkeley Symposium on Mathematical Statistics and Probability, pp. 281–297. University of California Press, Okland(1967)
9. Mansour, E.M., Ahmadi, A.: A novel clustering algorithm based on fully-informed particle swarm. In: 2019 IEEE Congress on Evolutionary Computation (CEC), pp. 713–720 (2019)

10. Pedregosa, F., et al.: Scikit-learn: machine learning in python. J. Mach. Learn. Res. **12**, 2825–2830 (2011)
11. Xu, R., Wunsch, D.C.: Clustering. Wiley-IEEE Press, Piscataway (2009)
12. Yang, L., Li, K., Zhang, W., Ke, Z., Xiao, K., Du, Z.: An improved chaotic ACO clustering algorithm. In: 2018 IEEE 20th International Conference on High Performance Computing and Communications; IEEE 16th International Conference on Smart City; IEEE 4th International Conference on Data Science and Systems (HPCC/SmartCity/DSS), pp. 1642–1649 (2018)
13. Zhu, W., Luo, W., Ni, L., Lu, N.: Swarm clustering algorithm: let the particles fly for a while. In: 2018 IEEE Symposium Series on Computational Intelligence (SSCI), pp. 1242–1249 (2018)

Interactive Process Drift Detection Framework

Denise Maria Vecino Sato[1,2](✉) ⓘ, Jean Paul Barddal[1] ⓘ,
and Edson Emilio Scalabrin[1] ⓘ

[1] Pontifical Catholic University of Paraná (PUCPR),
Imac. Conceição. 1155, 80215-901 Curitiba, Brazil
{denise.sato,jean.barddal,scalabrin}@ppgia.pucpr.br
[2] Federal Institute of Paraná (IFPR), João Negrão. 1285, 80230-150 Curitiba, Brazil
denise.sato@ifpr.edu.br

Abstract. This paper presents a novel tool for detecting drifts in process models. The tool targets the challenge of defining the better parameter configuration for detecting drifts by providing an interactive user interface. Using this interface, the user can quickly change the parameters and verify how the process evolved. The process evolution is presented in a timeline of process models, simulating a "replay" of models over time. One instantiation of the framework was implemented using a fixed-size sliding window, discovering process maps using directly-follows graphs (DFGs), and calculating nodes and edges similarities. This instantiation was evaluated using a benchmarking dataset of simple and complex drift patterns. The tool correctly detected 17 from the 18 change patterns, thus confirming its potential when an adequate window size is set. The user interface shows that replaying the process models provides a visual understanding of the changing process. The concept drift is explained by the similarity metrics' differences, thus allowing drift localization.

Keywords: Process drift · Concept drift · Drift detection · Evolving environment

1 Process Mining in Evolving Environments

Process Mining (PM) is gathering more enthusiasts in recent years. The growing interest can be explained by the current availability of process data recorded by the informatics systems (big data) and the increasing development of tools to provide easy access to different PM techniques. The primary input of any PM technique is event data, which contains information about business process executions, and can be accessed in the form of event logs (historical information about the process) or event streams (continuous flow of events associated with processing instances). The event data must include at least an identifier of the process instance (case), the event (indicating the occurrence of activity), and the timestamp in which the event occurred.

Supported by Coordenação de Aperfeiçoamento de Pessoal de Nível Superior-Brasil (CAPES)-Finance Code 001, Grant No.: 88887.321450/2019-00.

There are three types of PM: discovery, conformance checking, and enhancement [1]. Discovery aims at learning what is happening with the processes by automatically discovering process models from the event logs without any *a priori* knowledge. Conformance checking compares a discovered or designed process model to the event log to pinpoint possible deviations. The goal is to help business analysts to understand problems or even unexpected behavior to support improvements. In the enhancement, an existent process model is extended (including new perspectives, e.g., performance) or improved using information from the event logs. Business analysts can then change the process model to reflect better reality based on the information provided by discovering or conformance. Discovery, conformance checking, and enhancement are usually applied offline by analyzing an event log, but PM techniques can also perform online analysis.

Regardless of the type of analysis, the business processes are not static. Companies and their analysts are always trying to improve the processes to minimize costs or maximize customer satisfaction. The processes also change when there is a new regulation or in case of unexpected events. Recently, the whole world had to adapt roughly every process to address the COVID-19 pandemic situation. So, it is naïve to assume that models do not change when business processes are placed in evolving environments. Besides, most state-of-the-art PM techniques consider the processes to be in a "steady-state", i.e., thus assuming that an event log contains information about a single version of the process. This assumption does not consider the existence of process drifts.

A process drift, or concept drift in processes, occurs when the business process changes while being analyzed [7]. The correct identification of the process drifts is critical when conducting process analysis in evolving environments. By detecting process drifts, the analysts should better understand the actual process model without facing a mixed model from different versions of the process. Process drift detection can improve reporting and diagnosis analysis, predictions, recommendations, and operational support by assuring that a more adherent version of the process model is considered. In online analysis, the process drift detection can help discovery or conformance checking techniques to maintain an up-to-dated process model, without the need to continually rediscover it.

Process drift was one of the challenges cited in the Process Mining Manifesto [7]. Since 2011, researchers have developed different process drift detection tools, such as ProM ConceptDrift plugin [5,6,10], Apromore ProDrift [8,9,12–14], and other experimental tools [3,11,15,16,18–21]. However, specific issues make it difficult to compare the tools affecting the adoption in real scenarios.

We identified the following experimental tools for process drift detection: Process Drift Detector Plugin (PDD) for ProM [5,6,10], Tsinghua Process Concept Drift Detection (TPCDD) [21], Concept-Drift in Event Stream Framework (CDESF) [3,11], Visual Drift Detection (VDD) [18,19], Dynamic Outlier Aggregation (DOA) [20], Online Trace Ordering for Structural Overviews (OTOSO) [15]. These tools share the drawback of ProM and Apromore, i.e., drift detection accuracy is highly affected by the hyperparameter configuration.

Some of them are essentially affected by the window size: TPCDD, CDESF (the window size is defined by the time-horizon parameter, controlling model updates), and VDD. The approaches based on clustering, i.e., TPCDD, CDESF, VDD, DOA, and OTOSO, are also sensitive to the clustering hyperparameters. The PDD applies an adaptive window approach aiming to solve the window size choice, but the adaptation is limited to user-given upper and lower bounds.

This paper proposes the Interactive Process Drift Detection (IPDD) Framework to provide a practical tool for process drift detection that can be easily applied in real scenarios. Section 2 describes process drifts and explores the available tools. In Sect. 3, the new framework is described. Section 4 presents results obtained with an instantiation of the proposed framework. Finally, Sect. 5 concludes the paper and indicates next steps for IPDD enhancement.

2   Process Drift

A process drift indicates a point in time where the process changes for a reason. The changes can be planned and documented or unexpected. A change in the process reflects a change in its process model, meaning a new one replaces the existing process model's version. The new process model can affect the ongoing process instances in different ways: sudden or gradual [5,6].

In a sudden drift, all the ongoing process instances start to emanate from the new process model when the drift occurs. It can occur when a new regulation should be followed or even within an epidemic situation. In a gradual drift, the new process starts emanating process instances, but both versions coexist for some time, indicating a gradual replacement of the model. The authors in [5,6] also describe two different dynamics for the changes: recurring and incremental. In the recurring drift, the current model is replaced by a new one, but the new model is replaced by the previous one after some time. The incremental drift represents minor incremental changes implemented during some time. The recurring drift can indicate seasonal changes, for instance. An incremental drift can occur in companies to minimize risks for significant changes. In each process model transition, we can have a sudden or a gradual drift.

Process drifts can occur at different time granularities. For instance, we can have a recurring drift that occurs every season, e.g., summer, winter, fall, and spring; and another drift occurring at the last week of the month, e.g., for specific accounting tasks. Both drifts coexist, yet they occur at different time granularities. In [10], the authors named this situation as multi-order dynamics, reinforcing that any drift detection mechanism should deal with different time granularities when identifying process drifts.

A process drift can affect one or more perspectives of the model, which are partially overlapping. The most common perspectives described in [1] are:

1. **Control-flow.** It represents the process model's behavior based on the structure of activities (sequential, parallel, choice, loops).
2. **Organizational.** Resources related to the process activities, which can be people, systems, departments, or others.

3. **Data.** It is related to the information relevant to the process associated with the case, e.g., supplier, or to a specific activity, e.g., a machine.
4. **Time.** Time and frequency of activities.

Tools and methods form handling process drifts should consider the types and perspectives of change and can also address different problems [5,6]:

1. **Change point detection.** Identify the point in time (timestamp) that the drift occurred. The change point can be reported by the case index, the event index, or the date/time where the change starts. Change point detection is the most common problem addressed by the available tools.
2. **Change localization.** Report the process model region which has changed, e.g., between activities A and B. This problem partially overlaps with "unravel process evolution", however, ProM and Apromore addressed this challenge without providing the process evolution. This problem is more about local changes and not the global picture of the drift.
3. **Change characterization.** Specify the type of change and in which perspective it occurred. This problem is less explored because few methods detect different process drifts or consider different perspectives.
4. **Unravel process evolution.** Relate and explore the former discoveries, putting everything together to understand the process's evolution over time. We did not identify any tool that explored this problem.

We focus on problems 1, 2, and 4. We propose IPDD, a framework for detecting change points and visually presenting its localization and process model evolution. IPDD deals with sudden drifts in the control-flow perspective. Incremental, recurring drifts, or multi-order dynamics are addressed by the user interface, allowing to check the detected drifts with different parameters.

2.1   Process Drift Detection Tools

Process drift detection tools can be divided into two categories: academic and experimental tools. We did not find any commercial tool with a process drift detection mechanism. We only report the tools that provide the source code or an executable interface due to space limitations.

The first identified tool for process drift detection is the Concept Drift plugin in ProM[1]. Different approaches have been implemented in this plugin [5,6,10] to address change point detection for sudden and gradual drifts and multi-order dynamics. The user can also search for change localization by selecting a pair of activities; in this case, the tool checks if there is a change between the selected activities. The user has to select many options for using the plugin: log configuration (join logs, split the log or not), feature to be compared (global or local), parameters of the feature, window strategy (fixed or adaptive, sliding over traces or time periods) with parameters, type of drift to detect (gradual or sudden),

---

[1] ProM is an open source framework that provides a big set of tools for the discovery and analysis of process models from event logs: http://www.processmining.org.

and the statistical test applied along its parameters. The user is required to set many configurations before even know if there is potential drift in the event log. Furthermore, the two types of drifts detected (sudden and gradual) must be separately checked. The plugin is designed for offline analysis, and it is not working if any global feature is selected, thus raising an exception. The results are highly sensitive to the parameters chosen, and there is no user-friendly interface to compare the results from different configurations. It is not possible to obtain the accuracy of the method, e.g., *F-score*, for synthetic logs in the plugin's interface.

Apromore[2] is another academic tool that provides a plugin for concept drift detection reporting change points, change characterization, and localization (ProDrift). ProDrift has different approaches implemented [8,9,12–14] and can detect drifts from a stream of traces (based on runs, which is an abstraction for the traces) or event streams. The tool can detect sudden and gradual drifts from event streams in an integrated way, i.e., the user does not have to specify the type of drift to be detected. ProDrift also has different types of parameters: approach (runs or events), windowing strategy (fixed or adaptive), and window size. The event-based approach includes a noise filter threshold, drift detection sensitivity, characterization (on/off), characterization method (activity-based or fragment-based), and characterization noise filter threshold. The accuracy of the detection method is highly sensitive to the chosen parameter values. An advantage of Apromore is that it has default values for each parameter. Yet, a drawback is that it is not possible to quickly check the differences in the process model before and after a detected change point. The adaptive window approach uses an initial window size value as input, but the plugin applies an algorithm to initialize it if the user does not specify. However, this initial size definition is not explained in the papers supporting the implementation [8,9,12–14].

The identified experimental tools for process drift detection are: Process Drift Detector Plugin for ProM [16], Tsinghua Process Concept Drift Detection (TPCDD) [21], Concept-Drift in Event Stream Framework (CDESF) [3,11], Visual Drift Detection (VDD) [18,19], Dynamic Outlier Aggregation (DOA) [20], Online Trace Ordering for Structural Overviews (OTOSO) [15]. These tools share the drawback of ProM and Apromore, i.e., drift detection accuracy is highly affected by the hyperparameter configuration. Some of them are essentially affected by the window size: TPCDD, CDESF (the window size is defined by the time-horizon parameter, controlling model updates), and VDD. The approaches based on clustering, i.e., TPCDD, CDESF, VDD, DOA, and OTOSO, are also sensitive to the clustering hyperparameters. The Process Drift Detector Plugin for ProM applies an adaptive window approach aiming to solve the window size choice, but the adaptation is limited to user-given upper and lower bounds.

The remaining issue for the available tools is hyperparameter setup. The choice of the window size is critical in any drift detection approach because a small window size may lead to false positives, and a large one may lead to false

---

[2] Apromore is a collaborative business process analytics platform with distinct editions. The ProDrift is an experimental plugin: https://apromore.org/platform/tools/.

negatives making it challenging to pinpoint the exact location of the drift [9]. Hyperparameter tuning is essential for providing a tool for detecting drifts in real-world scenarios. The current tools do not provide accuracy metrics neither reveal process evolution, turning the hyperparameter tuning an arduous task. By default, the tools focus on reporting the change points and leaving the user to understand the process model's change by splitting the log and applying discovery techniques in all the resulted sub-logs. Our proposal (IPDD) fills this gap by providing an interactive process drift detection tool, easy to use, and a reduced number of parameters. The interface allows the user to quickly verify the detection results by visualizing the process models over time.

3   Interactive Process Drift Detection Framework

The proposed framework (IPDD) aims at overcoming the reported issues of the available tools: a not so user-friendly interface, the difficulty in comparing results obtained by different parameter configurations (not allowing hyperparameter tuning), complex configuration, and not reporting the accuracy metrics in a common manner. Figure 1 shows an overview of the proposed framework.

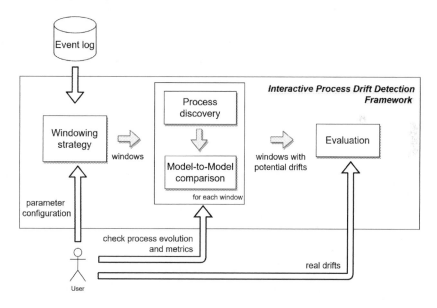

Fig. 1. Overview of interactive process drift detection framework.

The definition of the parameters' values is challenging when starting a process drift analysis because the user has to determine *a priori* what will be a "good" parameter for each situation. We propose to solve this issue by providing an interactive and easy user interface, leaving the user free to test different parameters quickly. The commercial tools for process discovery (e.g., Disco) inspired this format, as they usually present a simplified version of the process model in which the user can navigate by zooming in or out to understand the process. The goal is to allow the user to navigate between different granularities of change and inspect the process evolution. The hyperparameter tuning is allowed by the user interface, which provides the process evolution for each tested parameter configuration. The user can also verify the chosen parameters' accuracy by checking the evaluation module.

3.1 Windowing Strategy

An event log in XES format[3] can be uploaded into the IPDD. Next, by applying a windowing strategy, the events are split into separated windows. The window strategy can consider the event log as time series of traces, i.e., a stream of traces, by ordering the traces by the first event's timestamp. The windows can also be defined based on events, sorted by their timestamp, i.e., time series of events. The size of the window can be fixed or adaptive. Using fixed-size windows, the user must specify the size as the number of traces or event or as a time window, e.g., hours, days. The windows can be overlapping or non-overlapping, continuous, or non-continuous. All of the identified options can be implemented in the windowing strategy step of the framework. The window slots containing a set of traces/events are forward to the next step, named *process discovery*.

3.2 Process Discovery

For each window slot provided in the first step, IPDD applies a process discovery algorithm to derive a process model. Several algorithms have been proposed for process discovery, and any discovery technique has its own representational bias, i.e., the process model that can be discovered [1]. The resulted process models can be declarative or imperative, and several notations are available [1]. Another option is to use a DFG, also named process map, for simplicity and scalability. This option simplifies the mined process models but still shows relevant insights about the process paths with metrics.

The discovery algorithm's choice is related to the resulted process model, its ability to filter noise, its performance, etc. In the *process discovery* step, any implemented discovery algorithm can be called, like a black-box. The derived models are forward to the next step. An advantage of using the process model is that the framework can quickly show the process changes over time to the user.

3.3 Model-to-Model Comparison

In the previous step, IPDD derives a process model for each window slot. Comparing models from adjacent windows allow identifying differences. Any difference can be a potential drift, and the type of change that can be detected is related to the metric chosen for this comparison. There are several metrics for

---

[3] See www.xes-standard.org for detailed information about the standard.

comparing process models [4], and they are related to the notation of the process model mined. In this step, IPDD calculates the implemented similarity metrics between adjacent models.

The metric should be adherent to the process model derived by the chosen discovery algorithm. There is no limitation about the number of metrics that can be included, and the user can select one or more of the available metrics. IPDD implements a timeout mechanism for avoiding freezing the user interface when calculating the metrics. If the timeout is reached, only the finished metric results are shown. The result of this stage is reporting each metric's value. If a metric has a value bounded in the [0, 1] interval, with one indicating that the two models are similar, a potential drift can be considered when this metric value is below one. Any metric with a value indicating dissimilarity triggers an update in the user interface, marking the window as a potential drift.

We choose to use a model-to-model comparison instead of trace comparison scheme, e.g., applying statistical or clustering approaches, because we can identify drifts that affect the process model. The chosen process model notation can provide the detection of different drifts. For instance, if we choose to generate a model with the frequencies annotated in the path between activities, IPDD can implement a metric for comparing these frequencies. In this example, IPDD can handle control-flow drifts that do not affect the structure of the process model but affects the routing of cases. It is also possible to localize the drift in the process model by checking the similarity metrics' differences.

3.4   Evaluation

IPDD receives the real drift position as trace/event indexes or date/time values. It can then calculate an accuracy metric to measure if the windows reported as potential drifts include the real drifts. We identified two metrics for measuring the accuracy of concept drift in process models: *F-score* and *mean delay*, reported in [8–10, 12–14, 16, 21]. IPDD reports a drift by indicating the window that renders a model dissimilar to the one obtained in the previous window, so the *F-score* can be applied to evaluate the accuracy of the detected drifts.

The *F-score* represents the harmonic mean of recall and precision, calculated based on the true positives (TP), false positives (FP), and false negatives (FN). It is critical to define that a TP should consider an interval of indexes because the detection mechanism cannot detect the drift by the time it has occurred. In other words, if the real drift initiates in the i-th trace, a TP occurs when the detection method reports a drift in the interval $[i, i + \text{et}]$, where et indicates an error tolerance and should be configured. The FPs and FNs should also be consistent with the definition of the TP. The *mean delay* represents the distance between the occurrence of the real drift and the drift flagged. This distance relates to the windowing strategy adopted. For instance, if the instantiation used a window over traces, this distance is the difference between the trace index of the real drift and the detected one.

4   Results

We implemented a prototype to validate the IPDD and its user interface[4]. This prototype is an instantiation of the framework that encompasses the following:

1. **Windowing strategy.** Non-overlapping and continuous windows of traces (ordered as time series, based on the first activity's timestamp). The windows have a fixed size of traces defined by the user.
2. **Process discovery.** We applied the Pm4Py[5] framework to discover the DFG, with the frequencies of activities and paths.
3. **Model-to-model comparison.** We calculated the node similarity (NS) and edge similarity (ES) scores between two consecutive process maps (P and Q). NS is calculated using Eq. 1 [2], where n_p and n_q are the number of activities in process maps P and Q, respectively, and n_{cs} indicates the number of common activities between P and Q. ES is calculated using Eq. 2, which is similar to NS, however using: e_p is the number of edges in P, e_q is the number of edges in Q, and e_{cs} indicates the number of common edges in both P and Q.

$$NS = 2 * n_{cs}/(n_p + n_q) \tag{1}$$

$$ES = 2 * e_{cs}/(e_p + e_q) \tag{2}$$

The prototype calculates both metrics, and if one or both is less than zero, IPDD marks the current window as a drift.
4. **Evaluation.** There is no evaluation metric already implemented, but we have the *F-score* metric defined to measure the detected drifts' accuracy. Because of the window strategy choice, a TP represents a window reported as a drift containing a trace inputted as a real drift. An FP should be counted when a window reporting a drift does not contain any of the traces inputted as real drifts. Finally, an FN should be incremented when a window that does not report a drift contains any traces inputted as real drifts.

Figure 2 shows a snapshot of the prototype. After setting the window size, the prototype mine the models and calculate the similarity metrics between adjacent models (NS and ES). The window with a similarity metric value below one is marked in red, indicating a drift. The user can check the process mined for this window and verify the metric value and the differences. In Fig. 2 (displayed in the lower-left corner), the edges from activity "Assess eligibility" to "Send acceptance pack" and to "Send home insurance" are included, indicating what has changed in the model. In this example, the activities "Prepare acceptance pack" and "Check if home insurance quote is requested" are optional after the drift; the new edges allow skipping both activities after performing "Assess eligibility".

To validate the tool's usage, we apply the drift detection in some of the logs publicized in [9]. The dataset contains 72 logs with different change patterns

---

[4] Available at https://github.com/denisesato/InteractiveProcessDriftDetectionFW.
[5] PM4Py is a python open source PM platform: https://pm4py.fit.fraunhofer.de/.

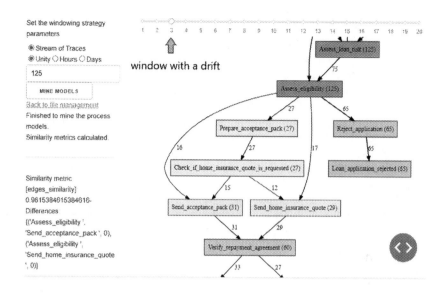

Fig. 2. Snapshot of the prototype implementation.

Table 1. Experiments description.

Change pattern	Category	Drifts detected?	Metric that detects the drifts
cb: make fragment skippable/non-skippable	O	Yes	ES
cd: synchronize two fragments	R	Yes	ES
cf: make two fragments conditional/sequential	R	Yes	ES
cm: move fragment into/out of the conditional branch	I	Yes	ES
cp: duplicate fragment	I	Yes	ES
fr: change branching frequency	O	No	-
lp: make two fragments loopable/not-loopable	O	Yes	ES
pl: make two fragments parallel/sequential	R	Yes	ES
pm: move fragment into/out of parallel branch	I	Yes	ES
re: add/remove fragment	I	Yes	NS and ES
rp: substitute fragment	I	Yes	NS
sw: swap two fragments	I	Yes	ES
IOR	IOR	Yes	NS and ES
IRO	IRO	Yes	NS and ES
OIR	OIR	Yes	NS and ES
ORI	ORI	Yes	NS and ES
RIO	RIO	Yes	ES
ROI	ROI	Yes	NS

and inter-drift distances (distance between each injected drift). The authors use a business process for assessing loan applications containing 15 activities and different control-flow structures as the base model. Next, they injected different types of control-flow changes, simulating sudden recurring drifts (9 drifts in each log). The base model was changed using 12 simple change patterns (described in [17]) and 6 complex patterns (the composition of 3 simple patterns). Each change pattern is injected using 4 inter-drift distances (250, 500, 750, and 1,000). We select all change patterns with the inter-distance of 500 to validate IPDD (Table 1). For the change patterns lp and re, we used the files named 2.5k as these files contain inter-drift distance of 500. Each pattern is categorized as Insertion (I), Resequentialization (R), and Optionalization (O). For the complex pattern, the authors randomly apply one pattern from each category in a nested way.

Table 1 shows that the implemented instantiation for IPDD correctly detects drifts for 17 out of the 18 patterns using NS and ES. The detection is possible when configuring a window size equal to the inter-drift distance (500). The fr pattern is not detectable because there is no structural difference between the models; the drift only changes the frequencies in one branch of the process. The choice of the window size is still an issue. If the window size is configured with a value higher than the inter-drift distance, the drift will not be detected.

5   Conclusion

IPDD has been validated by a prototype implementation that demonstrates its use for detecting sudden drifts in the control-flow perspective. The main contribution of the novel approach is the interactive user interface, which provides the user with a tool for quickly checking different values for window sizes. The drift can be visually checked by analyzing the process models from adjacent windows and the metrics' value describing the detected differences. The implemented metrics (NS and ES) can detect 17 (from 18) change patterns in the public dataset. The window size choice is still a challenge, but the interactive user interface provides an easy way of testing different values. We plan to extend IPDD by including new instantiations implementing the defined evaluation metric (F-$score$) and a similarity metric related to the frequencies between activities.

References

1. van der Aalst, W.M.: Process Mining: Data Science in Action. Springer, Heidelberg (2016). https://doi.org/10.1007/978-3-662-49851-4
2. Akkiraju, R., Ivan, A.: Discovering business process similarities: an empirical study with SAP best practice business processes. In: Maglio, P.P., Weske, M., Yang, J., Fantinato, M. (eds.) ICSOC 2010. LNCS, vol. 6470, pp. 515–526. Springer, Heidelberg (2010). https://doi.org/10.1007/978-3-642-17358-5_35
3. Barbon Junior, S., Tavares, G.M., da Costa, V.G.T., Ceravolo, P., Damiani, E.: A framework for human-in-the-loop monitoring of concept-drift detection in event log stream. In: WWW 2018: Companion Proceedings of the The Web Conference 2018, vol. 2, pp. 319–326. Association for Computing Machinery (ACM) (2018)

4. Becker, M., Laue, R.: A comparative survey of business process similarity measures. Comput. Ind. **63**(2), 148–167 (2012)
5. Bose, R.P.J.C., van der Aalst, W.M., Žliobaite, I., Pechenizkiy, M.: Dealing with concept drifts in process mining. IEEE Trans. Neural Netw. Learn. Syst. **25**(1), 154–171 (2014)
6. Bose, R.P.J.C., van der Aalst, W.M.P., Žliobaitė, I., Pechenizkiy, M.: Handling concept drift in process mining. In: Mouratidis, H., Rolland, C. (eds.) CAiSE 2011. LNCS, vol. 6741, pp. 391–405. Springer, Heidelberg (2011). https://doi.org/10.1007/978-3-642-21640-4_30
7. van der Aalst, W., et al.: Process mining manifesto. In: Daniel, F., Barkaoui, K., Dustdar, S. (eds.) BPM 2011. LNBIP, vol. 99, pp. 169–194. Springer, Heidelberg (2012). https://doi.org/10.1007/978-3-642-28108-2_19
8. Maaradji, A., Dumas, M., Rosa, M.L., Ostovar, A.: Detecting sudden and gradual drifts in business processes from execution traces. IEEE Trans. Knowl. Data Eng. **29**(10), 2140–2154 (2017)
9. Maaradji, A., Dumas, M., La Rosa, M., Ostovar, A.: Fast and accurate business process drift detection. In: Motahari-Nezhad, H.R., Recker, J., Weidlich, M. (eds.) BPM 2015. LNCS, vol. 9253, pp. 406–422. Springer, Cham (2015). https://doi.org/10.1007/978-3-319-23063-4_27
10. Martjushev, J., Bose, R.P.J.C., van der Aalst, W.M.P.: Change point detection and dealing with gradual and multi-order dynamics in process mining. In: International Conference on Business Informatics Research, pp. 1–15 (2015)
11. Mora, D., Ceravolo, P., Damiani, E., Tavares, G.M.: The CDESF toolkit: an introduction. In: ICPM Doctoral Consortium and Tool Demonstration Track 2020, vol. 2703, pp. 47–50 (2020). CEUR-WS.org
12. Ostovar, A., Leemans, S.J.J., Rosa, M.L.: Robust drift characterization from event streams of business processes. ACM Trans. Knowl. Discovery from Data **14**(3), 1–57 (2020)
13. Ostovar, A., Maaradji, A., La Rosa, M., ter Hofstede, A.H.M.: Characterizing drift from event streams of business processes. In: Dubois, E., Pohl, K. (eds.) CAiSE 2017. LNCS, vol. 10253, pp. 210–228. Springer, Cham (2017). https://doi.org/10.1007/978-3-319-59536-8_14
14. Ostovar, A., Maaradji, A., La Rosa, M., ter Hofstede, A.H.M., van Dongen, B.F.V.: Detecting drift from event streams of unpredictable business processes. In: Comyn-Wattiau, I., Tanaka, K., Song, I.-Y., Yamamoto, S., Saeki, M. (eds.) ER 2016. LNCS, vol. 9974, pp. 330–346. Springer, Cham (2016). https://doi.org/10.1007/978-3-319-46397-1_26
15. Richter, F., Maldonado, A., Zellner, L., Seidl, T.: OTOSO: online trace ordering for structural overviews. In: Leemans, S., Leopold, H. (eds.) ICPM 2020. LNBIP, vol. 406, pp. 218–229. Springer, Cham (2021). https://doi.org/10.1007/978-3-030-72693-5_17
16. Seeliger, A., Nolle, T., Mühlhäuser, M.: Detecting concept drift in processes using graph metrics on process graphs. In: Proceedings of the 9th Conference on Subject-oriented Business Process Management, S-BPM ONE 2017, vol. Part F1271 (2017)
17. Weber, B., Reichert, M., Rinderle-Ma, S.: Change patterns and change support features - enhancing flexibility in process-aware information systems. Data Knowl. Eng. **66**(3), 438–466 (2008). ISSN 0169023X
18. Yeshchenko, A., Di Ciccio, C., Mendling, J., Polyvyanyy, A.: Comprehensive process drift detection with visual analytics. In: Laender, A.H.F., Pernici, B., Lim, E.-P., de Oliveira, J.P.M. (eds.) ER 2019. LNCS, vol. 11788, pp. 119–135. Springer, Cham (2019). https://doi.org/10.1007/978-3-030-33223-5_11

19. Yeshchenko, A., Mendling, J., Ciccio, C.D., Polyvyanyy, A.: VDD: a visual drift detection system for process mining. In: ICPM Doctoral Consortium and Tool Demonstration Track 2020 (2020). CEUR-WS.org
20. Zellner, L., Richter, F., Sontheim, J., Maldonado, A., Seidl, T.: Concept drift detection on streaming data with dynamic outlier aggregation. In: Leemans, S., Leopold, H. (eds.) ICPM 2020. LNBIP, vol. 406, pp. 206–217. Springer, Cham (2021). https://doi.org/10.1007/978-3-030-72693-5_16
21. Zheng, C., Wen, L., Wang, J.: Detecting process concept drifts from event logs. In: Panetto, H., et al. (eds.) OTM 2017. LNCS, vol. 10573, pp. 524–542. Springer, Cham (2017). https://doi.org/10.1007/978-3-319-69462-7_33

Mining of High-Utility Patterns in Big IoT Databases

Jimmy Ming-Tai Wu[1], Gautam Srivastava[2(✉)], Jerry Chun-Wei Lin[3], Youcef Djenouri[4], Min Wei[1], and Dawid Polap[5]

[1] Shandong University of Science and Technology, Qingdao, China
`wmt@wmt35.idv.tw`
[2] Department of Math and Computer Science, Brandon University, Brandon, Canada
`SRIVASTAVAG@brandonu.ca`
[3] Western Norway University of Applied Sciences, Bergen, Norway
`jerrylin@ieee.org`
[4] SINTEF, Trondheim, Norway
`Youcef.Djenouri@sintef.no`
[5] Faculty of Applied Mathematics, Silesian University of Technology, Kaszubska 23, 44-100 Gliwice, Poland
`Dawid.Polap@polsl.pl`

Abstract. In general data mining, HUIM also known as high-utility itemset mining is an offshoot of frequent item set mining (FIM). HUIM is known to give more emphasis to many factors which can give HUIM a distinct edge over FIM. PHIUM, or Potential high-utility item set mining has been created to give intrinsic patterns in databases that tend to be uncertain. Despite most previous methods being highly effective and powerful miners, PHUIM needs to work fast. Most current mining techniques do not handle databases with extremely large number of records when performing HUIM. In this paper, we make the assumption that the dataset is bigger than a direct load into RAM could handle. Furthermore, the dataset is not of the size where modification or duplication is possible, and as such a MapReduce framework is created that can be used to handle datasets that fall into these categories. One of the main objectives of this research is to be able to reduce the frequency of database scans while simultaneously maximizing parallel processing. Using experimental analysis, our Hadoop based algorithm performs well to mine high utility itemsets from big databases.

Keywords: IoT data analytics · Utility patterns · Data mining

1 Introduction

The essential goal of KDD (knowledge discovery from databases) is to have the option to infer an important and helpful message from enormous assortments of information. Association rule mining, known as ARM, just as FIM, or frequent itemset mining [1] have become key parts for KDD that, when appropriately applied to different situations and applications, are successful apparatuses.

© Springer Nature Switzerland AG 2021
L. Rutkowski et al. (Eds.): ICAISC 2021, LNAI 12855, pp. 205–216, 2021.
https://doi.org/10.1007/978-3-030-87897-9_19

Various varieties of ARM-based or FIM-based calculations have been generally evolved to recover the necessary information (i.e., Association rules (ARs) or frequentitemsets (FIs)) from an exact data set, and most algorithms were based on level-wise/pattern-growth approaches.

High utility mining (HUIM) [2] uses two components: unit profit and quantity to decide "important" patterns from the data set. We can call an itemset a high-utility itemset (HUI) when the utility worth of an itemset is more noteworthy than a pre-characterized minimum utility count. Accordingly, it shows the most beneficial itemsets to the retailer. The crucial HUIM idea was presented by Chen [2], which brought together a system of HUIM later characterized by Yao et al. [3]. Be that as it may, the nonexclusive HUIM model does not give the downward closure (DC) property for proficiently mining the set of HUIs; in this manner, the calculation is exorbitant to track down the necessary data in a tremendous massive space.

There can be uncertainty created when data is gathers from sources synonymous with noise like wireless sensors, GPS, RFID, etc. Realitcally, any real-world data source may introduce noise. State of the art pattern-mining approaches like ARM or FIM are not able to be used to find the needed knowledge for making decisions while simultaneously being able to handle missing and unreliable data. There have been many approaches that attempt to handle these uncertain data mining applications like (CUFP)-tree [7], (UFP)-growth algorithm [6], and UApriori [4].

UApriori is known to be one of the first algorithms that adopted a level-wise approach into the concept of uncertainty using a generate-and-test approach as well as the breadth-first search method to create required knowledge. UFP-growth was the designed using FP-growth and the UFP-tree structure to discover rules. Furthermore, CUFP-tree gives a data structure that is compressed that keeps necessary information using in the mining processes. For handling big data, Lin et al. made an algorithm that is able to reveal HUIs in uncertain databases [8]. In [11], Zhang furthered this concept making HUI-probability sequential patterns from uncertain databases. In [13], Ahmed used utility and uncertainty as the 2 factors used for mining HUI patterns also leaning on evolutionary computation. Srivastava et al. [10] developed a method high in efficiency to mine HUI-sequential patterns from uncertain databases. To date, there are many more algorithms still in development that can be used in uncertain databases for knowledge discovery in uncertain situations [5].

In real life Internet of Things applications, data is often absent or partially present while also being joined to an existential probability, especially information gathered from measurements or sensors. In general, the utility of data usually reflects the end-users rank of its usefulness. In uncertain databases, even the implementations with high practicality can consider utility. In any market basket analysis the customer transactions can often be less than precise. For example, several items may be linked to sensors, RFID, or smart device information gathering [5]. As an example, take { W:3 X:2 Y:5 Z:3, 85%} which shows that the items (W X Y Z) are bought at the same time (same customer) in the quantities $(3, 2, 5, 3)$ in a single transaction with an existence probability of 85%. The risk in an avent can as well be regarded as a probability of an event in the area of risk prediction. In {X:2, Y:4, Z:3, 90%}, shows that the three objects {X Y Z} that have an occurrence frequency of {2, 4, 3}, respectively have an existence probability = 90%. Taking into account the utility factor, we can consider many factors for real situations like weight, interestingness, importance, and others so as to show that the utility concept can easily be amalgamated with lots of domains in pretty much any scenario.

Our main contributions are:

1. Propose EUHUP that is based on EFIM for revealing PHUIs while keeping a close performance to EFIM.
2. Proposed HUHUP by extending EUHUP to handle big data.
3. Both EUHUP as well as HUHUP can use a similar pruning strategy to EFIM.
4. HUHUP attempts at the reduction of dataset scans.

2 Problem Statement

The goal of PHUIM is to reveal all of the PHUIs in an uncertain database. The formal problem statement of PHUI is described below.

> **Input:** an uncertain (transaction) database, D,
> a unit profit table for each item, p,
> a predefined minimum utility threshold, ε,
> a predefined minimum potential probability threshold, μ
>
> **Output:** a set of PHUIs in D

The input data that is needed to solve the PHUIM issues usually includes some uncertain database, some threshold for minimum utility, a unti profit table, and another threshold for minimum potential probability. The utility information is contained in the uncertain database for each and every record and there is also probability information for each item in a given record. We use the unit profit table for indicating the unit utility profit of every item. Calculations can be done further on the transaction values utility or the itemset utility. The thresholds that are predefined for minimum potential probability and minimum utility can be set by users or through the application requirements.

2.1 Downward Closure Property

We discuss the downward closure properties in this section. It is often used in PHUI mining algorithms. Based on the downward closure property, PHUI mining algorithms can limit the searching space scale to assist in the derivation of PHUIs.

Theorem 1 (*Downward Closure Property of High Probability Itemsets*). *If the itemset X is not an HPI in the uncertain database D, all of the supersets of X is not an HPI. Thus, if X is an HPI, all of the subsets of X is HPI.*

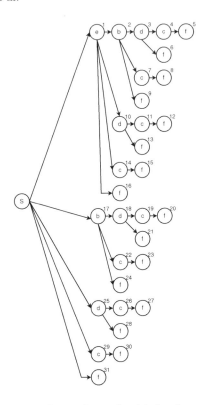

Fig. 1. A search graph with five items.

By applying the downward closure property, the searching process can be designed effectively to generate the candidates of HPIs and reduce the size of the searching space. The downward closure property using in high-utility itemset mining is described below.

Definition 1 (*Transaction-Weighted Utility, TWU*). *The transaction-weighted utility of the itemset X in the dataset D is defined as:*

$$TWU(X) = \sum_{X \subseteq T \in D} tu(T) \tag{1}$$

Definition 2 (*High Transaction-Weighted Utilization Itemset, HTWUI*). *The itemset X is a high transaction-weighted utilization itemset if and only if:*

$$TWU(X) \geq TU \times \varepsilon, \tag{2}$$

where ε is the minimum utility threshold.

Theorem 2 *(Downward Closure Property of High Transaction-Weighted Utilization Itemsets, TWDC). If the itemset X is not an HTWUI in the database D, all of the supersets of X is not an HTWUI. Thus, if X is an HTWUI, all of the subsets of X is HTWUI.*

Obviously, $TWU(X) \geq u(X)$. Therefore, if $TWU(X) < TU \times \varepsilon$, $u(X) < TU \times \varepsilon$. That is to say, if the itemset X is not an HTWUI, it is also not a HUI. Because HUIs do not hold the downward closure property, the downward closure property of HTWUI is used to reveal the HUIs in a database. To find all of the PHUIs in a database, the downward closure property of PHUIs and the related itemset definition are shown below.

Definition 3 (*High Transaction-Weighted Probabilistic and Utilization Itemset, HTWPUI*). *The itemset X is a high transaction-weighted probabilistic and utilization itemset if and only if X is both HTWUI and HPI.*

Theorem 3 (*Downward Closure Property of High Transaction-Weighted Probabilistic and Utilization Itemset, HTWPUI*). *If the itemset X is not a HTWPUI in the uncertain database D, all of the supersets of X is definitely not a HTWPUI. Thus, if X is a HTWPUI, all of the subsets of X is definitely a HTWUI.*

Theorem 4 (*PHUIs $\subseteq$ HTWPUIs*). *According to the definitions of PHUI and HTWPUI, PHUIs is obviously a subset of HTWPUIs. Therefore, if an itemset is not a HTWPUI, it is definitely not a PHUI.*

Due to no direct downward closure property of PHUI and HUI, the previous algorithms applied Theorem 3 and Theorem 4 are used to keep a set of candidate itemset of HTWPUIs and further to reveal PHUIs by HTWPUIs. To put it another way, HTWPUIs is an upper bound of PHUIs and can be used to reduce the searching space in the process of mining PHUIs.

Definition 4 (*k-High Transaction-Weighted Probabilistic and Utilization Itemset, k-HTWPUI*). *Assume the itemset X includes k items and it is also a HTWPUI, it is called as k-HTWPUI.*

In the previous Apriori-based methods, the mining process will generate k-HTWPUIs as candidates and reveal PHUIs in the k-th round. In the proposed efficient uncertain high utility pattern mining (EUHUP) and Hadoop uncertain high utility pattern mining (HUHUP), 1-HTWPUIs are used to establish the searching graph for the mining process. The detailed descriptions are shown in the following section.

3   Proposed Uncertain High Utility Pattern Mining Algorithms

Before performing the proposed algorithms, a search graph is first built that can be used to apply the pruning strategies and the proposed algorithms. The proposed algorithms discover all the PHUIs (potential high-utility itemsets) in a dataset when the proposed algorithms estimate or prune all the nodes in a graph. Based on the developed Theorem 3 and 4, the search graph contains only 1-HTWPUIs in a transactional database. After obtaining the 1-HTWPUIs, the proposed algorithms sort the 1-HTWPUIs (using the value of transaction-weighted utility for each item) by descending order. After that, a routing graph will be generated from this ordered list. Details are then described in Algorithms 1 and 2.

Algorithm 1. Construct searching graph

Input: an ordered item list l.
Output: a searching graph G.

1: set $G = \emptyset$;
2: put the starting point s into G;
3: run BUILDCHILDNODES(G, s, l);
4: **return** G;

Algorithm 1 and Algorithm 2 are a recursive process to establish a searching graph. This recursive process ensures all of the possible PHUIs existed in the generated searching graph. Figure 1 is an example of five items. In this graph, a node (item) represents an item in the database. In Fig. 1, the constructing process will follow the (depth-first) numerical increasing order to build the graph (1, 2, 3, ..., 31).

Algorithm 2. Build child nodes algorithm

1: **function** BUILDCHILDNODES(G, s, l)
2:      **for** each item i in l **do**   ▷ select item i by l order.
3:          $G \leftarrow i$;
4:          generate a directed link from s to i;
5:          **if** item i is not the last one in list l **then**
6:              set a list l_i is a sub list of l after item s;
7:              run BUILDCHILDNODES(G, i, l_i);
8:          **end if**
9:      **end for**
10: **end function**

3.1   The Node's Expression for Itemset

According to the above process, a search graph is generated before EUHUP and HUHUP. Therefore, each node in the search graph indicates a specific itemset which can be estimated whether it is a PHUI. The expression of a node is the traveling log between the starting node to this node. For example, the corresponding itemset of node 23 in Fig. 1 is $\{b, c, f\}$. The detailed searching process and pruning strategies are given below.

4   Experimental Evaluation

This section estimates the performance of the proposed EUHUP, HUHUP, and the previous traditional Apriori algorithm and its MapReduce version (called as Apriori(M)) [9]. The apriori-based algorithms will apply HTWPUIs and its downward closure property. The experiments were conducted in computing nodes equipped with the Intel Core i7-6700 CPU @ 3.40 GHz * 8 and 32 GB assigned RAM, running Linux Ubuntu 16.04 LTS. To achieve parallelization, the experiment was run under the `hadoop2.5.1` cluster with one master node and five data nodes. The number of nodes can be adjusted according to the size of the input data. First, the following session will show the different runtime of these four algorithms in the different sizes of databases. Generally, a single computer algorithm can obtain better performance in a small size dataset and a Hadoop-based framework can handle well in a huge size dataset. Then, to compare these two MapReduce algorithms, the memory usage and runtime will be compared in the different numbers of nodes and the different sizes of databases. These experiments can show the influence of different numbers of nodes.

Table 1. Database information

Dataset	Trans.	Items	Avg. trans. length
BMS	59,601	497	4.8
Mushroom	8,124	119	23.0
Accident	340,183	468	33.8
Connect	67,557	129	43.0
Chess	3,196	75	37.0
Foodmart	4,141	1,559	4.4

4.1 Data Information and Preparation

The databases [12] used in the experiments include many transactions with items, unit utilities, and probabilities. To facilitate calculation, firstly, the experiment initializes the database and take the line number as TID. Each transaction contains four parts: the first part is the items, the second part is the utility of each item, and the third part and four-part are the *twu* and the probability of the transaction. By analyzing these data sets, reveal all of the PHUIs in the input uncertainty database. To analyze the different performance between the three algorithms more comprehensively and show the performance differences between different databases, six standard datasets for evaluating the traditional HUIM algorithms are selected as input data (namely BMS, Mushroom, Accident, Connect, Foodmart, and Chess), and they are shown in Table 1. Besides, pre-defined probability information is also provided to calculate the probability of each itemset. For sure, the above datasets are all not IoT datasets. That is because it needs to spend a lot of resources to establish a real Internet of Things cluster. The experiments, thus, applied these datasets to do some simulations and obtain similar results. The correspondence of the simulation datasets and a real Internet of Things dataset is shown below. A transaction indicates an Internet of Things device in a real Internet of Things network. The items in a transaction indicate the attribute values collected by a sensor in a device. The setting probability of each item means the state or credibility of a sensor. For example, from a real scenario, Internet of Things devices might include several sensors, and these sensors collect different kinds of data separately. Due to the sensor state or communication network situation, each record from Internet of Things data has different reliability (probabilities). It, therefore, can not apply the traditional high-utility mining algorithms directly. Moreover, to verify the advantages and disadvantages of the algorithm in big data, the experiments are divided into three groups, namely small, medium, and large of the conducted datasets. After preparing these databases, upload them to HDFS for the initialization for performing the Hadoop-based HFUPM algorithm.

Runtime. As shown in Fig. 2, the runtime of EUHUP and Apriori is slightly less than that of HUHUP and Apriori (M) by taking the first group of smaller size databases. In this figure, The x-coordinate represents the different databases within the category, and the y-coordinate represents the runtime of the compared algorithms. From the experimental results, it can be seen that Hadoop-based frameworks have no advantage when they come to handling small datasets. It is reasonable as it takes a certain amount of time to start the Hadoop cluster, and also prohibitive is the cost of communication between different nodes. When the first category of databases (small size) is utilized, the size of the six databases for the HUHUP and Aproiro (M) algorithms is not up to the size of a block in HDFS. The process is, therefore, only performed on one node. Nonetheless, the two algorithms will start the entire Hadoop cluster using the master node scheduling the whole operations, but the EUHUP and Apriori will complete in one machine only. Therefore the performance of single-machine algorithms is better than Hadoop-based frameworks when using this set of data. Furthermore, the performance difference between EUHUP and Apriori (M) at low data volumes is not significant since the calculation amount is small.

The next experiments take the second group of medium-size databases, and the results are displayed in Fig. 3. From the analysis, one can see that the runtime of the single machine algorithms is still less than the frameworks based on Hadoop. For this experiment, the size of the input data exceeded the block size, and the parallel computation in the Hadoop-based algorithms was implemented. The input data was split into separate nodes for calculation. Even parallel computation save a lot of computational costs. However, as already stated, when using Hadoop's MapReduce architecture, the cost of communication between different nodes is still very high. Therefore, because EUHUP and Apriori algorithms operate on a single machine, their performance requirements on the computer are beginning to increase as the size of the dataset becomes greater. And the output gap between HUHUP and Apriori (M) is evident as the amount of computation increased. The explanation is that the upper-bounds are efficient in reducing the search space; they can save a lot of time in the process of calculation.

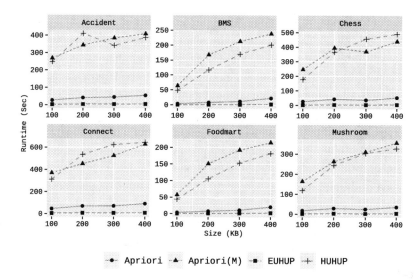

Fig. 2. The runtimes of small databases.

The input data in the third experiment is the large datasets, and the results of this experiment are shown in Fig. 4. The results of this experiment showed how vital the proposed HFUPM is. The runtime of both Hadoop-based algorithms has increased significantly in this experiment, but the runtime of the Apriori (M) algorithm is growing faster. What's more, the single computer can no longer support the EUHUP and Apriori algorithms until the storage size exceeds 1.5

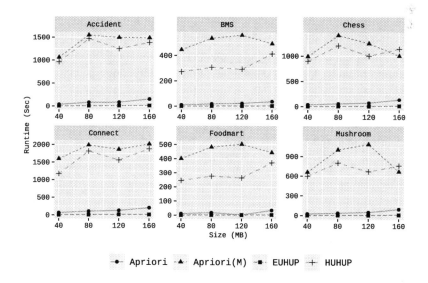

Fig. 3. The runtimes of middle databases.

GB. The final results can still be obtained for the HUHUP and Apriori (M) algorithms. Even, they take a long time to perform the MapReduce system in the large scale databases, we also can see more clearly that the performance of HUHUP is better than the Apriori (M) in the large databases.

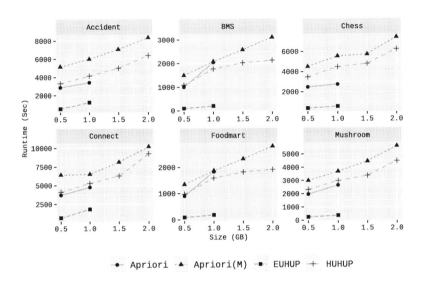

Fig. 4. The runtimes of large databases.

Recently, due to the flourishing development of 5G, the size of an Internet of Things cluster network is getting bigger. That is to say, the scale of an Internet of Things dataset usually achieves petabytes level. Even though the proposed Hadoop framework cannot get better performance than the previous single-machine method in a small dataset. The ability of the proposed framework to handle a big dataset causes that HUHUP has more value to be applied in a real application.

Memory Usage. According to the previous works, the Apriori (M) generates a lot of intermediate data during computation, which takes up more storage space. The designed HUHUP can thus solve this limitation. In these experiments, we compared the different sizes of temp data (memory usages) generated by HFUPM and Apriori (M) while they were performed and executed in various sizes of databases. The experimental results are provided in Fig. 5.

To minimize the impact of different factors on the performance of the comparative algorithms, the number of items stays stable while adjusting the size of the database. The size of the transaction will be increased to evaluate the algorithms. For the experimental results, the Apriori (M) can be shown to have better memory usage when performing it in a tiny database. That's because to minimize the unpromising candidate itemset, and the proposed HUHUP needs

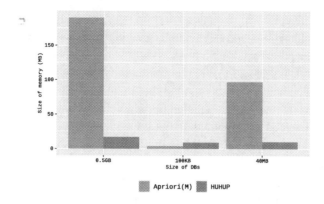

Fig. 5. The memory usages of two Hadoop frameworks in different DB size.

to conduct a very complicated process in the search process. While the database size was increasing, however, HFUPM shows excellent memory usage compared to the Apriori (M); it kept the memory size in a suitable rage, but Apriori(M) suffered the memory explosion crisis [14–17].

5   Conclusion

Commercial dataset availability has grown rapidly. It has become a necessity of industry to be able to attain knowledge from datasets to aid in decision making processes. However, for the most part, traditional classic mining algorithms were not made to handle big data. In our work here, we have designed and implemented HFUPM based on Hadoop. In the future, we hope to implement a similar framework using Spark to be able to handle fuzzy datasets. We also wish to consider more constraints as they have been shown to give significant benefits in pattern mining. We can also look to involve machine and deep learning algorithms to help predict fuzzy high utility patterns in sequential databases.

References

1. Agrawal, R., Imielinski, T., Swami, A.: Mining association rules between sets of items in large database. In: ACM SIGMOD International Conference on Management of Data, pp. 207–216 (1993)
2. Chan, R., Yang, Q., Shen, Y.D.: Mining high utility itemsets. In: IEEE International Conference on Data Mining, pp. 19–26 (2003)
3. Yao, H., Hamilton, H.J., Butz, C.J.: A foundational approach to mining itemset utilities from databases. In: SIAM International Conference on Data Mining, pp. 211–225 (2004)
4. Chui, C.K., Kao, B., Hung, E.: Mining frequent itemsets from uncertain data. In: The Pacific-Asia Conference on Knowledge Discovery and Data Mining, pp. 47–58 (2007)

5. Wu, T.Y., Lin, J.C., Yun, U., Chen, C.H., Srivastava, G., Lv, X.: An efficient algorithm for fuzzy frequent itemset mining. J. Intell. Fuzzy Syst. **38**(5), 5787–5797 (2020)
6. Leung, C.K.S., Mateo, M.A.F., Brajczuk, D.A.: A tree-based approach for frequent pattern mining from uncertain data. In: The Pacific-Asia Conference on Knowledge Discovery and Data Mining, pp. 653–661 (2008)
7. Lin, C.W., Hong, T.P.: A new mining approach for uncertain databases using CUFP trees. Expert Syst. Appl. **39**(4), 4084–4093 (2012)
8. Wu, J.M., Srivastava, G., Wei, M., Yun, U., Lin, J.C.: Fuzzy high-utility pattern mining in parallel and distributed Hadoop framework. Inform. Sci. **1**(553), 31–48 (2021)
9. Lin, Y.C., Wu, C.W., Tseng, V.S.: Mining high utility itemsets in big data. In: Pacific-Asia Conference on Knowledge Discovery and Data Mining, pp. 659–661 (2015)
10. Lin, J.C., Srivastava, G., Li, Y., Hong, T.P., Wang, S.L.: Mining high-utility sequential patterns in uncertain databases. In: 2020 IEEE International Conference on Big Data (Big Data) 10 Dec 2020, pp. 5373–5380. IEEE (2020)
11. Zhang, B., Lin, J.C., Fournier-Viger, P., Li, T.: Mining of high utility-probability sequential patterns from uncertain databases. PloS ONE **12**(7), e0180931 (2017)
12. Fournier-Viger, P., et al.: The SPMF open-source data mining library version 2. In: Berendt, B., et al. (eds.) ECML PKDD 2016. LNCS (LNAI), vol. 9853, pp. 36–40. Springer, Cham (2016). https://doi.org/10.1007/978-3-319-46131-1_8
13. Ahmed, U., Lin, J.C., Srivastava, G., Yasin, R., Djenouri, Y.: An evolutionary model to mine high expected utility patterns from uncertain databases. IEEE Trans. Emerg. Top. Comput. Intell. **5**(1), 19–28 (2020)
14. Srivastava, G., Lin, J.C.-W., Zhang, X., Li, Y.: Large-scale high-utility sequential pattern analytics in Internet of Things. IEEE Internet Things J. **8**(16), 12669–12678 (2021). https://doi.org/10.1109/JIOT.2020.3026826
15. Singh, R., Dwivedi, A.D., Srivastava, G., Wiszniewska-Matyszkiel, A., Cheng, X.: A game theoretic analysis of resource mining in blockchain. Cluster Comput. **23**(3), 2035–2046 (2020). https://doi.org/10.1007/s10586-020-03046-w
16. Wu, J.M.-T., et al.: Mining of high-utility patterns in big IoT-based databases. Mob. Netw. Appl. **26**(1), 216–233 (2021). https://doi.org/10.1007/s11036-020-01701-5
17. Raj, E.D., Manogaran, G., Srivastava, G., Wu, Y.: Information granulation-based community detection for social networks. IEEE Trans. Comput. Soc. Syst. **8**(1), 122–133 (2020)

Dynamic Ensemble Selection
for Imbalanced Data Stream
Classification with Limited Label Access

Paweł Zyblewski$^{(\boxtimes)}$ ⓘ and Michał Woźniak ⓘ

Department of Systems and Computer Networks, Wrocław University of Science
and Technology, Wybrzeże Wyspiańskiego 27, 50-370 Wrocław, Poland
{pawel.zyblewski,michal.wozniak}@pwr.edu.pl

Abstract. Real data streams often, in addition to the possibility of
concept drift occurrence, can display a high imbalance ratio. Another
important problem with real classification tasks, often overlooked in the
literature, is the cost of obtaining labels. This work aims to connect three
rarely combined research directions i.e., data stream classification, imbalanced data classification, and limited access to labels. For this purpose,
the behavior of the DESISC-SB framework proposed by the authors in
earlier works for the classification of highly imbalanced data stream was
examined under the scenario of limited label access. Experiments conducted on synthetic and real streams confirmed the potential of using
DESISC-SB to classify highly imbalanced data streams even in the case of
low label availability.

Keywords: Data stream · Classifier selection · Active learning

1 Introduction

Due to the fact that in real classification problems, data often arrive continuously
over time, the classification of data streams has recently become one of the
more interesting research topics [5]. Designing effective data stream analysis
algorithms means the need to take into account important problems, such as the
need to quickly classify upcoming objects and limitations in resources such as
memory, storage, or computing power. However, the most defining characteristic
of the data stream is the possibility of a phenomenon called concept drift, which
may cause changes in the problem's decision boundary, and thus lead to the
degeneration of the classification model. Therefore, data stream classification
algorithms, in addition to memory and computational limitations, must also be
able to adapt appropriately to continuous changes that may occur in the data
distribution [4].

Another extremely important problem, though still poorly represented in literature, is the potentially highly imbalanced nature of the real data streams. In
this case, apart from the possibility of a concept drift occurrence, the designed
classification algorithms must also deal with bias in relation to the majority

© Springer Nature Switzerland AG 2021
L. Rutkowski et al. (Eds.): ICAISC 2021, LNAI 12855, pp. 217–226, 2021.
https://doi.org/10.1007/978-3-030-87897-9_20

class, which results from a large discrepancy in the cardinality of the problem classes [3]. Here, the most popular group of algorithms are approaches based on classifier ensemble combined with data preprocessing techniques. Existing methods for mining imbalanced data streams work in two distinctive modes, i.e., the data is arriving in chunks or incoming samples are processed online. Wang et al. proposed an ensemble approach, where the *k-Means* based *undersampling* is performed on each data chunk before the training phase [16]. Ditzler and Polikar proposes two modification of the well known *Learn++* algorithm, designed for imbalanced data stream classification, namely *Learn++*.NIE and *Learn++*.CDS [2]. Another interesting approach based on classifier ensemble, which takes into account the possibility of dynamic changes in the *Imbalance Ratio*, was proposed by Sun et al. in [14]. As an example of the incremental methods, the interesting proposition by Wang et al. may be cited [15]. Here, authors proposes the *Sampling-based Online Bagging*, employing both *undersampling* and *oversampling*. Grzyb et al. proposed the *Hellinger Distance Weighted Ensemble*, which employs the *Hellinger* distance in order to prune the maintained ensemble [6]. Ksieniewicz [8] proposed the *Prior Imbalance Compensation* algorithms, which modifies the obtained predictions based on the *prior* class probabilities.

Finally, an additional problem when dealing with real data streams is access to labels. Many works in the area of data streams ignore the fact that instances may arrive too quickly to enable full labeling. Another important element in the case of a real classification problem is the cost of obtaining the labels. Their acquisition very often involves the work of a human expert, which not only carries high financial costs but also takes time. One way to deal with this problem is an approach known as *active learning*, which allows for selecting only the so-called *interesting* objects for labeling [11]. Among the *active learning* solutions for data stream classification, the query strategies based on instance weighting, such as the one proposed by Bouguelia et al. [1], can be distinguished. Krawczyk et al. classified drifting data streams using the *query by committee* strategy [7]. Mohamad et al. proposed *Stream Active Learning* (SAL) methodology for classifying data streams with unknown class number [10]. *Active learning* was combined with random labeling approach by Shan et al. [12]. Zhang et al. proposed a hybrid labeling strategy, combining both *uncertainty* and *imbalance* strategies, known as the *Reinforcement Online Active Learning Ensemble* [17].

The purpose of this paper is to extend the DESISC-SB framework [19] previously introduced by the authors for the classification of highly imbalanced data sets with an active learning module to deal with the limited access to labels scenario.

In brief, the main contributions are as follows:

- Extension of the previously proposed DESISC-SB framework for highly imbalanced data stream classification with *active learning* strategies for dealing with limited label access.
- Exhaustive experimental evaluation of the DESISC-SB framework paired with various labeling approaches for synthetic and real highly imbalanced drifting data streams.

2   Methods

This work focuses on extending the DESISC-SB imbalanced data stream classification framework with an active learning module. This is to asses the compatibility of the proposed batch approach with active labeling methods and to evaluate its behavior when dealing with restricted access to labels. The schema of the expanded framework is presented in the Fig. 1. DESISC-SB uses a stratified version of bagging to generate a new learner on the current data chunk. This classifier is then added to the pool, which is then used in the Dynamic Ensemble Selection process. The previous preprocessed data chunk is used as Dynamic Selection Dataset. When the predetermined maximum size of the classifier pool is exceeded, the worst model in terms of Balanced Accuracy is removed from it. A detailed description of how the framework, along with an analysis of its computational complexity, can be found in [19].

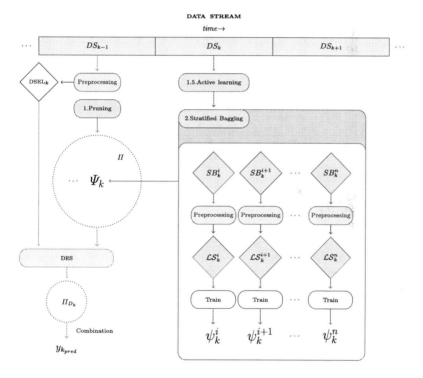

Fig. 1. The framework extended with the active learning module for training base classifiers and to prepare a DSEL for dynamic selection process. Here, T_k is the training data produced by preprocessing (*Preproc*) data chunk DS_k and Ψ_k is the base classifier trained on the kth data chunk. Π denotes the classifier pool.

As labeling methods, two approaches described in the previous work of the authors are used [18]. The first is *Budget Labeling* (BLS) which trains each new base classifier on a fixed percentage of randomly selected problem instances from the current chunk. The second method is the ALS that has been modified. As before, this algorithm selects patterns that are within a certain distance from the problem's decision boundary defined by the threshold t, but this time it can also be given the budget b, which defines the percentage of these patterns we want to label. ALS pseudocode is presented in Algorithm 1.

As the framework is supposed to work with highly imbalanced problems, another modification has been made to labeling methods. If all the labeled instances come from the same class, a new model is not added to the classifier pool.

Algorithm 1. Pseudocode for the modified ALS

Input:
  $Stream = \{\mathcal{DS}_1, \mathcal{DS}_2, \ldots, \mathcal{DS}_k, \mathcal{DS}_{k+1}, \ldots\}$ – data stream,
  Ψ – classification algorithm,
  t – threshold,
  b – budget.

1: **for each** $k, \mathcal{DS}_k = \{x_k^1, x_k^2, \ldots, x_k^N\}$ in $Stream$ **do**
2:     **if** $k == 0$ **then**
3:         $\Psi \leftarrow$ UPDATECLASSIFIER$(\Psi, \mathcal{DS}_k)$ ▷ Update the classifier using whole chunk
4:     **else**
5:         $\mathcal{X}_k =$ ACTIVELEARNING$(t, b, \mathcal{DS}_k)$ ▷ Select instances using *active learning*
6:         $\mathcal{LS}_k =$ GETLABELS$(\mathcal{X}_k)$
7:         $\Psi \leftarrow$ UPDATECLASSIFIER$(\Psi, \mathcal{LS}_k)$ ▷ Update the classifier
8:     **end if**
9: **end for**

---

3   Experimental Evaluation

The experiments were designed to answer the question, whether the batch-based DESISC-SB framework for imbalanced data stream classification is compatible with *active learning* methods. The aim of the conducted experiment is to see how the use of a data labeling strategy affects the results achieved by the DESISC-SB framework.

Experimental Set-Up
The analysis was based on six types of synthetic streams generated using *stream-learn* module [9], replicated 10 times for stability of the achieved results. All experiments can be replicated according to the *Python* source code, available on *GitHub* repository[1]. The detailed characteristics of the generated streams are described below:

– *Concept drift* types – *sudden, gradual* and *incremental*,
– Approaches to repetitive concepts – *recurrent* and *non-recurrent concept drift*,
– Data stream size – 50000 instances (200 data chunks, 250 instances each)

---

[1] https://github.com/w4k2/icaisc21-al-stream.

- Number of concept drift per stream – 9,
- Global label noise – 5%,
- Imbalance Ratio – 19.

Additionally, experiments were carried out on 3 real data streams from the INSECTS database [13], the characteristics of which are presented in the table 1.

Table 1. Real data streams characteristics.

Data stream	#Samples	#Features	IR
INSECTS-abrupt_imbalanced_norm	300 000	33	19
INSECTS-gradual_imbalanced_norm	100 000	33	19
INSECTS-incremental_imbalanced_norm	380 000	33	19

The experimental evaluation was carried out in accordance with the *Test-Then-Train* methodology [5]. The DESISC-SB framework parameters (i.e. dynamic selection method and preprocessing technique) were selected based on the results of the experiments performed previous works:

- Base classifier – *Naïve Bayes Classifier*,
- *Dynamic Ensemble Selection* – KNORA-U at the level of bagging classifiers,
- Data preprocessing – *Random Oversampling*,
- Fixed classifier pool size – 5 bagging classifiers, 10 base classifiers each (50 models in total).

Comparative methods:

- Whole - model updated using all available data,
- BLS-15 - 15% of random budget,
- ALS-15 - 15% of instances closes to the decision boundary,
- ALS - all instances within distance of 0.2 from the decision boundary.

The methods' parameters were selected based on the experience gained during research on the BALS algorithm and also taking into account the batch approach and base classifier used.

Experiment – The Impact of Active Learning on the DESISC-SB Framework
Figure 2 shows the results of using the proposed framework in the case of sudden concept drifts occurrence. The first thing that stands out is that the BLS-15 result is similar to that of the random classifier. This is due to the high imbalance ratio (5% of minority class) in the data stream. Because of that BLS selects only instances belonging to the majority class and the new model is not added to the pool. At the same time, we can see that both ALS-15 and ALS are doing relatively well. Both in the case of recurring and non-recurring drift, ALS is better at identifying the minority class, due to the lack of a set budget. Thanks to this, it maintains a high generalization capacity and in some cases is able to perform similarly to the model learned on all available data.

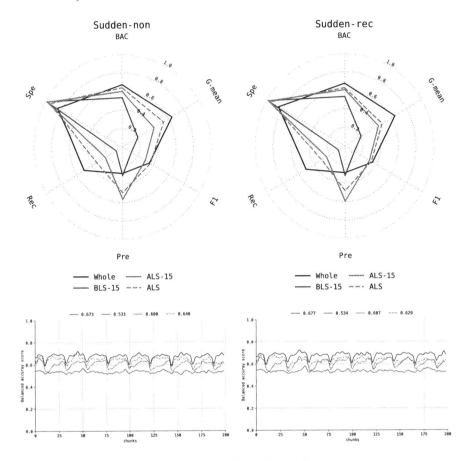

Fig. 2. Results for sudden drift.

Figure 3 shows the results obtained in the case of gradual drift, characterized by slower dynamics of change and the occurrence of instances from both concepts at the same time. In this case, for non-recurring drift, we can observe a progressive deterioration of the generalizing ability of ALS. This may be due to a small number of instances located within a given distance from the decision boundary, which in turn leads to underfitting in the face of a constant concept change. On the other hand, in the case of recurring gradual drift, the ALS and ALS-15 remain on a similar level, because the ensemble always includes models that remember the old concept.

In the case of the of incremental drift occurrence (Fig. 4) the observations are similar to those regarding the gradual concept drift. The difference is that whether the drift is recurring or non-recurring, the ALS-15 and ALS methods achieve nearly identical results. This may be the result of more instances available to ALS as one concept blends seamlessly into another.

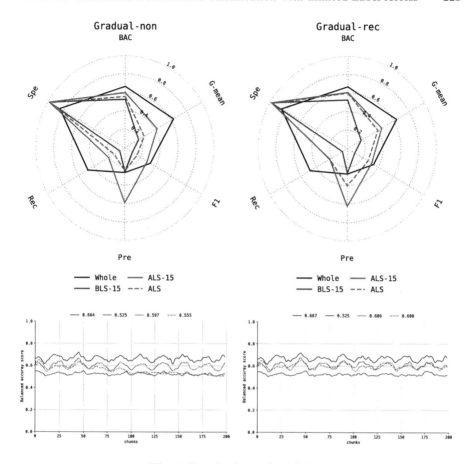

Fig. 3. Results for gradual drift.

Figure 5 shows the results of combining the DESISC-SB framework with active learning methods in a classification task of five real imbalanced data streams. Radar charts show values of six metrics averaged over the entire length of the stream, while the runs are shown for the $Gmean_s$ metric. Due to the use of the batch-based data stream classification approach and the GNB classifier as the base model, it was decided to abandon the BLS approach, which in this case would remain at the level of the random classifier.

The observations related to the classification of INSECTS streams, presenting three defined types of concept drift, are particularly interesting. In the case of sudden drift, the model learned using the ALS approach achieves the generalization ability at the level of the full model. It may be caused by low classification certainty, and thus a large number of patterns located at a given distance from the decision boundary. The model using the ALS-15 approach achieves slightly lower results than ALS, which is a direct result of the smaller number of training patterns available. In the case of gradual drift, all three approaches have very

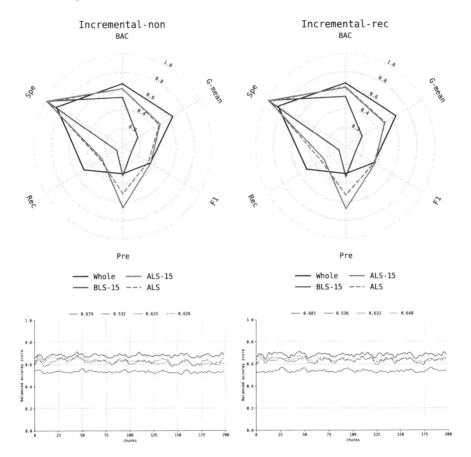

Fig. 4. Results for incremental drift.

similar performance. This is due to the drift characteristics and proves that only a small number of instances closest to the decision limit is sufficient for building a useful model. When dealing with incremental concept drift, the model learned using the ALS-15 approach displays a correspondingly lower generalization capacity, resulting from the smaller number of patterns used for updating the classifier. At the same time, however, this model is relatively stable compared to the classifier trained using the ALS method, which demonstrates greater degeneration in the event of concept drift occurrence. This is due to the changes in the support space and the lack of patterns that can be used during the training process.

Lessons Learned
Based on the results obtained, it can be concluded that the proposed batch-based framework for imbalanced data stream classification is compatible with active learning methods. ALS works especially well in the case of sudden drift,

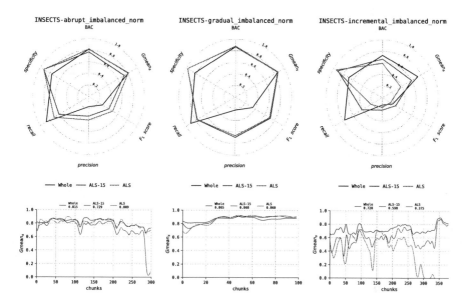

Fig. 5. Results for real data streams.

where about 25% of the instances closest to the decision boundary are sufficient to achieve results similar to the model trained using all instances of the problem.

Research on real data streams has shown that the ALS approach - using all patterns within a given distance from the problem's decision boundary - cannot be used in its current form for every data stream. This is due to the high sensitivity of the method to the distribution of patterns in the support space, which, if the classification is too certain, leads to the lack of patterns that can be used in the model training process. To deal with this problem, threshold t should not be set as a fixed parameter, but rather optimized for each data chunk, to ensure that models using this approach always get a training set containing patterns useful in the training process.

4   Conclusions

The goal of this work was to evaluate the performance of the DESISC-SB framework when combined with *active learning* method based on selecting patterns located at a certain distance from the decision boundary. The conducted research confirmed the usefulness of the framework under a high imbalance ratio and limited access to labels for both synthetic and real data streams.

Future works may include conducting a broader experimental evaluation of the DESISC-SB framework for the highly imbalance data stream classification under a scenario of restricted access to labels.

Acknowledgment. This work was supported by the *Polish National Science Centre* under the grant No. 2017/27/B/ST6/01325.

References

1. Bouguelia, M., Belaïd, Y., Belaïd, A.: An adaptive streaming active learning strategy based on instance weighting. Pattern Recogn. Lett. **70**, 38–44 (2016)
2. Ditzler, G., Polikar, R.: Incremental learning of concept drift from streaming imbalanced data. IEEE Trans. Knowl. Data Eng. **25**(10), 2283–2301 (2013)
3. Fernández, A., García, S., Galar, M., Prati, R.C., Krawczyk, B., Herrera, F.: Learning from Imbalanced Data Sets. Springer, Cham (2018). https://doi.org/10.1007/978-3-319-98074-4
4. Gama, J., Žliobaitė, I., Bifet, A., Pechenizkiy, M., Bouchachia, A.: A survey on concept drift adaptation. ACM Comput. Surv. (CSUR) **46**(4), 1–37 (2014)
5. Gomes, H.M., Barddal, J.P., Enembreck, F., Bifet, A.: A survey on ensemble learning for data stream classification. ACM Comput. Surv. (CSUR) **50**(2), 1–36 (2017)
6. Grzyb, J., Klikowski, J., Woźniak, M.: Hellinger distance weighted ensemble for imbalanced data stream classification. J. Comput. Sci. **51**, 101314 (2021)
7. Krawczyk, B., Pfahringer, B., Wozniak, M.: Combining active learning with concept drift detection for data stream mining. In: IEEE International Conference on Big Data, Big Data 2018, Seattle, WA, USA, 10–13 December 2018. pp. 2239–2244. IEEE (2018)
8. Ksieniewicz, P.: The prior probability in the batch classification of imbalanced data streams. Neurocomputing **452**, 309–316 (2020)
9. Ksieniewicz, P., Zyblewski, P.: Stream-learn-open-source python library for difficult data stream batch analysis. arXiv preprint arXiv:2001.11077 (2020)
10. Mohamad, S., Sayed-Mouchaweh, M., Bouchachia, A.: Active learning for classifying data streams with unknown number of classes. Neural Netw. **98**, 1–15 (2018)
11. Settles, B.: Active Learning. Morgan & Claypool Publishers (2012)
12. Shan, J., Zhang, H., Liu, W., Liu, Q.: Online active learning ensemble framework for drifted data streams. IEEE Trans. Neural Netw. Learn. Syst. **30**(2), 486–498 (2019)
13. de Souza, V.M.A., Silva, D.F., Batista, G.E.A.P.A.: Classification of data streams applied to insect recognition: initial results. In: 2013 Brazilian Conference on Intelligent Systems, pp. 76–81 (2013). https://doi.org/10.1109/BRACIS.2013.21
14. Sun, Y., Tang, K., Minku, L.L., Wang, S., Yao, X.: Online ensemble learning of data streams with gradually evolved classes. IEEE Trans. Knowl. Data Eng. **28**(6), 1532–1545 (2016)
15. Wang, S., Minku, L.L., Yao, X.: A systematic study of online class imbalance learning with concept drift. CoRR abs/1703.06683 (2017)
16. Wang, Y., Zhang, Y., Wang, Y.: Mining data streams with skewed distribution by static classifier ensemble. In: Chien, B.C., Hong, T.P. (eds.) Opportunities and Challenges for Next-Generation Applied Intelligence. Studies in Computational Intelligence, vol 214, pp. 65–71. Springer, Heidelberg (2009). https://doi.org/10.1007/978-3-540-92814-0_11
17. Zhang, H., Liu, W., Liu, Q.: Reinforcement online active learning ensemble for drifting imbalanced data streams. IEEE Trans. Knowl. Data Eng. (2020)
18. Zyblewski, P., Ksieniewicz, P., Woźniak, M.: Combination of active and random labeling strategy in the non-stationary data stream classification. In: Rutkowski, L., Scherer, R., Korytkowski, M., Pedrycz, W., Tadeusiewicz, R., Zurada, J.M. (eds.) ICAISC 2020. LNCS (LNAI), vol. 12415, pp. 576–585. Springer, Cham (2020). https://doi.org/10.1007/978-3-030-61401-0_54
19. Zyblewski, P., Sabourin, R., Woźniak, M.: Preprocessed dynamic classifier ensemble selection for highly imbalanced drifted data streams. Inf. Fusion **66**, 138–154 (2021)

Various Problems of Ariticial Intelligence

Applying and Comparing Policy Gradient Methods to Multi-echelon Supply Chains with Uncertain Demands and Lead Times

Julio César Alves[1,3]() , Diego Mello da Silva[2,3] ,
and Geraldo Robson Mateus[3]

[1] Federal University of Lavras, Lavras, Brazil
`juliocesar.alves@ufla.br`
[2] Federal Institute of Minas Gerais, Formiga, Brazil
`diego.silva@ifmg.edu.br`
[3] Federal University of Minas Gerais, Belo Horizonte, Brazil
`mateus@dcc.ufmg.br`

Abstract. In the present work, we have applied and compared Deep Reinforcement Learning techniques to solve a problem usually addressed with Operations Research tools. State-of-the-art Policy Gradient methods are used in a production planning and product distribution problem, considering a four-echelon supply chain with uncertain lead times, to minimize total operating costs while meeting uncertain seasonal demands. Two-phases experiments are conducted regarding eight different scenarios. Firstly, A2C, DDPG, PPO, SAC, and TD3 are used considering fixed training budgets. In the second phase, the best two algorithms from the first phase are tuned considering different stopping criteria. Results show that PPO and SAC perform better for the problem addressed. They achieve comparable learning behavior, final performance, and sample complexity, while PPO is faster in terms of wall-clock time.

Keywords: Supply chain · Uncertainty · Deep reinforcement learning · Policy gradient methods

1 Introduction

Nowadays there is a fast-growing interest in applying Machine Learning (ML) algorithms to solve problems in many fields. Supervised and Unsupervised learning had a boom in the last decade due to the increasing amount of available data, the revival of neural networks, and the hardware evolution that allowed to handle bigger and bigger problems. Reinforcement Learning (RL) is another branch of ML that started to gain attention later, mainly after works that solved game problems at unprecedented levels using deep neural networks (like Atari games

This research has been supported by the following Brazilian institutions: Minas Gerais Research Funding Foundation (FAPEMIG) and National Council for Scientific and Technological Development (CNPq).

L. Rutkowski et al. (Eds.): ICAISC 2021, LNAI 12855, pp. 229–239, 2021.
https://doi.org/10.1007/978-3-030-87897-9_21

[13] and Go [19]). Many groups from academia have been looking for this type of technique to solve sequential decision-making problems under uncertainty. Problems like that from the Operations Research (OR) area usually have much more actions to be taken on each step than problems commonly addressed by the RL community. Nevertheless, in this work, we have successfully applied and compared state-of-the-art Deep RL algorithms to solve a production planning and product distribution problem usually handled by OR techniques. There are two main types of RL techniques: based on function values or based on policies [20]. In value-based RL an agent learns values for each state (or state/action pair) and uses them to find a good policy. In policy-based RL an agent learns a parameterized policy directly, and if it uses a gradient method to maximize the policy performance, the method is called policy-gradient. This type of algorithm uses deep neural networks to approximate the policy and is currently the state-of-the-art for problems with continuous state and action spaces, which is the case of the problem addressed here.

We consider a four-echelon, centrally controlled, supply chain with uncertain seasonal demands and lead times. The first echelon is composed of suppliers that produce raw materials and send them to the factories. Factories process raw materials to make products and send these to the wholesalers. Wholesalers send products to retailers that, in turn, need to attend to uncertain demands from customers. All nodes of the chain have local capacitated stocks, and supplying, processing, and transport are also limited. There are lead times to produce raw materials and in the transportation of materials along the chain. The target is to meet the clients' uncertain demands for a 360-period time horizon with the lowest total operation cost.

In recent years, Deep RL techniques have been used in similar (but usually less complex) problems than that we address here. Some of the works deal with two-echelon supply chains with small action space and have applied algorithms like REINFORCE, A2C/A3C, DQN, and DDPG [6,8–10,15]. Oroojlooyjadid [14] have applied DQN in a multi-echelon supply chain, but only one of the nodes is controlled by the RL agent. Geevers [5] has achieved unstable results with PPO applied in a two-echelon supply chain with non-seasonal demands and 9-dimensional action space. Perez et al. [16] use PPO in a four-echelon supply chain, but they consider non-capacitated stocks, non-seasonal demands, and the state and action spaces are discrete. It is important to notice that none of those works have considered stochastic lead times. We have built upon our previous work [1] in which we have applied PPO in a four-echelon supply chain with seasonal demands and stochastic lead times, including also capacitated stocks and factories. Therefore, to the best of our knowledge, there is no comparison between Deep RL techniques for this type of multi-echelon supply chain problem. Thus the main contribution of this work is to properly apply and compare state-of-the-art Policy Gradient methods on solving a four-echelon supply chain problem with two nodes per echelon and uncertain seasonal demands and lead times. Other contributions are the use of a different stopping criterion in the training and Welch's t-tests for performance comparison and to select algorithms.

The remainder of this text is structured as follows. Section 2 presents the Markov Decision Process (MDP) formulation of the problem. Experimental methodology and test scenarios are presented in Sect. 3. Section 4 presents experiments and results analysis, whereas conclusions are presented in Sect. 5.

2   Problem Formulation

MDP is a classical formalization of sequential decision-making problems [20], and it is the first step to apply RL. We have proposed the MDP formulation for the problem addressed in [1], and it is given as follows.

The **state space** is given by values scaled in the interval $[-1, 1]$, which is useful to get good results with Deep RL methods [17]. A state is composed of A: demands for each retailer; B: current stock level of each node; C: the amount of material that will be available on the next time steps on each node; and D: number of remaining time steps to complete the episode. Range scaling takes into account the minimum and maximum bounds of each field. For A, the bounds come from the scenario, and for D they are 0 and 360. The bounds for B are zero and the stock capacity of the node. Regarding material to be available (C), the minimum is zero and the maximum is the supplying capacity, in case of suppliers, or the sum of the stock capacities of the sources, in case of the other nodes. Still on C, there are $l^{avg} + 1$ values for each node, where l^{avg} is the average lead time. So, the first l^{avg} values represent the material that will arrive on the next l^{avg} periods, and the last value is the sum of material that will arrive later than that (scaled properly considering the number of steps). For the scenarios addressed here, with $l^{avg} = 2$, the space state has 27 dimensions.

Regarding **action space** there are two types of decisions: how much material to be supplied by each supplier (a_n^p) and to be sent from one node to its possible destinations (a_{nm}^t). Material to be kept in stock can be indirectly defined, and there are no decisions for retailers since they attend to clients' demands whenever possible. The actions are values in the $[-1, 1]$ interval which is scaled to $[0, 1]$ before being used in the environment. The amount of material to be supplied by node n is given by $a_n^p b_n^p$, where b_n^p is the supply capacity of the node n. Regarding the amount of material to be transported, we sort the action values for each origin in a way that each value is viewed as a cut in the current amount of available material. Thus for the least action value, the material to be transported is $x_1 = a_{nm}^t s_n$ (where s_n is the current stock level of the node n), and for the other action value the material to be transported is $x_2 = a_{nm'}^t (s_n - x_1)$. The left material ($s_n - x_1 - x_2$) is kept in stock. Considering the scenarios addressed here, the action space has 14 dimensions.

The transition function is not known by the agent, since we are applying model-free RL methods. So it is replaced by the simulation of the **environment's dynamics**. As the formulated action space only generates viable decisions, the simulation always follows the decision of the agent. Stochastic demands and lead times are sampled for each time step, and materials that exceed stock or processing capacities are discarded.

The **reward** is the negative of the operational costs incurred on each step since we want to minimize the total operation cost. Therefore it is composed of supply, stock, transport, and product processing costs. Besides that, we also have penalization costs per unit of material missing in meeting demands and discarded by exceeding stock or processing capacities.

3   Experimental Methodology

Our main objective is to apply and compare Policy Gradient algorithms (A2C, DDPG, PPO, SAC, TD3 [4,7,11,12,18]) to the proposed supply chain problem. In the first phase, the methods are used in eight different test scenarios using fixed training budgets. The best two algorithms are selected and used in the second phase, in which they are tuned considering new stopping criteria and applied in the same eight scenarios (also with the new stopping criteria).

The **test scenarios** refers to a four-echelon supply chain with two nodes per echelon. They were selected from [1] to have variety in terms of demand types (seasonal and regular), lead times (stochastic and constant), and perturbations. The differences between the scenarios are shown in Table 1. Let $\mathcal{N}$ and $\mathcal{U}$ be the normal and uniform distributions, respectively. Uncertain seasonal demands for each time step t is given by $f(t) + \mathcal{N}(0, p)$, where f is a sinusoidal function with four peaks, minimum 100 and maximum 300, and p is the standard deviation given by each scenario. The regular demands are given by $200 + \mathcal{N}(0, p)$ or $200 + \mathcal{U}(-p, p)$. Constant lead time is equal to two and stochastic lead times are given by $\min(Poisson(1) + 1, 4)$. Others parameters (like costs, capacities, etc.) are the same for all scenarios and can be found in [1].

Table 1. Test scenarios and their differences

Scenarios	N20	N60	N20cl	N60cl	rN50	rU200	rN50cl	rU200cl
Lead time	stoch	stoch	const	const	stoch	stoch	const	const
Demands	seas	seas	seas	seas	reg	reg	reg	reg
Pert. fun.	$\mathcal{N}$	$\mathcal{N}$	$\mathcal{N}$	$\mathcal{N}$	$\mathcal{N}$	$\mathcal{U}$	$\mathcal{N}$	$\mathcal{U}$
p	20	60	20	60	50	200	50	200

In the **first-phase experiments**, the algorithms are trained with 3 predefined random seeds, considering 10 thousand episodes (or 3.6 million time steps). In each training run, the model is evaluated on every 50 episodes, considering the average return (accumulated rewards) of 10 episodes. The resulting model of each run is the one with the best evaluation throughout the training. A common way to compare Deep RL algorithms is to analyze the behavior of the algorithms during training using learning curves. We do this, but as we are interested in the performance of the methods regarding a specific problem, it is also interesting to compare the final results. This is done by calculating the average return of

100 episodes (using 10 predefined random seeds for the environment) for the 3 best models (considering deterministic actions) of each algorithm.

Besides comparing the algorithms, we are also interested in selecting the best two in performance to be used in the next phase. To take such a decision we also use Welch's t-tests [3] considering the evaluation data during training runs. We do multiple comparisons considering 95% confidence and Bonferroni correction. The main reason to not use all five algorithms in the second phase is the time needed to run all experiments, especially for off-policy methods (DDPG, SAC, TD3). One could also note that, somehow, PPO can be viewed as an evolution of A2C, and SAC an evolution of TD3 (which, in turn, is an improved DDPG). Therefore, it would be expected that the newer algorithms would perform better.

In **second-phase experiments**, we consider a different stopping criterion, in which training is stopped if there is no improvement in the model after N consecutive evaluations. The idea is to allow an algorithm to run longer when the model is still getting improved, and, on other hand, save machine time by stopping training when the algorithm cannot find better solutions after many time steps. Experiments from the first phase are used to choose the value of N. Once this is done, a hyperparameter tuning is conducted for each of the selected algorithms. The tuning consists of running 100 training trials with different combinations of values of the hyperparameters for the $N20$ scenario. The very first attempt uses the default values for the algorithm since they are tuned considering several continuous benchmark problems [17]. From the second to the 20th attempt, the values are randomly chosen from predefined intervals. TPE algorithm [2] and pruning by median mechanism are used for the next 80 attempts to, respectively, choose the next parameter values and to early-stop unpromising attempts. In each attempt, the agent is evaluated on every 50 episodes, by averaging the returns of 5 episodes. If it is not pruned, an attempt is ended after N consecutive evaluations without improvement or after 10 thousand episodes of training. The best values of the hyperparameters are chosen from the attempt with the best result.

Afterward, the training phase consists of using the best hyperparameters values found to execute 5 training runs, with predefined random seeds, of each algorithm in each of the eight scenarios. A training run is stopped after N consecutive evaluations without improvement or after 20 thousand episodes. As before, the model is evaluated on every 50 episodes, considering the average return of 10 episodes, and the resulting model is the one with the best evaluation throughout the training. The same evaluation of the best models and analysis tools used in the first phase of the experiments are used in this phase, but now considering 5 training runs per scenario and algorithm.

4   Experiments and Results

For the experiments, we have used a computer with a 6 GB GPU, 2.9 GHz × 6 processor, and 32 GB of RAM, and implementations in Python 3.6.10. We have used the environment's implementation we have done in [1], and algorithm's versions from the Stable Baselines 3 (SB3) library [17]. Automatic reward normalization was used because it proved beneficial for all algorithms in exploratory experiments. Experiments and results of the first and second-phase are presented in Sects. 4.1 and 4.2, respectively.

4.1   First-Phase Experiments

The average learning curves of the experiments, considering the evaluations during training, are presented in Fig. 1. They show that A2C was not able to learn, since it has not converged in any of the scenarios. DDPG had good results in two scenarios but had a large variance among different seeds in the others. TD3 had good results in many scenarios, but presented high variance in **rN50cl** and **rU200cl** scenarios. PPO and SAC had the best results in terms of learning because the curves are well-behaved in all tested scenarios.

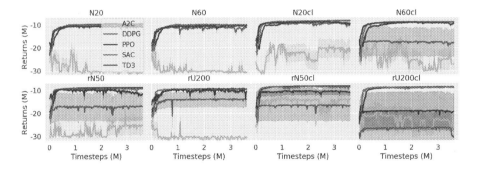

Fig. 1. First-phase experiments: average learning curves by scenario and algorithm, based on the evaluations during training. The values are in millions and shaded area represents standard error.

We also have done a multiple comparison analysis considering training data to select the best algorithms. The performance of the algorithms was summarized in four intervals (each one with 50 evaluation points), using Welch's t-tests, with 95% confidence and Bonferroni correction, to identify the algorithm with the lowest average returns in each interval. The idea was to find the best two algorithms with statistical confidence, considering stability throughout training. The results is shown in Fig. 2. In the first round, we considered all five algorithms and, as expected, A2C had the lowest averages in some scenarios. Looking at the learning curves, one would expect that A2C had the lowest averages also in others scenarios, but this did not happen because of the high variance of DDPG results. In the second round, we repeated the analysis excluding A2C, and DDPG had the lowest averages in some scenarios. The same process was repeated, without DDPG, and TD3 had the lowest averages. Therefore, the analysis reinforced that PPO and SAC were the algorithms with better performance in the experiments.

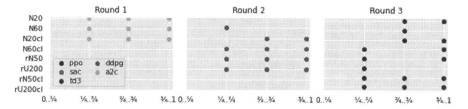

Fig. 2. Multiple comparison analysis by training intervals and scenario. Markers indicate algorithms with lowest average returns, considering Welch's t-tests. In each round, the algorithm with lowest average is excluded from the evaluation.

The evaluation of the final results (best models) is presented in Table 2. The columns present the scenarios and each line has the results for an algorithm. For PPO, average total operating costs (negative of the returns), are presented. On the other lines, the values are the relative differences to the PPO result. The results are compatible with the behavior of the learning curves. SAC and PPO had lower operational costs and were comparable to each other. TD3 had relatively comparable results in several scenarios but very high costs in some others. DDPG had poor results in several scenarios and A2C in all of them.

Table 2. First-phase experiments: evaluation of the best models. For PPO, values are the average total operating costs, for the others, the relative difference to PPO.

Alg/Scen	N20	N60	N20cl	N60cl	rN50	rU200	rN50cl	rU200cl
PPO	9,534 k	10,031 k	8,259 k	8,862 k	8,827 k	9,315 k	8,228 k	8,647 k
A2C	+76.2%	+76.5%	+94.6%	+55.3%	+94.3%	+109.6%	+46.6%	+56.8%
DDPG	+8.4%	+6.8%	+6.6%	+91.9%	+90.5%	+47.1%	+98.9%	+204.2%
SAC	+1.6%	−3.4%	+1.1%	−1.4%	+0.3%	−0.4%	−0.0%	−1.3%
TD3	+7.1%	+3.1%	+3.9%	+0.7%	+5.9%	+3.1%	+23.7%	+117.4%

We have chosen PPO and SAC for the second-phase experiments since they performed better considering the learning curves, Welch's t-tests per training intervals, and the final results regarding the best models. Analyzing the average wall-clock time per training run, PPO took 1.7 h and SAC 9.3 h (the values for A2C, DDPG, and TD3 were 0.7 h, 5.8 h, and 5.6 h, respectively).

4.2   Second-Phase Experiments

The first step in this phase was to define the value of the parameter N related to the proposed stopping criterion. Using the evaluations done during the training runs of the first-phase experiments, we have calculated the maximum value of N that would lead to a maximum loss of 1% in the evaluation of the best models, in at least 90% of the training runs. We have found $N = 38$, and looking at data, we realized that $N = 40$ would lead to a maximum loss of 1.01%, so we have decided to use the round value 40. As the evaluations are done on every 50 episodes, $N = 40$ means that the training is stopped if there are no model improvements after 2 thousand episodes (or 720 thousand training time steps). The next step was the hyperparameter tuning of PPO and SAC that followed the proposed methodology. The predefined intervals of the hyperparameters of each algorithm and the best values found are not presented here for the sake of space. The tuning process for SAC took 16 days, even using the stopping criterion and two parallels processes, whereas for PPO spent less than 48 h.

The training phase was executed considering 5 runs of each algorithm, with predefined random seeds and stopping criteria of 40 consecutive evaluations without improvement (or a maximum of 20 thousand episodes), for the eight test scenarios. The average learning curves of PPO and SAC, considering the evaluations during training, are presented in Fig. 3. It is important to note that due to the stopping criteria, the training size of each training run can be different. So, the average curve considers this detail, and each algorithm can stop the curve at different points. We can observe that PPO and SAC have comparable performance in terms of learning since the curves grow fast and stabilize at similar levels for all scenarios. It is interesting to note that although in many problems SAC has better sample efficiency than PPO, this is not observed for our problem. In fact, in some scenarios, the PPO learning curve grows faster than SAC.

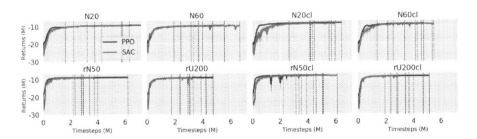

Fig. 3. Average learning curves from the second-phase experiments, based on the evaluations during training. The values are in millions and shaded area represents standard error. Curves stop early due to the stopping criterion and dashed lines show when each training run has finished.

The final results, considering the best models, are shown in Fig. 4. The left graph shows the average total operating costs for scenarios with regular demands and the right graph for those with seasonal demands. We can see that, for both algorithms, the greater the uncertainty of the scenario, the greater the costs. Considering Welch's t-tests, PPO performed better in scenarios $rN50cl$ and $N20cl$, while SAC did better in the other three scenarios with seasonal demands. In terms of wall-clock time, each training run of PPO and SAC took, on average, 0.9 h and 10.7 h, respectively.

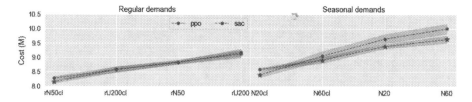

Fig. 4. Average total operating costs by scenario from evaluation of the best models (2nd-phase experiments). The values are in millions and shaded area represents standard error. Marker ⋆ indicate whether there is difference between the algorithms, considering Welch's t-tests.

We have evaluated the average costs over the planning horizon found by the solutions of PPO and SAC, in each scenario. Both algorithms were able to build stocks with seasonality, but PPO built slightly lower stocks than SAC in all scenarios. SAC was better than PPO in decreasing production at the end of the episode. Supplying and transport material from SAC solutions usually had more variance than PPO. Finally, the unmet demands seem to had been the main reason for one algorithm performs better than the other in each scenario.

Although we have done a fair comparison between the algorithms concerning the stopping criterion, in practical terms, PPO needed $1/12$ of the time spent by SAC (per training run on average). So, we have decided to run new experiments with PPO considering a fixed budget of 20k episodes, to compare the final results and wall-clock time. The results have shown that, in this case, and considering Welch's t-tests, PPO had better performance in scenarios *N20cl, N60cl, rN50*, and *rN50cl*, while SAC performed better only in scenario *N60*. In the other three scenarios, there was no significant difference. Regarding wall-clock time PPO took 1.7 h per training run, which was still $1/6$ of the time needed by SAC.

5   Conclusions

In this work, we have experimented with five Policy Gradient algorithms in a four-echelon supply chain problem with two nodes per echelon and uncertain seasonal demands and lead times. In the first phase, the algorithms were trained considering fixed training budgets in eight representative scenarios. We have used learning curves, statistical multiple comparisons by training intervals, and final results with the best models, to select two algorithms that performed better (PPO and SAC). Their hyperparameters were tuned with a stopping criterion that considers a maximum number of consecutive evaluations without the model's improvement. The methods were trained in the same scenarios and have achieved comparable learning behaviors, final results, and sample efficiency. In seasonal demands scenarios, SAC performs slightly better in three of the four scenarios, in terms of final results. However, PPO is faster regarding wall-clock time with $1/8$ of the time required by SAC for tuning, and $1/12$ per training run. Additional experiments with PPO, with a fixed training budget, show that it achieves slightly better performance than SAC in the final results, and still using only $1/6$ of wall-clock time. Therefore, the experimental results suggest that PPO may be a good choice to be used in practice for the problem addressed.

References

1. Alves, J.C., Mateus, G.R.: Multi-echelon supply chains with uncertain seasonal demands and lead times using deep reinforcement learning. Submitted (2021)
2. Bergstra, J., Bardenet, R., Bengio, Y., Kégl, B.: Algorithms for hyper-parameter optimization. In: Proceedings of the 24th International Conference on Neural Information Processing Systems. NIPS 2011, Red Hook, NY, USA, pp. 2546–2554. Curran Associates Inc. (2011)
3. Colas, C., Sigaud, O., Oudeyer, P.Y.: A Hitchhiker's guide to statistical comparisons of reinforcement learning algorithms. In: ICLR Worskhop on Reproducibility, Nouvelle-Orléans, United States, May 2019. https://hal.archives-ouvertes.fr/hal-02369859
4. Fujimoto, S., van Hoof, H., Meger, D.: Addressing function approximation error in actor-critic methods. In: Dy, J., Krause, A. (eds.) Proceedings of the 35th International Conference on Machine Learning. Proceedings of Machine Learning Research, 10–15 Jul 2018, vol. 80, pp. 1587–1596. PMLR (2018). http://proceedings.mlr.press/v80/fujimoto18a.html
5. Geevers, K.: Deep reinforcement learning in inventory management. Master's thesis, University of Twente, December 2020. http://essay.utwente.nl/85432/
6. Gijsbrechts, J., Boute, R.N., Van Mieghem, J.A., Zhang, D.: Can deep reinforcement learning improve inventory management? performance on dual sourcing, lost sales and multi-echelon problems. SSRN (2020). https://doi.org/10.2139/ssrn.3302881
7. Haarnoja, T., Zhou, A., Abbeel, P., Levine, S.: Soft actor-critic: off-policy maximum entropy deep reinforcement learning with a stochastic actor. In: Dy, J., Krause, A. (eds.) Proceedings of the 35th International Conference on Machine Learning. Proceedings of Machine Learning Research, 10–15 Jul 2018, vol. 80, pp. 1861–1870. PMLR (2018). http://proceedings.mlr.press/v80/haarnoja18b.html
8. Hachaïchi, Y., Chemingui, Y., Affes, M.: A policy gradient based reinforcement learning method for supply chain management. In: 2020 4th International Conference on Advanced Systems and Emergent Technologies (IC_ASET), pp. 135–140 (2020). https://doi.org/10.1109/IC_ASET49463.2020.9318258
9. Hutse, V.: Reinforcement learning for inventory optimisation in multi-echelon supply chains. Master in business engineering, Gent University (2019). http://lib.ugent.be/catalog/rug01:002790831
10. Kemmer, L., von Kleist, H., de Rochebouët, D., Tziortziotis, N., Read, J.: Reinforcement learning for supply chain optimization. In: European Workshop on Reinforcement Learning, vol. 14 (2018). https://ewrl.files.wordpress.com/2018/09/ewrl_14_2018_paper_44.pdf
11. Lillicrap, T.P., et al.: Continuous control with deep reinforcement learning. arXiv preprint arXiv:1509.02971 (2015)
12. Mnih, V., et al.: Asynchronous methods for deep reinforcement learning. In: Balcan, M.F., Weinberger, K.Q. (eds.) Proceedings of The 33rd International Conference on Machine Learning. Proceedings of Machine Learning Research, 20–22 Jun 2016, New York, USA, vol. 48, pp. 1928–1937. PMLR (2016). http://proceedings.mlr.press/v48/mniha16.html

13. Mnih, V., et al.: Human-level control through deep reinforcement learning. Nature **518**(7540), 529–533 (2015). https://doi.org/10.1038/nature14236
14. Oroojlooyjadid, A.: Applications of machine learning in supply chains. Ph.D. thesis, Lehigh University (2019). https://preserve.lehigh.edu/etd/4364
15. Peng, Z., Zhang, Y., Feng, Y., Zhang, T., Wu, Z., Su, H.: Deep reinforcement learning approach for capacitated supply chain optimization under demand uncertainty. In: 2019 Chinese Automation Congress (CAC), pp. 3512–3517 (2019). https://doi.org/10.1109/CAC48633.2019.8997498
16. Perez, H.D., Hubbs, C.D., Li, C., Grossmann, I.E.: Algorithmic approaches to inventory management optimization. Processes **9**(1) (2021). https://doi.org/10.3390/pr9010102
17. Raffin, A., Hill, A., Ernestus, M., Gleave, A., Kanervisto, A., Dormann, N.: Stable baselines3 (2019). https://github.com/DLR-RM/stable-baselines3
18. Schulman, J., Wolski, F., Dhariwal, P., Radford, A., Klimov, O.: Proximal policy optimization algorithms. arXiv preprint arXiv:1707.06347 (2017)
19. Silver, D., et al.: Mastering the game of go with deep neural networks and tree search. Nature **529**(7587), 484–489 (2016). https://doi.org/10.1038/nature16961
20. Sutton, R., Barto, A.: Reinforcement learning, second edition: an introduction. In: Adaptive Computation and Machine Learning series. MIT Press (2018). https://mitpress.mit.edu/books/reinforcement-learning-second-edition

Formants Analysis of L2 Arabic Short Vowels: The Impact of Gender and Foreign Accent

Ghania Droua-Hamdani[✉]

Centre for Scientific and Technical Research on Arabic Language Development (CRSTDLA), Algiers, Algeria
gh.droua@post.com, g.droua@crstdla.dz

Abstract. The paper examines the formant of short vowels in Modern Standard Arabic (MSA) language produced by native and non-natives speakers. The experiment displays variations in MSA vowel quality when the mother tongue of L2 speakers is English. The analysis was conducted on F1, F2, and F3 formants computed from 145 Arabic sentences of the West Point corpus. Statistical analyses were applied on formant values to reveal the impact of foreign accent and the gender on Arabic short vowel quality produced by L2 speakers. Results show a significant effect of gender (female/male) and foreign accent of speakers on the formant frequencies.

Keywords: Formants · Statistical analysis · Modern Standard Arabic · Native speakers · Non-native speakers · Gender

1 Introduction

Recognizing native speakers from non-native speakers is often easier for a human being, but this task becomes a challenging problem when it comes to an automatic system. The distinction of speakers' accents in automatic recognizers and classifiers needs to be trained and tested using several kinds of speech features such as MFCC, rhythm metrics, etc. [1–6]. Formants that refer to the frequency resonance of the vocal tract are important acoustic features that are widely studied in speech processing as well as in: sounds production comparison within languages, second language (L2) acquisition, speech pathology studies, etc. [3, 7–12]. The present study examines vowel variation quality (the first, second, and third formants, hereafter F1, F2, and F3) within L1 and L2 Arabic language. The objective is to put forward formant variation in vowel production in Modern Standard Arabic (MSA) spoken by native vs. non-natives speakers using statistical analyses. Thus, we examined speakers' foreign accent, gender, and variations in the articulation of Arabic short vowels formant within connected sentences produced by Arabic and American participants.

MSA is a Semitic language that is endowed by six vowels: three short vowels (/a/, /u/ and /i/) vs. three long vowels (/a:/, /u:/ and /i:/). However, the English language has fifteen vowel sounds. The investigation examines only short vowels extracted from the speech material.

© Springer Nature Switzerland AG 2021
L. Rutkowski et al. (Eds.): ICAISC 2021, LNAI 12855, pp. 240–246, 2021.
https://doi.org/10.1007/978-3-030-87897-9_22

The paper is organized as follows. Section 2 exposes speech material and participants used in the study. Section 3 describes the measurement of the formants dataset. Section 4 shows experiments and findings. Section 5 gives the concluding remarks based on the analysis.

2 Speakers and Speech Material

Recordings of 29 speakers (15 natives/14 non-native) were used in the study. Speech material was taken from the West Point corpus that was dedicated to recording MSA texts by Arabic and American speakers [13]. The recordings were collected at a normal speech rate, a sampling frequency of 22.05 kHz. Text material included five sentences from scripts 1 that were read by all speakers. A total of 145 recordings were used in the analysis. Table 1 shows the number and gender of speakers in the sample.

Table 1. Distribution of native and non-native speakers per gender

Native speakers		Non-native speakers	
Male	Female	Male	Female
5	10	6	8
Total	15	14	

3 Measurement

Formants are distinctive frequency features of the speech signal that refer to frequency resonances of the vocal tract cavities. F1, F2,..., F5 express local maxima in the signal spectrum. To compute speech formants, an experimented annotator segmented manually all speech material i.e. 145 recordings of the dataset onto their different segmental units (vowels and consonants) using Praat software. Formants values of the short vowels were calculated using Linear Predictive Coding Coefficients (LPCC). Most often, the two first formants, F1 and F2, are sufficient to identify the kind of vowel. Nevertheless, in the study, we exploited three formants (F1, F2, and F3) values for each vowel to reveal a maximum variation in pronunciation. The data were submitted to a MANOVA to test for significant differences depending on the foreign accent (L1/L2), gender, and kind of vowels.

4 Results

4.1 L1/L2 Formant Analysis

Figure 1 shows the density of vowels spreading on (F1, F2) plan for each vowel for both native and non-native speakers. The outcomes reveal a wide distribution of vowels on

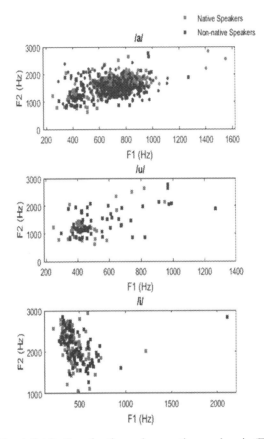

Fig. 1. Vowel distribution of native and non-native speakers in (F1, F2) plan

(F1, F2) plan especially /a/ and /u/ both speakers groups. Regarding the vowel /i/, we can see from the figure that formants computed for natives and non-native speakers are gathered in one consolidated area.

The assessment of the mean values of all formants for both corpora was conducted regardless of the gender of speakers (male/female). Table 2 expresses the average and standard deviation values of F1, F2 and F3 measured from L1 and L2 formants for each short vowel /a/, /u/, and /i/.

The outcomes performed for the vowel /a/: the average value of F1 of native speakers is close to that calculated for non-native ones. The same observation is valid for F3. However, the results express a deviation in the case of F2 formant. It can be noticed that the standard deviation measured for all formants for both categories (L1/L2 speakers) is nearby to each other. Regarding findings of /u/ analysis, the most significant difference between L1 and L2 formants scores is distinguished in F2 value. For the vowel /i/, the higher value is reached in the case of F3. Moreover, we point out, for /u/ and /i/, a slight variation in formant standard deviations between L1 and L2.

Table 2. Means and standard deviation of F1, F2 and F3 (Hertz) of native (L1) and non-native (L2) speakers

		/a/ M	SD	/u/ M	SD	/i/ M	SD
F1	L1	756,50	125,39	456,54	130,13	435,19	116,58
	L2	722,95	130,50	544,68	200,89	487,36	217,41
F2	L1	1643,09	244,87	1284,66	441,14	2139,58	357,23
	L2	1514,58	264,37	1420,32	472,96	2089,83	368,68
F3	L1	2735,39	325,26	2712,75	253,23	2929,33	289,58
	L2	2702,39	275,37	2785,36	356,50	2819,84	301,39

A statistical analysis (One Way ANOVA) was directed on the data formant set for each L1/L2 vowels (/a/; /u/ and /i/) with $\alpha = 0.05$. The outcomes display a significant effect of L2 accent in both F1 and F2 formant for the vowel /a/. The results found are F1: $F(1,658) = 11.32\ p = 0.01$; second formant F2: $F(1,658) = 42\ p = 0.00$ and for the third formant F3: $F(1,658) = 1.95\ p = 0.163$. Likewise, findings obtained from /u/ and /i/ analyses show a significant effect L1/L2 accent on F1 for both vowels and only on F3 in the case of /i/. The results are F1: $F(1,103) = 6.58\ p = 0.012$; second formant F2: $F(1,103) = 2.024\ p = 0.13$; third formant F3 $F(1,103) = 1.35\ p = 0.24$; F1: $F(1,199) = 4.77\ p = 0.03$; F2: $F(1,199) = 0.9\ p = 0.33$; F3 $F(1,199) = 6.80\ p = 0.01$ respectively. We can observe from the results that all first formants F1 for all vowels were significant to foreign accent. Regarding formants F2 and F3, the effect depends on the vowel type. We can suggest that the difference in generating Arabic vowels between Arabic speakers and their counterpart Americans maybe occurs in the first level of production of the segment i.e. at the F1 formant stage. The statement is valid for all Arabic short vowels. In the case of the vowel /a/, a difference is also noticed in the second formant. As the analysis excludes the physiological study of the vocal cavities and as regards the vowel diagram, the outcomes can be explained as follow: F1, which refers to the position of the tongue on a vertical axis and ranges from open to close, reveal a deviation in F1 when L2 speakers produced all Arabic vowels. Regarding F2, which refers to the position of the tongue on a horizontal axis in the vowel diagram, shows also a deviation in the case of /a/ pronounced by L2 subjects. These variations can be explained by the English mother tongue influence on L2 vowels production.

4.2   Gender L1/L2 Formant Analysis

The second experiment concerns the assessment of the mean values of all formants regarding L1 and L2 speakers' gender (male/female). The analysis aims to express possible foreign accent influence in vowel production regarding both the gender and the origin of speakers. Tables 3 and 4 present the average and standard deviation values of F1, F2, and F3 measured for female and male speakers for each short vowel.

Table 3. Means and standard deviation of F1, F2 and F3 (Hertz) for MSA native and non-native female short vowels

		/a/		/u/		/i/	
		M	SD	M	SD	M	SD
F1	L1	790,37	114,39	462,90	139,74	443,76	131,60
	L2	747,34	124,14	505,68	131,04	474,50	114,90
F2	L1	1701,24	223,40	1223,66	393,33	2227,62	329,61
	L2	1615,99	260,06	1369,23	387,40	2172,78	371,36
F3	L1	2810,87	329,06	2763,83	167,85	3004,11	285,92
	L2	2747,26	309,52	2705,47	296,18	2865,76	294,14

Table 4. Means and standard deviation of F1, F2 and F3 (Hertz) of MSA native and non-native male speakers

		/a/		/u/		/i/	
		M	SD	M	SD	M	SD
F1	L1	688,74	119,21	443,80	111,83	417,34	74,94
	L2	691,91	132,23	599,96	263,49	507,39	319,53
F2	L1	1526,78	245,35	1406,68	516,90	1955,71	346,23
	L2	1385,57	208,17	1491,83	572,84	1960,51	329,60
F3	L1	2584,41	329,06	2610,57	355,52	2773,69	231,86
	L2	2645,29	212,30	2897,18	407,20	2748,25	302,85

As it can be seen from Tables 3 and 4 all formants values computed for L1 speakers (females and males), in the case of the vowel /a/, are relatively higher than those measured for L2 speakers (females and males). The only exception is noticed in the F1 score of L2 male participants where the tendency is inversed. Moreover, findings of formants /u/ vowel show increased F1, F2, and F3 values for L2 males in comparison to their L1 counterparts; and higher F1 and F2 in L2 female speakers. For the/i/ vowel, the results vary depending on gender and formants.

Statistical analyses (One Way ANOVA) were applied separately on the data formant set for female speakers then for male speakers for each short vowels (/a/; /u/ and /i/). The results achieved for females: vowel /a/ showed a significant effect of L2 accent on formant values (F1, F2 and F3). The findings for first formant F1 are $F_{(1, 405)} = 13,08$ $p = 0.0$, for the second formant F2 are $F_{(1, 405)} = 12,58$ $p = 0.0$ and for the third formant F3 are $F_{(1, 405)} = 3,90$ $p = 0.049$. For the vowels /u/ and /i/; the findings are for F1: $F_{(1, 63)} = 1,58$ $p = 0.21$; F2: $F_{(1, 63)} = 2,24$ $p = 0.13$ and for F3: $F_{(1, 63)} = 0,91$ $p = 0.037$; F1: $F_{(1, 128)} = 1,90$ $p = 0.17$, F2: $F_{(1, 128)} = 0,78$ $p = 0.37$ and F3: $F_{(1, 128)} = 7,17$ $p = 0.08$ respectively. The results above showed significant effects of speaker origin for female group in all formants of the vowel /a/ and only on F3 formant of /i/ vowel. The same analysis was conducted on male formant group. The outcomes indicated a significant effect of foreign accent on male formant measures F2 and F3 in case of /a/: F1, $F(1, 251) = 0.4$ $p = 0.84$; F2, $F(1, 251) = 24.52$ $p = 0.0$; F3: $F(1, 251) = 4.21$ $p = 0.04$. For /u/, there is significant effect only on F1 and F3. The results are: F1, $F(1, 38) = 4.7$ $p = 0.03$; F2: $F(1, 38) = 0,22$ $p = 0.64$ and finally F3: $F(1, 38) = 5,09$ $p = 0.03$. Thus, we can conclude that the quality of Arabic vowels pronounced by L2 speakers depends on gender and the kind of vowel.

5 Conclusion

A formant analysis was directed on Modern Standard Arabic (MSA) language produced by native and non-natives speakers. The experiment displays variations in MSA vowel quality between L1 and L2 speakers. Thus, we examined speakers' foreign accents, gender, and variations on short Arabic vowels. The study was conducted basing on 145 speech files recorded by 15 speakers. Three formants were computed from a set of vocalic segments (/a/, /u/, and /i/) of L1 and L2 speech material. Two experiments were done on the dataset. The first consisted on studying L1 and L2 formant regardless of the speaker's gender. The average values and statistical analysis were performed for F1, F2, and F3 and all vowels. Results expose that the difference in generating Arabic vowels between L1 speakers and their counterpart L2 speakers occurs in the F1 formant stage. The statement is valid for all Arabic short vowels. In the case of the vowel /a/, a variation in the F2 formant value is also noticed. The second experiment concerned the assessment of the mean values of all formants regarding L1 and L2 speaker's gender (male/female), followed by two MANOVA executed on the L1/L2 female group than on the L1/L2 male group. The outcomes state that Arabic vowels pronounced by L2 speakers in the speech material do not answer to a unique pattern of MSA short vowel quality. The production depends on the gender of speakers and the kind of pronounced vowel.

References

1. Nicolao, M., Beeston, A.V., Hain, T.: Automatic assessment of English learner pronunciation using discriminative classifiers. In: IEEE International Conference on Acoustics, Speech and Signal Processing (ICASSP), Brisbane, QLD, pp. 5351–5355 (2015). https://doi.org/10.1109/ICASSP.2015.7178993
2. Marzieh Razavi, M., Magimai Doss, M.: On Recognition of Non-native Speech Using Probabilistic Lexical Model INTERSPEECH 2014 15th Annual Conference of the International Speech Communication Association, Singapore, 14–18 September (2014)
3. Alotaibi, Y.A., Hussain, A.: Speech recognition system and formant based analysis of spoken Arabic vowels. In: Lee, Y.-H., Kim, T.-H., Fang, W.-C., Ślęzak, D. (eds.) FGIT 2009. LNCS, vol. 5899, pp. 50–60. Springer, Heidelberg (2009). https://doi.org/10.1007/978-3-642-105 09-8_7
4. Droua-Hamdani, G., Selouani, S.A., Boudraa, M.: Speaker-independent ASR for modern standard arabic: effect of regional accents. Int. J. Speech Technol. **15**(4), 487–493 (2012)

5. Droua-Hamdani, G., Sellouani, S.A., Boudraa, M.: Effect of characteristics of speakers on MSA ASR performance. In: IEEE Proceedings of the First International Conference on Communications, Signal Processing, and their Applications (ICCSPA 2013), pp. 1–5 (2013)
6. Droua-Hamdani, G.: Classification of regional accent using speech rhythm metrics. In: Salah, A., Karpov, A., Potapova, R. (eds.) SPECOM 2019. LNCS, vol. 11658, pp. 75–81. Springer, Cham (2019). https://doi.org/10.1007/978-3-030-26061-3_8
7. Droua-Hamdani, G.: Formant frequency analysis of MSA vowels in six algerian regions. In: Karpov, A., Potapova, R. (eds.) SPECOM 2020. LNCS, vol. 12335, pp. 128–135. Springer, Cham (2020). https://doi.org/10.1007/978-3-030-60276-5_13
8. Farchi, M., Tahiry, K., Soufyane, M., Badia, M., Mouhsen, A.: Energy distribution in formant bands for Arabic vowels. Int. J. Elect. Comput. Eng. Yogyakarta 9(2), 1163–1167 (2019)
9. Mannepalli, K., Sastry, P.N., Suman, M.: Analysis of emotion recognition system for Telugu using prosodic and formant features. In: Agrawal, S.S., Dev, A., Wason, R., Bansal, P. (eds.) Speech and Language Processing for Human-Machine Communications. AISC, vol. 664, pp. 137–144. Springer, Singapore (2018). https://doi.org/10.1007/978-981-10-6626-9_15
10. Korkmaz, Y., Boyacı, A.: Classification of Turkish vowels based on formant frequencies. In: International Conference on Artificial Intelligence and Data Processing (IDAP), Malatya, Turkey, 2018, pp. 1–4 (2018). https://doi.org/10.1109/IDAP.2018.8620877
11. Natour, Y.S., Marie, B.S., Saleem, M.A., Tadros, Y.K.: Formant frequency characteristics in normal Arabic-speaking Jordanians. J. Voice 25(2), e75–e84 (2011)
12. Rusza, J., Cmejla, R.: Quantitative acoustic measurements for characterization of speech and voice disorders in early-untreated Parkinson's disease. J. Acoust. Soc. Am. 129, 350 (2011). https://doi.org/10.1121/1.3514381
13. Linguistic Data Consortium LDC. http://www.ldc.upenn.edu

Cluster Analysis of Co-occurring Human Personality Traits and Depression

Marta Emirsajłow[(✉)] [iD]

Department of Computer Engineering, Wrocław University of Science
and Technology, Wybrzeże Wyspiańskiego 27, 50-370 Wrocław, Poland
marta.emirsajlow@pwr.edu.pl

Abstract. Methods of artificial intelligence are widely used in medicine
and psychology. Their development gives hope to find various intercon-
nections between human characteristics and states that would not be
possible without the use of advanced computer science technologies. The
aim of this paper is to use unsupervised machine learning to investigate
whether there are any relationships between human personality traits
and susceptibility to depression. It is well known that diseases like depression are associated with many factors and it is very difficult to isolate
them. Using the tool of clusterization, an analysis of personality traits co-
occurring with depressive states was carried out, using publicly available
data base containing reliable questionnaire studies.

Keywords: Methods of artificial intelligence · Clusterization ·
Personality traits · Depression

1 Introduction

Taking into account the complexity of human behaviour and the difficulties
that arise when looking for relationships between human personality features
and states or behaviours, one should look at the possibilities offered by artificial
intelligence. It is a very dynamically developing field in research and applications.
By combining psychological knowledge and computer science technology, you
can get completely new possibilities of data analysis, and thus also of human
features and behaviours, see [1,4,9,12,14]. When reflecting on the determinants
of depression, it is hard not to notice that there may be a relationship between the
set of personality traits of an individual and an increased tendency to manifest
depressive symptoms [5,7].

This article examines the relationship between the combination of personality traits and the results on the depression scale. The data that was used for
the research came from students and were obtained in the **PHQ-9 depression
scale** and **Big Five personality traits** official questionnaires [6,8]. Data analysis was performed using clustering. It is a term used in data mining, and more
specifically, it relies on a patternless classification. This means that the classification is done unsupervised [13]. It is a grouping of elements into relatively sets.

© Springer Nature Switzerland AG 2021
L. Rutkowski et al. (Eds.): ICAISC 2021, LNAI 12855, pp. 247–256, 2021.
https://doi.org/10.1007/978-3-030-87897-9_23

In most algorithms, this is done on the basis of the similarity between the elements of the sets, determined by a function. Thanks to clustering, it is possible to discover the data structure and to extract classes and apply generalizations.

2   Method Description

This paper presents the compilation of data into relatively homogeneous groups called clusters. Based on the analysis of the scientific literature on clustering, the Matlab environment was selected and the methods available in its *Statistics and Machine Learning Toolbox* [10] were used. The methods used are the routine *cluster*, which works in connection with the routines *linkage* and *pdist* as follows

$$y = pdist(x, \text{'metric'});$$
$$z = linkage(y, \text{'grouping method'});$$
$$t = cluster(z, \text{'number of groups'});$$

The *cluster* routine performs grouping of the elements from the whole collection. Before applying it, it is necessary to define the metric to be used and to select the method on which clustering will be performed. The result of this procedure is a vector with a length equal to the number of points to be grouped. The elements of the vector are numbers ranging from 1 to the number of groups. These numbers represent the numbers of the cluster to which a given point will be assigned.

The choice of the metric we want to use for calculations is done in the *pdist* procedure. The coordinates of the points to be grouped are placed in the x array, while the second input parameter is the selected metric that will be used to calculate the distance between the points. In this study, the metric *euclidean* was used. It consists in calculating the Euclidean distance described by the usual formula:

$$d(X, Y) = \left(\sum_{i=1}^{n} (x_i - y_i)^2 \right)^{\frac{1}{2}} \tag{1}$$

where X, Y denote the points of the set in n-dimensional space, and x_i and y_i are the coordinates of these points.

Another procedure used in this paper is called *linkage*. It is used to specify the method on which clustering is based. The input parameters include the vector y, obtained from the *pdist* procedure, and the method of clustering. In this paper, the method *'complete'* is chosen and it turns out to be the most effective for the analyzed data. This is a hierarchical cluster analysis method, also known as the full binding method. The distance is calculated as the greatest distance between the elements in the set and the external elements. For the point P, which is not included in the cluster $X = \{X_i\}$, the distance between the cluster and the point P is described by the formula:

$$d(P, X) = \max_i d(P, X_i) \tag{2}$$

where X_i stand for the elements of X in n-dimensional space.

For the two sets X and Y the distance $d(X, Y)$ between them is calculated as the distance between the two most distant elements as follows

$$d(X, Y) = \max_{i,j} d(X_i, Y_j) \tag{3}$$

where $X_i \in X$, $Y_j \in Y$.

The methods used in the approach presented in the paper have been chosen on the basis literature and own research and have been found to be most effective with the data used in this work.

3   Research Description

In order to carry out the research in this paper we used data from the project *StudentLife: Using Smartphones to Assess Mental Health and Academic Performance of College Students* [15], realized by a group of researchers from Dartmouth College in the United States. This research was done on a group of respondents studying computer science programming at Dartmouth College and its raw data base was made publicly available at [16].

During that project, the respondents were examined using a dedicated smartphone application for many different factors influencing their behavior, status and academic performance. The tools used also included a number of psychological questionnaires examining, among others, personality traits and the level of depression. 48 people took part in the research, organized during a 10-week trimester of study, including 10 women and 38 men. The data we were interested in, namely, the **Big Five** questionnaire and the **PHQ-9** questionnaire answers, were obtained as part of this project and made publicly available at [16]. We included 46 complete sets of results and excluded some which we incomplete.

3.1   Results Description

The results from the Big Five and PHQ-9 questionnaires based on the respondents' answers were counted to raw scores by a number of programs written in JavaScript. The obtained raw results from the depression level questionnaire were divided into three levels. According to the official scale, the PHQ-9 questionnaire differentiates the respondents into 5 levels of depression (none-to-minimal, mild, moderate, moderately severe, severe), while for the purposes of this study, the results of the respondents were divided into three levels: *none-to-minimal, mild, moderate-to-severe*. The above division was made to properly balance the data set, because a very small percentage of the population suffers from severe depression and often, with such a severe course of the disease, it is not possible to actively participate in university classes. For this reason the collections in which the respondents would have such high scores on the scale of depression would be too unrepresentative. In the Big Five questionnaire, each individual was assigned every of the five personality traits: *extraversion, neuroticism, openness, conscientiousness, agreeableness*, with a certain intensity.

For each individual the **vector of five personality traits** was created, and then the results were normalized so that each of the personality traits factor was described on the same normalized scale. In some comments we refer to the original scale of scores and then we use the name *raw points* or simply *points*. The respondents were divided into **three groups depending on their result in the PHQ-9 questionnaire**, and then a table containing vectors of personality traits was created for each group.

Table 1. Table of trait vectors for the level *none-to-minimal*

Extraversion	Neuroticism	Openness	Conscientiousness	Agreeableness
0.4250	0.7000	0.6400	0.6000	0.8444
0.6250	0.5500	0.7800	0.7556	0.8889
0.4750	0.8250	0.5800	0.6222	0.5333
0.4750	0.4250	0.6600	0.9111	0.5333
0.3500	0.6750	0.7200	0.2667	0.9556
0.5500	0.5750	0.6800	0.7111	0.7333
0.6500	0.4250	0.6600	0.7333	0.7333
0.5500	0.6750	0.8600	0.7111	0.4889
0.6500	0.5500	0.6000	0.7111	0.6222
0.7250	0.5750	0.7600	0.6889	0.9111
0.5000	0.6250	0.6600	0.7778	0.5333
0.8500	0.3500	0.7400	0.8444	0.8667
0.6500	0.5750	0.6600	0.8667	0.6222
0.7000	0.3500	0.7000	0.8222	0.7556
0.6750	0.6000	0.7000	0.6667	0.6222
0.6250	0.5000	0.7800	0.8444	0.4889
0.6750	0.4500	0.7200	0.7333	0.7556
0.5750	0.5000	0.8600	0.8444	0.7333
0.6750	0.5500	0.8000	0.7556	0.7556
0.3250	0.4000	0.8200	0.8222	0.6667
0.3750	0.4250	0.5400	0.8000	0.6444
0.3750	0.7750	0.7200	0.8667	0.7111

The Table 1 shows the normalized vectors of personality traits for the *none-to-minimal* depression level. For each table, the average personality trait vector was calculated, so that it was possible to determine the average values of the traits in a given group, assigned to the depression level. People from the *none-to-minimal* level were characterized by an average level of extraversion and neuroticism, only one point below the average level of these traits in relation to the entire group of respondents (25 raw points). The level of openness and agreeableness was also around the average (respectively, 36 raw points and 33 raw points), while the level of conscientiousness was 3 points higher than the average obtained (31 raw points).

Table 2. Table of trait vectors for the level *mild*

Extraversion	Neuroticism	Openness	Conscientiousness	Agreeableness
0.4250	0.6750	0.5400	0.6444	0.7778
0.6250	0.5500	0.6600	0.7556	0.8889
0.5250	0.6500	0.6600	0.7556	0.8444
0.6250	0.3750	0.4600	0.8444	0.7556
0.3000	0.8500	0.7600	0.6667	0.7111
0.3250	0.7250	0.6400	0.6222	0.6889
0.7750	0.4750	0.7200	0.8667	0.6667
0.7000	0.6250	0.7000	0.5333	0.4667
0.9000	0.4000	0.7200	0.8222	0.8889
0.5250	0.5250	0.6800	0.7778	0.7111
0.4750	0.5000	0.6600	0.8222	0.8444
0.6250	0.6250	0.7400	0.7333	0.6889
0.8750	0.4000	0.7600	0.6444	0.6889
0.7750	0.5000	0.7000	0.6667	0.7111
0.4500	0.9000	0.8800	0.8000	0.5778
0.8000	0.5750	0.7000	0.8222	0.5111

The Table 2 shows the personality traits vectors for the level *mild* of depression. Respondents assigned to this group were characterized by average, in relation to the whole population, extraversion and neuroticism, as well as average openness and agreeableness and increased conscientiousness.

Table 3. Table of trait vectors for the level *moderate-to-severe*

Extraversion	Neuroticism	Openness	Conscientiousness	Agreeableness
0.7500	0.6250	0.9000	0.5778	0.6000
0.5250	0.6250	0.9000	0.7778	0.6667
0.5250	0.7500	0.5000	0.4444	0.6444
0.5250	0.7500	0.7200	0.7333	0.6000
0.4750	0.6250	0.7400	0.7778	0.5778
0.5500	0.7250	0.9200	0.6667	0.3778
0.6250	0.7000	0.8400	0.6222	0.7111
0.4500	0.9000	0.8800	0.8000	0.5778

The last analyzed group were individuals with the *moderate-to-severe* level of depression, i.e. people with a significant depression. The Table 3 shows vectors corresponding to their personality traits. People in this group differed the most from the mean in terms of their average personality trait vector. Extraversion was slightly reduced, neuroticism was significantly higher than average, openness to experience was also higher than average, while agreeableness was 7 points lower than the average in the study population. Only conscientiousness did not differ from the average.

The next step was to carry out clustering and the dendrogram was shown in the Fig. 1 (using the Matlab procedure *dendrogram*).

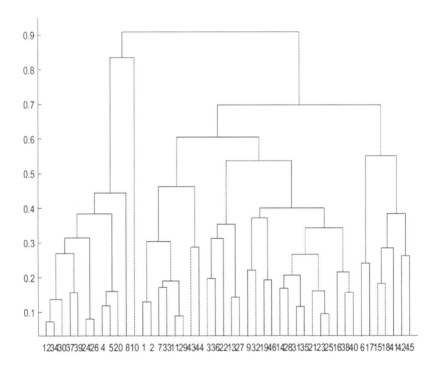

Fig. 1. Clusters for personality traits.

Based on the analysis of the dendrogram it was decided to perform clustering by dividing the data into five clusters which corresponds to the distance $d = 0.55$. It can be seen that out of the five created clusters, one contains only one individual, so it is omitted in the analysis. For each of the remaining four clusters, a table of personality trait vector was created and the average personality trait vector was calculated. Considering the legitimacy of the division into such four clusters, it can be noticed that in the first cluster there were people with relatively low extraversion, 6 raw points lower than the average, and the result of other features close to the average. In the second cluster, the mean extraversion was increased by 2 raw points from the average among the respondents. Neuroticism was around average, openness and conscientiousness were increased relative to the mean by 3 and 4 raw points, respectively, while agreeableness was decreased by 5 points. People with low extraversion were sent to the third cluster, 8 raw points lower than the average in the whole group of respondents. High neuroticism 9 points above average, close to average openness and conscientiousness, and agreeableness 4 points down from the mean. The fourth cluster collected the respondents with the results of high extraversion, 6 points above the average, reduced neuroticism by 5 raw points, average openness and higher conscientiousness and agreeableness by 4 raw points.

In order to analyze the relationship between the scores on the personality trait scale and the assignment to a given depression level, the maximal distances between clusters $\{X_i\}_{i=1}^{i=4}$ and sets $\{Y_j\}_{j=1}^{j=3}$ of individuals belonging to each of the depression level were calculated. Matlab functions max and $pdist2$, were used for this purpose. The function $pdist2$ computes the matrix of pairwise distance between two sets and then the function max selects the largest distance. The maximal distances between sets were calculated according to the formula

$$di_{\max} = \max(pdist2(X_i, Y_j)) \quad \text{for} \quad j = 1, 2, 3. \tag{4}$$

The Table 4 presents a comparison of the obtained distances between the farthest elements in the sets representing three levels of depression and four clusters of personality traits.

Table 4. Comparison of maximal distances between clusters and depression level sets

Clusters	None-to-minimal	Mild	Moderate-to-severe
1	$d1_{\max} = 0.6916$	$d1_{\max} = 0.6245$	$d1_{\max} = 0.6389$
2	$d2_{\max} = 0.8480$	$d2_{\max} = 0.7440$	$d2_{\max} = 0.6044$
3	$d3_{\max} = 0.8928$	$d3_{\max} = 0.9085$	$d3_{\max} = 0.6018$
4	$d4_{\max} = 0.8353$	$d5_{\max} = 0.7873$	$d5_{\max} = 0.7585$

The distance analysis shows that the *none-to-minimal* level is closest to the first cluster because the distance between the farthest points in both sets is the shortest one. The distances between the first level and the other three clusters are clearly larger. This means that due to the similarity of personality trait levels, people with the lowest level of depression and people in the first cluster are the closest. A similar situation occurs with the *mild* level of depression. The respondents from the first cluster have the most similar personality traits, i.e. those who are characterized by low extraversion and average level of other personality traits. The next cluster in terms of closeness, but still further, is the second cluster. The smallest closeness of features is between the *mild* level and the third cluster. The last, highest level is *moderate-to-severe*. It includes people who score high on the depression scale. The distances between this level and the first three clusters are the smallest. However, this especially applies to the second and third clusters, where the shortest distances from all the measured distances were obtained. The highest level of depression is the farthest from cluster 4, which includes individuals with increased extroversion, lower than average neuroticism, medium openness, and above average conscientiousness and agreeableness.

3.2 Conclusions

On the basis of the obtained results, it can be observed that all three levels are in similar proximity to the first cluster. This means that it cannot be concluded that the personality traits of people in the first cluster differentiated the respondents in terms of the level of depression. However, it can be seen that the distance obtained decreases slightly with increasing levels of depression. The biggest difference, although still small, is between the non-depressed subjects and the other two levels.

Another interesting observation is the distance of the third depression level from the first three clusters. As the first cluster was relatively close to the remaining levels, it should not be associated with depression, while the other two clusters, the second and the third, were in such proximity only to the highest level of depression. This may suggest that the personality traits found in these clusters may be related in some way to the disease. The second cluster was characterized by average scores on extraversion and neuroticism, increased scores on openness and conscientiousness, and reduced agreeableness. This means that such a combination of personality traits may be associated with high levels of depression. The third cluster included people with low extroversion, high neuroticism, moderate openness and conscientiousness, and reduced agreeableness. Researchers from Istanbul Billim University in the work [5] showed a relationship between extroversion and a higher quality of life of patients, as well as the relationship between high neuroticism and depression, which is confirmed by the obtained results associating the third cluster, in which the respondents showed high neuroticism and low extraversion, with high depression. On the other hand, scientists in the study [7] show moderate to large associations between depression and personality traits such as neuroticism, extraversion and conscientiousness.

Summing up, the obtained data show that both combinations of personality traits occurring in the second and third clusters, and especially in the third cluster, are associated with high depression. These results comply with some existing literature findings.

4 Final Remarks

The results of this study are based on data from questionnaire answers. It is therefore necessary to raise the issue of the reliability of the results obtained with such tools. It is known that questionnaire research is prone to the natural and occurring desire of self-presentation in every human being. It also happens that the respondents are not able to accurately define their own characteristics, it is possible to give random answers in the case of a question that is too difficult to construct. Nevertheless, questionnaire surveys are the most popular form of research and provide the greatest opportunities for large-scale research. Additionally, the PHQ-9 and Big Five questionnaires are highly reliable tools, which reduces the chance of obtaining false results.

The study conducted by the author of the paper showed the existing links between personality traits and susceptibility to the occurrence of symptoms of depression. It was noticed that the highest level, and therefore the most symptoms of depression, was observed in people with high neuroticism scores and low scores on the extraversion and agreeableness continuum. It should also be noted that the more the results were different from the average, the greater the disturbance in the sense of life satisfaction in the form of higher scores on the depression scale.

All these observations were obtained through the analysis of data in the field of psychology with the use of IT tools. Such a combination may expand the possibilities of conducting research on personality traits and depression, and as a result, provide a better understanding of the impact of a combination of personality traits on the risk and severity of depression, which may lead to the individualization of the selection of therapy and treatment for this psychological disorder.

Therefore, undoubtedly, the work on the analysis of the application of information technology tools in psychology may contribute to the development of both the field of computer science and psychology and create new possibilities for research on personality traits. The obtained results open the possibility of extending the use of IT tools, such as clustering in social sciences. Undoubtedly, the obtained results encourage further research and focus on the analysis of much larger data sets and the development of tools towards the analysis of human characteristics (see also [2,3,11]).

References

1. Chetouani, M., Cohn, J., Salah, A.A. (eds.): HBU 2016. LNCS, vol. 9997. Springer, Cham (2016). https://doi.org/10.1007/978-3-319-46843-3
2. Cohen, S., Kamarck, T., Mermelstein, R.: A global measure of perceived stress. J. Health Soc. Behav. **5**, 385–396 (1983)
3. Diener, E., et al.: New well-being measures: short scales to assess flourishing and positive and negative feelings. Soc. Ind. Res. **97**(2), 143–156 (2010)
4. Emirsajłow, M.: Pattern recognition methods in human behaviour analysis. M. Sc. thesis, Faculty of Electronics, Wroclaw University of Science and Technology, Wrocław (2019)
5. İzci, F., et al.: Impact of personality traits, anxiety, depression and hopelessness levels on quality of life in the patients with breast cancer. Eur. J. Breast Health **14**(2), 105–111 (2018)
6. John, O.P., Srivastava, S.: The big five trait taxonomy: History, measurement, and theoretical perspectives. In: Handbook of personality: Theory and Research, vol. 2, pp. 102–138 (1999)
7. Klein, D.N., Kotov, R., Bufferd, S.J.: Personality and depression: explanatory models and review of the evidence. Ann. Rev. Clin. Psychol. **7**, 269–295 (2011)

8. Kroenke, K., Spitzer, R.L., Williams, J.B.: The PHQ-9. J. Gen. Intern. Med. **16**(9), 606–613 (2001)
9. Pentland, A., Liu, A.: Modeling and prediction of human behavior. Neural Comput. **11**(1), 229–242 (1999)
10. Statistics and Machine Learning Toolbox^{TM} User's Guide, Matlab R2019a. The MathWorks (2019)
11. Russell, D.W.: UCLA loneliness scale (version 3): Reliability, validity, and factor structure. J. Pers. Assess. **66**(1), 20–40 (1996)
12. Salah, A.A., Gevers, T. (eds.): Computer Analysis of Human Behavior. Springer, London (2011). https://doi.org/10.1007/978-0-85729-994-9
13. Shalev-Shwartz, S., Ben-David, S.: Understanding Machine Learning: From Theory to Algorithms. Cambridge University Press, New York (2014)
14. Salah, A.A., Kröse, B.J.A., Cook, D.J. (eds.): HBU 2015. LNCS, vol. 9277. Springer, Cham (2015). https://doi.org/10.1007/978-3-319-24195-1
15. Wang R., et al.: StudentLife: assessing mental health, academic performance and behavioral trends of college students using Smartfons, In: Proceedings of ACM Conference on Ubiquitous Computing (2014)
16. Database of project StudentLife. http://studentlife.cs.dartmouth.edu/. Accessed 14 June 2019

Polynomial Algorithm for Solving Cross-matching Puzzles

Josef Hynek[⊠]

Faculty of Informatics and Management, University of Hradec Kralove, Rokitanskeho 62,
500 03 Hradec Králové, Czech Republic
josef.hynek@uhk.cz

Abstract. The aim of this paper it to analyze the cross-matching puzzle and to propose a fast and deterministic algorithm that can solve it. Nevertheless, there is a bigger goal than designing an algorithm for a particular problem. We want to show that while AI researchers constantly look for new constraint-satisfaction problems that could be utilized for testing various problem-solving techniques it is possible to come up with the problem that can be solved by much simpler algorithms. We would like to stress that there is an important misconception related to NP class that a huge number of potential solutions to the specific problem almost automatically implies that the relevant problem belongs to the class of NP. Such a misunderstanding and misclassification of the particular problem leads to false impression that there is no chance to design a simple and fast algorithm for the problem. Therefore, various heuristics or general problem-solving techniques are unnecessarily employed in order to solve it. And moreover, the wrong impression that the problem is difficult is further supported. We believe that our paper can help to raise the awareness that not all the problems with immense search spaces are hard to be solved and the polynomial algorithm to tackle the cross-matching puzzle that is described here is a good example of such an approach.

Keywords: Cross-matching puzzle · Efficient algorithm · Time complexity

1 Introduction

It is a very common approach that various games and puzzles are utilized in order to demonstrate the power of specific problem-solving techniques. There is a great advantage hidden in the fact that simple rules that can be easily understood could generate a huge space of potential candidate solution. There are many traditional types of puzzle like, for example, 15-puzzle (or Loyd's puzzle), Sudoku or jigsaw puzzles, while tic-tac-toe, Othello (reversi), checkers, chess or some specific tasks involving individual pieces of chess (like knights or queens) are often used to present specific problem-solving techniques or algorithms. For example, the classic 9×9 Sudoku is well-known and extremely popular amongst general public. The general problem of solving $N \times N$ Sudoku has been proved to be NP-complete [1] while there are really fast algorithms solving Sudoku of small sizes including 9×9 grid (see, for example, [2]). Furthermore,

© Springer Nature Switzerland AG 2021
L. Rutkowski et al. (Eds.): ICAISC 2021, LNAI 12855, pp. 257–266, 2021.
https://doi.org/10.1007/978-3-030-87897-9_24

various of these puzzles and games are also very often used by teachers and lectures in programming courses as it is quite easy to define the problem precisely and then to show how to tackle it algorithmically (see, for example, [3]).

On the other hand, AI researchers have always looked for new challenges and new constraint-satisfaction or constraint-optimization problems that could be utilized as testbeds for various approaches and techniques. If there is not at hand some real and practical problem to be solved, it is a nice challenge to design an artificial problem and then show the way how to solve it. On one side this approach is understandable as there are, for example, too many papers devoted to jigsaw puzzles or the safe placement of N-queens on the chessboard, while the newly designed problem might look not too common, more attractive and it could also possess some specific features that make the search for the solution somehow different or even more difficult. On the other hand, the design of new artificial problems brings along numerous questions or even risks. First of all, the new problem could be exactly the same one as the already known and well described but this time it is only defined in a different way. Secondly, while the classic problems are well-known and correctly classified as belonging to the class of NP problems, the new problem can be easily misunderstood and misclassified as being more difficult that it actually is. Finally, if some general problem-solving technique is unnecessarily employed in order to solve it, it is then clearly a worthless waste of computational resources while, at exactly the same time, the wrong impression that the problem is difficult is further supported.

2 Problem Description

We have decided to illustrate the problem on a simple cross-matching puzzle that was described by Kesemen and Özkul in [4] and the again in [5]. Their cross-matching puzzles consist of three tables with the size of M × N. For the sake of simplicity, we will consider squared tables of the size N × N in this paper but the algorithm presented below works with the size M × N as well.

The cross-matching puzzle is represented by three tables that are shown in Fig. 1. The table in the center is the solution table whose content is to be found. The table to the right is the detection table and the control table is located below. We adopted the same terminology as is used in [5]. The detection and control table represent the constraints under which the solution is to be sought.

The easiest way to describe the cross-matching puzzle is to show how to create it. At the beginning of this process, the solution table is filled by randomly generated symbols (letters or numbers). Then the symbols of the i-th row of the selection table are sorted and put into i-th row of the detection table.

We can see in Fig. 2a that the first row of the solution table containing the letters {D,M,I,B,E} was transformed into the sorted set {B,D,E,I,M} that is placed in the first row of the detection table. The same principle is applied to the creation of the control table and that is why the first column of the solution table comprising symbols {D,E,L,F,N} was converted into {D,E,F,L,N} in the first column of the control table. As soon as the detection as well as the control table are filled in, all the symbols in the solution table are erased and the puzzle is ready to be solved (Fig. 2b). In order to not confuse the reader, we have utilised exactly the same example (assignment of letters) as it was presented in [5] and this puzzle will be used throughout this paper to show how to solve it by the algorithm we are going to propose here.

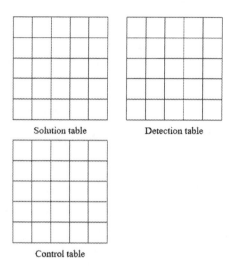

Solution table   Detection table

Control table

Fig. 1. The cross-matching puzzle tables.

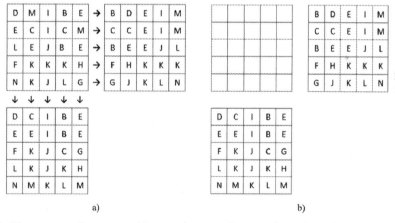

a)   b)

Fig. 2. The process of cross-matching puzzle generation – a) formation of the detection and control table, b) created cross-matching puzzle.

The aim of the puzzle is to re-construct the original solution table by using the clues given by the content of the detection and control tables. Should S be a $N \times N$ matrix representing the solution table, it is obvious that the sought symbol S_{ij} has to be present in the i-th row of the detection table and the j-th column of the control table. More specifically, if we convert the symbols of the i-th row of the detection table into a set of symbols and label it as D_i and if we do the same with the j-th column of the control table and label it as C_j, it is clear that the symbol we are looking for must lie in the intersection of these two sets. Formally written:

$$S_{ij} \in D_i \cap C_j \tag{1}$$

Based on this formula, it is evident that if the puzzle consists of N^2 symbols that are different from each other, the cardinality of the intersection (1) is always equal to one and the solution of the puzzle is absolutely straightforward. Should there be a repeated occurrence of some symbols there, the cardinality of the intersection (1) is greater than one and there is more than one candidate symbol for the placement into the given position. Then we have to wait for the other positions to be filled in order to get a clue which symbol should be selected there.

Finally, the random generation of symbols during the process of the puzzle creation does not guarantee that there is a unique solution of the puzzle. There must be at least one but depending on the frequency of the letters and their specific position there might very easily exist multiple solutions. We will address this issue later when discussing the functioning and performance of our algorithm.

Of course, the puzzle can be solved using backtracking when the positions of the solution table are consecutively assigned starting from right-top corner and trying all the possibilities from the appropriate row in the detection table while checking the constraints given by the relevant column in the control table. This algorithm provides the exact solution but its time complexity is $O(N^N)$. Therefore, due to the combinatorial explosion it can be used for small instances of the cross-matching puzzles only. Kesemen and Özkul in [5] accepted this fact as an argument that the solution of the puzzle belongs to NP-class and hence a genetic algorithm (or another stochastic search method) is needed to tackle it. We are not going to give more details on their approach using multi-layer genetic algorithm here as the details can be found in their paper and we will directly skip to their so called intelligent genetic algorithm [5].

Their main improvement there is based on their observation that it is possible to fix some elements of the solution table because in some cases the intersection of the relevant row and the column provides only one symbol to be placed there. Using this simple idea they were able to generate partial solutions to the cross-matching puzzle where some positions were fixed (the wanted symbol has been found) and the other positions were assigned randomly using the remaining available symbols. This approach has been utilized to generate the initial population of individuals and thus all of these individuals presented partial solutions where the "already known" positions were fixed.

The only remaining concern for Kesemen and Özkul [5] was to make sure that the genetic operators employed there (crossover and mutation) would not damage the already fixed parts of the partial solution represented by chromosomes. They managed to solve this obstacle easily and as they significantly reduced the size of the search space (because of the already fixed positions) their intelligent genetic algorithm works rather nicely. However, they still reported that the algorithm needed nearly 6 s to solve 10×10 cross-matching puzzle and larger instances were not attempted.

Their paper raised our curiosity and the certain similarity of cross-matching puzzles with Sudoku inspired us to analyze the problem in order to devise a heuristic algorithm that would be capable of solving it quickly. We have realized that such an algorithm exists, it is really fast and very simple.

3 Proposed Solution

It is a well-known fact that a huge number of potential solutions to the specific problem does not automatically mean that the relevant problem is hard to be solved and that it belongs to the class of NP [6]. For example, the problem of computing the shortest path between two vertices in a complete graph with positive edge weights can be solved in polynomial time despite the fact that there are exponentially many possible paths between two vertices in a complete graph. The same applies to the minimal spanning tree problem and many others. The trick is that there is no need to asses all potential candidate solutions. Utilizing the specific features of the problem we can design an efficient algorithm that finds the optimal solution without having to traverse the whole search space. Therefore, if the representation of the candidate solutions is wrongly designed causing that the search space is even bigger than it is necessary (as it was discussed, for example, in [7]) and then a brute-force approach to check the whole search space is employed, it cannot be taken as a proof that the problem is impossible to be solved efficiently by a polynomial algorithm.

It was exactly the argument concerning the huge search-space of the cross-matching puzzle that attracted our interest. Moreover, as Kesemen and Özkul [5] realized, the search-space could be rather easily narrowed by the fixation of the symbols that were unique for the particular position within the solution table. Therefore, the first step of the algorithm we would like to present here is also the calculation of the intersections between the relevant row and the relevant column using the formula (1) above. Taking step by step the symbols from the row of the detection table and browsing for them within the control table, we will reach the stage depicted in Fig. 3.

Using the example presented above we can see that there are fifteen positions in the solution table where the cardinality of the just executed intersection is one and these cells are ready to be fixed. Nevertheless, we can see that there are several more places in the table, where the cardinality of the intersection is higher than one but some symbols appear repeatedly there.

For example, the content of the cell on the second row and the second column is {C,C,M} which can be simplified to {C,M}, because only these two symbols are eligible to stand here. Similarly, the content of the cell on the fourth row and the second column is {K,K,K} which without any doubt can be simplified to {K} only. This repeated occurrence of some symbols is due to the procedure that was used to perform the intersection operation and we can either tailor it or simply check the output for uniqueness of the symbols contained within each cell. Then we reach the situation depicted in Fig. 4. There are N^2 cells, the intersection can be done in $O(N^2)$, the uniqueness of each cell in $O(N^2)$ and therefore the overall time of reaching the initial stage of the algorithm shown in Fig. 4 is $O(N^4)$.

{D,E}	{E,M}	{I}	{B}	{E,M}
{E}	{C,C,M}	{I}	{C,C}	{M}
{L}	{E,E}	{J}	{B}	{E,E}
{F}	{K,K,K}	{K,K,K}	{K,K,K}	{H}
{N}	{K}	{J}	{L}	{G}

B	D	E	I	M
C	C	E	I	M
B	E	E	J	L
F	H	K	K	K
G	J	K	L	N

D	C	I	B	E
E	E	I	B	E
F	K	J	C	G
L	K	J	K	H
N	M	K	L	M

Fig. 3. The puzzle after the application of the intersection operator.

Now it is the right time to fix all the positions where the set containing only one candidate symbol exists. Whenever a symbol is fixed in cell S_{ij} we have to delete this symbol from the i-th row in the detection table as well as from the j-th column in the control table. Keeping in our mind that some symbols occur there multiple times it is necessary to make sure that only one symbol is deleted from the respective row and column each time. There are N^2 cells and therefore the fixation including the removal of the fixed symbol from the row and the column can be done in $O(N^3)$.

{D,E}	{E,M}	{I}	{B}	{E,M}
{E}	{C,M}	{I}	{C}	{M}
{L}	{E}	{J}	{B}	{E}
{F}	{K}	{K}	{K}	{H}
{N}	{K}	{J}	{L}	{G}

B	D	E	I	M
C	C	E	I	M
B	E	E	J	L
F	H	K	K	K
G	J	K	L	N

D	C	I	B	E
E	E	I	B	E
F	K	J	C	G
L	K	J	K	H
N	M	K	L	M

Fig. 4. The puzzle after the application of the intersection operator including the uniqueness of symbols in individual cells.

We can see in Fig. 5 that in our illustrative example nearly all the positions within the solution table are fixed right now (21 out of 25). Moreover, we can see that the rest of the puzzle will be solved quickly using the same process based on the row and column intersections that will be performed only for the positions that are unfixed yet.

Naturally, symbol D will occupy the top left corner position in the solution table as the result of intersection between {D} and {D,E,M}, the next position has to be M, because {D,E,M} ∩ {C,M} = {M}, etc. Therefore, in this specific case it is clear that after the second cycle of computing the intersections and fixing the positions where only one candidate symbol exists, the algorithm will reach the solution in Fig. 6.

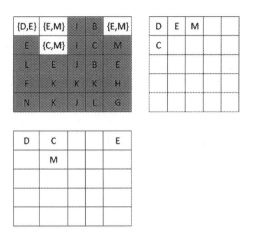

Fig. 5. The puzzle after the fixation of the already known positions.

Of course, depending on the size of the cross-matching puzzle and the number of symbols utilised there could be more cycles needed. If at the end of the cycle all the positions of the solution table are known (it means fixed), we have reached the solution and the program can terminate. The other option is to check the content of the detection or the control table, because at this stage of the computation both of them must be empty (all the symbols were placed and thus deleted from these tables). Secondly, for each cycle we calculate the number of symbols that were fixed within the cycle (*Fixed*). Positive value of *Fixed* indicates that at least one symbol was fixed and removed from the detection and control table and we can safely continue with another cycle. If there is an unique (single) solution to the cross-matching puzzle, our algorithm finds it and terminates.

However, there are situations (especially when the cross-matching puzzle has been generated randomly) that there are several solutions there. In Fig. 7a) we can see the assignment that leads to multiple solution. Utilising the algorithm described above we will reach after two cycles the situation depicted in Fig. 7b). From this stage of calculation no further improvement is possible because the intersection in all four corners always contains two symbols {A,C}. Nevertheless, as no further symbol can be fixed, the value of the above defined indicator *Fixed* is equal to zero and algorithm terminates as well. The output here is a partial solution as one described in Fig. 7b) and in this particular case it indicates that there are two solutions to the cross-matching puzzle depending whether the symbol A or symbol C is selected for the upper left corner in the solution table. However, it is easy to find a solution (or even to generate all the solutions) as whenever we select the particular symbol from the set, this symbol is deleted from the relevant row as well as the relevant column and the maximum number of choices to be made is less or equal to $(N^2-N)/2$, where N is the size of the puzzle. We can see that in our example from Fig. 7 only one decision is needed in order to obtain one of the two existing solutions.

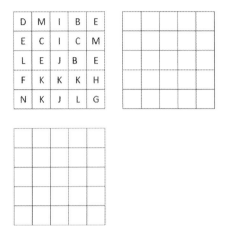

Fig. 6. The solution of the cross-matching puzzle.

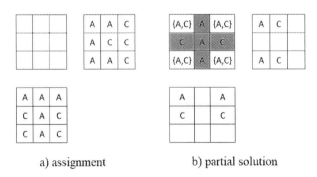

a) assignment   b) partial solution

Fig. 7. The cross-matching puzzle with two different solutions.

We can conclude that we have designed a simple fast polynomial deterministic algorithm that solves cross-matching puzzle efficiently. If there is a unique solution to the puzzle, our algorithm will find it. If there are multiple solutions the algorithm will reach the stage when all the positions that could be determined are fixed and the remaining cells are assigned the relevant set of possible symbols. The solution is then to be find by making the choices for these cells one-by-one and by deleting the chosen (and therefore fixed) symbols from the detection and decision tables. Once again, this process will terminate when all the symbols are fixed. The pseudocode of our algorithm is given in Table 1.

Table 1. Cross-matching puzzle algorithm.

```
Input:  N > 0   ... size of the puzzle
        S(N,N)  ... solution table (empty)
        D(N,N) ... detection table (given)
        C(N,N) ... control table (given)
Output: S(N,N) ... filled solution table
```

```
for i := 1 to N do
  for j := 1 to N do
     GotIt(i,j):=false;   % no fixed symbol in this cell yet (auxiliary table)
repeat
  Fixed:=0;
  for i := 1 to N do
    for j := 1 to N do
      begin
          if GotIt(i,j)=false then    % still unknown/unfixed cell
          begin
            ISC(i,j):=(i-th row of D) ∩ (j-th column of C);    % intersection
            if ISC(i,j)={X} then    % single option only
              begin
                S(i,j):=X;
                GotIt(i,j):=true;
                remove X from the i-th row of D;
                remove X from j-th column of C;
                Fixed := Fixed + 1;
              end
          end
      end
  until Fixed = 0;        % there was no change (no further symbol was fixed)
  return S;
```

4 Conclusion

The deterministic algorithm for solving the cross-matching puzzle was presented in this paper. The algorithm works in polynomial time and therefore even large instances of the cross-matching puzzle can be tackled and solved. If there is a unique solution only, this solution is found deterministically. Should there be multiple solutions, the algorithm presented here will terminate at the situation where all the symbols that could be fixed unambiguously were placed in the respective position while there are the sets of the candidate symbols amongst which is it necessary to make the choice elsewhere. Therefore, even in this situation a solution to the cross-matching puzzle (or all of them) can be easily found.

Nevertheless, the main aim of this paper was not to design an algorithm for a specific problem. We wanted to show that while AI researchers constantly look for new constraint-satisfaction problems that could be utilized for testing various problem-solving techniques it is possible to come up with the problem that can be solved by much simpler algorithms. Moreover, it is important to repeat that one of the top misconceptions related to NP class is that a huge number of potential solutions to the specific problem does not automatically mean that the relevant problem inevitably belongs to the class of NP. Such a misunderstanding and misclassification of the particular problem leads to false impression that there is no chance to design a simple and fast algorithm for such a problem. Consequently, various heuristics or general problem-solving techniques are unnecessarily employed in order to solve it and the wrong impression that the problem is difficult is further supported. There are too many problems in the class of NP anyway and so there is no need to waste our effort as well as unnecessary computational resources on problems that could be solved using polynomial deterministic algorithms.

References

1. Yato, T., Seta, T.: Complexity and completeness of finding another solution and its application to puzzles. IEICE Trans. Fundam. Electron. Commun. Comput. Sci. **86**, 1052–1060 (2003)
2. Chatterjee, S., Paladhi, S., Chakraborty, R.: A Comparative Study on the performance characteristics of Sudoku solving algorithms. IOSR J. Comput. Eng. **1**, 69–77 (2014)
3. Slabý, A., Ševčíková, A.: Chess as a motivational tool in education. In: 29th Annual Conference of the European Association for Education in Electrical and Information Engineering, EAEEIE, pp. 1–6 (2019)
4. Kesemen, O., Özkul, E.: Solving crossmatching puzzles using multi-layer genetic algorithms. In: First International Conference on Analysis and Applied Mathematics, 18–21 Oct 2012 Gumushane (2012)
5. Kesemen, O., Özkul, E.: Solving cross-matching puzzles using intelligent genetic algorithms. Artif. Intell. Rev. **49**(2), 211–225 (2016)
6. Mann, Z.A.: The top eight misconceptions about NP-hardness. Computer **50**, 72–79 (2017)
7. Hynek, J.: Genetic algorithms for the N-queens problem. In: Arabnia, H.R., Mun, Y. (Eds.): Proceedings of the 2008 International Conference on Genetic and Evolutionary Methods, pp. 64–68. CSREA Press (2008)

Credit Risk Assessment in the Banking Sector Based on Neural Network Analysis

Vera Ivanyuk[1,2]([✉]) [iD], Egor Slovesnov[1], and Vladimir Soloviev[1]

[1] Financial University Under the Government of the Russian Federation,
Moscow, Russia
[2] Bauman Moscow State Technical University, Moscow, Russia

Abstract. The present research explores the possibility of using neural networks to predict credit risk in the banking sector through a case study of a database of one of the American banks. Scoring is a mathematical or statistical model used by a bank to determine, based on the credit history of "past" clients, how likely it is that a particular potential borrower will repay the loan on time. A scoring model is a weighted sum of certain characteristics. The result is an integrated parameter (score); the higher it is, the more reliable the client is, and the bank can order the clients according to their level of creditworthiness in increasing order.

The integrated parameter of each client is compared with a certain numerical threshold, or boundary line, which is essentially a break-even line and is obtained from the reckoning of the average number of clients paying on time needed to compensate for losses from a single debtor. Clients with an integrated parameter above this line are given credit, while clients with an integrated parameter below this line are not.

Theoretical aspects of the neural network application were considered. A basic table of real data on the bank's clients was studied. Based on the results of the study, conclusions were made that helped solve the problem of building a neural network.

Keywords: Credit risk · Neural network · Assessment

1 Introduction

To assess credit risk, the borrower's creditworthiness is analyzed. In banking practice, creditworthiness is interpreted as a desire combined with the ability to repay the issued obligation in a timely manner. According to this definition, the main goal of scoring is not only to find out whether the client is able to pay back the loan or not but also to examine the degree of client's reliability and commitment.

In the banking system, when a person applies for a loan, the bank may have the following information to analyze:

- the questionnaire filled out by the borrower
- information on this borrower from the credit bureau, an organization that stores the credit histories of the entire adult population of the country
- the borrower's account history, if he or she is the bank's client.

© Springer Nature Switzerland AG 2021
L. Rutkowski et al. (Eds.): ICAISC 2021, LNAI 12855, pp. 267–277, 2021.
https://doi.org/10.1007/978-3-030-87897-9_25

Credit analysts use the following concepts: clients' "attribute-characteristics" (in terms of mathematical logic—variables, factors) and "grade-values" that a variable takes. In the questionnaire that the client fills out, the characteristics are represented by the questions (age, marital status, profession), and the grade-values are the answers to these questions.

In its simplest form, a scoring model is a weighted sum of certain characteristics. The result is an integrated parameter (score); the higher it is, the more reliable the client is, and the bank can order the clients according to their level of creditworthiness in increasing order.

The integrated parameter of each client is compared with a certain numerical threshold, or boundary line, which is essentially a break-even line and is obtained from the reckoning of the average number of clients paying on time needed to compensate for losses from a single debtor. Clients with an integrated parameter above this line are given credit, while clients with an integrated parameter below this line are not.

Currently, it is customary to distinguish four areas of scoring:

1. Application scoring—models for evaluating the financial status of an entity to decide on the feasibility of a transaction
2. Behavioural scoring—models for evaluating the financial status of an entity in the process of implementing a transaction
3. Collection scoring—models for building relationships with entities in high-risk transactions
4. Fraud scoring—models for building relationships with entities to minimize non-financial (in particular, legal) transaction risks.

The first credit scoring models were developed by Fair Isaac Corporation more than half a century ago. The scores resulting from these models are named after the company—FICO. Now the FICO score is widely known and massively used in the United States and Canada when making decisions about issuing loans. The FICO score is calculated based on information from the three largest national credit bureaus: Experian, Equifax, and TransUnion. Depending on the credit bureau whose data is used for calculation, the credit score varies slightly.

The FICO score ranges from 300 to 850. A higher score, as in most other models, corresponds to lower risks. It should be noted that determining the threshold for screening applications that will not be satisfied requires additional efforts. There is no strictly defined procedure and the choice of this feature depends on the bank's strategy: what risks the bank is willing to accept, how much it seeks to expand its loan portfolio, etc.

2   Credit Risk Assessment in the Banking Sector

Currently, the main algorithms used in credit scoring models are:

- logistic regression
- neural networks

- decision trees (and their ensembles such as random forests and gradient boosting).

The use of various machine learning methods, such as neural networks, logistic regression, random forests, etc. in credit scoring has been considered in many papers.

So in the paper West, D. [1] the use of 5 different neural network architectures for credit scoring tasks was considered and benchmarked against classical statistical methods including discriminant analysis, logistic regression and non-parametric methods. The results of the study showed that the use of artificial neural networks could significantly improve the quality of classification.

Boguslauskas, V., Mileris, R. concluded in their work [2] that neural networks and logistic regression appeared to be the most effective models for solving the problem of credit scoring. Their analysis showed that artificial neural networks were superior to other methods in terms of prediction accuracy.

The paper Pawel Plawiak, Moloud Abdar, Joanna Plawiak, Vladimir Makarenkov [3] also considered the credit scoring application of neural networks. The authors used a genetic algorithm of a deep learning neural network on a small data set, which resulted in outperforming traditional scoring algorithms.

The purpose of the present research is to study and summarize theoretical and practical issues of using neural networks in the banking sector for credit risk analysis.

The information base is a table of clients of an American bank with data for the period 2011–2015. The database contains information about more than 800,000 clients including 74 characteristics for each client (see Fig. 1).

id	member_id	loan_amnt	funded_amnt	funded_amnt_inv	term	int_rate	installment	grade	sub_grade
0 1077501	1296599	5000.0	5000.0	4975.0	36 months	10.65	162.87	B	B2
1 1077430	1314167	2500.0	2500.0	2500.0	60 months	15.27	59.83	C	C4
2 1077175	1313524	2400.0	2400.0	2400.0	36 months	15.96	84.33	C	C5
3 1076863	1277178	10000.0	10000.0	10000.0	36 months	13.49	339.31	C	C1
4 1075358	1311748	3000.0	3000.0	3000.0	60 months	12.69	67.79	B	B5

5 rows × 74 columns

Fig. 1. Information base.

The objective of the study is to build a neural network using real data to assess credit condition (good, bad).

Formation of requirements to the model. Below we list some important clients' attribute-characteristics:

1. ID. Borrower's identification number
2. Loan_amnt. Amount of the loan requested by the borrower
3. Funded_amnt. Amount of the loan issued

4. Term. Period for which the loan was issued
5. Int_rate. Interest rate of the loan
6. Installment. Amount of regular payment
7. Grade and Sub_grade. Score of the borrower's reliability
8. Emp_length. Borrower's employment length
9. Home_ownership. The form of housing tenure of the borrower (own, rent, mortgage)
10. Annual_inc. Borrower's annual income
11. Loan_status. Current status of the loan (current, fully paid, late).
12. Issue_d. Loan issue date
13. Purpose. The purpose provided by the borrower for the loan request (car, business, educational).

Let us look at the amounts of loans requested by the clients and what loans were issued overall and by year (see Fig. 2).

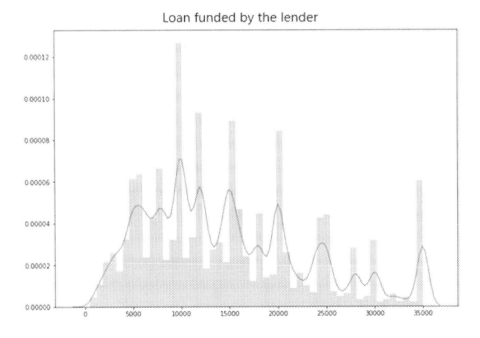

Fig. 2. Amount of loans issued

We also consider the average amount of loans issued by year (see Fig. 3):

According to the charts, most of the loans issued were in the range of $10,000 to $20,000. We can also note a steady increase in the average amount of the loans issued.

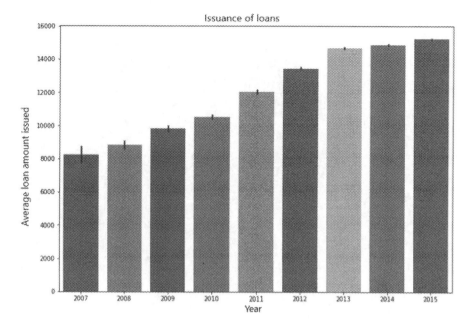

Fig. 3. Average amount of loans issued by year.

Let us consider the number of loans by their status:

1. Current – 601779
2. Fully paid – 207723
3. Charged off – 45248
4. Late (31–120 days) – 11591
5. Issued – 8460
6. In Grace Period – 6253
7. Late (16–30 days) – 2357
8. Does not meet the credit policy. Status: Fully Paid – 1988
9. Default – 1219
10. Does not meet the credit policy. Status: Charged Off – 761.

We shall consider Charged Off, Default and Late (in any stage) as bad loans. Now let us look at the ratio of good loans to bad loans, as well as their number by year (see Fig. 4 and 5).

Two important conclusions can be drawn from these charts. First, bad loans comprise only 7.6% of all loans issued. Second, it is important to remember that the database contains a lot of current loans, which may become bad and somewhat affect the quality of the network.

Let us consider the importance of the borrower's credit score. In order to understand exactly how the grade of the credit score affects the final risk, we need to consider the number of bad loans against the borrower's score. Let us plot the number of loans issued depending on the borrower's credit score (see Fig. 6).

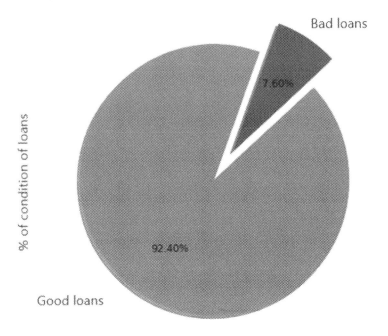

Fig. 4. The ratio of good loans to bad loans

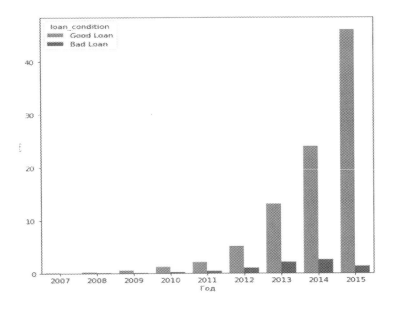

Fig. 5. Percentage of good and bad loans by year

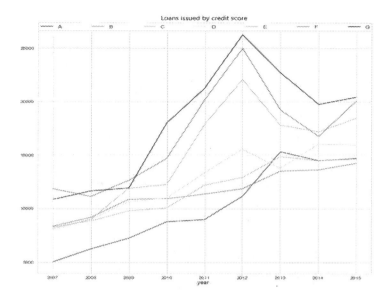

Fig. 6. Number of loans issued

Conclusions on the impact of credit score:

- The scores that had a lower grade received a larger amount of loans in comparison to the higher grade of credit score. This contributes to a higher level of risk for the bank as a whole.
- The interest rate increases as the grade deteriorates.
- Most bad loans were issued to borrowers with a grade of "B".

Let us explore the reasons why a loan becomes bad. Logically, it can be assumed that the borrower's credit score and annual income will have the greatest impact on the level of credit risk. We will identify factors that increase the risk of loan default, such as low annual income, high interest rate, and low grade of the credit score. Let us build a correlation heatmap based on numerical variables (see Fig. 7):

Let us plot the amounts of bad loans with a breakdown by condition (see Fig. 8).

According to this plot, bad loans tended to decline by 2015.

3   Analysis of the Results Obtained

Let us describe the structure of the neural network that will be used for prediction [4–6]. The neural network will consist of input neurons, two output layers, and two hidden layers, with 66 neurons in each. For the research, a feedforward network will be used. The activation function will be the ReLu function [7,8].

It is the most convenient function which often performs better than others. Schematically, the neural network has the form (see Fig. 9):

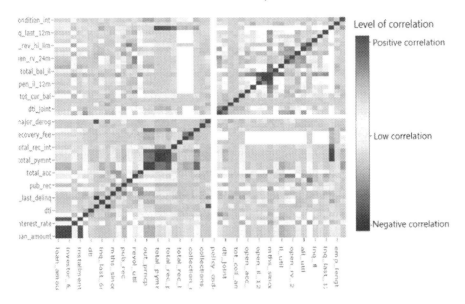

Fig. 7. Correlation heatmap.

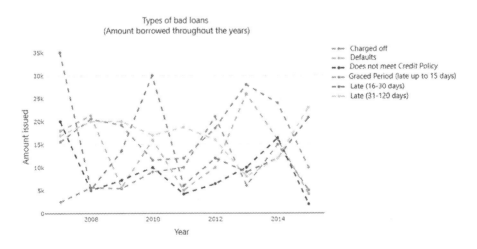

Fig. 8. Amount of bad loans by condition.

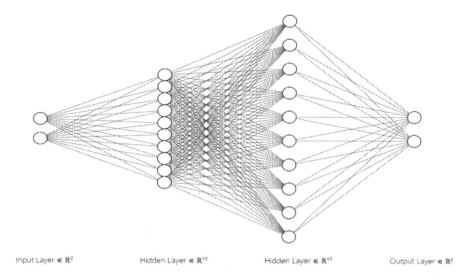

Fig. 9. Neural network scheme.

The scheme shows only 10 neurons in hidden layers for visual clarity. Also, the number of inputs in the scheme is two, but in fact, there will be more of them, equal to the number of attribute-characteristics.

Since there is only 9% of bad loans in the source data, the dataset can be considered unbalanced. Specifically for such cases, an algorithm named Synthetic Minority Over-Sampling Technique (SMOTE) was developed to improve the accuracy of predictions. We explain the principle of operation by giving an example. Assume that the total number of loans is D_0, then the number of good loans is S_0 and the number of bad loans is B_0. Consequently, $D_0 = S_0 + B_0$. Since the data is highly unbalanced, i.e. $S_0 \gg B_0$, we will increase the percentage of bad loans, as shown in Table 1.

Table 1. Example of SMOTE operation.

	"Good loans"		Bad loans		Total
	Quantity	Percentage ($\times 100\%$)	Quantity	Percentage (100%)	Quantity
Original data	S_0	S_0/D_0	B_0	B_0/D_0	D_0
SMOTE	$S_1 = S_0$	S_0/D_1	$B_1 = S_0$	S_0/D_1	D_1

It is important to note that instead of using existing data, SMOTE generates new rows by combining the characteristics of the target class with those of its neighbours.

The number of layers, the number of neurons per layer, and the learning rate were selected experimentally [9]. The learning rate is a setting parameter in the optimization algorithm that determines the step size at each iteration when moving to the minimum of the loss function [10, 11]. The loss function is a function that, in statistical decision theory, characterizes the loss associated with incorrect decision-making based on observed data.

Using the Tensorboard library, we will create the final scheme of the neural network (see Fig. 10).

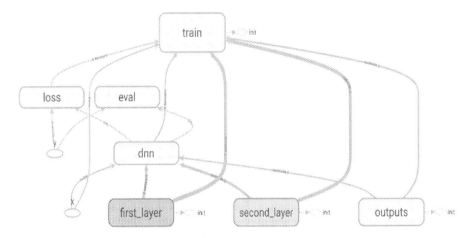

Fig. 10. Neural network scheme.

Assessment of model prediction accuracy. To assess the accuracy of the model, we compare it with the prediction obtained using logistic regression on the same data. The performance of the developed model is compared with the results of alternative models below (see Table 2)

Table 2. Performance comparison of different prediction methods

Prediction method		Prediction accuracy (%)
Parametric statistical methods	Logistic regression	84,5
	Linear discriminant analysis [12]	88,1
Nonparametric statistical methods	Decision tree (CRT) [13]	71,2
	Decision tree (CHAID) [13]	70,1
Support vector (SNN)	SVM in IBM SPSS modeler [13]	72,1
	Neural network with 1 hidden layer [13]	71,2
Deep neural networks (DNN)	Neural network with 2 hidden layers (our study)	92,4

The table shows that the accuracy of the neural network exceeds the accuracy of other prediction methods.

4   Conclusion

The study was conducted using a neural network with 2 hidden layers, with 66 neurons in each layer 22 categorical and numeric attribute-characteristics were used to create the model, and loans were divided into two classes: good and bad. Since categorical data cannot be used to build a neural network, it was converted to numeric data using the One-Hot Encoding algorithm 887379 observations were used to train the model. Due to the domination of good loans among all observations, SMOTE was used to increase the number of bad loans and thus balance the data. After training the neural network, the prediction accuracy comprised 0.92, which exceeds the results of other prediction methods.

References

1. West, D.: Neural network credit scoring models. Comput. Oper. Res. **27**(11–12), 1131–1152 (2000)
2. Boguslauskas, V., Mileris, R.: Estimation of credit risk by artificial neural networks models. Eng. Econ. **64**(4) (2009)
3. Plawiak, P., Abdar, M., Plawiak, J., Makarenkov, V., Acharya, U.R.: DGHNL: a new deep genetic hierarchical network of learners for prediction of credit scoring. Inf. Sci. **516**, 401–418 (2020)
4. Eliana, A., di Tollo, G., Roli, A.: A neural network approach for credit risk evaluation. Q. Rev. Econ. Financ. **48**(4), 733–755 (2008)
5. Ivanyuk, V., Tsvirkun, A.: Intelligent system for financial time series prediction and identification of periods of speculative growth on the financial market. IFAC Proc. Vol. **46**(9), 1128–1133 (2013)
6. Koroteev, M.V., Terelyanskii, P.V., Ivanyuk, V.A.: Approximation of series of expert preferences by dynamical fuzzy numbers. J. Math. Sci. **216**, 5692–695 (2016)
7. Chuang, C.-L., Huang, S.-T.: A hybrid neural network approach for credit scoring. Expert Syst. **28**(2), 185–196 (2011)
8. Lee, T.-S., et al.: Credit scoring using the hybrid neural discriminant technique. Expert Syst. Appl. **23**(3), 245–254 (2002)
9. Khemakhem, S., Said, F.B., Boujelbene, Y.: Credit risk assessment for unbalanced datasets based on data mining, artificial neural network and support vector machines. J. Model. Manage. **13**, 932–951 (2018)
10. Oreski, S., Oreski, D., Oreski, G.: Hybrid system with genetic algorithm and artificial neural networks and its application to retail credit risk assessment. Expert Syst. Appl. **39**(16), 12605–12617 (2012)
11. Wang, S., Yin, S., Jiang, M.: Hybrid neural network based on GA-BP for personal credit scoring. In: 2008 Fourth International Conference on Natural Computation, vol. 3. IEEE (2008)
12. Feis, A., et al.: P2P loan selection. Stanford Univesity Algorithmic Trading and Big Financial Data MS&E, p. 448 (2016)
13. Jin, Y., Zhu, Y.: A data-driven approach to predict default risk of loan for online peer-to-peer (P2P) lending. In: 2015 Fifth International Conference on Communication Systems and Network Technologies. IEEE (2015)

Neural Network Model for the Multiple Factor Analysis of Economic Efficiency of an Enterprise

Vera Ivanyuk[1,2](✉) [ID] and Vladimir Soloviev[1]

[1] Financial University Under the Government of the Russian Federation,
Moscow, Russia
[2] Bauman Moscow State Technical University,
Moscow, Russia

Abstract. The paper proposes a neural network model for assessing the impact of financial instruments and organizational forms on the growth of efficiency within the industry based on the case study of such a high-technology company as the Rosatom State Atomic Energy Corporation. A large holding that is a monopoly state corporation (Rosatom SC) manages more than 300 large enterprises which it owns (either fully or partially, through joint ventures, such as JSCs), or controls directly, such as FSUEs (Federal state unitary enterprises) and FSBIs (Federal state budgetary institutions). Objective: To explain the degree of impact of financial instruments and their groups on the overall economic efficiency using a non-recurrent neural network-based analysis, and to build a neural network-based profit generation model. The main criterion for the economic efficiency of the head enterprise of Rosatom group is its combined profit for the year. Since 2007, Rosatom group has used EBITDA as the main indicator of the company's performance. The Rosatom's order portfolio exceeds $133 billion, which is 67% of the global nuclear power plant construction market. The present paper suggests a methodology for evaluating the economic efficiency of existing organizational forms, financial instruments and support institutions for Rosatom. The paper proposes an algorithm for building a neural network model for evaluating an enterprise's efficiency.

Keywords: Neural network model · Algorithm · Artificial intelligence · Multiple factor analysis

1 Introduction

We will assess the performance of Rosatom group companies and evaluate what factors and financial instruments have influenced the growth of the holding's economic efficiency. To do this, we propose a method for evaluating the economic efficiency of an enterprise.

The methodology for evaluating the economic efficiency of an enterprise involves the following aspects:

© Springer Nature Switzerland AG 2021
L. Rutkowski et al. (Eds.): ICAISC 2021, LNAI 12855, pp. 278–289, 2021.
https://doi.org/10.1007/978-3-030-87897-9_26

1. **Formalization of the efficiency measure.** At this stage, the form and type of efficiency measurement are determined. The form of measurement should be determined by a mathematical formula that expresses the ratio of positive and negative aspects of the evaluated parameter. Let us suggest a formula for calculating technological efficiency. Technological efficiency can be measured by such indicators as 1) the number of patents; 2) economic impact of the introduction of new technology that reflects the positive aspect of this parameter and 3) R&D costs, 4) technology introduction costs and 5) organizational costs that reflect the negative aspect of this parameter. Let us derive the ratio of these indicators, which will characterize the measure of technological efficiency (TE).

$$TE = \frac{\text{number of patents} \times \frac{\text{economic impact of the introduction}}{\text{number of introductions}}}{\text{R\& D costs} + \text{introduction costs} + \text{organizational costs}} \quad (1)$$

To measure economic performance, specialists often use such indicators as EBITA and EBITDA that exclude the tax and political components as well as depreciation and amortisation.

2. **Formalization of evaluation components.** At the next stage, we need to formalize the components of the evaluation by determining their type, data sources, and, if necessary, methods of normalization and reduction to a unified scale.

3. **Deductive analysis. Building a tree of components and factors.** At this stage, the process of forming the evaluation factor is deduced to the level of finite elements related to organizational forms, financial instruments and support institutions, after which a deductive tree is constructed.

4. **Formalization of the deductive tree of components.** After building a deductive tree, it is necessary to formalize all its components including formula descriptions and group allocation with respect to organizational forms and financial instruments, as well as to determine the sources, types, and completeness of data along with methods of its normalization and reduction to a unified economic scale.

5. **Collection and normalization of data.** At the fifth stage, it is necessary to obtain the requisite data and verify its reliability with the eventual application of significance coefficients, and then normalize it to reach a unified economic scale, e. g. XDR in the case of monetary funds, or another generally accepted measure in other cases.

6. **The choice and construction of the model.** At the next stage, depending on the volume of data (sample size), preferred deviations (absolute or relative), and established methodologies, it is necessary to select and apply a standardized method for building a linear statistical, variance-analytical, or neural mathematical model. The selection stage is followed by a formal mathematical description of the model, including all its components, dependencies and constraints of its organizational structures and parameters.

7. **Selection of optimization criteria.** In this case, one of three typical criteria is selected: Bayesian (the smallest integral of the difference between

model results and empirical data), Fischer's (the highest frequency of agreement between model results and empirical data), or the Neyman-Pearson approach, which determines the minimum MSE (mean square error).

8. **Calculation of the optimal coefficients of the model.** Depending on the selected criterion, we determine the optimal coefficients for the equation described in paragraph 6, using the gradient Newton method, or combined stochastic gradient methods (Newtonian Monte Carlo).

9. **Empirical testing of the model quality.** After determining the optimal coefficients, we wait for further results, check them, and, if necessary, adjust the model.

2   Objective Setting. Neural Network Model for the Multiple Factor Analysis of Economic Efficiency of an Enterprise

While applying neural network modelling as a method of the multiple factor analysis of an enterprise's economic efficiency, it is necessary to define some key issues related to the characteristics of the factors under study:

- To determine the network architecture, it is necessary to know exactly the number of groups and subgroups, the factors under study, and their cross-impact structure. All the factors used must be strictly formalized, and the cross-impact must be represented in the form of a tree structure.

- There are factors of economic inertia. It is necessary to determine the possibility of the autocorrelation of the actual performance indicators analyzed. If the presence of autocorrelation or inertia is determined explicitly or at least partially, an autocorrelation branch must be added to the generated structure.

- It is necessary to account for the eventual impact of the studied factors on the ultimate efficiency value in cases where their weights differ significantly and play various roles in shaping the final result. In this case, it is necessary to determine the boundary values of the output coefficients related to these factors of neurons, in order to avoid their unjustified increase for factors that actually make a small contribution, but have a good correlation with the resulting efficiency value.

- If the natural fluctuations of the resulting value are inconsiderable, and there are no significantly correlating factors, instead of taking the actual normalized values of the initial factors as the initial analytical data, it is recommended to use their differentials, which will increase the sensitivity of the network not so much to the measure of the standard deviation, but to the general direction of growth or decline in economic efficiency.

- Another important step is to choose a technique for evaluating the quality of the trained network, both for overfitting and for compliance of the network model dynamics with the dynamics of real performance data. In some cases, it is necessary to introduce additional estimates that correspond to both the degree of overfitting based on the assumed mean values and the degree of Bayesian likelihood. For a common evaluation of the network quality, we will use a generalized parameter that is the product of the standard deviation by the sum of differences between the differentials of the model and real data [1–3].

- In some cases, it is necessary to introduce additional estimates that correspond to both the degree of overfitting based on the assumed mean values and the degree of Bayesian likelihood. For a common evaluation of the network quality, we will use a generalized parameter that is the product of the standard deviation by the sum of differences between the differentials of the model and real data [4,5].

As an example, let us consider the generation of an analytical neural model of economic efficiency for a real enterprise.

As input data, let us take an enterprise whose economic efficiency is presumably influenced by two groups of factors, two factors in each. It is understood that the enterprise is profitable and invests a part of its profit in its own fixed and intangible assets, which means it has economic inertia. To construct a simple non-recurrent convolutional network, we define a tree structure of cross-impact of efficiency elements for the two-year depth of impact (see Fig. 1)

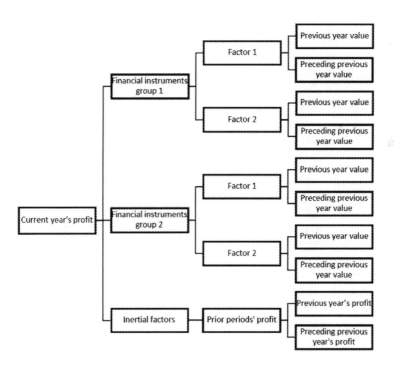

Fig. 1. Hierarchy of the system.

Having this structure, we can define the architecture of the neural network (see Fig. 2).

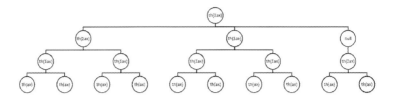

Fig. 2. Structure of the neural network.

Next, having the complete structure of the neural network, we formalize the calculation of the economic efficiency function, expressed as the current year's profit:

$$E = a \sum_{g=1}^{g} a_g \operatorname{th} \sum_{s=1}^{s} a_s \operatorname{th} a_n x, \tag{2}$$

where

E — output of neural network (current year's profit);

a — coefficient for the highest level neuron;

a_g — coefficients of the second level neurons;

a_s — coefficients of the third level neurons;

a_n — coefficients of neurons of the subsequent levels;

th — activation function of neural network $\operatorname{th} = \frac{e^x - e^{-x}}{e^x + e^{-x}}$.

As an optimization criterion for this network, we select a generalized indicator of the form:

$$\text{Total network error} \quad (S) = MSE * \sum_{x=x_1}^{x_n} \left| \frac{sy_m}{dx} - \frac{dy_d}{dx} \right|, \tag{3}$$

where $\frac{dy_d}{dx}$ — the differential for the output parameter (d) of the neural network;

$\frac{dy_m}{dx}$ — the differential for the input parameter (m) of the neural network;

MSE — mean square error of the neural network at time t;

The total error of the neural network (S) is a parameter of neural network optimization that needs to be minimized.

If the difference of derivatives $\frac{dy_m}{dx} - \frac{dy_d}{dx}$ is close to one, then the total network error (S) will tend to the value of the stand-ard error (MSE). Therefore, the greater the difference between the derived input and output parameters of the neural network, the smaller should be the coefficient of impact of the network's input parameter on the output of the network. Conversely, the smaller the difference between the derivatives of the input and output parameters of a neural network, the greater the coefficient of impact of the input parameter of the network on the output parameter of the network [6,7].

When optimizing the total network error to a minimum, using any of conventional methods, we get a trained neural network, the weights of coefficients in which are distributed according to the degree of impact of factors and their groups on the final economic efficiency [8,9]. However, it should be borne in mind that insignificant factors may have a good correlation with the output data and therefore artificially limit the output coefficients of neurons as and ag according to the assumed share of the factor' contribution to the final result (usually 3–5 times the weight).

3 Development of a Neural Network Model for the Multiple Factor Analysis of Economic Efficiency Based on a Case Study of Rosatom

Let us propose an algorithm for building a neural network model for evaluating an enterprise's efficiency:

- Objective statement. We define the input and output elements of the system to build a model of financial flows representing the movement of finance from input to output.
- Grouping elements by factors. For instance, an enterprise's financial instruments influencing its economic efficiency can be divided into two groups: financial instruments related to joint-stock ownership (JSC) and financial instruments related to state ownership (federal state unitary enterprise).
- Analysis of the hierarchy of factors influencing economic efficiency.
- Collecting raw information from public sources.
- Creating a neural network based on a grouped tree. Development of a mathematical model.
- Determination of optimal model coefficients.
- Estimation and interpretation of the network coefficient values.

Objective Statement. To explain the degree of impact of financial instruments and their groups on the overall economic efficiency of the enterprise using a non-recurrent neural net-work-based analysis, and to build a neural network-based profit generation model. The main criterion for the economic efficiency of the head enterprise of Rosatom group is its combined profit for the year (EBITDA). Thus, the output elements of the system are the EBITDA data for 2007–2018. The collected data is formalized, cleared of inflation, and brought to XDR to ensure a unified measurement scale. As input elements, we will use three financial indicators that affect the enterprise's total profit (EBITDA), such as a) subsidies; b) loans; c) security issue.

Grouping Elements by Factors. Let us describe the component tree. The component tree consists of three levels. At the first level, we highlight one component — the EBITDA performance indicator. At the second level, we distinguish two components (organizational forms of incorporation):

- federal ownership (Rosatom SC, as well as FSUEs and FSBIs affiliated with Rosatom);
- collective (joint-stock) ownership (JSCs affiliated with Rosatom).

At the third level, we allocate three components (three financial instruments): a) instruments of state influence (subsidies); b) lending; c) issue of securities.

Each of these instruments can be related to both federal and joint-stock forms of ownership.

The general model of the problem can be described as finding the relationship between the profitability of the enterprise and each of the independent groups (organizational forms), as well as subgroups in each group (financial instruments).

Analysis of the Hierarchy of Factors Influencing Economic Efficiency. Let us build a hierarchy diagram of the impact of financial instruments on the enterprise's profit (see Fig. 3).

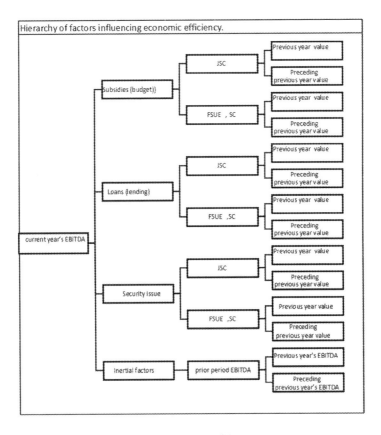

Fig. 3. Hierarchy of factors.

Collecting Raw Information From Open Sources. Data for building the model are obtained from official public sources. Data on subsidies received by Rosatom in the period 2007–2018 were taken from the website of the Ministry of Finance of the Russian Federation.

Since Rosatom group incorporates more than 360 enterprises including nuclear weapons enterprises, research establishments, and the nuclear icebreaker fleet, we classify these enterprises by sector. Table 1 shows data on subsidies for 2017.

Table 1. Subsidies received by Rosatom holding in 2017.

Rosatom holding enterprises	Form of ownership	bln rubles
Rosatom	Federal	2.883644
Operation	Federal	0.25
Supervision	Federal	4,841428
Export	Federal	7,107074
Operation	Federal	3,608192
R&D	Federal	0
R&D	Federal	1,64655
Operation	Joint-stock	0.0668
R&D	Joint-stock	1.3187
Production	Joint-stock	0
Operation	Joint-stock	0
Production	Joint-stock	0
Operation	Joint-stock	0
Operation	Joint-stock	6,039757

Table 2 presents aggregated data on subsidies to enterprises of the Rosatom holding for the years 2007–2018.

Table 3 shows data on the issue of bonds of Rosatom group.

After collecting, processing, and summarizing data from public sources, the in-formation for each of the model elements is expressed in XDR. Thus, we reduce the impact of fluctuations and plummeting of exchange rates. Table 4 shows con-solidated indicators for building the model, expressed in XDR.

Development of Mathematical Model. Since the numbers of groups of ownership forms and subgroups of financial instruments are clearly defined, this problem can be solved using a convolutional neural network [10]. In this case, the coefficients of convolutional neurons will correspond to the efficiency weights of groups of ownership forms and subgroups of financial instruments (see Fig. 4).

Determination of Optimal Model Coefficients. Table 5 shows the results of the neural network. The neural network selects coefficients that determine the degree of impact of financial instruments on the company's profit.

Table 2. Subsidies for Rosatom.

	Subsidies for joint-stock companies of Rosatom group (bln rub)	Subsidies for federal enterprises of Rosatom group (bln rub)
2007	7,425257	20,336888
2008	60,0267	109,719681
2009	66,800266	185,730887
2010	8,294913	167,364135
2011	34,67569	161,36264
2012	7,808726	196,333049
2013	7,265003	168,561916
2014	1,748965	161,36407
2015	5,016797	195,062353
2016	16,288606	95,747481
2017	15,821427	91,394533
2018	3,790242	111,973298

Table 3. Issue of bonds of Rosatom State Corporation.

Date	Amount, bln rub	Rosatom holding enterprises	Form of ownership
May 2007	1.5	Atomstroyexport	JSC
November 2009	30	Atomenergoprom	JSC
November 2009	30	Atomenergoprom	JSC
November 2009	5	Atomenergoprom	JSC
August 2010	10	Atomenergoprom	JSC
November 2011	12.5	Atomredmetzoloto	JSC
December 2011	16.5	Atomredmetzoloto	JSC
August 2013	12.5	Atomredmetzoloto	JSC
July 2015	15	Rosatom SC	FED
July 2015	10	Atomenergoprom	JSC
July 2015	10	Atomenergoprom	JSC
December 2015	10	Atomenergoprom	JSC
December 2015	10	Atomenergoprom	JSC
November 2016	15	Atomenergoprom	JSC
December 2016	15	Atomenergoprom	JSC

Table 4. Summary data for the model.

	Financial instruments						Profit
Year	JSC subsidies	FED subsidies	JSC issue	FED issue	JSC loans	FED loans	EBITDA
2007	0,192	0,524	0,039	0	1,140	0	2,69
2008	1,313	2,399	0	0	0	0	2,6
2009	1,407	3,913	1,369	0	0	0	2,87
2010	0,178	3,582	0,214	0	0	0	3,87
2011	0,704	3,275	0,589	0	13,099	0	3,24
2012	0,167	4,193	0	0	0	0	3,01
2013	0,144	3,333	0,247	0	7,403	0,316	3,07
2014	0,021	1,98	0	0	4,522	4,908	2,46
2015	0,05	1,927	0,39	0,15	18,2	0,079	2,4
2016	0,20	1,178	0,369	0	20,1	0	3,21
2017	0,193	1,117	0	0	25,86	0,789	3,74
2018	0,039	1,159	0	0	3,33	0	3,65

Table 5. Calculation of neural network coefficients.

Neurons	Output neuron	JSC input	Joint-stock ownership			FED input	Federal ownership			
			Subsidies	Issue	Loans		Subsi	Issue	Loans	
Network Coeff	6.64	1,31	8,42	0,6	0,0	0,7	0,02	41,19	1,8	Preced, Previous year
	77.7	55,82	13,4	0,0	0,0	20,6	0,0	0,002	0,0	Previous Year

Table 6. Interpretation of neural network coefficients.

Neurons	Output neuron	JSC in	Joint-stock ownership			FED input	Federal ownership			
			Subsidies	Issue	Loans		Subsidies	Issue	Loans	
Network coeff	6,64	67%	93%	7%	0%	33%	0,04%	95,73%	4,24%	Preced previous year (2017)
	77.7	73%	100%	0,0%	0%	27%	0,0%	100,0%	0%	Previous year (2018)

Estimation and Interpretation of Neural Network Coefficient Values.
To interpret the coefficients of the neural network, we express them as a percentage. Table 6 provides the network's estimated coefficients as a percentage.

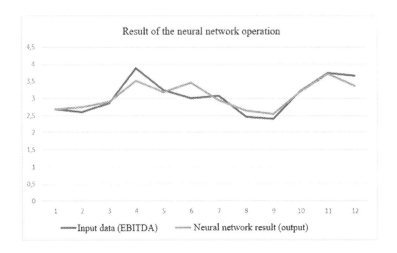

Fig. 4. Output of neural network.

4   Conclusion

The results showed that in 2018, the group of financial instruments of joint-stock enterprises of Rosatom group affected the holding's profit (EBITDA) to a degree of 73%, while the degree of impact of financial instruments of federal enterprises of the holding equalled to 27%. Hence, it may be noted that financial instruments (subsidies, loans, issue of securities) allocated to joint-stock companies of the Rosatom group had a greater impact on the profit of the group compared to financial instruments allocated to federal enterprises of Rosatom group. The same results were obtained when we evaluated the impact of financial instruments on the profit of Rosatom group for the year 2017.

Let us consider which financial instruments have the greatest impact on the company's total profit (EBITDA). The evaluation of financial instruments of joint-stock enterprises showed that the greatest contribution (impact) to the profit of Rosatom group is made by subsidies. The degree of impact of subsidies in the financial instruments group comprised 93% in the year 2017 and 100% in the year 2018. Thus, we can conclude that subsidies from the federal budget granted to joint-stock companies of Rosatom group make the greatest impact on the total profit (EBITDA) of Rosatom State Corporation.

References

1. Fu, Y.W., Yin, H., Yang, G.B.: Application of BP neural network in evaluating enterprise operation performance. Operations Research and Management Science, vol. 4 (2006)
2. Ivanyuk, V., Tsvirkun, A.: Intelligent system for financial time series prediction and identification of periods of speculative growth on the financial market. IFAC Proc. Volumes **46**(9), 1128–1133 (2013)

3. Elizarov, M., Ivanyuk, V., Soloviev, V., Tsvirkun, A.: Identification of high-frequency traders using fuzzy logic methods. In: 2017 Tenth International Conference Management of Large-Scale System Development (MLSD), pp. 1–4. IEEE (2017)
4. Wen, W., Chen, Y.H., Chen, I.C.: A knowledge-based decision support system for measuring enterprise performance. Knowl. Based Syst. **21**(2), 148–163 (2008)
5. Koroteev, M.V., Terelyanskii, P.V., Ivanyuk, V.A.: Approximation of series of expert preferences by dynamical fuzzy numbers. J. Math. Sci. **216**(5), 692–695 (2016)
6. Chang, I.C., Hwang, H.G., Liaw, H.C., Hung, M.C., Chen, S.L., Yen, D.C.: A neural network evaluation model for ERP performance from SCM perspective to enhance enterprise competitive advantage. Expert Syst. Appl. **35**(4), 1809–1816 (2008)
7. Fenglan, L.: Evaluating competitive edge for logistics enterprises based on BP neural network. Journal of Theoretical and Applied Information Technology, vol. 46, no. 1 (2012)
8. Jiang, H., Ruan, J.: Investment risks assessment on high-tech projects based on analytic hierarchy process and BP neural network. J. Netw. **5**(4), 393 (2010)
9. Staub, S., Karaman, E., Kaya, S., Karapınar, H., Güven, E.: Artificial neural network and agility. Procedia-Soc. Behav. Sci. **195**, 1477–1485 (2015)
10. Zhining, Y., Yunming, P.: The genetic convolutional neural network model based on random sample. Int. J. u- e-Serv. Sci. Technol. **8**(11), 317–326 (2015)

A Study of Direct and Indirect Encoding in Phenotype-Genotype Relationships

Clyde Meli[1]([✉]) [ID], Vitezslav Nezval[1], Zuzana Kominkova Oplatkova[2] [ID],
Victor Buttigieg[3] [ID], and Anthony Spiteri Staines[1]

[1] Department of Computer Information Systems, University of Malta, Msida, Malta
{clyde.meli,tony.spiteri-staines}@um.edu.mt, vnez@cis.um.edu.mt
[2] Faculty of Applied Informatics, Department of Informatics and Artificial Intelligence,
Tomas Bata University in Zlín, Zlín, Czech Republic
oplatkova@utb.cz
[3] Department of Communications and Computer Engineering, University of Malta,
Msida, Malta
victor.buttigieg@um.edu.mt

Abstract. This paper examines phenotype and genotype mappings that are biologically inspired. These types of coding are used in evolutionary computation. Direct and indirect encoding are studied. The determination of genotype and phenotype relationships and the connection to genetic algorithms, evolutionary programming and biology are examined in the light of newer advances. The NEAT and HyperNEAT algorithms are applied to the 2D Walker [41] problem of an agent learning how to walk. Results and findings are discussed, and conclusions are given. Indirect coding did not improve the situation. This paper shows that indirect coding is not useful in every situation.

Keywords: Indirect encoding · Direct coding · Genotype · Phenotype · Genetic algorithms · Evolutionary programming · Neuroevolution · NEAT · HyperNEAT

1 Introduction to Phenotype and Genotype Mappings

This paper deals with the study of the influence of direct and indirect encoding for the artificial neural network design by evolutionary computation, i.e. neuroevolution. Evolutionary computation meets terminology as phenotype (behaviour, physiology, the morphology of an organism) and genotype (genetic coding of the organism). The relationship between both terms has been the subject of various investigations, including [1–3]. Such phenotype mapping has been used to predict disease-related genes. Van Driel et al. [1] say that phenotypes can be used to predict biological interactions, which are the effect which two genes have on each other.

Many evolutionary computation systems, such as genetic programming (GP) work with direct encodings (sometimes the term coding is used). Program trees representing a computer program evaluated recursively are used as genotypes. They translate directly into solutions. Some hybrid systems operate on graphs so there may be a problem with

© Springer Nature Switzerland AG 2021
L. Rutkowski et al. (Eds.): ICAISC 2021, LNAI 12855, pp. 290–301, 2021.
https://doi.org/10.1007/978-3-030-87897-9_27

devising direct genetic operators. In a direct encoding [4], the genotype specifies every neuron and connection explicitly. Not all authors differentiate between direct encodings and structural encodings like [5]. Structural encoding implies that the encoding also holds information on connection weights. This is used when a genetic algorithm (GA) evolves Artificial Neural Network (ANN) parameters. In such an encoding, there are few constraints to the GA's exploration [5].

The minimal alphabet principle in Gas [6] specifies the selection of the smallest alphabet that permits a natural problem expression. This holds for direct encoding but for indirect encoding the search space can be reduced [7].

1.1 Direct Encodings for ANNs

Montana [8] reviewed how GAs can be used to represent and train ANNs. The genetic representation must include the network topology, real-valued weights associated with every link and real-valued bias associated with every node.

Specifically, an example of a direct and structural encoding of an ANN, as found in [8] is given in Fig. 1. Typically, the layout, such as the number of nodes in each layer and the number of hidden layers, is fixed, and not part of the encoding.

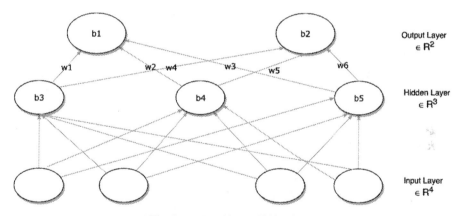

Fig. 1. ANN with one hidden layer

The ANN in Fig. 1 would be encoded as the following chromosome $(w_1,w_2,w_3,w_4,w_5,w_6,b_1,b_2,w_7,w_8,w_9,w_{10},w_{11},w_{12},w_{13},w_{14},w_{15},w_{16},w_{17},w_{18},b_3,b_4,b_5)$.

This representation scheme degrades its performance of convergence as the network size is enlarged [9]. It is noted that using direct encodings, it is normally not possible to represent the ANN graphical structure geometrically [10]. This lack of geometry restricts ANNs from evolving brain-like structures.

1.2 Indirect Encodings for ANNs

[9] devised a genetic algorithm representation for ANNs, which represents connections and network topology. The existence of a connection is represented by '1' and its absence

by a '0'. An extension of L-Systems [11] called Graph L-System, is used to generate graphs. Starting from an axiom of a 2 × 2 matrix, a set of deterministic rewriting rules are applied for edges and nodes represented by symbols. The connectivity matrix will enlarge every generation and the last generation will contain '1' and '0''s representing feed-forward network connections as an upper-right triangle; only the bold connections are used, the rest are discarded. An example representation for the XOR problem [12] is given in Fig. 2 where the symbol S is a starting axiom and symbols A, B, C, D stand for nonterminal for the grammar graph generation based on Graph L-System. The details of the concept are described in [9] including rewriting rules to final '1' and '0''s.

For the first line, '01100' represents a network with the top node not having a connection from itself, and only the next two nodes connecting to the top node.

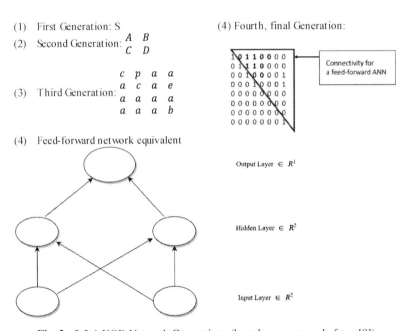

(1)   First Generation: S

(2)   Second Generation: $\begin{matrix} A & B \\ C & D \end{matrix}$

(3)   Third Generation: $\begin{matrix} c & p & a & a \\ a & c & a & e \\ a & a & a & a \\ a & a & a & b \end{matrix}$

(4)   Feed-forward network equivalent

(4) Fourth, final Generation:

```
1 0 1 1 0 0 0 0
0 1 1 1 0 0 0 0
0 0 1 0 0 0 0 1
0 0 0 1 0 0 0 1
0 0 0 0 1 0 0 0
0 0 0 0 0 1 0 0
0 0 0 0 0 0 1 0
0 0 0 0 0 0 0 1
```

Connectivity for a feed-forward ANN

Output Layer $\in R^1$

Hidden Layer $\in R^2$

Input Layer $\in R^2$

Fig. 2. 2-2-1 XOR Network Generations (based on an example from [9])

1.3   The Evolutionary Search Space of Indirect Encodings

Pigliucci [13] quotes Hartmann et al. [7] as having determined that indirect encoding dramatically reduced the evolutionary search space in evolutionary computation as opposed to the classical direct encoding. However, on a closer look, this is actually not something that Hartmann et al. claim as fact in all cases and situations. Indeed, they only claimed this holds for evolved digital circuits. So, further research may be required in this area to see whether indirect encoding would be beneficial in certain cases. The importance of this can be seen in the field of genetics, where [13] assumed that indirect encoding was found to be beneficial in all cases by an analogy that biological systems should behave like simulated evolutionary models. On that unproven argument, it is claimed in [13] that the old metaphors of genetic blueprints and genetic programs are misleading or inadequate. [13] claims developmental or indirect encoding must be a promising basis to understand evolvability and the Genotype-Phenotype (G-P) mapping problem.

Interestingly, developmental encoding itself was inspired by biological development [14]. More recently, Clune et al. [15] showed an indirect encoding outperforming direct encoding. On the contrary, Harding and Miller [16] showed that for lower complexity (in the sense of Kolmogorov) encoding patterns, direct encoding sometimes performs worse than indirect encodings.

Other studies utilising generative encodings include Clune and Lipson [17], Meli [18], Jacob and Rozenberg [11] (using the grammar by Lindenmayer [19]) and Kitano [9].

Kwasnicka and Paradowski [20] found that in a few generations, indirect encoding gave excellent solutions though the evolved networks were larger than the equivalent directly encoded ones. Da Silva et al. also demonstrated the benefit of indirect encoding [21] where the quality of the Particle Swarm Optimisation (PSO), GA and GP-based solutions for web services using using indirect encoding was higher than the equivalent baseline direct encoding approach for twelve out of thirteen datasets. Also Hotz [22] in a case study of lens shape evolution schemes showed that indirect encoding converged faster than direct ones.

Compared to the advantages stated in the above-mentioned papers, [23] found in a TETRIS problem that HyperNEAT, a tool for indirect encoding in ANNs, is superior to NEAT (a tool for direct encoding in ANNs) early in evolution, but this fizzles out eventually and NEAT performs better. Authors showed that HyperNEAT was better for raw but NEAT performed better for hand-designed features. The aim of this paper is to compare direct and indirect encoding on a problem to analyse the performance of evolution.

2   Indirect Encoding

Encodings in evolutionary programming are typically binary, real-valued, graph-based, computer code [24]. The selected representation affects the effectiveness of a genetic algorithm [25] and probably even other binary-coded evolutionary computation. Direct encoding is claimed to be ineffective [14]. Indirect encoding can vary. One form, developmental encoding, employs gene reuse. Used in evolutionary computation, this is referred to as embryogeny. This is called Artificial Embryogeny [14]. Some exciting research in indirect encoding involves ANNs. An ANN is represented as a binary tree with grammar rules generating the ANN as seen in [9].

In GP, one finds the use of indirect encodings like Cartesian GP (CGP) [26] and Grammatical Evolution [27]. They can also help in "encoding neutrality". *Neutrality* is defined as a situation when small genotype mutations (*neutral mutations*) do not have an effect on the fitness of the expressed phenotype. Miorandi et al. [28] explain that a neutral mutation gives a substantial advantage for state space exploration. This can potentially increase the evolvability of a population providing further robustness.

Indirect encoding representation can allow optimisation to occur without restrictions, since "functional constraints are subsequently enforced during the decoding step" [21].

An issue which Ronald [29] finds with the developmental form of indirect encoding used mainly in hybrid systems is that some of these encodings do not address the entire search space. On the other hand, an alternate representation may address the entire search space, as the Millipede [18] representation does by combining or folding several points in the search space into one. Similarly, Della Croce et al. [30] in their GA, which solves the Traveling Salesperson Problem (TSP) employ a lookahead representation based on Falkenauer and Bouffoix's Linear Order Crossover (LOX). Another GP representation uses a block-oriented representation instead of the usual direct one [31]. It describes how a block diagram is built rather than the directly coded structure.

3   Summary of Research on Indirect Encoding

Research on indirect encoding has been summarized in two tables, Table 1 involving evolutionary computation with neural networks and Table 2 which involved the use of other evolutionary computation techniques.

Table 1. Research involving indirect encoding with ANN or other neural networks

Paper	Indirect coding used
Clune et al. [15]	Evolution of Compositional Pattern Producing Networks (CPPNs) via Hypercube-based NeuroEvolution of Augmenting Technologies (HyperNEAT) algorithm. The latter outperformed direct encodings in three problem domains
Clune et al. [32]	Comparison of HyperNEAT and FT-NEAT algorithms, the former outperformed the latter in a quadruped problem
D'Ambrosio and Stanley [33]	Multiagent HyperNEAT outperformed multiagent Sarsa(λ)
Gillespie et al. [23]	HyperNEAT and NEAT comparison in TETRIS, NEAT overtakes HyperNEAT eventually
Hussain and Browse [5]	Evolving Neural Networks using Attribute Grammars
Kitano [9]	Artificial Neural Network (ANN) with GA and Graph L-System
Kwasnicka and Paradowski [20]	Neural Network with Direct and Indirect encoding, the latter does not reach the global optimum but gives good ANNs in few generations compared to direct encoding

Table 2. Research involving indirect encoding with other evolutionary computation techniques

Paper	Indirect coding used
Aickelin [35]	Indirect GA with a hill climber algorithm
Aickelin and Dowsland [36]	Indirect GA with a heuristic decoder and hybrid crossover operator
Brucherseifer et al. [31]	GP with block-oriented encoding
da Silva et al. [21]	PSO, GA and GP with four indirect representations
Haj-Rachid et al. [37]	GA comparisons of Indirect and Direct coding. Best direct encoding performance with PMX crossover and Inversion Mutation. Best indirect encoding performance with OX crossover with inversion mutation
Hartmann et al. [7]	Cartesian Genetic Programming using a symbolic netlist representation
Hotz [22]	Evolution strategies with indirect coding
Jacob and Rozenberg [11]	Genetic L-System Programming (GLP) Paradigm for development of L-Systems
Meli [18]	Genetic Algorithm (GA) with Millipede encoding
Thangavelautham and D'Eleuterio [38]	Artificial Neural Tissue (ANT) GP architecture with introns

3.1 Basic Processing of HyperNEAT

HyperNEAT is also known as Hypercube-based NEAT [10], where NEAT stands for NeuroEvolution of Augmented Topologies. The main idea in HyperNEAT is that it is possible to learn relationships when the solution is represented indirectly. It is a generative description of the connectivity of the ANN rather than searching for and tuning the connection weights of the ANN itself. Such approach is very important for evolution of large scale neural networks with huge amount of nodes a and many more of connections.

HyperNEAT uses Compositional Pattern Producing Networks (CPPNs) which can produce augmented structures including possibility to use any activation functions via evolutionary process of genetic algorithm. Figure 3 briefly shows the steps involved in the HyperNEAT algorithm.

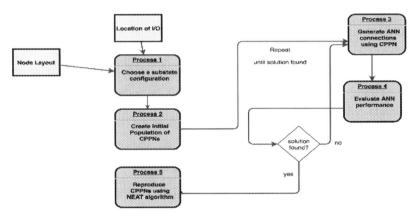

Fig. 3. HyperNEAT processing

4   Experiment

The experiment compared two direct and indirect encoding algorithms used for ANNs. The direct algorithm chosen was the NEAT algorithm [34], and the indirect algorithm used was HyperNEAT [10]. Recall that NEAT is used to evolve dense ANN network nodes and connections specifically. It uses a direct encoding because of the difficulty in obtaining extensive knowledge of how such encoding will be used and, such indirect search can be biased [34]. HyperNEAT extends the NEAT algorithm. It utilizes Compositional Pattern Producing Networks (CPPNs) to evolve ANNs using principles from the NEAT algorithm. These algorithms were preferred due to the availability of easily usable software, SharpNEAT[1].

4.1   Experimental Parameters

The open-source SharpNEAT neuroevolution C#.NET implementation was compiled and used unmodified to compare the two algorithms, NEAT and HyperNEAT, applied to the 2D Walker problem [41]. Settings were taken from [39], which compared the two algorithms applied to the T-Maze learning problem [39, 40] used HypersharpNEAT instead of SharpNEAT. The 2D Walker problem was chosen as it was the only problem available in SharpNEAT using both NEAT and HyperNEAT algorithms. The problem was investigated by [41], who used their implementation of NEAT, Covariance Matrix Adaptation Evolution Strategy (CMAES) and other deep reinforcement learning algorithms. The 2D Walker task involves an agent learning how to walk. [32] investigated a similar quadruped gait problem and concluded that HyperNEAT performed better than FT-NEAT, a directly encoded algorithm. They used the ODE physics simulator with a small population of 150, 1000 generations and 50 runs. The fitness function for 2D Walker is the mean hip position (if it is positive, otherwise 0) over five trials squared.

Every run consisted of 500 generations, the population size was 500, and 10% elitism was used. 50% of sexual offspring were not mutated. The asexual offspring "had 0.94

---

[1] Available at https://sharpneat.sourceforge.io/.

probability of link weight mutation, 0.03 chance of link addition, and 0.02 chance of node addition" [39]. Initial connections proportion was set to 0.05. Mutate connection weights was set to 0.89, mutate delete connection to 0.025. Connection weight range was five.

4.2 Results

Figure 4 below shows the results of running the NEAT and HyperNEAT algorithms using the SharpNEAT application for 500 generations, charting mean fitness.

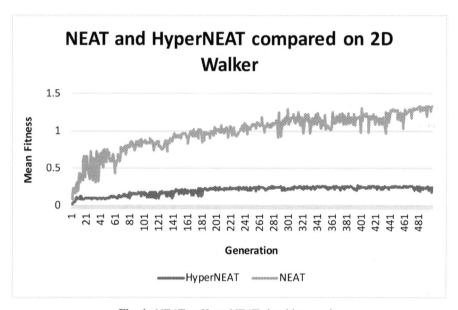

Fig. 4. NEAT vs HyperNEAT algorithm results

Interestingly, NEAT performed well on the task. It can be seen how the HyperNEAT algorithm using indirect coding does not manage to get good mean fitness values. Indeed all values are smaller than 0.3. On the other hand, the NEAT algorithm manages to get close to 1.3 mean fitness. By generation 200 the mean fitness for HyperNEAT is 0.239891, whereas the equivalent for NEAT is 0.96752. The largest mean fitness reached by HyperNEAT is 0.264622 at generation 396. This is never reached again by the algorithm. The largest mean fitness reached by NEAT is that of 1.333688, at generation 493.

Table 3 shows the mean and maximum fitnesses reached in the last generation 500 averaged across runs for NEAT and HyperNEAT.

The Mann-Whitney U test was performed across the final generation runs for mean fitness since the data is not normally distributed. This showed that the difference is significant ($U = 0$; z-score is 5.39649, p-value is $<.00001$ and the result is significant at $p < 0.05$).

Table 3. Evaluation of mean and average of runs (Mean and Maximum fitnesses at Generation 500 for NEAT and HyperNEAT)

	Mean NEAT	Mean HyperNEAT	Max NEAT	Max HyperNEAT
Mean	1.1381	0.2925	2.5640	0.4874
Std Dev	0.4296	0.0701	1.6208	0.0588

The results clearly ascertain how indirect coding in the Walker problem does not help. Indeed, it appears to have stifled the evolution. Some other tests were made to see if HyperNEAT could be improved, e.g. using the larger number of generations used by [32] or interspecies crossover, however there was no improvement.

5   Conclusion

Most current research into indirect encoding involves GP, GAs or ANNs. This paper has covered the issues involving indirect coding which are important aspects of evolutionary computation. We looked at differences involving indirect and direct representation. Significant progress has been made in the theoretical and practical developments. More research is needed involving indirect coding with neural networks and other evolutionary computation techniques, resulting in better representations.

These negative results do not show that indirect encoding is better than direct coding for the 2D Walker problem. This shows an example where indirect encoding is hard to apply successfully and is less effective, reminiscent of [23].

Further research should clarify whether indirect encoding is better than direct encoding in some areas, as well as which categories of problems might be convenient for either type of encoding. It definitely cannot be claimed that indirect coding is always better. This will also ascertain the veracity of Pigliucci's claim that "old metaphors of genetic blueprints and genetic programmes are misleading or inadequate" [13].

Acknowledgments. This work was supported by financial support of research project NPU I No. MSMT-7778/2014 by the Ministry of Education of the Czech Republic, by the European Regional Development Fund under the Project CEBIA-Tech No. CZ.1.05/2.1.00/03.0089 and by resources of A.I. Lab research group at Faculty of Applied Informatics, Tomas Bata University in Zlin (ailab.fai.utb.cz).

References

1. Van Driel, M.A., Bruggeman, J., Vriend, G., Brunner, H.G., Leunissen, J.A.: A text-mining analysis of the human phenome. Eur. J. Hum. Genet. **14**, 535 (2006)
2. Meli, C.: Using a GA to determine genotype and phenotype relationships. In: European Simulation and Modelling Conference 2007. Westin Dragonara, St Julians (2007)

3. Fogel, D.B.: Phenotypes, genotypes, and operators in evolutionary computation. In: Proceedings of the 1995 IEEE International Conference on Evolutionary Computation (ICEC 1995), pp. 193–198 (1995)
4. Galushkin, A.I.: Neural Networks Theory. Springer, Heidelberg (2007). https://doi.org/10.1007/978-3-540-48125-6
5. Hussain, T.S., Browse, R.A.: Evolving neural networks using attribute grammars. In: 2000 IEEE Symposium on Combinations of Evolutionary Computation and Neural Networks. Proceedings of the First IEEE Symposium on Combinations of Evolutionary Computation and Neural Networks (Cat. No. 00), pp. 37–42 (2000). https://doi.org/10.1109/ECNN.2000.886217
6. Goldberg, D.E.: Genetic Algorithms in Search, Optimization and Machine Learning. Addison-Wesley (1989)
7. Hartmann, M., Haddow, P.C., Lehre, P.K.: The genotypic complexity of evolved fault-tolerant and noise-robust circuits. Biosystems. **87**, 224–232 (2007)
8. Montana, D.J., Davis, L.: Training feedforward neural networks using genetic algorithms. In: IJCAI, pp. 762–767 (1989)
9. Kitano, H.: Designing neural networks using genetic algorithms with graph generation system. Complex Syst. **4**, 461–476 (1990)
10. Gauci, J., Stanley, K.O.: Autonomous evolution of topographic regularities in artificial neural networks. Neural Comput. **22**, 1860–1898 (2010)
11. Jacob, C.: Genetic L-system programming. In: Davidor, Y., Schwefel, H.-P., Männer, R. (eds.) PPSN 1994. LNCS, vol. 866, pp. 333–343. Springer, Heidelberg (1994). https://doi.org/10.1007/3-540-58484-6_277
12. Zhao, Y., Deng, B., Wang, Z.: Analysis and study of perceptron to solve XOR problem. In: The 2nd International Workshop on Autonomous Decentralized System, 2002, pp. 168–173 (2002) https://doi.org/10.1109/IWADS.2002.1194667
13. Pigliucci, M.: Genotype-phenotype mapping and the end of the 'genes as blueprint' metaphor. Philos. Trans. R. Soc. Lond. B Biol. Sci. **365**, 557–566 (2010)
14. Stanley, K.O., Miikkulainen, R.: A taxonomy for artificial embryogeny. Artif. Life. **9**, 93–130 (2003)
15. Clune, J., Stanley, K.O., Pennock, R.T., Ofria, C.: On the performance of indirect encoding across the continuum of regularity. IEEE Trans. Evol. Comput. **15**, 346–367 (2011)
16. Harding, S., Miller, J.F.: A comparison between developmental and direct encodings. Presented at the GECCO 2006 (Updated version) (2006)
17. Clune, J., Lipson, H.: Evolving three-dimensional objects with a generative encoding inspired by developmental biology. In: ECAL, pp. 141–148 (2011)
18. Meli, C.: Millipede, an extended representation for genetic algorithms. In: International Journal of Computer Theory and Engineering. IACSIT PRESS, Rome, Italy (2013)
19. Lindenmayer, A.: Mathematical models for cellular interactions in development. I. Filaments with one-sided inputs. J. Theor. Biol. **18**, 280–299 (1968)
20. Kwasnicka, H., Paradowski, M.: Efficiency aspects of neural network architecture evolution using direct and indirect encoding. In: Ribeiro, B., Albrecht, R.F., Dobnikar, A., Pearson, D.W., Steele, N.C. (eds.) Adaptive and Natural Computing Algorithms, pp. 405–408. Springer, Vienna (2005). https://doi.org/10.1007/3-211-27389-1_98
21. da Silva, A.S., Mei, Y., Ma, H., Zhang, M.: Evolutionary computation for automatic web service composition: an indirect representation approach. J. Heuristics. **24**, 425–456 (2018)
22. Hotz, P.E.: Comparing direct and developmental encoding schemes in artificial evolution: a case study in evolving lens shapes. In: Proceedings of the 2004 Congress on Evolutionary Computation (IEEE Cat. No. 04TH8753), vol. 1, pp. 752–757 (2004). https://doi.org/10.1109/CEC.2004.1330934

23. Gillespie, L.E., Gonzalez, G.R., Schrum, J.: Comparing direct and indirect encodings using both raw and hand-designed features in tetris. In: Proceedings of the Genetic and Evolutionary Computation Conference, pp. 179–186. Association for Computing Machinery, Berlin (2017). https://doi.org/10.1145/3071178.3071195

24. Kicinger, R., Arciszewski, T., De Jong, K.: Evolutionary computation and structural design: a survey of the state-of-the-art. Comput. Struct. **83**, 1943–1978 (2005)

25. Caruana, R.A., Schaffer, J.D.: Representation and hidden bias: Gray vs. binary coding for genetic algorithms. In: Machine Learning Proceedings 1988, pp. 153–161. Elsevier, Amsterdam (1988)

26. Miller, J.F., Thomson, P.: Cartesian genetic programming. In: Poli, R., Banzhaf, W., Langdon, W.B., Miller, J., Nordin, P., Fogarty, T.C. (eds.) EuroGP 2000. LNCS, vol. 1802, pp. 121–132. Springer, Heidelberg (2000). https://doi.org/10.1007/978-3-540-46239-2_9

27. Ryan, C., Collins, J.J., Neill, M.O.: Grammatical evolution: evolving programs for an arbitrary language. In: Banzhaf, W., Poli, R., Schoenauer, M., Fogarty, T.C. (eds.) EuroGP 1998. LNCS, vol. 1391, pp. 83–96. Springer, Heidelberg (1998). https://doi.org/10.1007/BFb0055930

28. Miorandi, D., Yamamoto, L., De Pellegrini, F.: A survey of evolutionary and embryogenic approaches to autonomic networking. Comput. Netw. **54**, 944–959 (2010). https://doi.org/10.1016/j.comnet.2009.08.021

29. Ronald, S.: Robust encodings in genetic algorithms: a survey of encoding issues. Presented at the (1997). https://doi.org/10.1109/ICEC.1997.592265

30. Della Croce, F., Tadei, R., Volta, G.: A Genetic algorithm for the job shop problem. Comput. Oper. Res. **22**, 15–24 (1995). https://doi.org/10.1016/0305-0548(93)E0015-L

31. Brucherseifer, E., Bechtel, P., Freyer, S., Marenbach, P.: An indirect block-oriented representation for genetic programming. In: Miller, J., Tomassini, M., Lanzi, P.L., Ryan, C., Tettamanzi, A.G.B., Langdon, W.B. (eds.) EuroGP 2001. LNCS, vol. 2038, pp. 268–279. Springer, Heidelberg (2001). https://doi.org/10.1007/3-540-45355-5_21

32. Clune, J., Beckmann, B.E., Ofria, C., Pennock, R.T.: Evolving coordinated quadruped gaits with the HyperNEAT generative encoding. In: 2009 IEEE Congress on Evolutionary Computation, pp. 2764–2771 (2009). https://doi.org/10.1109/CEC.2009.4983289

33. D'Ambrosio, D.B., Stanley, K.O.: Scalable multiagent learning through indirect encoding of policy geometry. Evol. Intell. **6**, 1–26 (2013). https://doi.org/10.1007/s12065-012-0086-3

34. Stanley, K.O., Miikkulainen, R.: Evolving neural networks through augmenting topologies. Evol. Comput. **10**, 99–127 (2002). https://doi.org/10.1162/106365602320169811

35. Aickelin, U.: An indirect genetic algorithm for set covering problems. J. Oper. Res. Soc. **53**, 1118–1126 (2002)

36. Aickelin, U., Dowsland, K.: An indirect genetic algorithm for a nurse scheduling problem. Comput. Oper. Res. **31**, 761–778 (2008). https://doi.org/10.1016/S0305-0548(03)00034-0

37. Haj-Rachid, M., Ramdane-Cherif, W., Chatonnay, P., Bloch, C.: Comparing the performance of genetic operators for the vehicle routing problem. IFAC Proc. **43**, 313–319 (2010). https://doi.org/10.3182/20100908-3-PT-3007.00068

38. Thangavelautham, J., D'Eleuterio, G.M.T.: A coarse-coding framework for a gene-regulatory-based artificial neural tissue. In: Capcarrère, M.S., Freitas, A.A., Bentley, P.J., Johnson, C.G., Timmis, J. (eds.) ECAL 2005. LNCS (LNAI), vol. 3630, pp. 67–77. Springer, Heidelberg (2005). https://doi.org/10.1007/11553090_8

39. Risi, S., Stanley, K.O.: Indirectly encoding neural plasticity as a pattern of local rules. In: Doncieux, S., Girard, B., Guillot, A., Hallam, J., Meyer, J.-A., Mouret, J.-B. (eds.) SAB 2010. LNCS (LNAI), vol. 6226, pp. 533–543. Springer, Heidelberg (2010). https://doi.org/10.1007/978-3-642-15193-4_50

40. Soltoggio, A., Bullinaria, J.A., Mattiussi, C., Dürr, P., Floreano, D.: Evolutionary advantages of neuromodulated plasticity in dynamic, reward-based scenarios. In: Proceedings of the 11th International Conference on Artificial Life (Alife XI), pp. 569–576. MIT Press (2008)
41. Zhang, S., Zaiane, O.R.: Comparing deep reinforcement learning and evolutionary methods in continuous control (2017). https://arxiv.org/abs/1712.00006

Study of the Influences of Stimuli Characteristics in the Implementation of Steady State Visual Evoked Potentials Based Brain Computer Interface Systems

José Luis Murillo López[1,2], Johanna Carolina Cerezo Ramírez[1,2], and Sang Guun Yoo[1,2(✉)] (iD)

[1] Departamento de Informática y Ciencias de la Computación, Escuela Politécnica Nacional, Quito, Ecuador
{jose.murillo01,johanna.cerezo,sang.yoo}@epn.edu.ec
[2] Smart Lab, Escuela Politécnica Nacional, Quito, Ecuador

Abstract. The characteristics of the stimulus can influence the strength of the SSVEP responses. However, a complete study of the influence of different characteristics of SSVEP stimuli was not considered in previous works. The present work has performed different experiments including the most important characteristics of a visual stimulus that can affect the results of a BCI system based on SSVEP. For the present study, we have used a methodology called action research to perform different experiments and get new learnings in each experiment. At the end of the experimentation, it was possible to find the optimal combination of stimulus' characteristics that can allow us to implement a more precise SSVEP systems. For the experimentation, the traditional CCA algorithm and the proposed algorithm called RCA were used for signal classification. The results indicated that a white color circular visual stimulus of 7 to 10 Hz generated by an LED with a diffuser filter reaching 6 lx delivered the best SSVEP responses. With this combination of characteristics, it was possible to build CCA based BCI prototypes having up to 4 simultaneous stimuli with a precision higher than 82% and a response time of 0.5 s. In a similar way, it was also possible to generate an RCA based systems of up to 7 simultaneous stimuli with a precision higher than 90% and a response time of 2.5 s.

Keywords: Brain computer interface · Steady state visually evoked potentials · Stimulus characteristics · SSVEP response · CCA · EEG

1 Introduction

Brain-Computer Interface (BCI) is a technology that allows a direct connection between the Central Nervous System (CNS) of a person and an external device [1]. This is a new Augmentative and Alternative Communication (AAC) technology that is very useful for people with severe motor disabilities [2]. For the implementation of a BCI system, the

© Springer Nature Switzerland AG 2021
L. Rutkowski et al. (Eds.): ICAISC 2021, LNAI 12855, pp. 302–317, 2021.
https://doi.org/10.1007/978-3-030-87897-9_28

acquisition of brain waves is essential, and this is the reason why a large number of techniques have been developed to achieve this purpose. In general, brain waves acquisition techniques have been classified into invasive and non-invasive techniques [3]. Invasive techniques use electrodes that are placed inside of the head of the person and includes techniques such as Electrocorticogram (ECoG) and Intracortical Neuron Registry (INR) [4], which allow to have high quality signals, but they are very risky in terms of health. On the other hand, non-invasive techniques do not use surgical implants or require any surgery; that is the reason why this method is more widely used in research activities. One of the most common non-invasive techniques is the electroencephalography (EEG), which requires the placement of electrodes on the skin of the head to capture the electrical activity of the cerebral cortex [5]. Among the different non-invasive techniques, EEG is the most studied because it provides faster responses, requires less expensive equipment, and is simpler to implement [6]. Furthermore, it is the technique that covers the large number of brain signal patterns. Among those patterns, the most used are: P300, which is an Event-Related Potential (ERP), Slow Cortical Potentials (SCPs), Steady-State Visual Evoked Potentials (SSVEPs) and Sensorimotor Rhythms (SMRs) [4]. Among the aforementioned patterns, SSVEP is the one that is gaining the most popularity in recent years. This is because it is one of the patterns that provides the fastest and most reliable communication [6]. Furthermore, it is easy to implement since it requires very short or no training periods for users, and because it requires very few electrodes to start working [7]. In an SSVEP based BCI, the classification precision is related to the strength of the SSVEP response, Signal-to-Noise Ratio (SNR) and the different properties of the stimulus [8].

There are many studies that have successfully implemented SSVEP based BCI systems by improving SNR or adjusting a classification algorithm to the SSVEP response. However, there are very few works that focus on the study of the different characteristics of the stimulus. Faced with this situation, this work has been motivated to carry out an analysis of how the different characteristics of the stimulus could improve the effectiveness of SSVEP that allow the implementation of an effective BCI system.

2   State of Art

According to [9], characteristics of a stimulus can influence highly on the SSVEP response, so they must be carefully considered when designing BCI systems. The most used characteristics are the type of the source of the visual stimulus, blinking frequency, light color, size of the stimulus, and number of simultaneous stimuli.

The blinking frequency (or just frequency) of the visual stimulus is generally correlated with the number of simultaneous stimuli, and these two characteristics are the most commonly used in SSVEP implementations. The frequency stimulus can evoke an SSVEP response at a variety of amplitudes, classified as low-band (5–12 Hz), mid-band (12–25 Hz), and high-band (25–50 Hz) frequencies [10], being the most used, the low and medium bands [9]. Many studies have also analyzed the influence of the color of the stimulus, having as a common point the usage of primary colors of the RGB model (see Table 1) and the white color [11–15]; few works also analyzed violet [16] and yellow [12]. In all these works, the authors analyzed the influence of the color in the result of SSVEP responses; however, the level of illuminance of the stimulus were not mentioned in all of those works, except for [12]. Additionally, the influence of the color on chromatic stimuli has also been measured in [17]; however, this work does not make a comparison of its results with the monochromatic stimuli.

The size and shape of the stimulus were also studied in [12]. Additionally, in [18], the influence of the source type of the visual stimulus was studied, finding that frosted LEDs generated the best SSVEP responses by having a certain type of diffusion.

As we can see in the analyzed works, during the last years, the study of the influence of different characteristics of the SSVEP responses has been carried out. However, those works analyzes the characteristics one by one in a separate manner. In other words, there is no such structured study in which the most important characteristics of the visual stimuli are analyzed together, correlating them, and using the results to obtain a complete and effective BCI implementation. In this situation, the present work has decided to analyze the different characteristics of the SSVEP stimuli to define the optimal combination of characteristics for generating an effective BCI system.

Table 1. SSVEP stimuli characteristics used in previous works.

Reference	Frequency (Hz)	Source	Size	Color	Background	Illuminance	# of stimuli
[11]	7.5, 8, 8.57, 9.23, 10, 10.9, 12, 13.33, 15, 17.14	LCD 120 Hz	–	Red Blue Green White Gray	Black	–	5
[12]	14, 17, 25, 30	Diode LED	~0.99 cm ~2.6 cm ~4.5 cm ~6.5 cm	Blue Red Green White Yellow	Black	Blue: 4 lx Red: 12 lx Green: 20 lx White: 30 lx Yellow: 30 lx	4
[13]	13, 14, 15, 16, 17	LCD	–	White	–	–	5
[14]	6.67, 7.5, 8.57, 10, 12, 15	LCD 60 Hz	–	White	Black	–	6
[15]	5.45, 8.57, 12, 15	LED 60 Hz	~6.2 cm (diameter)	White	Black		4
[16]	7, 9, 11, 13	LCD 60 Hz	4 cm	Green Red Blue Violet	Black		4
[17]	Low (< 12) Medium (12–30) High (>30)	Diode LED	–	White-gray Green-red Green-blue	–	750 mcd 250 mcd	1

(*continued*)

Table 1. (*continued*)

Reference	Frequency (Hz)	Source	Size	Color	Background	Illuminance	# of stimuli
[18]	10, 10.5, 11, 11.5, 12, 12.5, 13, 13.5, 14	LCD Diode LED Frosted LED	4 × 4 LEDs 111 × 111 px	Red	Black Cardboard	–	9
[19]	6.32, 6.67, 7.06, 7.5, 8, 8.57, 9.23, 9.32, 10, 10.91	LCD 120 Hz	175 × 175 px	–	–	–	4
[20]	7, 8, 9, 10	Ring COB LED	130 mm (radius)	Green	–	357 lx 715 lx 1072 lx 1430 lx	1
[21]	8, 10, 12, 14	Panel LED	3 × 3	Green	Black	–	4
[22]	12, 13, 14, 15	Diode LED	2.6 mm (radius)	Red	–	7100 mcd 2525 mcd	1

3   Methods

Action Research is a methodology that is based on a strong relationship between experimentation and the generation of knowledge [23]. Given the iterative process it represents, the generality of its phases and its prioritization in the generation of knowledge, action research has begun to be considered as a general empirical research methodology [24]. In this work, the cyclical model and the empirical approach of Action Research are adjusted to the objective of finding the characteristics of the stimulus that can generate the best classification results in a SSVEP based BCI system. To reach the objective of the present research work, we have proposed the generation of a prototype, which covers the acquisition, processing, and classification of brain signals (see Fig. 1). This system will be used to evaluate the precision of each important characteristic of the stimulus used in SSVEP. Each evaluation (experiment) will be a cycle of the action research methodology which will generate some knowledge for the next cycle.

Before starting a new experiment, the results of different values of the stimulus' characteristics will be analyzed statistically by using the non-parametric Friedman test to determine if there are significant differences between the evaluated parameters. If so, a Friedman test of two factors by ranges will be applied to find the values of the characteristic with best results (which will be used for the next experiments) and worst results (which will be excluded for the next experiments). Once the best results have been obtained, they will also be discriminated considering for the next experiments only those that reach a precision higher than 80%. All statistical analysis will be carried out with a confidence level of 95%. Once this is done, the new experiment (cycle of action research) will start with the best results of the last cycle combining with a new characteristic of the SSVEP stimulus.

3.1 Environment for the Experimentation

In all experiments, the OpenBCI Cyton board will be used, which will be connected to the user by means of gold cup electrodes in the Oz, O1 and O2 channels, with A1 as Ground and A2 as reference, based on the international 10–20 system. The user will be located at 60 cm from the stimulus approximately. There will be no distracting noises of any kind, but the room in which the tests are taken place will not be isolated from uncontrolled noise either. All tests will be done in a moderately illuminated room. The experiment involved 16 adults aged 22.4 ± 1.4 years (4 females and 12 males) with normal or corrected vision.

The experiment will consist of creating a scenario that meets the conditions proposed in each phase i.e. the application of the characteristics of the visual stimulus which will be analyzed. When a start signal is given, the user will focus on the stimulus for 20 s until the delivery of an end signal; then, the user will rest for 15 to 25 s. During the break, the value of the characteristic being evaluated (e.g. frequency, color, illuminance, etc.) will be changed; and if the user is ready, the start signal will be given to begin the new test.

The brain signals will be transmitted at a sampling rate of 250 Hz via Bluetooth to a computer. Such data will be captured in real-time by using a software called Open-ViBE [25] (see Fig. 1). For the signal processing process, a Butterworth type Band-Pass temporal filter of order 4 was used with cutoff frequencies of 1 Hz and 40 Hz and with a maximum ripple of 0.5 dB. The signal is cut into 3 s fragments that move in a 0.5 s window. These fragments are classified using the Canonical Correlation Analysis (CCA) method and the classification returned by the algorithm is recorded for further analysis.

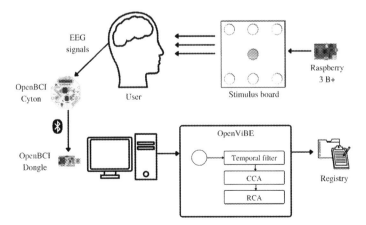

Fig. 1. Experimentation environment

CCA is a multivariate statistical method used to explore the relationships between two sets of variables. With CCA, the SSVEP response can be detected by finding the maximum correlation between two signals [26]. One of these signals is the acquired brain signal from the user and the other can be a reference signal. This reference signal is generated through a pure sinusoidal signal with the frequency of interest [9].

The signal acquisition, processing, and classification flow generates an output every 0.5 s approximately, not counting the initial 3 s after running the experiment. This way of gathering data can be useful for applications that require a quick response e.g. vehicle control. However, not having an inactivity control can significantly affect the usability of the system since the system can generate unstable signals when the user does not see any stimulus or when his/her sight changes from one stimulus to another (in multi stimuli SSVEP systems). For this reason, beside of using the CCA classification algorithm, an optimized algorithm was developed that retains the CCA responses and generates an output only when the frequency of a response exceeds a certain percentage of acceptance. For practical purposes, we have called this algorithm as Response Control Algorithm (RCA). The diagram how the proposed RCA works is shown in Fig. 2.

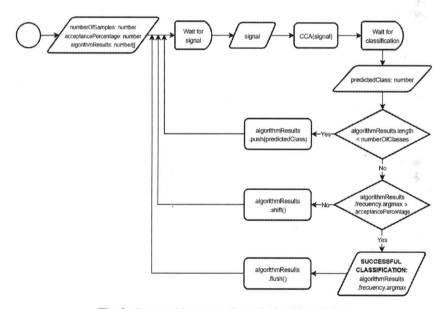

Fig. 2. Proposed Response Control Algorithm (RCA)

As indicated in Fig. 2, RCA make use of CCA outputs and two variables i.e. the number of samples and the percentage of acceptance. Each time CCA generates an output, it is stored in the memory until the number of stored outputs reaches the pre-configured number of samples. When this condition is met, the number of recurrence of each response is calculated, and if any response's recurrence is equal or greater than the acceptance percentage, this result becomes the RCA output and the stored responses are deleted; on the other hand, if the samples does not reach the acceptance percentage, the oldest response is removed, a new CCA response is saved, and the number of recurrence of all responses is recalculated again; this process is executed repeatedly until having the preconfigured percentage of acceptance. It is important to indicate that, in all experiments, an RCA with 5 samples and a percentage of acceptance of 80% will be used.

3.2 Description of Experiments

In each experiment, we firstly define the characteristic to be analyzed, the limitation presented in previous works, and the details of the experiment performed in this work to deliver a conclusion for solving the described limitation.

3.2.1 Frequency of the Stimulus

- **Limitation of Previous Works:** Most solutions use a low range frequency as shown in Table 1. This may be because the SSVEP response is often significantly more difficult to detect at high frequencies due to its low amplitude [17]. However, there is not a study analyzing the different frequencies to verify their efficiency.
- **The experiment:** This experiment has the purpose of analyzing the results of 20 frequencies between 7 and 16.5 Hz jumping 0.5 Hz. For this process, a system composed of Raspberry Pi 3B+ (including a Python script) and a customized stimulus board with RGB LEDs will be used to generate the stimulus signal (see Fig. 1).

3.2.2 Source of Stimulus

- **Limitation of Previous Works:** Using LEDs in microcontrollers or open board computers as the source of the stimulus can be an accurate way to generate the signal. But building these systems requires certain electronic skills and involves difficulties integrating with an application's GUI. On the other hand, if the stimuli were generated on an LED screen, it would facilitate the work of generating the stimulus and integrating with the GUI of any application. This antecedent raises the following question: can a LED screen generate a visual stimulus for SSVEP without having a significant loss of precision in the results?
- **The experiment:** An LED screen with a refresh rate of 60 Hz will be tested using the frequencies with the best results in the previous experiment. The stimuli will be white squares with a size of 100 × 100 pixels over a black background.

3.2.3 Color and Illuminance of the Stimulus

- **Limitation of Previous Works:** The color of the stimulus is not a standard and how such characteristic was selected in previous works are not justified. Only some of works provides some information about the influence of the color in the SSVEP response [11, 16, 17], indicating that bright colors have better results, although they are more annoying for the user [12]. In case of illuminance, not only is it not usually justified, but the vast majority of work ignore this characteristic or do not provide this data. Some previous studies such as [22] have shown that illuminance can influence the results.

- **The experiment:** The color of the stimuli will be varied in each test to measure if there are differences between them when having SSVEP responses. The colors used will be White, Red, Blue, and Green. This range will allow us to cover the primary colors of light (RGB) as well as the result of its additive synthesis. Generally, the RGB LEDs have its own maximum values of candela, which are different one from another in order to cover a greater range of colors. This would mean that the conditions under which each color is evaluated are different from others; this is one of the great limitations of [12] where the illuminance of the best-performing stimuli was up to 7 times greater than that of the worst. As the margin is so large, it is natural to question whether the results are really better because of its color or if it was because of the illuminance.

To solve this question, this work has decided to measure the illuminance generated by the stimulus with a lux meter to balance the conditions of each color. And at the same time, the present work has decided to vary the illuminance to discover how much they affect the results. The maximum values of the RGB LED used in the experiment are summarized in Table 2. Since red presents the lowest level of illuminance with 6 lx, it was decided to use this as the "Low illuminance level" and vary the duty cycle of the rest of colors to equal this level. The duty cycle that each color will work with is found in Table 2. For the maximum values, it was necessary to use a different red LED (capable of reaching a maximum illuminance of 27 lx) and to adjust the duty cycle of the RGB LEDs to reach this illuminance in the rest of the colors. The duty cycles used in each color to reach the "High illuminance level" are indicated in Table 2.

Table 2. Duty cycle and illuminance values per color.

Color	Illuminance (lux) at 100% duty cycle	Duty cycle (%) to reach 6 lx	Duty cycle (%) to reach 27 lx
White	95	10	35
Red	6	100	100 (using a different LED)
Blue	39	25	67
Green	46	20	69

3.2.4   Diffuse (Frosted) Light

- **Limitation of Previous Works:** Even with the optimal color and illuminance settings, the stimuli can be tiring for the user if it is watched over long period of time. That is why some works make use of LEDs with opaque (frosted) glass, which indicates that the frosted stimulus gives better results than LEDs with transparent glass [18]. This improvement was attributed to the comfort of the stimuli which allows better level of concentration. However, there is no studies of the efficiency difference between the transparent light stimuli and frosted ones in SSVEP responses.

- **The experiment:** Up to this point of the present work, clear glass LEDs have been used. For this experiment, the LED will be enclosed in a circular tube of 9.2 cm long and 2.5 cm in radius with a diffuser filter at the end. At this point, the lux meter was used again to compare how much the luminosity has been attenuated using the diffuser. The illuminance of the diffused LED was 6 lx.

3.2.5 Size of the Stimulus

- **Limitation to Solve:** Another study suggested that the amplitude of the SSVEP response is related to the size of the stimulus [12]. A larger size can achieve a greater amplitude, and therefore better classification results. But this parameter must be balanced, because it could reduce the usability or portability of the system.
- **The experiment:** In the previous experiment, a diffuse light of 2.5 cm radius was used. In this experiment, a stimulus with a radius of 5 cm will be used to verify if a bigger stimulus has an important effect over the results of SSVEP responses.

3.2.6 Number of Simultaneous Stimuli

- **Limitation to solve:** Up to this point, each stimulus has been tested individually (a single stimulus blinking). However, this condition has a low level of usability since most of applications require multiple operating commands i.e. multiple stimuli.
- **The experiment:** Since most of BCI applications require multiples stimuli, we will test how many simultaneous stimuli the system can have without having a significant loss in the results. The experiment will consist of choosing 2 stimuli and climbing to a maximum of 7 simultaneous stimuli. At this point, the stimuli will be generated using the best combination of characteristics selected in previous experiments.

4 Results

4.1 Frequency of the Stimulus

The results of varying the frequencies of the stimulus are presented in Fig. 3. It shows that the precision achieved by the system decreases as the frequency increases. The statistical analysis on the CCA classification results indicates that the frequencies ranging from 7 to 10.5 Hz has the best results. Statistical analysis was also done on the RCA classification results; such analysis determined that, in addition to the 7 to 10.5 Hz frequencies, 11 Hz also generate results that stand out significantly.

4.2 Source of Stimulus

The results using a signal generated by an LCD screen are shown in Table 3. Statistical analysis indicates that there is no significant difference between the samples in this phase. However, all the results were lower than those obtained with the LED (from previous experiment). For this reason, the use of LCD screen was discarded for the next experiments. The reason for this result is analyzed in the next section.

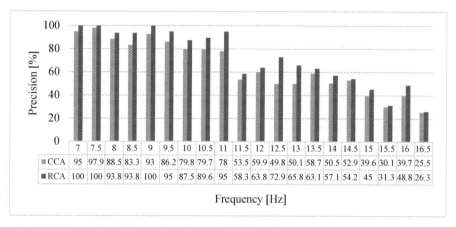

Fig. 3. System's precision using stimulus of different frequencies generated by an LED.

Table 3. System's precision using different frequencies of an LCD screen.

Classification algorithm	Frequency (Hz)							
	7	7.5	8	8.5	9	9.5	10	10.5
CCA precision (%)	33.57	30.26	19.24	20.10	15.99	3.60	17.46	11.41
RCA precision (%)	37.92	23.75	20.42	25.00	20.00	0.00	11.25	18.75

4.3   Color and Illuminance of the Stimulus

The results obtained after analyzing the 8 frequencies which were selected in the first experiment combining with different colors and illuminance are presented in Fig. 4. The stimuli of all frequencies and colors with low illuminance were significantly lower, except for Red with 7.5, 8 and 10 Hz. While for the high illuminance stimuli, the only ones that were significantly low accurate were Red and Green of 10.5 Hz, and the Blues of 8, 8.5, 9, 10 and 10.5 Hz. With this analysis, for the next experiments, the CCA results that exceeded the minimum threshold of 80% will be considered. Being more specific, the high illuminance stimuli corresponding to Red from 7 to 9 Hz, Green from 7 to 10 Hz, Blue from 7 to 7.5 Hz and White from 7 to 10 Hz will be considered for the next experiments.

4.4   Diffuse (Frosted) Light

The results after adding a diffuser filter are shown in Fig. 5. The statistical analysis has revealed that there is no significant difference without a diffuser filter. However, the filter provides the benefit of being more comfortable and less tiring to users. Based on this analysis, we have decided to use the diffuser filter for the next experiment.

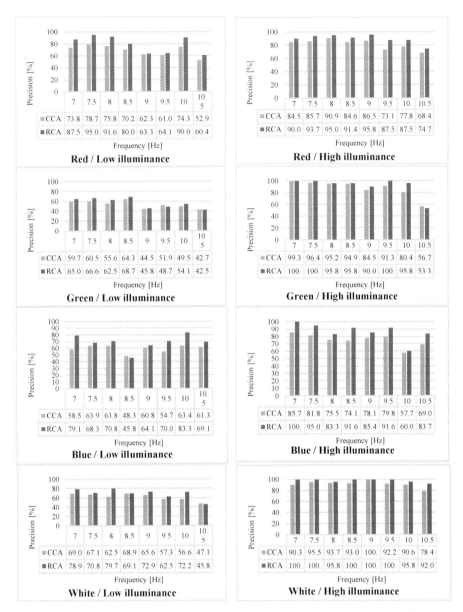

Fig. 4. Precision achieved with an LED of different colors, frequencies, and illuminance.

4.5 Size of the Stimulus

The results of increasing the size of the stimulus are presented in Fig. 6. Compared to the results of the previous experiment, the difference is minimal; the statistical analysis also indicate that the difference is not significant. In this experiment, the red color stimulus had the best result, which improved its precision by approximately 4%. Since in most of the stimuli, the difference was not significant, it was decided to continue with the stimuli used for the previous experiment. Furthermore, as the last experiment requires multiples (simultaneous) stimuli of different frequencies to represent different commands of a BCI system, only the white color stimuli has been chosen since it presents the greater number of frequencies with the highest classification precision.

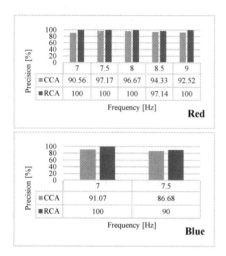

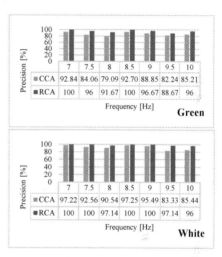

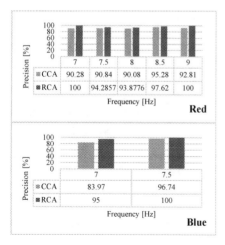

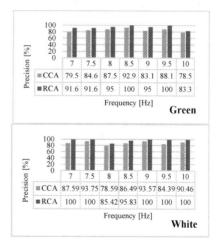

Fig. 5. Precision achieved with a diffuser filtered LED with different colors.

Fig. 6. Precision achieved with big stimuli generated by a diffuser filtered LED with different colors.

4.6   Number of Simultaneous Stimuli

Given that a total of 7 frequencies have achieved the best results up to this point, they were used to design 6 BCI systems: the first with 2 simultaneous stimuli, the second with 3, the third with 4 and so on. The distribution of the stimuli in the board is presented in Fig. 7. The results of the best stimuli are shown in Fig. 8. The average precisions are also presented in Table 4. At this point, it is important to indicate that the proposed RCA algorithm has a remarkable improvement over the CCA algorithm.

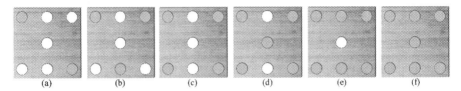

Fig. 7. Design of the stimulus boards: (a) 2 stimuli (7 and 10 Hz), (b) 3 stimuli (7, 8.5 and 10 Hz), (c) 4 stimuli (7, 8, 9 and 10 Hz), (d) 5 stimuli (7, 7.5, 8, 8.5 and 9 Hz), (e) 6 stimuli (7, 7.5, 8, 8.5, 9 and 9.5 Hz) and (f) 7 stimuli (7, 7.5, 8, 8.5, 9, 9.5 and 10 Hz).

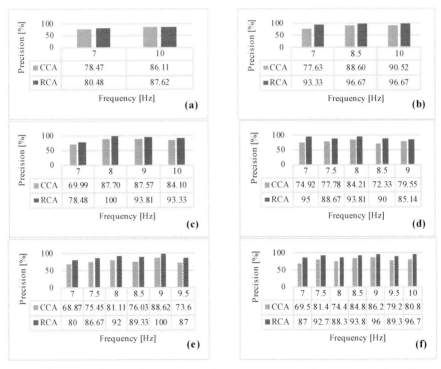

Fig. 8. Stimulus precision of the systems created in this study. The data correspond to a BCI with (a) 2 stimuli, (b) 3 stimuli, (c) 4 stimuli, (d) 5 stimuli; (e) 6 stimuli, (f) 7 stimuli.

Table 4. Average precision obtained by BCI systems by the experimented algorithms.

Number of stimuli	CCA	RCA	Number of stimuli	CCA	RCA
2	82.3%	84.05%	5	77.76%	90.52%
3	85.59%	95.56%	6	77.28%	89.17%
4	82.34%	91.4%	7	79.49%	91.97%

5 Discussion

After analyzing the previous works, it can be indicated that this study is one of the leading works that propose the generation of SSVEP systems that seeks the most optimal combination of characteristics of the stimuli.

Starting with the objective of creating an optimal BCI system, the selection of the frequency of the stimulus was surely the most important point. In the first experiment, a relationship between the frequency and the precision of CCA and RCA was found. As other studies hinted, as the frequency increased, the precision decreased significantly. The second experiment presented a surprising result indicating that an LCD screen gave results with low precision; it was a surprising result since many previous works made use of an LCD screen. One of the reasons of this result was since LCD screens have a great limitation regarding the frequencies that can be generated in it. In an LCD screen, based on its refresh rate, only frequencies that comply with the following relationship [9] can be used:

$$F = \frac{f}{n} \quad n = 2, \ldots, f \tag{1}$$

where F is the frequency of the stimulus, f is the refresh rate, and n is an integer which value is lower than the refresh rate. Therefore, for a 60 Hz LCD monitor (like the one used in this experiment), only stimuli of 6, 6.67, 7.5, 8.57, 10, 12, 15, 20 and 30 Hz could be used. Additionally, vertical synchronization must be enabled for ensuring better response [8]. Although 7.5 and 10 Hz were present in the experiment, vertical synchronization was not considered. Therefore, the tests performed in this experiment did not included the best conditions, which will be solved in future works.

The third experiment could show how increasing the intensity of illuminance, the precision of SSVEP responses improves significantly. In fact, the results using a low illuminance were disappointing. The present work could also show that the white color is the one that gives the best results, while the blue color gives the worst ones.

The last three experiments allowed to show how the usage of the diffuser filter allowed to have a great improvement in the aspect of the user's comfort while having a slight improvement in the results. Additionally, such experiments showed how the size of the stimulus did not generate an improvement in the results in a general way. In addition, the usage of the best frequencies generated by diffused white color LEDs allowed to have a BCI system with a greater number of simultaneous stimuli. It is evident that the precision achieved with multiple simultaneous stimuli drops markedly compared to individual stimulus. Even so, it was possible to create a BCI system of up to 4 simultaneous stimuli with a good level of precision (precision higher than 80%) and systems with more than 4 stimuli with an acceptable precision (higher than 77%) using CCA. It is also important to indicate that the proposed RCA algorithm had an average improvement of 9.44% in BCI systems of up to 4 simultaneous stimuli and 15.83% in BCI systems of more than 4 stimuli (compared to CCA).

As indicated before, it is important to indicate that one of the most important contributions of the present work is the proposal of the RCA algorithm, which throughout all the experiments gave the best results than the traditional CCA.

6   Conclusions

Through the performed experiments, it was possible to analyze the influence of the different characteristics of a visual stimulus in the construction of a BCI system based on SSVEPs. Using CCA, one of the most used classifiers, SSVEP response precision decreased as the stimulus frequency increased, finding the best results between 7 and 10.5 Hz. Additionally, it was possible to verify the advantages of generating the stimulus with an LED diode instead of an LCD screen that is limited to its refresh rate. In addition, the presented work showed that a small white stimulus with a diffuser filter having an illuminance of 6 lx gave the best results. With the optimal characteristic combination, it was possible to build BCI systems of 2, 3 and 4 simultaneous stimuli with an average precision greater than 82% and with an average response time of 0.5 s. Additionally, the present work implemented a new algorithm i.e. RCA, to control the responses emitted by the system and to handle in a certain way the lack of an idle class in CCA. This algorithm not only significantly improved the results of each stimulus, but also allowed the creation of BCIs of up to 7 simultaneous stimuli with an average precision of around 90% with an average response time of 2.5 s.

Acknowledgement. The authors gratefully acknowledge the financial support provided by the Escuela Politécnica Nacional, for the development of the project PVS-2018-022 – "Silla de ruedas eléctrica controlado por ondas cerebrales".

References

1. He, B., et al.: Brain–computer interfaces. In: Neural Engineering, pp. 87–151 (2013)
2. Wolpaw, J.R., et al.: Brain-computer interface technology: a review of the first international meeting. IEEE Trans. Rehabil. Eng. **8**(2), 164–173 (2000)
3. Graimann, B., Allison, B., Pfurtscheller, G.: Brain-Computer Interfaces. Springer, Heidelberg (2010). https://doi.org/10.1007/978-3-642-02091-9
4. Nam, C.S., Nijholt, A., Lotte, F.: Brain-Computer Interfaces Handbook, 1st edn. CRC Press, New York (2018)
5. Rao, R.P.N.: Brain-Computer Interfacing: An introduction, 1st edn. Cambridge University Press, Cambridge (2013)
6. Schomer, D.L., Lopes da Silva, F.H.: Niedermeyer's Electroencephalography, 7th edn. Oxford University Press, Oxford (2017)

7. Poveda, S., Murillo, J.L., Ortíz, K., Yoo, S.: Review of steady state visually evoked potential brain-computer interface applications: technological analysis and classification. J. Eng. Appl. Sci. **15**(2), 659–678 (2019)
8. Zhu, D., Bieger, J., Garcia, G., Aarts, R.M.: A survey of stimulation methods used in SSVEP-based BCIs. Comput. Intell. Neurosc. **2010**, 702357 (2010)
9. Évain, A.: Optimizing the use of SSVEP-based brain-computer interfaces for human-computer interaction. https://tel.archives-ouvertes.fr/tel-01476185. Accessed 16 May 2021
10. Wu, Z., Lai, Y., Xia, Y., Wu, D., Yao, D.: Stimulator selection in SSVEP-based BCI. Med. Eng. Phys. **30**(8), 1079–1088 (2008)
11. Cao, T., Wan, F., Mak, P., Mak, P., Vai, M., Hu, Y.: Flashing color on the performance of SSVEP-based brain-computer interfaces. In: Proceedings of the Annual International Conference of the IEEE Engineering in Medicine and Biology Society, EMBS, IEEE, USA, pp. 1819–1822 (2012)
12. Duszyk, A., et al.: Towards an optimization of stimulus parameters for brain-computer interfaces based on steady state visual evoked potentials. PLoS ONE **9**(11), 1–11 (2014)
13. Allison, B., et al.: BCI Demographics: How many (and what kinds of) people can use an SSVEP BCI? IEEE Trans. Neural Syst. Rehabil. Eng. **18**(2), 107–116 (2010)
14. Zerafa, R., Camilleri, T., Camilleri, K.P., Falzon, O.: The effect of distractors on SSVEP-based brain-computer interfaces. Biomed. Phys. Eng. Exp. **5**(3), 035031 (2019)
15. Işcan, Z., Nikulin, V.V.: Steady state visual evoked potential (SSVEP) based brain-computer interface (BCI) performance under different perturbations. PLoS ONE **13**(1), 1–17 (2018)
16. Singla, R., Khosla, A., Jha, R.: Influence of stimuli colour in SSVEP-based BCI wheelchair control using support vector machines. J. Med. Eng. Technol. **38**(3), 125–134 (2014)
17. Floriano, A., Diez, P.F., Bastos-Filho, T.F.: Evaluating the influence of chromatic and luminance stimuli on SSVEPs from behind-the-ears and occipital areas. Sensors. **18**(2), 615 (2018)
18. Mu, J., Grayden, D.B., Tan, Y., Oetomo, D.: Comparison of Steady-State Visual Evoked Potential (SSVEP) with LCD vs. LED Stimulation. In: 2020 42nd Annual International Conference of the IEEE Engineering in Medicine and Biology Society, IEEE, Monreal, Canada, pp. 2946–2949 (2020)
19. Volosyak, I., Gembler, F., Stawicki, P.: Age-related differences in SSVEP-based BCI performance. Neurocomputing **250**, 57–64 (2017)
20. Mouli, S., Palaniappan, R.: Eliciting higher SSVEP response from LED visual stimulus with varying luminosity levels. In: 2016 International Conference for Students on Applied Engineering ICSAE, IEEE, UK, pp. 201–206 (2016)
21. Pathiranage, S., Paranawithana, I., Perera, M., De Silva, A.C.: An in-depth study of SSVEP signals against stimulus frequency and distance to the stimulus. In: Moratuwa Engineering Research Conference (MERCon), IEEE, Sri Lanka, pp. 60–65 (2018)
22. Wu, C.H., Lakany, H.: The effect of the viewing distance of stimulus on SSVEP response for use in brain-computer interfaces. In: Proc. 2013 IEEE International Conference on Systems, Man, and Cybernetics, IEEE, UK, pp. 1840–1845 (2013)
23. Susman, G.I., Evered, R.D.: An assessment of the scientific merits of action research. Adm. Sci. Q. **23**(4), 582–603 (1978)
24. Staron, M.: Action Research in Software Engineering: Theory and Applications. Springer, Cham (2020). https://doi.org/10.1007/978-3-030-32610-4
25. Renard, Y., et al.: OpenViBE: an open-source software platform to design, test, and use brain-computer interfaces in real and virtual environments. Presence: Teleoper. Virtual Environ. **19**(1), 35–53 (2010)
26. Kwak, N.S., Müller, K.R., Lee, S.W.: A convolutional neural network for steady state visual evoked potential classification under ambulatory environment. PLoS ONE **12**(2), 1–20 (2017)

Promises and Challenges of Reinforcement Learning Applications in Motion Planning of Automated Vehicles

Nikodem Pankiewicz[1,2(✉)] , Tomasz Wrona[1,2] , Wojciech Turlej[1,2] ,
and Mateusz Orłowski[1,2]

[1] AGH University of Science and Technology, Krakow, Poland
[2] Aptiv, Krakow, Poland
{nikodem.pankiewicz,tomasz.wrona, wojciech.turlej,
mateusz.orlowski}@aptiv.com

Abstract. As automated driving development progresses forward, novel methods are required to handle the vastness of possible road situations and to face end user's high demands. Trying to solve the problem of motion control involving decision making and trajectory planning it is reasonable to take into consideration reinforcement learning as a viable approach. In this paper, we present the promises reinforcement learning can bring to an automated driving domain and the list of challenges we encountered during our work. We address the issues related to the environment definition, sample efficiency, safety and explainability.

Keywords: Automated driving · Behavior planning · Reinforcement learning · Real-world

1 Introduction

Applying reinforcement learning (RL) methods to solve complex tasks is becoming increasingly popular and progressing research constantly delivers new state-of-the-art algorithms [15, 23, 30]. In some of these applications, RL-based solutions exceed even the human performance [10]. However, the vast majority of the considered problems were placed in the virtual domain, which usually takes the form of a computer game or a simulation, while the progress in solving problems in the real world domain is not that advanced. This immaturity is a concern for the industry, where the application potential is significant.

One of the potential areas of reinforcement learning methods application is behavior and trajectory planning of the automated vehicles. Due to the problem's complexity, rules-based deterministic algorithms may not be sufficient to achieve desirable system's performance. Therefore, a need for data-driven methods emerges and reinforcement learning seems to be a promising approach.

As a research group involved in the development of automated vehicle technology, we present our motivation and problems behind the application of reinforcement learning to this area. In this paper, we refer to the publication [13],

© Springer Nature Switzerland AG 2021
L. Rutkowski et al. (Eds.): ICAISC 2021, LNAI 12855, pp. 318–329, 2021.
https://doi.org/10.1007/978-3-030-87897-9_29

in which the authors outlined nine major problems of applying reinforcement learning in a real world setting. We describe the problems we encountered and investigated during the implementation of the behavior planning functionality in a close to production ready system.

1.1 Our Work

In this section we present an architecture of our system involving reinforcement learning agent in the place of behavior planning module of an automated driving stack.

Motion Planning System. Because of the reasons described in more detail in Sect. 3.4, we decided to create a hybrid system consisting both of reinforcement learning and model-based algorithms. The motion planning system is fed by a perception stack, which provides a representation of an ego car's surroundings, such as other road users and a road itself. The system consists of four main entities: the behavior planning module, the safety framework, the trajectory planning module and the control block. The architecture is presented in Fig. 1. In our case, the behavior planning module, providing a car with a high-level description of an intention, like a maneuver to execute or a speed recommendation, has a form of an reinforcement learning agent. The returned behavior is then realized by model-based methods responsible for trajectory planning and control, which are under the supervision of the safety framework. The safety framework itself, heavily based on the ideas presented in [31], along with an additional set of deterministic rules based on traffic law, assures safety of the AI-based behavior planning as well as the trajectory planning. The behavior planning module can also be additionally supplied with signals from the mission planning module, responsible for producing sub-goals necessary to follow the defined path.

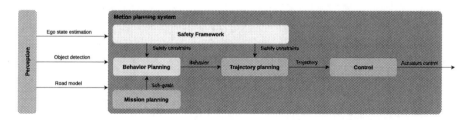

Fig. 1. The motion planning architecture: the RL-based behavior planning module (in red) and the model-based trajectory planning and control modules (in blue). (Color figure online)

Reinforcement Learning Environment. To train a reinforcement learning agent responsible for behavior planning we design a custom environment based on a proprietary closed-loop simulation tool [6]. The environment consists of multiple blocks, which we listed below.

– **Behavior decoding** decodes an action (outputted from a policy neural network) to a semantic value, like a maneuver to execute or a velocity set point (see Sect. 3.3).

– **Trajectory planning** establishes a specific realization of a behavior and returns a continuous trajectory to be followed by the agent.

– **Control** produces low-level control signals (e.g. throttle, braking, steering wheel angle) to keep the car on its reference trajectory.

– **Simulation** delivers a closed-loop simulation of vehicle dynamics and road users interactions.

– **Perception simulation** disturbs the perfect ground truth obtained from the simulation using high-level sensor models (see Sect. 3.5).

– **Observation encoding** encodes the state of the environment to a form easily interpretable by the policy network. The design of this transformation is crucial for the policy to be able to generalize well in various situations (see Sect. 3.3).

– **Reward calculation** defines a numerical reward signal for the agent, designed to promote the agent for keeping a cruising velocity and reaching predefined lane-based goals, while maximizing comfort and safety of passengers (see Sect. 3.6).

By defining the environment in such way, we are able to control the current challenge for the agent as well as experiment with different versions of each module.

Training Setup. Having the environment defined, we utilize standard reinforcement learning algorithms, such as DQN [23], PPO [30] or SAC [15] to train the behavior planning agent. For better generalization and more efficient training, these algorithms were scaled up to support a distributed setup. Our best performing setup is a combination of PPO algorithm and multiple state-of-the-art methods available in the literature. In [28] we presented our previous results with DQN algorithm, along with the more detailed description of the environment definition, including the observation and action spaces and the reward shaping mechanism (Fig. 2).

2   Promises

Using data-driven methods over other methods in behavior and trajectory planning modules brings several promises. The most obvious one is the transfer of the responsibility for the design of driving policy rules and actions from the group of human experts to the computer program. This design might be done based on a large amount of data collected with the help of a realistic virtual simulation. A good example might be Waymo, a self-driving company, which states that their systems drive 20 million miles in a simulation each day [4]. Still, the choice between supervised learning and reinforcement learning methods is an open question.

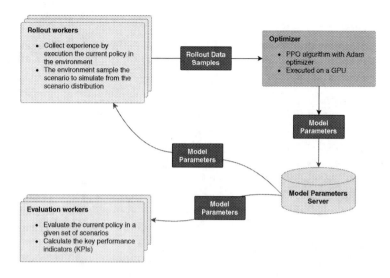

Fig. 2. The architecture of the behavior planning agent training setup.

Supervised learning methods, such as e.g. behavior cloning [33], might be more interesting due to their higher stability of the training process, but they require human interaction in the data collection process, thus making this process significantly slower and more expensive [1]. Furthermore, when the trained policy is found to be unsatisfactory, the data has to be collected again. In the case of reinforcement learning, for some algorithms, the data can be reused by recalculating the rewards only.

The more important difference, however, is with the trained policy itself. With supervised learning, we are limited to a single policy demonstrated by a driver. Moreover, it is unlikely that the driver's policy is the optimal one. This is due to the fact that it is usually hard to decide, which action is the best decision at a given moment, and how this action will affect the future states. This problem is known as the credit assignment problem, and many reinforcement learning algorithms tackle this problem by default. Yet another problem we observed is the difference between the perception system of a human and an automated vehicle. The drivers' decisions are affected by human factors such as lack of attention, cognitive distraction [21,36] or aggressiveness level [22]. That and other not listed discrepancies in perception causes the distributional shift between learning and evaluation domain of automated vehicle. Finally, an exploration of corner cases (accidents, dangerous or unusual situations), even when carried out in the virtual simulation, is more efficient with reinforcement learning methods, as humans are willing to risk less, than computer programs.

3   Challenges

3.1   Training Off-Line from the Fixed Logs of an External Behavior Policy

Fixed logs from driving are valuable assets in the entire learning process. Such logs can be used in a supervised manner to train an initial policy with behavior cloning [33], as a part of imitation learning process [34,35] or as an additional exploration guiding for policy optimization [20].

A good example of successful logs utilization is the work [11] where the team from Waymo used 30 million samples of car's trajectories to train a neural network policy, which was responsible for generating a future trajectory for a vehicle. As the authors admitted, the model was good enough to drive a vehicle successfully, however it was not fully competitive with the deterministic motion planning despite using a considerable amount of data.

Besides the amount of data, the samples should be also of sufficient quality. Erroneous actions resulting from driver's mistakes are misleading during training. Additionally, data should include states in which a reasonable driver will never be situated. These states provide, however, valuable inputs to the learning process as they alleviate distributional shift between the learning and the operating domains.

3.2   Learning on the Real System from Limited Samples

The process of developing functionalities, capable of operating in the real world, based solely on simulation is burdened with the simulation to reality (sim-to-real) gap issue. Therefore, a logical step to close this gap is to fine-tune the policy on public roads. In this process, a most recent driving policy is used to interact with the target environment, explore it and collect data for the next training iteration. The relatively slow rate of data collection, thus the limited number of samples, and willingness of data re-usage suggest utilization of off-policy methods, such as DQN [23] or SAC [15], which latter's efficiency in training the physical robots was demonstrated in [16].

The complication which arises in such an approach is a risk resulting from exploiting a deficient policy in a vulnerable environment. The application of such a solution requires ensuring the safety of the environment influenced by the agent. In our work, this approach is viable because the agent's action is secured by a deterministic trajectory planning and its safety module (discussed in Sect. 3.4).

In case of recognizing performance drops in some situations during evaluation, the process enables counteracting them by involving more vehicles in those problematic scenarios, which can be later used to improve the policy in those cases.

3.3   High-Dimensional Continuous State and Action Spaces

While designing state spaces for robotic applications it is reasonable to convert raw sensor inputs to another, meaningful representation, e.g. camera images to a set of detected objects or a lidar points cloud to a spatial map of surroundings. That way we decrease the amount of black-box models in the motion planning module, which significantly increases its explainability. Moreover, the specific representation of the input data contributes to a better generalization of a neural network and speeds up its training. We use this approach in our work, where we represent the environment with the high-level properties of the ego vehicle (e.g. orientation, velocity), the target vehicles, and the road model.

A reasonable action space for an automated vehicle motion planning module must consist of at least one control signal for each of the two spatial dimensions: longitudinal (velocity, acceleration, throttle, braking) and lateral (heading, steering angle). Another approach is to determine a set of waypoints, which describe consecutive features of a trajectory: velocity and acceleration values sampled along a reference path.

As mentioned in Sect. 1.1, we split the control decision component into two modules: behavior planning and trajectory planning. The behavior planning module is further combined with strategic and tactical planners. The former outputs one maneuver (e.g. follow lane, change lane left, prepare to change lane left) from the list of the currently available maneuvers. The state of the list depends on the state of the current situation on the road and is strictly controlled by the safety framework. The latter produces a combination of additional guiding signals for the trajectory planning module, such as a velocity set point or a lane bias. The real values are discretized with a resolution found empirically, having in mind the trade-off between the model complexity and the maneuvers' flexibility.

3.4   Satisfying Safety Constraints

All components of a modern vehicle must meet strict safety requirements. Both hardware and software are subject to validation accordingly to standards such as ISO 26262 (Functional Safety) or ISO/PAS 21448 (Safety of the Intended Functionality) and therefore are heavily tested before they reach an end customer. With the advent of advanced systems, which take more and more responsibility for control of a vehicle, safety considerations for highly automated vehicles is a complex topic in itself.

Looking for the required performance with regards to safety it is reasonable to refer to the human drivers. While car accidents are caused to over 1.3 million deaths annually [27], a fatal accident rate of a single vehicle is relatively low, reaching on average 1 fatality per 100 million miles in US [8]. Since statistics on the road-related accidents are shaping expectations regarding maximum failure rates of automated vehicles, a rarity of such events poses a great challenge to companies developing such systems. Showing with reasonable confidence that a system has a significantly lower failure rate in terms of fatal accidents through end-to-end test drives would require billions of miles of test drives [19].

A relative rarity and importance of severe accidents create a challenge not only for testing and validation but also for the development of such systems. Learning an end-to-end RL-based planning system to drive with reasonably low severe collision rate requires modeling of reward function with trade-off between efficiency and safety or using constrained RL [7]. However, to liberate the RL part from the problem which could be formulated in a deterministic way, we decided that the reinforcement learning system should not be formally responsible for the safety aspects of driving.

Another observation may be made with respect to traffic regulations. Training an agent to be compliant with them with the use of a reward signal will always end up in their approximations, which is unacceptable. Fortunately, it is straightforward to design a model-based system obeying such rules by definition.

The points above suggest that end-to-end planning systems based on reinforcement learning present several serious issues. It seems that the desired system should be rather a hybrid approach, composed of model-based and machine learning modules. The interface between the reinforcement learning module and the deterministic part of the system has to satisfy two contradicting objectives. On the one hand, the interface has to be constrained enough to not put safety requirements onto the reinforcement learning part. On the other, it has to be open enough to benefit from data-driven methods in planning and assure a sufficient level of policy transparency, allowing efficient and explainable learning.

An interesting approach intended for providing safety guarantees in path planning systems with reinforcement learning elements was proposed in [31]. Authors formulate a set of safety rules that are used to derive deterministic constraints for a path planning algorithm. The rules describe a safety envelope, i.e. a constrained area of a state space, which guarantees the existence of collision-free trajectories in all reasonable situations. The reasonable situations are the ones in which road users follow traffic laws and formalized common-sense traffic rules. Early detection of safety envelope violations, caused by a controlled agent, allows to execute predefined emergency responses in time and avoid collisions. Similar concepts were proposed also in [26] and [29].

The described approach can fulfill two major roles in the RL-based path planning system: limiting the available action space and triggering execution of emergency trajectories in unsafe situations (that may be caused by both ego's actions and other agents' behaviors). The safety envelope provides transparent, testable safety constraints that protect an agent against performing unsafe maneuvers and enforces proper responses when the constraints are violated.

This approach alone provides a certain level of protection against collisions, but still, overall system performance in terms of avoiding unsafe situations can be improved in a training process. Since emergency actions typically consist of severe braking, which is sub-optimal in terms of passengers' comfort and driving efficiency, a properly trained agent will proactively avoid potentially unsafe actions. An example of such behavior may be to switch a lane away from a merge-in lane to avoid potentially unsafe situations involving slow vehicles merging into highway traffic.

Proactive safety behaviors emerging in an RL-based system may have particular significance in a context of a realistic sensor stack with limited performance. Simulating typical sensors' error patterns and occlusions in a training setup may significantly decrease the impact of perception issues on overall system safety. An agent trained in such environments may learn to avoid situations that are correlated with hazardous perception errors, as well as to perform human-like visibility-enhancement actions, such as biasing on its lane to unveil occluded areas of the environment.

3.5   Partial Observability and Non-stationarity

In the automated driving domain, both partial observability and non-stationarity are major issues significantly affecting the trained policy's performance. They are a result of different factors, from the physical world barriers to software and hardware limitations. In this section, we summarize what type of problems we observed during our research, what solutions we applied, or what solutions we propose.

Occluded Objects. The first problem is occluded objects: other vehicles, pedestrians, traffic signs and lights, lane markings, and other elements of infrastructure. One possible general solution might be to involve V2V or V2X communication [18] to exchange the information between the objects, even when they are not visible to each other. Another option is to assume worst-case scenarios, as proposed in [31] and discussed in Sect. 3.4, and act appropriately, but this might lead to overly protective policies. We implemented this by injecting worst-case object hypotheses on occlusion edges into perception results and allow agents to learn avoiding potentially unsafe regions to minimize effects on driver's comfort or total efficiency.

Perception Issues. The second problem is the imperfection of a sensing stack. A typically automated vehicle prototype is nowadays equipped with a combination of multiple cameras, radars, and lidars. Each type of the sensors has its advantages and disadvantages, thus the common approach is to fuse data from multiple sources to obtain a high fidelity representation of the vehicle's surroundings. In practice, false positives (ghost objects), false negatives (missing objects), or significant measurement uncertainties are still present in the fused representation. To address these problems in simulation, sensor models can be introduced to imitate the behavior of the real-world sensing suite. A lot of work has already been done in the area of detailed sensor simulation [14,24]. However, due to the detailed simulation, the proposed solutions are usually computationally expensive, what significantly extends the simulation time, thus slow down the learning process. To alleviate this issue, we propose the usage of simple, high-level sensor models, which skip understanding the underlying physical principles of a given sensor (e.g. how radar wave is reflected from an object and how it is later tracked), and instead utilize statistical models for each common problem (ghost objects, uncertainty, etc.; Fig. 3).

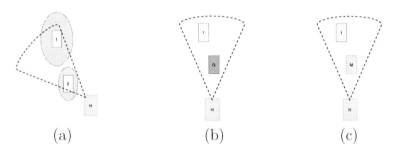

Fig. 3. The high-level sensor models: a) a measurement uncertainty (in red) is being applied to target objects (in gray); b) a ghost object (in green) is being inserted into the sensor's FOV; c) a missing object (in red) is being removed from the sensor's FOV. (Color figure online)

Intentions. The third problem is missing information about other road users' intentions. For the same observation of a state, the consecutive states might differ between episodes, depending on the *hidden state* of other drivers. One solution might be to add a pipeline responsible for the behavior or trajectory prediction of road users. This introduces an extra computing overhead and is a challenging task in itself, but significantly increases the model's explainability. Another solution is to let a policy's network spot the interesting connections on its own and store them in memory blocks.

In the previous paragraphs, we mentioned memory blocks as a solution for partial observability and non-stationarity of the environment. The obvious solution is to use modules such as LSTM [17] cells, but they lack interpretability. Frame stacking or putting historical data into the observation seems to be more suitable for safety-critical systems, however, in order to achieve satisfying policies, they might need to generate their own signals (e.g. information about a long-term decision taken in the past) and pass them between consecutive timesteps. Ideally, to achieve high interpretability, these signals have to be identified and defined by a human in advance. This, however, limits the capabilities being possible to be trained by the policy.

3.6   Unspecified and Multi-objective Reward Functions

Designing a reward function is another challenge that determines the success of the implemented solution. Drivers while making their decisions, take into account many aspects, which should be identified and represented during the reward function design.

Therefore, the reward function should be a compromise between pro-active safety assurance, performance on reaching destination targets, efficiency (time, fuel), comfort, and a general impact on a traffic flow. However, reward hacking [9] is a known problem in reinforcement learning, which causes an agent to not behave as expected, even though it achieves high rewards. We observed this in our work multiple times, e.g. when the agent was performing too frequent changes of its current maneuver. It is important to remember that the reward function designer must adequately prioritize and balance the individual parts of the reward signal or use meta-learning techniques.

3.7 Explainability

Explainability is a crucial requirement when working with safety-critical systems such as automated vehicles. In case of a system failure, whether in a form of an accident or a dangerous situation, a testing team must have a tool to understand why the situation happened, which decision was wrong, and what signals attributed most to this decision. Fortunately, the topic is being extensively researched recently and a good survey on available methods was presented in [25], as well as open-source tools [2,5] were released.

In this paper, we focus on neural network-based policies, where the input signals are high-level signals outputted from other systems (perception, localization). This is in contrast to end-to-end methods (e.g. [12]), where the input data is usually raw data and its interpretation is rather an uneasy task. In our situation, the input signals are labeled, and we can use simple methods such as e.g. Integrated Gradients [32] to calculate each input neuron's attribution to the output signal and provide meaningful insight on the policy's motivation.

With this methodology, we can navigate through a complete episode and investigate, what caused a problematic decision at any given time step, and how the decision would change, if we change the observation. We found this approach successful for some tasks and open-sourced our initial version of the tool [3]. We, however, encountered the problem with understanding neurons' state in LSTM cells. As mentioned in Sect. 3.5, there is a need for more explainable memory blocks.

4 Summary

In this paper, we presented the promises we see standing behind the application of the reinforcement learning method in motion planning. We also listed the challenges we faced during our research on the behavior planning module and described our solutions for these issues. Based on our observations, we believe that the current state of the research allows us to successfully deploy reinforcement learning in real-world automated driving. However, some work has to be done yet to e.g., ensure better explainability of an artificial agent and to provide an easier comparison between different versions and methods of the trained agent. As a consequence of the unpredictability of RL agents, we presented our method of satisfying safety constraints, but this approach still requires exhaustive testing in an operating domain. Additionally, the formulation of reward function remains an open research topic. As an active research group, we will continue our research towards more challenging traffic scenarios such as occurring on urban roads. Our future research will also be devoted to the explainability problem of RL agents.

References

1. Autonomous Vehicle Data Annotation Market Analysis. https://www. researchandmarkets.com/reports/4985697/autonomous-vehicle-data-annotation-market-analysis
2. Captum. Model Interpretability for PyTorch. https://captum.ai/
3. GitHub - iamhatesz/rld: A development tool for evaluation and interpretability of reinforcement learning agents. https://github.com/iamhatesz/rld
4. Off road, but not offline: How simulation helps advance our Waymo Driver. https://blog.waymo.com/2020/04/off-road-but-not-offline-simulation27.html
5. sicara/tf-explain: Interpretability Methods for tf.keras models with Tensorflow 2.x. https://github.com/sicara/tf-explain
6. Traffic AI – Simteract. http://simteract.com/traffic-ai/
7. Achiam, J., Held, D., Tamar, A., Abbeel, P.: Constrained policy optimization. CoRR abs/1705.10528 (2017). http://arxiv.org/abs/1705.10528
8. Administration, F.H.: Highway statistics, 2018. Technical report, Washington, DC: US Department of Transportation (2019)
9. Amodei, D., Olah, C., Steinhardt, J., Christiano, P.F., Schulman, J., Mané, D.: Concrete Problems in AI Safety. CoRR abs/1606.06565 (2016). http://arxiv.org/abs/1606.06565
10. Badia, A.P., et al.: Agent57: Outperforming the Atari Human Benchmark (2020)
11. Bansal, M., Krizhevsky, A., Ogale, A.S.: ChauffeurNet: Learning to Drive by Imitating the Best and Synthesizing the Worst. CoRR abs/1812.03079 (2018). http://arxiv.org/abs/1812.03079
12. Bojarski, M., et al.: End to End Learning for Self-Driving Cars (2016). http://arxiv.org/abs/1604.07316
13. Dulac-Arnold, G., Mankowitz, D.J., Hester, T.: Challenges of Real-World Reinforcement Learning. CoRR abs/1904.12901 (2019). http://arxiv.org/abs/1904.12901
14. Dworak, D., Ciepiela, F., Derbisz, J., Izzat, I., Komorkiewicz, M., Wojcik, M.: Performance of LiDAR object detection deep learning architectures based on artificially generated point cloud data from CARLA simulator. In: 2019 24th International Conference on Methods and Models in Automation and Robotics, MMAR 2019, pp. 600–605. Institute of Electrical and Electronics Engineers Inc. (2019). https://doi.org/10.1109/MMAR.2019.8864642
15. Haarnoja, T., Zhou, A., Abbeel, P., Levine, S.: Soft Actor-Critic: Off-Policy Maximum Entropy Deep Reinforcement Learning with a Stochastic Actor (2018). http://arxiv.org/abs/1801.01290
16. Haarnoja, T., et al.: Soft Actor-Critic Algorithms and Applications. CoRR abs/1812.05905 (2018). http://arxiv.org/abs/1812.05905
17. Hochreiter, S., Schmidhuber, J.: Long short-term memory. Neural Comput. **9**(8), 1735–1780 (1997). https://doi.org/10.1162/neco.1997.9.8.1735
18. Jung, C., Lee, D., Lee, S., Shim, D.H.: V2x-communication-aided autonomous driving: system design and experimental validation. Sensors (Switzerland) **20**(10), 2903 (2020). https://doi.org/10.3390/s20102903, http://pmc/articles/ PMC7287954/?report=abstract www.ncbi.nlm.nih.gov/pmc/articles/PMC7287954/
19. Kalra, N., Paddock, S.M.: Driving to safety: how many miles of driving would it take to demonstrate autonomous vehicle reliability? Trans. Res. Part A Policy Pract. **94**, 182–193 (2016)

20. Kang, B., Jie, Z., Feng, J.: Policy optimization with demonstrations. In: 35th International Conference on Machine Learning, ICML 2018, vol. 6, pp. 3855–3869 (2018)
21. Kass, S.J., Cole, K.S., Stanny, C.J.: Effects of distraction and experience on situation awareness and simulated driving. Transp. Res. Part F: Traffic Psychol. Behav. **10**(4), 321–329 (2007). https://doi.org/10.1016/j.trf.2006.12.002
22. Lajunen, T., Parker, D.: Are aggressive people aggressive drivers? A study of the relationship between self-reported general aggressiveness, driver anger and aggressive driving. Accid. Anal. Prev. **33**(2), 243–255 (2001). https://doi.org/10.1016/S0001-4575(00)00039-7, https://linkinghub.elsevier.com/retrieve/pii/S0001457500000397
23. Mnih, V., et al.: Playing Atari with Deep Reinforcement Learning (2013). http://arxiv.org/abs/1312.5602
24. Molenaar, R., Van Bilsen, A., Van Der Made, R., De Vries, R.: Full spectrum camera simulation for reliable virtual development and validation of ADAS and automated driving applications. In: IEEE Intelligent Vehicles Symposium, Proceedings, vol. 2015-August, pp. 47–52. Institute of Electrical and Electronics Engineers Inc. (2015). https://doi.org/10.1109/IVS.2015.7225661
25. Molnar, C.: Interpretable Machine Learning (2019)
26. Nistér, D., Lee, H.L., Ng, J., Wang, Y.: The Safety Force Field. Technical report
27. World Health Organization et al.: Global status report on road safety 2018: Summary. World Health Organization, Technical report (2018)
28. Orłowski, M., Wrona, T., Pankiewicz, N., Turlej, W.: Safe and goal-based highway maneuver planning with reinforcement learning. In: Advances in Intelligent Systems and Computing, vol. 1196 AISC, pp. 1261–1274. Springer (2020). https://doi.org/10.1007/978-3-030-50936-1_105, https://link.springer.com/chapter/10.1007/978-3-030-50936-1_105
29. Pek, C., Althoff, M.: Computationally efficient fail-safe trajectory planning for self-driving vehicles using convex optimization. In: 2018 21st International Conference on Intelligent Transportation Systems (ITSC), pp. 1447–1454. IEEE (2018)
30. Schulman, J., Wolski, F., Dhariwal, P., Radford, A., Klimov, O.: Proximal Policy Optimization Algorithms (2017). http://arxiv.org/abs/1707.06347
31. Shalev-Shwartz, S., Shammah, S., Shashua, A.: On a Formal Model of Safe and Scalable Self-driving Cars (2017). http://arxiv.org/abs/1708.06374
32. Sundararajan, M., Taly, A., Yan, Q.: Axiomatic Attribution for Deep Networks. Technical report (2017)
33. Torabi, F., Warnell, G., Stone, P.: Behavioral cloning from observation. In: IJCAI International Joint Conference on Artificial Intelligence 2018-July(July), pp. 4950–4957 (2018). https://doi.org/10.24963/ijcai.2018/687
34. Via, E.: What would II do?: Imitation Learning via off-policy Reinforcement Learning, pp. 1–13 (2019)
35. Wu, Y.H., Charoenphakdee, N., Bao, H., Tangkaratt, V., Sugiyama, M.: Imitation Learning from Imperfect Demonstration. Technical report. https://www.basketball-reference.com/leagues/NBA_stats.html
36. YoungPaul, K.L., Salmon, M.: Examining the relationship between driver distraction and driving errors: a discussion of theory, studies and methods. Safe. Sci. **50**(2), pp. 165–174 (2012)

The Usage of Possibility Degree in the Multi-criteria Decision-Analysis Problems

Andrii Shekhovtsov, Bartłomiej Kizielewicz, Wojciech Sałabun$^{(\boxtimes)}$, and Andrzej Piegat

Research Team on Intelligent Decision Support Systems, Department of Artificial Intelligence and Applied Mathematics, Faculty of Computer Science and Information Technology, West Pomeranian University of Technology in Szczecin, ul. Żołnierska 49, 71-210 Szczecin, Poland
`{andrii-shekhovtsov,bartlomiej-kizielewicz,wojciech.salabun,`
`andrzej.piegat}@zut.edu.pl`

Abstract. More and more multi-criteria problems are being analysed in an uncertain environment where the decision-making attributes' exact values are not known. For this reason, new methods are also being developed that can assess alternatives in conditions of uncertainty. However, many methods evaluate alternatives not as an exact value but as a preference interval value. It raises the problem of how to rank the alternatives assessed as interval values finally.

In this paper, we propose a simple approach to ranking, where a matrix of the possibility degree values is created based on which the final ranking is obtained. Afterwards, we compare the rankings identified by using the proposed method with naive approaches. For this purpose, a short numerical example is presented, where seven different formulas of the possibility degree are involved. In this example, the interval assessment is obtained by using the COMET method and the obtained results are ranked and compared with naive approaches and reference ranking. The proposed approach is useful and straightforward for ranking alternatives under uncertain conditions.

Keywords: Decision-making · Intervals · Possibility degree · Rankings

1 Introduction

Multi-criteria Decision-Analysis (MCDA) methods belong to the rapidly growing branch of operational research. They are widely used and developed by many scientists around the world. Their popularity is associated with the need to solve increasingly complex decision-making problems. One of the sources of this complexity is that MCDA methods increasingly have to use uncertain data. Sometimes it is also associated with problems in which partly incomplete data occur [10,15].

The most straightforward approach to dealing with uncertain data is to use interval values instead of exact values. However, the solution to such a problem

© Springer Nature Switzerland AG 2021
L. Rutkowski et al. (Eds.): ICAISC 2021, LNAI 12855, pp. 330–341, 2021.
https://doi.org/10.1007/978-3-030-87897-9_30

usually remains the numerical interval. To obtain a ranking from these values, one has to wonder what the preference interval represents. This is because it defines the smallest and largest possible assessment that an alternative can receive. Therefore, comparing the two alternatives, we can only periodically determine the exact ranking if they are separable intervals. Many methods also use other forms of expression of uncertain values such as fuzzy numbers [18], hesitant fuzzy numbers [2], interval valued fuzzy numbers [3], q-rung orthopair fuzzy set [11] or intuitionistic 2-tuple linguistic sets [4]. However, this work is limited to the ranking of the interval values.

This paper's main contribution is a new approach to ranking a set of alternatives, where alternatives have been assessed in an interval form. The proposed method is based on the possibility degree of two intervals. In order to rank the alternatives, a matrix should be defined that contains all the possibility degrees. Our work compares seven different definitions of the possibility degree. The numerical example is presented to show the efficiency of the proposed method and comparing with naive approaches. For this purpose, we considered assessing ten electric vans, where the part of the data was presented as interval numbers. The Characteristic Object METhod (COMET) was used to obtain preference intervals. This method was used because it does not require the weight of criteria. Obtained results have been compared by using similarity coefficients.

The rest of the paper is organised as follows. The priority degree definitions are given in Sect. 2. The COMET method and similarity coefficients are presented in Sect. 3. In Sect. 4, we propose a new approach to rank alternatives in the decision-making domain. The numerical example is given in Sect. 5. In Sect. 6, we conclude the paper.

2   Preliminries

Let suppose that we have two intervals $A = [a^L, a^R]$ and $B = [b^L, b^R]$, where $a_L < a_R$ and $b_L < b_R$. Then according to [16], the possibility degree $A \geq B$ is defined as $P(A \geq B)$. In the literature, various mathematical definitions can be found (1–7) [1,6–8]. For example, Wang et al. [17] presented a simple equation (1) which provides the degree of possibility that one interval is greater than another. Currently, this approach seems to be most popular in the literature.

$$P_1(A \geq B) = \frac{\max\left(0, a^R - b^L\right) - \max\left(a^L - b^R, 0\right)}{a^R + b^R - b^L - a^L} \tag{1}$$

Equations (2–4) give the same results as method presented by Wang et al. [17], what was proved by Gao in [5]. The methods are presented on account of the different approach adopted in determining the formulas:

$$P_2(A \geq B) = \frac{\max\left\{0, a^R - a^L + b^R - b^L - \max\left(b^R - a^L, 0\right)\right\}}{a^R + b^R - b^L - a^L} \tag{2}$$

$$P_3(A \geq B) = \begin{cases} 1, b^R < a^L \\ \dfrac{a^R - b^L}{a^R - a^L + b^R - b^L}, b^L < a^R, a^L < b^R \\ 0, b^L > a^R \end{cases} \tag{3}$$

$$P_4(A \geq B) = \max \left\{ 1 - \max \left(\frac{b^R - a^L}{a^R - a^L + b^R - b^L}, 0 \right), 0 \right\} \tag{4}$$

Other methods, although less frequently used, are also an important element of our study and are presented as follow (5)–(7):

$$P_5(A \geq B) = \begin{cases} 1, b^R < a^L \\ \dfrac{\left(a^R - b^L\right)^2}{\left(a^R - b^L\right)^2 + \left(b^R - a^L\right)^2}, b^L < a^R, a^L < b^R \\ 0, b^L \geq a^R \end{cases} \tag{5}$$

$$P_6(A \geq B) = \frac{1}{2} \left(1 + \frac{\left(a^R - b^R\right) + \left(a^L - b^L\right)}{|a^R - b^R| + |a^L - b^L| + l_{AB}} \right), \tag{6}$$

where l_{AB} means the length of the overlap part of two intervals and can be calculated as (8);

$$P_7(A \geq B) = \begin{cases} 1, b^R < a^L \\ 1 - \dfrac{\left(b^L - a^R\right)^2}{2 l_A l_B}, b^L < a^L < b^R < a^R \\ \dfrac{a^L + a^R - 2b^L}{2 l_B}, b^L < a^L < a^R < b^R \\ \dfrac{2a^R - b^L - b^R}{2 l_B}, a^L < b^L < b^R < a^R \\ \dfrac{\left(a^R - b^L\right)^2}{2 l_A l_B}, a^L < b^L < a^R < b^R \\ 0, a^R < b^L \end{cases} \tag{7}$$

where l_A and l_B are the lengths of interval A and B respectively.

$$l_{AB} = \begin{cases} 0, & b^L > a^R \vee a^L > b^R \\ \min\left(a^R, b^R\right) - \max\left(a^L, b^L\right), & otherwise \end{cases} \tag{8}$$

3   Methods

3.1   The COMET Method

The COMET is a newly developed method for identifying a multi-criteria expert decision-making model to solve complex problems. This method is used in the numerical example to obtained interval preferences. The whole algorithm can be presented as five following steps and has been provided following [12].

Step 1. Define the space of the problem—an expert determines dimensionality of the problem by selecting number r of criteria, $C_1, C_2, ..., C_r$. Subsequently, the set of fuzzy numbers for each criterion C_i is selected, i.e., $\tilde{C}_{i1}, \tilde{C}_{i2}, ..., \tilde{C}_{ic_i}$. Each fuzzy number determines the value of the membership for a particular linguistic concept for specific crisp values. Therefore it is also useful for variables that are not continuous. In this way, the following result is obtained (9).

$$
\begin{aligned}
C_1 &= \{\tilde{C}_{11}, \tilde{C}_{12}, ..., \tilde{C}_{1c_1}\} \\
C_2 &= \{\tilde{C}_{21}, \tilde{C}_{22}, ..., \tilde{C}_{2c_1}\} \\
&\cdots\cdots\cdots\cdots\cdots\cdots\cdots\cdots \\
C_r &= \{\tilde{C}_{r1}, \tilde{C}_{r2}, ..., \tilde{C}_{rc_r}\}
\end{aligned}
\tag{9}
$$

where $c_1, c_2, ..., c_r$ are numbers of the fuzzy numbers for all criteria.

Step 2. Generate the characteristic objects—characteristic objects are objects that define reference points in n-dimensional space. They can be either real or idealized objects that cannot exist. The characteristic objects (CO) are obtained by using the Cartesian product of fuzzy numbers cores for each criteria as follows (10):

$$
CO = \{\{C(\tilde{C}_{11}), C(\tilde{C}_{12}), ..., C(\tilde{C}_{1c_1})\} \times ... \times \{C(\tilde{C}_{r1}), C(\tilde{C}_{r2}), ..., C(\tilde{C}_{rc_r})\}\}
\tag{10}
$$

As the result, the ordered set of all CO is obtained (11):

$$
\begin{aligned}
CO_1 &= \{C(\tilde{C}_{11}), C(\tilde{C}_{21}), ..., C(\tilde{C}_{r1})\} \\
CO_2 &= \{C(\tilde{C}_{11}), C(\tilde{C}_{21}), ..., C(\tilde{C}_{r2})\} \\
&\cdots\cdots\cdots\cdots\cdots\cdots\cdots\cdots\cdots\cdots \\
CO_t &= \{C(\tilde{C}_{1c_1}), C(\tilde{C}_{2c_2}), ..., C(\tilde{C}_{rc_r})\}
\end{aligned}
\tag{11}
$$

where t is a number of CO (12):

$$
t = \prod_{i=1}^{r} c_i
\tag{12}
$$

Step 3. Rank the characteristic objects—the expert determines the Matrix of Expert Judgement (MEJ). It is a result of pairwise comparison of the characteristic objects by the expert knowledge. The MEJ structure is as follows (13):

$$
MEJ = \begin{pmatrix}
\alpha_{11} & \alpha_{12} & ... & \alpha_{1t} \\
\alpha_{21} & \alpha_{22} & ... & \alpha_{2t} \\
... & ... & ... & ... \\
\alpha_{t1} & \alpha_{t2} & ... & \alpha_{tt}
\end{pmatrix}
\tag{13}
$$

where α_{ij} is a result of comparing CO_i and CO_j by the expert. The more preferred characteristic object gets one point and the second object get zero points. If the preferences are balanced, the both objects get half point. It depends solely on the knowledge of the expert and can be presented as (14):

$$\alpha_{ij} = \begin{cases} 0.0, & f_{exp}(CO_i) < f_{exp}(CO_j) \\ 0.5, & f_{exp}(CO_i) = f_{exp}(CO_j) \\ 1.0, & f_{exp}(CO_i) > f_{exp}(CO_j) \end{cases} \tag{14}$$

where f_{exp} is an expert mental judgement function. Afterwards, the vertical vector of the Summed Judgements (SJ) is obtained as follows (15):

$$SJ_i = \sum_{j=1}^{t} \alpha_{ij} \tag{15}$$

The number of query is equal $p = \frac{t(t-1)}{2}$ because for each element α_{ij} we can observe that $\alpha_{ji} = 1 - \alpha_{ij}$. The last step assigns to each characteristic object an approximate value of preference P_i by using the following Matlab pseudo-code:

```
1: k = length(unique(SJ));
2: P = zeros(t, 1);
3: for i = 1:k
4:     ind = find(SJ == max(SJ));
5:     p(ind) = (k - i)/(k - 1);
6:     SJ(ind) = 0;
7: end
```

In the result, the vector P is obtained, where i-th row contains the approximate value of preference for CO_i.

Step 4. The rule base—each characteristic object is converted into a fuzzy rule, where the degree of belonging to particular criteria is a premise for activating conclusions in the form of P_i. Each characteristic object and value of preference is converted to a fuzzy rule as follows detailed form (16). In this way, the complete fuzzy rule base is obtained, that approximates the expert mental judgement function $f_{exp}(CO_i)$.

$$IF\ C_1\ \tilde{}\ \tilde{C}_{1i}\ AND\ C_2\ \tilde{}\ \tilde{C}_{2i}\ AND\ ...\ THEN\ P_i \tag{16}$$

Step 5. Inference and final ranking—The each one alternative A_i is a set of crisp numbers a_{ri} corresponding to criteria $C_1, C_2, ..., C_r$. It can be presented as follows (17):

$$A_i = \{a_{1i}, a_{2i}, ..., a_{ri}\} \tag{17}$$

Each alternative activates the specified number of fuzzy rules, where for each one is determined the fulfilment degree of the complex conjunctive premise. Fulfilment degrees of all activated rules are summed to one. The preference of alternative is computed as the sum of the product of all activated rules, as their fulfilment degrees, and their values of the preference. The final ranking of alternatives is obtained by sorting the preference of alternatives, where one is the best result, and zero is the worst. More details can be found in [9].

3.2   Similarity Coefficients

For a samples of size N, the rank values x_i and y_i is defined as (18) for WS coefficient [13] and as (19) for weighted Spearman's rank correlation coefficient. For the WS coefficient, the given comparison value is determined by the relevance of the position relative to the first ranking. This ranking is referential, and the coefficient itself determines the similarity of the second-ranking to referential. Therefore, it is an asymmetric measure.

$$WS = 1 - \sum_{i=1}^{N} 2^{-x_i} \frac{|x_i - y_i|}{max(|x_i - 1|, |x_i - N|)} \tag{18}$$

In the second approach, the positions at the top of both rankings are more important than the rest positions. The weight of significance is calculated for each comparison. It is the element that determines the main difference to the Spearman's rank correlation coefficient, which examines whether the differences appeared and not where they appeared [14].

$$r_w = 1 - \frac{6 \sum_{i=1}^{N} (x_i - y_i)^2 ((N - x_i + 1) + (N - y_i + 1))}{N^4 + N^3 - N^2 - N} \tag{19}$$

4   The Proposed Approach

Let us assume that we have N alternatives that have been assessed using the appropriate MCDA method. A suitable MCDA method must be applicable, correctly selected according to [19] and returns the preference results in the form of intervals. In the following, the universal COMET method described in Sect. 3.1 will be used.

As a result of the evaluation, we obtained preference intervals for all alternatives, which can be written as $A_i = [A_i^L, A_i^R]$, where $i = 1...N$. Then, the Possibility Degree (PD) matrix with all values of the possibility degree should be determined as follow (20):

$$PD = [P(A_i \geq A_j)]_{N \times N} \tag{20}$$

where $i = 1...N$, $j = 1...N$, and P is used one of the equation (1)–(7). Then we count the cumulative probability vector PR in according to (21):

$$PR_i = \sum_{j=1}^{N} PD_{ij} \tag{21}$$

Finally, the alternatives are ranked from the highest to the smallest value of PR_i, where the highest value means the maximum cumulative probability degree. This approach will be compared in the next section with three naive approaches, i.e., ranking made up of A_i^L pessimistic version, A_i^R optimistic version and $\frac{A_i^L + A_i^R}{2}$ average version.

5   Comparative Study Case

This study case is based on data and initial results published in [20]. The decision problem is about obtaining a ranking of electronic vans according to selected nine criteria. We randomly selected ten electric vans, which we use to demonstrate the effectiveness of our approach. The description of all criteria is presented in Table 1.

Table 1. Description of the criteria.

C_i	Criterion name	Units	Direction
C_1	Carrying capacity	[kg]	max
C_2	Max velocity	[km/h]	max
C_3	Travel range	[km]	max
C_4	Engine power	[kW]	max
C_5	Engine torque	[Nm]	max
C_6	Battery charging time 100%	[h]	min
C_7	Battery charging time 80%	[min]	min
C_8	Battery capacity	[kWh]	max
C_9	Price	[thous. USD]	min

Table 2 shows all the vans selected at random and their performance concerning the analysed criteria. Some attributes are given as exact numerical values and some as intervals. This is due to the partial lack of data on engine torque, battery charging time 80%, battery capacity, and price.

Table 2. The performance table of the alternatives $A_1 - A_{10}$ in respect to nine criteria.

A_i	C_1	C_2	C_3	C_4	C_5	C_6	C_7	C_8	C_9
A_1	520	110	100	45	[80.0, 900.0]	8	[10.0, 180.0]	30.0	[12.9, 150.0]
A_2	1000	60	100	70	280	8	45	28.0	[12.9, 150.0]
A_3	1830	90	140	80	320	9	60	50.0	[12.9, 150.0]
A_4	750	110	400	[9.0, 200.0]	[80.0, 900.0]	8	[10.0, 180.0]	[2.7, 120.0]	[12.9, 150.0]
A_5	2000	80	160	70	300	8	[10.0, 180.0]	[2.7, 120.0]	32.3
A_6	600	60	150	[9.0, 200.0]	[80.0, 900.0]	6	[10.0, 180.0]	[2.7, 120.0]	14.1
A_7	695	110	170	49	200	7.5	30	22.5	[12.9, 150.0]
A_8	660	105	155	60	[80.0, 900.0]	8.5	[10.0, 180.0]	43.0	75.0
A_9	830	40	118	14	98	8	120	2.7	[12.9, 150.0]
A_{10}	650	130	170	44	226	8	[10.0, 180.0]	22.0	22.0

In order to present our ranking approach, a first assessment should be made using the COMET method. Table 3 gives each criterion's characteristic values that will be used to calculate the intervals of preference for each alternative.

Table 3. Characteristic values for each criterion.

C_i	Min	Mean	Max
C_1	340.00	1770.00	3200.00
C_2	40.00	95.00	150.00
C_3	100.00	250.00	400.00
C_4	9.00	104.50	200.00
C_5	80.00	490.00	900.00
C_6	2.00	7.00	12.00
C_7	10.00	95.00	180.00
C_8	2.70	61.35	120.00
C_9	12.90	81.45	150.00

The detailed results of the COMET interval assessments and the reference ranking derived from [20] are presented in Table 4. It should be borne in mind that this task is solved under uncertain conditions, which means that an exact solution cannot be expected in the sense of specific data. Some discrepancies are observed due to the burden of uncertain data.

Table 4. Interval preferences P and reference ranking.

A_i	ref	P
A_1	9	[0.0825, 0.5073]
A_2	10	[0.1770, 0.3368]
A_3	5	[0.2734, 0.4535]
A_4	1	[0.1754, 0.8711]
A_5	4	[0.2609, 0.5779]
A_6	2	[0.1616, 0.7127]
A_7	8	[0.2476, 0.4099]
A_8	3	[0.1627, 0.4908]
A_9	6	[0.0500, 0.1546]
A_{10}	7	[0.2836, 0.4399]

Based on the results obtained in column P of Table 4, the approach proposed in Sect. 4 is applied to calculate vector PR, which contain the cumulative possibility degree for each alternative. Table 5 presents detailed results for the different methods of determining the probability degree, according to (1)–(7) and naive approaches. As formulas (1)–(4) give the same results, only formula (1) is included in the table. In each analysed case, higher values mean a higher position in the ranking. The complete ranking for each approach is presented in Table 6. Most of the methods correctly indicated the first position of the ranking. However, none of the approaches accurately represent the reference ranking, what was easy to predict. All approaches have indicated A_9 as the worst alternative, ranked sixth in the reference ranking.

Table 5. The cumulative possibility degree and results by using naive approaches.

A_i	ref	P_1–P_4	P_5	P_6	P_7	A_i^L	A_i^R	$\frac{A_i^L + A_i^R}{2}$
A_1	9	4.5835	4.0524	4.2711	4.2623	0.0825	0.5073	0.2949
A_2	10	3.5182	2.5075	2.9900	2.3248	0.1770	0.3368	0.2569
A_3	5	5.7100	5.7826	5.7972	3.7674	0.2734	0.4535	0.3635
A_4	1	6.9773	7.9946	7.4617	15.8800	0.1754	0.8711	0.5232
A_5	4	6.4525	7.0695	6.8308	4.0052	0.2609	0.5779	0.4194
A_6	2	6.3358	6.9898	6.6316	13.0322	0.1616	0.7127	0.4372
A_7	8	5.0119	4.6151	4.8031	3.4127	0.2476	0.4099	0.3287
A_8	3	5.0756	4.7209	4.8620	6.3406	0.1627	0.4908	0.3267
A_9	6	0.6362	0.5243	0.5788	0.0585	0.0500	0.1546	0.1023
A_{10}	7	5.6991	5.7432	5.7736	3.5912	0.2836	0.4399	0.3617

Table 6. Rankings based on the cumulative possibility degree and naive approaches.

A_i	ref	P_1–P_4	P_5	P_6	P_7	A_i^L	A_i^R	$\frac{A_i^L + A_i^R}{2}$
A_1	9	8	8	8	4	9	4	8
A_2	10	9	9	9	9	5	9	9
A_3	5	4	4	4	6	2	6	4
A_4	1	1	1	1	1	6	1	1
A_5	4	2	2	2	5	3	3	3
A_6	2	3	3	3	2	8	2	2
A_7	8	7	7	7	8	4	8	6
A_8	3	6	6	6	3	7	5	7
A_9	6	10	10	10	10	10	10	10
A_{10}	7	5	5	5	7	1	7	5

Comparing individual places in the received rankings makes it quite challenging to indicate which ranking fits the reference better. Of course, only the ranking based on the formula 7 correctly indicated the second position in the ranking.

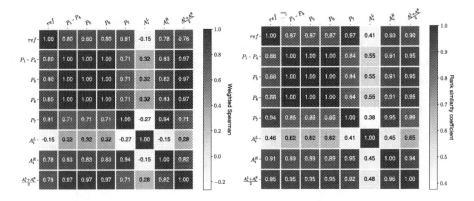

Fig. 1. Thermal maps of r_w and WS correlations

However, to comprehensively compare the obtained rankings, similarity coefficients of rankings r_w and WS described in Sect. 3.2 were calculated. Figure 1 shows thermal maps of both indicators calculated for all rankings. Analysing Fig. 1 we will focus on the first line. For the r_w coefficient, it is clear that the worst match is achieved by choosing a pessimistic solution (min column). Only for this ranking, a negative correlation was obtained. The other rankings are quite similar to the reference ranking, and the value of the index oscillates around 0.80. The best match has been suggested using the possibility degree according to the formula (7).

The situation is similar when we take the WS coefficient into account. However, we see a much better fit for the proposed approach using the formula 7. The resulting value of 0.97 indicates a very high similarity of this ranking to the reference ranking. These coefficients are mainly focused on the top of the ranking and not on its final part. As shown in the example shown, the proposed approach can effectively rank a set of alternatives assessed using interval values.

6 Conclusions

This paper proposes a new approach to the ranking of alternatives assessed by interval values. The proposed approach is based on the possibility degree. In this work, we have prepared seven possible formulas that can be used to calculate the cumulative possibility degree matrix. The numerical example demonstrated that, for the designated rankings, the proposed approach had returned rankings largely in line with the reference ranking. The proposed approach gave a better result on average than the three presented naive methods. More extensive tests should be carried out in future research directions to improve setting rankings based on interval values, and the proposed approach should be extended to include fuzzy numbers.

Acknowledgements. The work was supported by the National Science Centre, Decision number UMO-2018/29/B/HS4/02725.

References

1. Da, Q., Liu, X.: Interval number linear programming and its satisfactory solution. Syst. Eng. Theory Practice **19**, 3–7 (1999)
2. Faizi, S., Rashid, T., Sałabun, W., Zafar, S., Wątróbski, J.: Decision making with uncertainty using hesitant fuzzy sets. Int. J. Fuzzy Syst. **20**(1), 93–103 (2018)
3. Faizi, S., Sałabun, W., Ullah, S., Rashid, T., Więckowski, J.: A new method to support decision-making in an uncertain environment based on normalized interval-valued triangular fuzzy numbers and comet technique. Symmetry **12**(4), 516 (2020)
4. Faizi, S., Sałabun, W., Nawaz, S., ur Rehman, A., Wątróbski, J.: Best-worst method and hamacher aggregation operations for intuitionistic 2-tuple linguistic sets. Expert Syst. Appl. 115088 (2021). https://doi.org/10.1016/j.eswa.2021.115088
5. Fengji, G.: Possibility degree and comprehensive priority of interval numbers. Syst. Eng. Theory Pract. **33**(8), 2033–2040 (2013)
6. Gu, Y., Zhang, S., Zhang, M.: Interval number comparison and decision making based on priority degree. In: Cao, B.-Y., Wang, P.-Z., Liu, Z.-L., Zhong, Y.-B. (eds.) International Conference on Oriental Thinking and Fuzzy Logic. AISC, vol. 443, pp. 197–205. Springer, Cham (2016). https://doi.org/10.1007/978-3-319-30874-6_19
7. Lan, J.B., Cao, L.J., Lin, J.: Method for ranking interval numbers based on two-dimensional priority degree. J. Chongqing Inst. Technol. (Nat. Sci. Ed.) **10** (2007)
8. Li, D., Gu, Y.: Method for ranking interval numbers based on possibility degree. Xitong Gongcheng Xuebao **23**(2), 243 (2008)
9. Piegat, A., Sałabun, W.: Identification of a multicriteria decision-making model using the characteristic objects method. Appl. Comput. Intell. Soft Comput. **2014** (2014)
10. Rehman, A., Shekhovtsov, A., Rehman, N., Faizi, S., Sałabun, W.: On the analytic hierarchy process structure in group decision-making using incomplete fuzzy information with applications. Symmetry **13**(4), 609 (2021)
11. Riaz, M., Sałabun, W., Farid, H.M.A., Ali, N., Wątróbski, J.: A robust q-rung orthopair fuzzy information aggregation using einstein operations with application to sustainable energy planning decision management. Energies **13**(9), 2155 (2020)
12. Sałabun, W., et al.: A fuzzy inference system for players evaluation in multi-player sports: the football study case. Symmetry **12**(12), 2029 (2020)
13. Sałabun, W., Urbaniak, K.: A new coefficient of rankings similarity in decision-making problems. In: Krzhizhanovskaya, V.V., et al. (eds.) ICCS 2020. LNCS, vol. 12138, pp. 632–645. Springer, Cham (2020). https://doi.org/10.1007/978-3-030-50417-5_47
14. Sałabun, W., Wątróbski, J., Shekhovtsov, A.: Are MCDA methods benchmarkable? A comparative study of TOPSIS, VIKOR, COPRAS, and PROMETHEE II methods. Symmetry **12**(9), 1549 (2020)
15. Shekhovtsov, A., Kołodziejczyk, J., Sałabun, W.: Fuzzy model identification using monolithic and structured approaches in decision problems with partially incomplete data. Symmetry **12**(9), 1541 (2020)
16. Wan, S., Dong, J.: A possibility degree method for interval-valued intuitionistic fuzzy multi-attribute group decision making. J. Comput. Syst. Sci. **80**(1), 237–256 (2014)
17. Wang, Y.M., Yang, J.B., Xu, D.L.: A two-stage logarithmic goal programming method for generating weights from interval comparison matrices. Fuzzy Sets Syst. **152**(3), 475–498 (2005)

18. Wang, Y.J., Lee, H.S.: Generalizing topsis for fuzzy multiple-criteria group decision-making. Comput. Math. Appl. **53**(11), 1762–1772 (2007)
19. Wątróbski, J., Jankowski, J., Ziemba, P., Karczmarczyk, A., Zioło, M.: Generalised framework for multi-criteria method selection. Omega **86**, 107–124 (2019)
20. Wątróbski, J., Małecki, K., Kijewska, K., Iwan, S., Karczmarczyk, A., Thompson, R.G.: Multi-criteria analysis of electric vans for city logistics. Sustainability **9**(8), 1453 (2017)

Ant-Based Hyper-Heuristics for the Movie Scene Scheduling Problem

Emilio Singh[iD] and Nelishia Pillay[(✉)][iD]

Department of Computer Science, University of Pretoria, Pretoria, South Africa
u14006512@tuks.co.za, npillay@cs.up.ac.za

Abstract. The paper provides a study of the use of hyper-heuristics on the movie scene scheduling problem. In particular, the paper extends the definition of the movie scene scheduling problem to include a new method of calculating the solution quality. The study is also a novel application of hyper-heuristics to the movie scene scheduling problem and demonstrates one potential method for using hyper-heuristics as a solution method for the given problem. This includes the development of new low-level heuristics for the problem that are presented as well. The study showed that hyper-heuristics could be applied to the problem doing better than a random approach but that work would need to be done on improving the low-level perturbative heuristics. The study also showed that the new formulation would be tenable as a problem definition with little change to the underlying problem itself.

Keywords: Selection hyper-heuristics · Discrete combinatorial optimization · Ant algorithms · Movie scheduling problem

1 Introduction

In the modern era of movie production, studios face rising budgets for films as movie complexity increases [1]. With a wider range of production areas, it is becoming more important for movie studios to consider improvements to their production schedules to decrease the overall costs of production. Thus, the movie scene scheduling problem (MSSP) has emerged as an NP-hard scheduling problem that chiefly revolves around determining a schedule for all of the scenes in the movie such that the costs of the movie's production are minimised [3].

The MSSP is a scheduling problem to produce an ordering of n scenes from the set of all scenes S such that the production costs related to these scenes are minimised. The MSSP primarily considers the costs of transferring between scenes and some amount of wages paid to actors in these scenes. Cost reduction occurs when scenes that minimise actor downtime and scene transfer costs are scheduled in proximity.

The existing research into movie scene scheduling has focused on single location productions [7]. An obvious extension is adding multiple locations [7]. It is this formulation of the problem that this paper primarily considers and extends

© Springer Nature Switzerland AG 2021
L. Rutkowski et al. (Eds.): ICAISC 2021, LNAI 12855, pp. 342–353, 2021.
https://doi.org/10.1007/978-3-030-87897-9_31

by examining how the costs of moving between scenes could be aligned more closely with real-life conditions.

As a fairly new problem, there are many potential avenues in terms of research. Recent research has focused on using techniques like PSO and tabu search [7]. Therefore there is research potential in examining whether hyper-heuristics (HHs) could be applied to the MSSP. Hyper-heuristics are effective at solving discrete combinatorial problems [10]. Therefore it would be a novel application to apply them to the MSSP. Thus the main contributions of this paper are as follows:

- Reformulating the cost calculation of the MSSP to better model real-world conditions.
- A comparative study of hyper-heuristics on the MSSP.
- The derivation and presentation of low-level constructive and perturbative heuristics for the MSSP.

The rest of the paper is structured as follows. In Sect. 2, a background to the problem is presented. Section 3 presents the definition of the MSSP and the extensions considered in this paper. The novel HH method is presented in Sect. 4. The experimental setup which includes the dataset generation procedure is outlined in Sect. 5. The results of the subsequent experiments are presented in Sect. 6. Finally, Sect. 7 provides a conclusion to the paper with an overview of the findings and avenues for future research.

2   Background

In this section, an overview of relevant topics is presented. This will focus on the development of the MSSP and the existing body of research contained within with some discussion of hyper-heuristics as well. The MSSP was not entirely derived in isolation. In fact, the problem is related to other entertainment scheduling problems. An early example is the rehearsal problem which involved the scheduling of musicians who need to rehearse pieces of music [5]. The ordering of the musicians to pieces is similar to that of actors to scenes. The authors' used a model checking procedure. This was extended to unequal length musical pieces and a two-stage method was then applied [11].

One of the earliest examples of a movie scheduling problem started with fixed length scenes [3]. The authors focused entirely on variable actor wages. Later authors extended this research by applying a genetic algorithm (GA) to this problem with better results [9]. These early studies did not consider variable-length scenes which is an issue for real-world applications.

Later work increased the complexity by considering different scene durations and different actor wages [2]. The authors made use of a dynamic programming algorithm although it required bounding to function optimally. However, the demands of modern movie production typically require scheduling at multiple locations.

Naturally scheduling of scenes in different locations was be next [7]. This required additional complexities like transfer costs and transfer times which were factored into the scheduling process. The authors initially made use of a tabu-search method and a particle swarm optimisation (PSO) method. They extended the work to include ant colony optimisation (ACO) [8]. This formulation extended the MSSP further by mirroring the realities of real-world movie production. What is apparent when one looks at the development of the MSSP is that the general trend is towards increasing the complexity of the models and their definitions to better replicate the actual conditions of movie production. No model can perfectly replicate the exact conditions of movie production but some models can get close enough to serve as useful approximations and as such this will be the basis of the MSSP definition considered here.

What is apparent from this survey of related research is that there is still potential to develop the MSSP in terms of expanding the formulation of the problem as well as applying different approaches to solve the problem. The formulation of the MSSP model uses a time interval variable to calculate the wages owed to an actor between scenes m and n [7]. The issue with the calculation is that it is independent of the actual ordering of the scenes, only considering the time of all of the actors in scene m and the time needed to move between scene locations. Extending this to factor in the position of the scenes relative to one another will more accurately reflect the original problem. Given that previous research has focused on the MSSP solution space, an attempt to use the heuristic space to solve the MSSP would represent a significant advancement in the field.

3   Movie Scene Scheduling Problem

In this section, the MSSP is formally defined with its mathematical model, parameters and inputs. The extensions to the problem that are proposed by this paper are presented here along with a justification.

3.1   Problem Definition

Three elements define the basic components of the MSSP: the set of n scenes S, the set of m locations L and the set of o actors A. These come with some constraints. The first is that all scenes must be scheduled only once with no scenes overlapping each other and every scene assigned to only one location. Multiple scenes can be assigned to a single location but not at the same time. Every scene must have at least one actor assigned to it and every actor can be assigned to multiple scenes.

There are several secondary variables required for defining the MSSP:

- W: This defines a set of daily actor wages. Each actor's daily wage is in the range of [50,100].

- D: This defines the duration of each scene in days. Every scene once scheduled will be fully completed.

Each scene's duration is in the range of [1,10] days.

- O: This refers to the location assigned to scene i. Each scene is assigned a randomly determined location out of the list of locations L. Every location has an equal probability of selection and every scene must be assigned a location.
- T_{xy}: This variable is a matrix of transfer times (in days) between different locations. The transfer time to move from location x to x is 0. Each value is in the range [1,10] days.
- C_{xy}: This variable is a matrix of transfer costs between different scenes. The transfer cost to move from scene x to x is 0. Each value is in the range [100,999].
- AS: This quantifies the assignment of o actors to scene i. For each scene, a randomly shuffled list of all of the actors is generated. A random number of actors, in the range of [1,n_s], are removed from this list. The remaining actors are then assigned to scene i.

The task of MSSP, therefore, is to order the set S into a schedule R such that the costs of scheduling the scenes are minimised.

3.2   Problem Extensions

This section details the extensions and modifications made to the definition of the problem as formulated by [7].

Wage Calculations. The first component of the objective function was calculated by determining the individual time intervals for every actor p between all of their scenes m and n. This calculation was performed by summing the duration of scene m with the transfer time between m and n. However different orderings of R would not have changed the result of this calculation since the durations of all of the scenes and the transfer times between them were fixed beforehand.

The new calculation was performed as follows. For every actor p, an ordered sub-list ($\bar{R}$) was generated of all scenes from the schedule R. From that point, every pair of scenes (m, n) was used to calculate a new time interval between (m, n). This was done by summing the scene duration values of all of the scenes in R from m to n (but not inclusive of n). The duration of the last scene was added to the final sum along with the transfer costs between m and n. This final value was then multiplied by the wage of the actor p, W_p. This produced the cost of scheduling that actor in the production.

This new method of calculation followed the intent of the prior formulation except that the quality of schedule R depended on the individual ordering of the scenes of R.

For example, the case of $R = \{0, 1, 2\}$ with an actor A who appeared in scene 0 and 2. The time interval would have added the time for shooting scene 0 plus the actor's downtime during scene 1 before they were transferred to scene 2 for shooting. If $R = \{0, 2, 1\}$ then the costs of this interval would have been lowered since the actor had no downtime to add costs between their scenes.

4   Fast Ant Hyper-Heuristic

In this section, the fast ant hyper-heuristic (FA-HH) algorithm (and its variants) is presented. The basis of the FA-HH method is presented with relevant discussion provided regarding the variants that are considered in this paper. These specifically consider a selective constructive HH (FA-HH-C), a selective perturbative HH (FA-HH-P) and a hybridisation of the two (FA-HH-H).

4.1   Hyper-Heuristic Adaptations

The HHs used in this paper are based on the ant-based HH methodology [2,4]. However what makes the methodology presented here novel is that it is based on the fast ant (FANT) algorithm [4]. This adaptation keeps the same basic idea of an ant-based HH but utilising the FANT methodology as a way of simplifying the application of the ant algorithm. The basis of this algorithm is provided in Sects. 4.2–4.5.

4.2   Heuristic Search Procedure

At the heart of any ant-based HH is a search process that searches the heuristic space for heuristics that are applied for making or perturbing a solution. The ant starts with an empty path and gradually chooses heuristics to add to this path. The process for picking heuristics is based on the values of the corresponding heuristic pheromone accumulation vector which is explained in Sect. 4.3. FA-HH-P uses the chosen heuristics on a new random solution to perturb it whereas FA-HH-C uses the chosen heuristics to construct a new solution.

The size of the path depends on the specific kind of HH. A selection constructive HH will stop once it has created a fully complete and feasible solution. For a selection perturbative HH, the size of the path is based on the number of perturbative heuristics for a given problem. This is an appropriate limit as it scales to the size of the selection problem.

The heuristic selection process makes use of an ant algorithm probabilistic selection equation applied to the pheromone accumulation vector. Equation 1 is used in Line 3 of the algorithm to select a new heuristic to add to the path. The selection is based on the accumulated value of the pheromone stored at each index of the pheromone accumulation vector. Roulette wheel selection is used as the mechanism to make the selection [6].

The rule makes use of purely pheromone accumulation to guide the selection of heuristics, obviating the need to use a visibility heuristic that is common to other ant-based HHs. Visibility refers to the desirability of a given heuristic based on the computational time it requires to do its particular task.

$$p_i = \frac{\tau_i}{\sum_{i \in N_i^k} \tau_i} \qquad (1)$$

In Eq. 1, τ_i refers to the pheromone accumulated for a heuristic i. All heuristics can be considered at any point in the search process.

4.3 Pheromone Accumulation Vector

A normal pheromone graph represents pheromone deposited on specific edges between nodes in the graph. In the FA-HH algorithm, ants deposit their pheromone into indices in a vector to indicate the desirability of choosing that heuristic. As a heuristic accumulates pheromone in its place in the vector, it becomes more likely to be selected during the heuristic search. The accumulation process is given by Algorithm 1. In terms of terminology, ph_C and ph_P refer to a pheromone accumulation vector for the selection constructive HH and selection perturbative HH respectively.

This change to the way pheromone is used greatly simplifies the search procedure for an ant, reducing the complexity of their search space by a dimension. Instead the influence of a heuristic is better reflected as the aggregation of its presence in heuristic paths. The more often it is featured (reflecting its value in producing solutions), the more likely it is to be selected in the future.

4.4 Ant Updates

For the FA-HH-C and FA-HH-P algorithms it is necessary to update the underlying ant's pheromone accumulation vector with the information produced from the search once the search is completed. This is detailed by Algorithm 1:

Algorithm 1: Ant Update Procedure

Data: a pheromone accumulation vector ph_C or ph_P
Result: ph_C or ph_P is updated
1 **for** *each heuristic p* **do**
2 $\quad$ $\delta_1 = w_1 *$countFrequencyCurr(p);
3 $\quad$ $\delta_2 = w_2 *$countFrequencyBest(p);
4 $\quad$ $ph(p) += \delta_1 + \delta_2$;
5 ant.path.clear();
6 placeAnt();

The procedure is derived from the FANT update rule, but modified for the FA-HH algorithms. For each heuristic under consideration, a count of its frequency in the current and best heuristic path, is multiplied by w_1 and w_2 respectively. The values of w_1 and w_2 balance the exploration/exploitation potential of the algorithm.

The values of w_1 and w_2 are initially set to 1. This ensures all heuristics have an equal probability of selection. If the solution produced by the ant is equal to the best solution found, then w_1 is increased to encourage more exploration while w_2 remains unchanged.

Finally, the current path is cleared and the updated ant is randomly placed back in the heuristic space. The stopping condition is the number of iterations n_t. This is a standard stopping condition for ant-based algorithms.

4.5   Learning Framework and Hybridisation

The FA-HH-H algorithm combines the efforts of two selection HHs through the use of a learning schedule. It provides a general outline of when the respective processes (construction or perturbation) are allowed to execute and under what conditions. Such a framework is designed to reduce the overall computational costs of the algorithm by constraining the more expensive operations to parts of the execution where they are more likely to be useful. The schedule is divided into two phases: exploration and exploitation.

During each iteration if the algorithm's execution percentage is below α, it will generate a random number $r = U(0, 1)$. The execution percentage is simply the ratio between the iterations performed divided by the total number of iterations chosen for the algorithm run. If $r < p$ then it will perturb the current solution and update ph_P accordingly. Otherwise it will construct a new solution with the constructive heuristics, perturb it and then update ph_C and ph_P accordingly. After that cut-off point, for the rest of the run, it will set the current solution to the best solution and perturb it with the appropriate update to ph_P.

The exploration phase is meant to facilitate a search of the solution space and the exploitation phase refines solutions. The schedule itself makes use of two variables α and p. The variable p controls the probability of refining a solution with perturbative heuristics. Equation 2 is used to alter the value of p during the algorithm's run:

$$p(t) = p(0) - p(n_t)\frac{(n_t - t)}{n_t} + p(n_t) \tag{2}$$

where n_t is the maximum number of iterations, $p(0)$ is the initial value of p, $p(n_t)$ is the final value and $p(t)$ is the value of p at iteration t.

The variable α, with the range of (0,1), is used to determine the transition point since α represents a percentage of the execution of the algorithm that is dedicated to the exploration phase. Larger values of α will increase the computational costs of the algorithm since more of its execution will be dedicated towards the exploration process which involves new solutions being created.

5   Experimental Setup

This section several details pertinent to the experiments conducted during this research. This includes the dataset generation process, the heuristics available for the problems, algorithm parameters and experimental configurations.

5.1   Dataset Generation

The MSSP lacks standardised datasets. One reason is that real-world data is particularly difficult to come by given the non-disclosure agreements under which most movies are produced. As a consequence this paper makes use of generated datasets that use a set of generation rules to produce the necessary input data. The methods used to generate the instances of the problem are described below. There are three input parameters (S, A, L) that define the boundaries of the problem as defined in Sect. 3.1. These parameters are not randomly generated but rather chosen as part of the experimental procedure as shown in Sect. 5.4.

5.2 Heuristics

Given the novel application of HHs to the MSSP and that the MSSP is a relatively new problem, there are no available low-level heuristics to consider for the HH. Instead low-level heuristics were created for the task with [10] serving as a guideline. What follows is a list of the constructive heuristics considered in the paper where c_i denotes constructive heuristic i and p_j denotes perturbative heuristic j:

- c_1: Choose a random scene.
- c_2: Pick the next scene with the most number of actors assigned to it.
- c_3: Pick the next scene with the fewest number of actors assigned to it.
- c_4: Pick the next scene with the longest duration of any of the remaining scenes.
- c_5: Pick the next scene with the shortest duration of any of the remaining scenes.
- c_6: Pick the next scene with the smallest transfer cost from the prior scene that was scheduled.
- c_7: Pick the next scene which has the largest transfer cost from the prior scene that was scheduled.
- c_8: Pick the next scene with is chosen randomly from a list of scenes that share the same location as the prior scene. If no such scenes exist, the next scene is chosen randomly.
- p_1: The scene order is shuffled until an improvement in fitness occurs.
- p_2: The scene with the longest duration is moved to the front of the schedule.
- p_3: The scene with the shortest duration is moved to the front of the schedule.
- p_4: A random (excluding the end) scene is chosen. From the remaining scenes, the one which has the lowest transfer cost to the original chosen scene is determined. This scene is then put into the adjacent position next to the original random scene.
- p_5: The first and last scenes are interchanged with a corresponding randomly chosen scene from the schedule.
- p_6: The scene with the most number of attached actors is moved to a random position in the schedule.
- p_7: The scene with the least number of attached actors is moved to a random position in the schedule.
- p_8: All of the scenes are shifted up one position in the schedule. The scenes wrap around.
- p_9: All of the scenes are shifted down one position in the schedule. The scenes wrap around.

5.3   Parameters

There are two parameters pertinent to this paper: α and p. These variables are used by the hybrid, FA-HH-H, in its execution and control the hybridisation strategy itself. The non-hybrid methods, FA-HH-C and FA-HH-P, do not depend on additional parameters. The variable α is used in the learning schedule to control the transition from the exploration phase to the exploitation phase. The value of α falls in the range $(0,1)$ and represents a transition point during a run of the algorithm and is multiplied by the maximum number of iterations to get the transition point.

The values of p are in the range $[0.1, 0.9]$. This was based on the number of iterations that would be used in the experiments. This results in a gradual transition from using constructive and perturbative heuristics together to explore the space towards using only perturbative heuristics to refine the best solutions found during the run.

5.4   Experimental Process

In order to assess the application of HHs to the MSSP, an empirical testing process is used. The first step defines a number of different problem classes. These classes describe the MSSP complexity in terms of actors, scenes and locations. For this paper, there are four classes considered and their characteristics are described in Table 1:

Table 1. Attributes of the four problem classes

Class	1	2	3	4
Scenes	10	25	50	100
Actors	5	10	30	60
Locations	3	5	10	15

A class 1 problem would represent something like a student film whereas a class 4 problem would represent a big-budget Hollywood movie. The use of these problem classes also enables the scalability of the HHs to be assessed across problems of varying sizes. Five instances of each class are created through the process detailed in Sect. 5.1. In terms of the experiments, each variation of the FA-HH algorithm (FA-HH-C, FA-HH-P and FA-HH-H) was run on all of the instances. Additionally, a randomly generated solution was also considered as a way to examine the efficacy of the method in comparison to a non-HH method as well.

5.5   Assessment Metrics

This paper generates its own datasets but this comes with a potential problem. Namely, the optimal value of each dataset is difficult to find when the dataset is sufficiently large. The search space for a given MSSP problem, as defined, has a factorial growth based on the number of scenes. A small problem with a few scenes has a correspondingly small search space meaning it is possible to determine all possible values for all possible scene orders within a reasonable degree of time. However this becomes less viable as the problem scales in complexity. As such, this paper does not consider the results in absolute terms, rather it examines the performance (in terms of solution value) of the HHs in relation to one another and the given random solution. In particular, the assessment will focus on presenting the average for all of the runs of that instance. Therefore this paper can serve as a benchmark for future approaches applied to the same data, given that the generated datasets can be made publicly available.

5.6   Technical Specifications

For this research, a computing cluster provided by the University of Pretoria was used. The technical specifications of this cluster are: 377GB RAM, 56 cores at 2.40GhX (Intel Xeon CPU E6-2680 v4) and 1TB of Ceph Storage.

6   Results and Discussion

Table 2. Results for Class 0 and 1

Algorithm	Class 0					Class 1				
	I0	I1	I2	I3	I4	I0	I1	I2	I3	I4
FA-HH-C	27316	21968	23331	27465	20344	102086	129587	76793	65010	94923
FA-HH-P	26567	20559	22562	26173	19742	99552	130211	73936	62828	93155
FA-HH-H	**25958**	**20062**	**22309**	**25774**	**19345**	**94265**	**123686**	**70946**	**58227**	**87790**
Random	38446	28874	31146	33712	27872	124503	162811	97400	85039	120396
Time (Min)	0.02	0.01	0.01	0.01	0.01	0.13	0.10	0.07	0.07	0.07

Table 3. Results for Class 2 and 3

Algorithm	Class 2					Class 3				
	I0	I1	I2	I3	I4	I0	I1	I2	I3	I4
FA-HH-C	755180	644641	**789068**	**826922**	684463	**2979244**	3204273	3011135	3263095	3076266
FA-HH-P	778387	684099	842038	893995	707605	3164543	3395114	3102676	3429304	3240924
FA-HH-H	**726000**	**634453**	792466	832579	**658184**	2984098	**3176430**	**2889222**	**3240983**	**3029655**
Random	902263	789943	959215	1013800	815608	3415345	3691469	3391147	3718238	3529288
Time (Min)	0.79	0.85	0.0088	0.0094	0.71	3.94	4.58	3.63	4.36	4.66

In this section, the results are grouped into comparisons between the four problem classes. The results for each instance are presented which includes the average solution value over all of the 30 runs for that instance. The best results are indicated in bold. Additionally, the time (in terms of a single run) of the best performing algorithm in terms of the average score is included as well.

The results of the experiments, Tables 2 and 3, contain some interesting trends. In the majority of cases, the FA-HH-H algorithm obtains the best average value for the majority of the instances and classes with FA-HH-C being the best in some cases. One inference from this is that the FA-HH-H algorithm is better in a more general sense than the other algorithms for this kind of problem as it performs better on average but the FA-HH-C does find better solutions for the larger problems than FA-HH-H in some instances. This trend only manifests in the larger classes of problems which suggests that it is more a function of the problem complexity (and scale) than the individual instances themselves.

6.1   Discussion and Analysis

The results presented in Tables 2 and 3 demonstrate that HHs can be applied to solve the MSSP. In particular, some kind of HH can be applied (a hybrid or otherwise) and this application performs better than the random algorithm. In this way, one of the contributions of this paper was demonstrated. In terms of the MSSP, this paper demonstrated that the extension to the problem (reformulating the wage calculation) is a viable addition that more closely resembles real-world conditions. This paper also presented new low-level heuristics for the MSSP and in that regard, there are several insights to be gleaned. In particular, the low-level perturbative heuristics appeared to be insufficiently useful to problem-solving in comparison to the low-level constructive heuristics. This is demonstrated by the fact that the worst-performing HH was the selective perturbative one. This carried over into the hybrid given that there were some cases, the larger problem classes, where the selective constructive HH was the best performing HH. One potential issue with the perturbative heuristics as they are constructed is that they simply do not modify the underlying solution enough to meaningfully explore the space in their operation. This is more true of the larger problems as modifying the order of one or two scenes in a one hundred scene order is likely to produce little meaningful impact on the overall order.

7   Conclusion

This paper has advanced the development of the MSSP. It has done this through the extension of the problem definition such that it better resembles the underlying real-world problem. The next major contribution of this paper was the comparative study of HHs on the MSSP. The results obtained from the experiments, presented here, showcase a wide variety of HH approaches, of varying degrees of success, that demonstrate that while hyper-heuristics can be applied to the MSSP, further work is needed to refine the approach.

The effect of this research has pushed forward the field of MSSP but also opened up new avenues for research. In particular, research into formulating better perturbative heuristics, and indeed exploring how perturbative heuristics could be applied to the problem as a whole is necessary given the problems demonstrated by these perturbative heuristics in the paper. Another potential area of future research would be examining the use of additional constraints on the problem such as budgets and how these would affect the solving process. Finally, examining the influence of the parameter matrices and whether their symmetrical nature affects the problem is also worth considering.

Acknowledgements. This work was funded as part of the Multichoice Research Chair in Machine Learning at the University of Pretoria, South Africa. This work is based on the research supported wholly/in part by the National Research Foundation of South Africa (Grant Numbers 46712). Opinions expressed and conclusions arrived at, are those of the author and are not necessarily to be attributed to the NRF.

References

1. Aarseth, E.: The culture and business of cross-media productions. Pop. Commun. **4**(3), 203–211 (2006). https://doi.org/10.1207/s15405710pc0403_4
2. Garcia de la Banda, M., Stuckey, P., Chu, G.: Solving talent scheduling with dynamic programming. INFORMS J. Comput. **23**, 120–137 (2011). https://doi.org/10.1287/ijoc.1090.0378
3. Cheng, T.C.E., Diamond, J., Lin, B.: Optimal scheduling in film production to minimize talent hold cost. J. Optim. Theory Appl. **79**, 479–492 (1993). https://doi.org/10.1007/BF00940554
4. Gambardella, L.M., Dorigo, M.: Solving symmetric and asymmetric TSPS by ant colonies. In: Proceedings of IEEE International Conference on Evolutionary Computation, pp. 622–627 (1996)
5. Gregory, P., Miller, A., Prosser, P.: Solving the rehearsal problem with planning and with model checking. In: European Conference on Artificial Intelligence, vol. 16 (2004)
6. Lipowski, A., Lipowska, D.: Roulette-wheel selection via stochastic acceptance. Phys. A **391**(6), 2193–2196 (2012)
7. Liu, Y., Sun, Q., Zhang, X., Wu, Y.: Research on the scheduling problem of movie scenes. Discrete Dyn. Nat. Soc. **2019**, 1–8 (2019)
8. Long, X., Jinxing, Z.: Scheduling problem of movie scenes based on three meta-heuristic algorithms. IEEE Access **8**, 59091–59099 (2020)
9. Nordstrom, A., Tufekci, S.: A genetic algorithm for the talent scheduling problem. Comput. Ind. Eng. **21**, 927–940 (1994)
10. Pillay, N., Qu, R.: Hyper-Heuristics: Theory and Applications. Springer, Cham (2018). https://doi.org/10.1007/978-3-319-96514-7
11. Sakulsom, N., Tharmmaphornphilas, W.: Scheduling a music rehearsal problem with unequal music piece length. Comput. Ind. Eng. **70**, 20–30 (2014)

AI Alignment of Disaster Resilience Management Support Systems

Andrzej M. J. Skulimowski[1,2(✉)] and Victor A. Bañuls[3]

[1] Chair of Automatic Control and Robotics, AGH University of Science and Technology,
al. Mickiewicza 30, 30-059 Kraków, Poland
ams@agh.edu.pl
[2] International Centre for Decision Sciences and Forecasting, Progress & Business Foundation,
ul. Lea 12B, 30-048 Kraków, Poland
[3] Universidad Pablo de Olavide, Ctra. de Utrera, 1, 41013 Seville, Spain
vabansil@upo.es

Abstract. This paper presents an application of Artificial Intelligence (AI) prospective studies to determine the most suitable AI technologies for implementation in Disaster Resilience Management Support Systems (DRMSSs). The pivotal role in our approach is played by the security needs analysis in the context of most common natural disasters and their co-occurrence with other threats. The AI trends and scenarios to align with are derived according to foresight principles. We apply expert knowledge elicitation and fusion techniques as well as a control model of technology dynamics. The pre-assessments of security needs and technological evolution prospects are combined to rank and select the most prospective AI methods and tools. Long-term ex-ante impact assessment of future disaster resilience improvements resulting from different DRMSS implementations, allows for the identification of the most suitable AI deployment variant. The target market of the DRMSSs under study includes industrial corporations and urban critical infrastructures, to become part of their Industry 4.0 ecosystems. The models of protected area and its environment are continually updated with visual monitoring and other sensors embedded in the Industrial Internet-of-Things infrastructure. The software architecture of DRMSS focuses on model-based decision support that applies fuzzy-stochastic uncertainty and multicriteria optimization. The business processes behind the AI alignment follow these goals. In the conclusion, we will show that DRMSS allows stakeholders to reach social, technological, and economic objectives simultaneously.

Keywords: Decision Support Systems · Disasters resilience management · AI alignment · Technological evolution · Security process modelling

1 Introduction

The progress in AI provides opportunities to implement new powerful methods for autonomously processing big and vulnerable data in disaster resilience management

© Springer Nature Switzerland AG 2021
L. Rutkowski et al. (Eds.): ICAISC 2021, LNAI 12855, pp. 354–366, 2021.
https://doi.org/10.1007/978-3-030-87897-9_32

support systems (DRMSSs). These systems are a relatively new class of advanced software [22] that support decision makers in public administration and corporations not only in case of emergency, as common crisis management systems do, but also in allowing them to build resilience of their organizations in a systematic way. To achieve anticipated results of enhanced disaster resilience, the DRMSS developers must also take into account any potential social and technological AI threats. To overcome this challenge, we propose to analyze the AI forecasts and potential future hazards in the strategic alignment framework. This analysis allows us to derive optimal models of AI deployment, while indicating prior best practices in applying AI tools and methods in DRMSS. According to recent research trends on natural threats and AI, both constitute a major challenge to mankind in upcoming decades. Threats include hazards of purely natural origin, such as volcanoes, as well as mixed-origin threats such as climate change and pandemics. When investigating the synergies between these threats, the benefits of AI are clear since a great deal of AI-based technologies may be deployed to prevent or mitigate the impact of natural disasters. However, little is known about other incidences of AI/disaster interdependence, specifically those that might become relevant in the middle- or long-term future.

On the one hand, developers of specialized information systems such as DRMSSs or intelligent Decision Support Systems (DSSs) to be used for crisis management anticipate that AI will predominantly contribute to the growth in overall system quality. AI techniques are also expected to make autonomous robotic systems, such as search and rescue robots (SRR), a powerful tool that assists first responder teams in case of emergency. However, the increasing share of autonomous decision making processes required from AI-based systems creates a diversified spectrum of new problems that need to be solved. These range from operational security assurance to global impact modelling and universal AI ethics related to autonomous systems deployment. The ever-growing sophistication and scope of AI-based systems, devices, and software agents make the problems of explainable AI [2] and human-computer understanding increasingly relevant, especially in cases of resilience building activities.

1.1 The Resilience Management Support System Aims and Functionalities

The characteristic feature of prevention and resilience assurance measures, to be recommended or automatically controlled by DRMSS, is a high share of decisions that should be implemented immediately with specialized actuators, humans as first responders, as well as human-robot and robot teams. This, in turn, creates a need for endowing the DRMSSs with learning modules that process feedback information, gathered as observations of consequences of prior decisions. The efficiency of such learning schemes is ensured by the fusion of large amounts of information on similar operations collected from global databases and data streams and processed by state-of-the-art machine learning algorithms. Processing big data coming from environmental observations, such as satellite images, meteorological radar measurement, distributed precipitation data and precipitation-dependent soil porosity is another challenge.

While AI research will feed decision making, pattern recognition, and machine learning algorithm updates into the DRMSS evolution control module, current decision planning and implementation will strongly depend on the capabilities of rescue teams, sensor networks, and robotic systems. Motivated by real-life applications, this article is focused on industrial threat monitoring, alerting, and mitigation with DRMSS endowed with evolving AI tools. The system should also provide AI-supported coordination of mixed teams, composed of humans, and autonomous unmanned aerial and ground vehicles. For brevity's sake we refer to all of them as 'first responders'. The other key functionality of DRMSS is to provide decision support to enterprise managers who should be capable of quickly finding trade-off decisions with – communication dependent - restricted information exchange with regional crisis authorities. The DRMSS recommendations are simultaneously presented to all actors, taking into account the anticipated consequences of planned decisions.

The AI research screening will identify new technologies that are suitable for implementation into the decision support engine. Of particular interest are information fusion and multicriteria decision support methods as well as carefully selected machine learning and data mining techniques. Coordination of multi-robot and human-robot teams will require optimization of cyber-physical system activities, path planning, and simultaneous location and mapping (SLAM) of inspection and rescue robots. These aforementioned AI and robotics areas will be further focal points of the AI evolution model. This modelling tool will evolve itself, based on technology screening and reinforcement learning procedures. The overall DRMSS will use its sensors to detect and manage varying natural threats, such as landslides, falling rocks, and floods. The DRMSS will also monitor and prevent malicious human activities, including intrusions, violation of security rules by employees, etc.

1.2   The Aims and the Structure of this Paper

AI development trends observed over the past decades, when coupled with natural disaster and anthropogenic threat models, and confronted with real-life security needs should make it possible to create intelligent technology solutions to enhance disaster resilience. The primary aim of this paper is to provide a systematic approach to the design, implementation and deployment of intelligent DRMSSs endowed with a sufficient flexibility to align with AI development and capable of handling new emergency situations. The AI alignment problem to be solved differs from both, the heretofore general formulations relating AI to human values and ethics [10] as well as from the classical AI [21] or information technology (IT) business alignment [11]. Instead, we focus on an informed selection of the best AI tools and techniques to enhance DRMSSs as complex information systems to achieve best disaster resilience. The implemented methods should also make the system competitive and resilient against any adversarial AI support that can be used by intruders. We propose a general principle of aligning the DRMSSs to state-of-the-art AI solutions. Consequently, most of the background research is contained within intelligent decision support systems and autonomous decision making areas, both applied to the design of DRMSSs.

The structure of presentation follows the above formulated aim, starting from a survey of recent AI methods and tools applicable in DRMSSs, which is contained in Sect. 2. Section 3 formulates the technological AI alignment problem to be solved when designing a DRMSS that is continually updated according to the progress in AI. Section 4 presents the AI evolution modelling tool (AIEM), which supplies information about AI trends and scenarios to DRMSS developers. Then we outline the AIEM implementation, the DRMSS architecture and its links to the AI evolution model. The summary of our research findings is discussed in the final Sect. 5, together with general conclusions and future research plans.

2 Related Work

One of the first design proposals for emergency management DSS can be found in [6]. An overview of early developments in emergency response information systems and their future development prospects is provided in [22]. Multicriteria optimization and preference modelling approaches are commonly used in DRMSSs to include different resilience-related goals. An example of using them for location planning in disaster areas was given in [7]. Fundamental for the DRMSS implementation is research on decision support, with collaborative scenario modeling, to protect critical infrastructures. This group of methods can be used in DRMSS for project management [4].

The above background research, together with urgent real-life needs met a number of DRMSS implementations, both research prototypes and commercial systems. The THEMIS prototype applies an AI-based decision support system to be used by disaster managers and first responders. The architectures of DSS tailored to optimizing aid distribution during natural disasters are discussed in [15] and [9], including the assurance of supply chains and vehicle routing problems.

A great challenge posed to our research involves the integration of human and robot rescue teams under one coordination support system, a component of the DRMSS. Meanwhile, Search & Rescue is one of the leitmotivs and drivers of the autonomous mobile robotics development. The research includes a combination of advanced AI methods used predominantly to plan the search, identify the surrounding with pattern recognition and scene understanding approaches, assess the emergency situation and prioritize actions. Cyber–physical aspects of SRR are of utmost importance, including assistance to victims, cooperation with human rescue and repair teams, and communication with the management team that controls or supervises disaster mitigation activities. Rescue vehicle and repair crew scheduling was optimized with a combination of Mixed Integer Linear Programming and Ant Colony Optimization algorithms in [16]. Non-linear optimization for the solution of swarm SRR is described in [3]. An implementation of the simulation of an Industrial Internet-of-Things (IIoT) environment in a salt mine, where several mining inspection robots seek optimal anticipatory strategies to mitigate water and gas leaks, is presented in [20].

The problem of explainable AI, within the context of communicating robot action plans during S&R activities and human-robot interactions is discussed e.g. in [2]. The DRMSS design, based on AI alignment, proposed in this paper is aimed to bridge this gap, focusing on rescue action efficiency and on minimizing risks for the enterprise's personnel. To sum up, optimal embedding of robotic teams into overall disaster resilience building processes coordinated with a DRMSS remains an important research problem [9] to solve utilizing the knowledge on rapidly changing AI and robotics technologies. As we will show in Sect. 3, this knowledge can be gained from foresight exercises, specifically from Delphi surveys.

More information on the relationship between AI and disaster resilience management can be found in the survey [14] of related technological challenges such as early warning systems, navigation, geographic information systems (GIS), and diverse heuristics. Two real-life examples of DSS applications, including the real-time water quality management are also given. An attempt to provide taxonomy of disaster management support with AI methods, based on nine cases, is presented in [1]. The authors claim that about 60% of disaster management tools use AI methods, which coincides with our estimations. A survey on the AI-related methods of systems analysis and control with hints regarding future trends is presented in [13]. Another survey on the relations between AI and business models is contained in [8]. Finally, [23] studies quantitative disaster resilience indicators, which may be employed in DRMSSs.

3 AI Alignment and the Software Architecture of DRMSS

Progress in technology, specifically AI, provides a chance to win the race between growing disaster threats and the decreasing will to accept disaster-related losses. Hence, the design of disaster management support system architecture and its implementation should be aligned to state-of-the-art AI solutions, while taking into account a prior needs analysis, and the availability of suitable AI technologies. The resulting alignment problem to be solved can be formulated as follows.

Problem 1 (technological AI alignment in the DRMSS context)
Keep the AI implementation level in an enterprise information system (EIS) to standards ensuring at least an equally strong response of this EIS to the challenges and threats created by external agents with adversarial AI.

AI alignment in an enterprise with integrated information systems or a situation where all EISs will be aligned can be referred to as *enterprise AI alignment*.

Problem 2 (enterprise AI alignment in the DRMSS context)
Identify the best AI technologies to be used in an enterprise as a remedy to any external threats, caused by adversarial AI or not, and implement them in the enterprise.

Problems 1 and 2 can be resolved with the following iterative and continual system design procedure involving in-the-loop AI technology implementation.

Procedure 1
Start point: A digital innovation-aware enterprise is committed to use AI for its business security advantage. To achieve this goal, it will design and implement a DRMSS based on an AI technology assessment process.

Step 1. Determine the goals, use cases, key features, and DRMSS success factors in the context of a class of specific applications (DRMSS pre-design).

Step 2   Perform a technological needs analysis addressing the relevant risks and threats and relate them to the system pre-design categories.

Step 3. Scan the market of AI technologies, links to prototype solutions, and ongoing applied AI research. Retrieve those fitting the identified needs of the DRMSS related to the relevant problem to be solved (Problem 1 or 2 or both).

Step 4.   Rank the technologies in each needs category according to the price, compatibility with other key technologies, forecasted outcomes of the updated system, and other DRMSS or user performance criteria suitable for the real-life situations concerned.

Step 5.   Select the best technology portfolio and build the implementation plan.

Step 6.   Assess ex-ante the resulting system architecture and performance with impact prediction and multicriteria analysis methods, taking into account the additional preference information coming from the simulation of the system performance and expert judgments.

Step 7.   Implement the final technology selection in the DRMSS architecture.

Step 8.   Repeat Steps 1–7 during the system lifetime or until the stop condition holds.

The overall procedure can be assessed with aggregated performance criteria evaluated in Step 6. The procedure ends when a disruptive change in system goals or architecture occurs so that the subsequent changes cannot be regarded incremental.

In contrast to typical strategic IT alignment [11], Problems 1 and 2 touch upon the alignment of information systems to be built with state-of-the-art AI methods, and of their users to learn these methods. Moreover, the rapid progress in AI hinders a traditional system design approach, where the methods and tools are selected just once, at the inception stage of system planning, and then are incrementally updated during the system lifetime. Instead, AI-driven systems of strategic relevance, such as DRMSS, require a continual re-design process.

4   AI Evolution Modelling Tool

The solution to above AI alignment problems can be supported with a dedicated technological foresight tool, termed the AI evolution model (AIEM). A general scheme of AIEM deployment, conforming to the Procedure 1, may be presented as follows:

- The AI evolution model will be established to provide clues regarding the development of selected AI technologies, relevant to DRMSSs for the next 10–15 years. The technology development modelling may use methods of various degrees of sophistication, from judgmental forecasts [18] to sophisticated system dynamics.
- The AIEM is a scalable tool that may align just one class of DRMSSs or even a single system as well as a broad class of enterprise IS. It can scan progress in a specific AI technology, e.g. photogrammetric drones, or an AI technology portfolio.
- AIEM can be self-contained and provided as software as a service (SaaS) by a specialist company or it may be coupled with other DRMSS modules. The latter case is more likely when it scans a single technology or a bulk of related technologies. On the other hand, a built-in AIEM can evolve itself towards a self-contained foresight support system, capable of modelling the development of a variety of AI and related technologies while serving multiple DRMSSs. Its owner can create a specialist spin-out company and the first of the above business models will apply.
- Needs assessments are to be performed prior to system implementation and updated every year. The assessments deliver the requirements for AI technologies and IT to be

implemented in the DRMSS. In case of large industrial plant resilience, the relevant AI techniques may include information fusion, autonomous inspection robotics, or sensor networks within industrial internet of things.

- Needs confronted with technology forecasts make it possible to select most viable and robust software architectures, AI techniques, and an implementation time plan.
- Finally, the subsequent versions of DRMSS, aligned to state-of-the-art AI methods, will be implemented and validated during the system's lifetime.

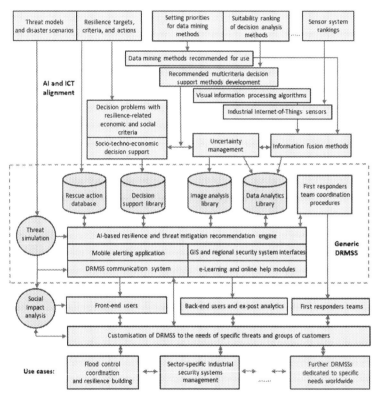

Fig. 1. A generic software architecture and uses scheme of the DRMSS coupled with an AIEM.

A scheme of the above-presented architecture in case of a single DRMSS coupled with an AIEM ensuring the AI alignment is shown in Fig. 1 above.

The AIEM building consists of four interdependent phases, interlaced with the implementation of selected methods in the core DRMSS. These phases will be repeated iteratively, based on learning from implementation outcomes.

Phase I.   Multi-level AIEM design, featuring expert information gathering and update exercises such as Delphi surveys interlaced with quantitative trend and qualitative scenario generation.

Phase II.   AI-related technological needs and impact model building. The natural disaster and anthropogenic threat characteristics will be optimally matched with the AI methods and tools to be deployed in the DRMSS. The impact modelling will be simulation-based and will stick to anticipatory network principles [19]. The impact assessment will cover at least three technology review periods.

Phase III. Implementation. The AIEM will gather sufficient information to implement the AI methods selected in Phase II into the core DRMSS.

Phase IV.   Testing and learning. The DRMSS's capabilities in solving real-life threat resilience problems, specifically those related to unauthorized intrusions and thefts, equipment damages related to fuel or gas leaks, landslides, and weather-dependent disasters such as floods, strong winds etc. will be tested and validated. The learning process will update the AIEM's technology selection procedures as well as the their implemention in the DRMSS.

The implementation order of specific DRMSS modules in Phase III should follow the ranking of the enterprise's security needs. For example, the system to be implemented may be focused on preventing and mitigating floods, the accompanying disastrous events, such as flood-related landslides and soil contamination, and on ensuring the resilience of endangered critical infrastructure. The most suitable AI methods and tools should be selected in compliance with the above scope. Specifically, these can include intelligent decision support technologies, team coordination and group decision making dedicated to managing threat prevention and mitigation operations, as well as the autonomous SRR. The forward-looking knowledge of evolution and development scenarios of AI methods can be a base of long-term flood resilience building.

4.1   The AIEM Implementation

The AI technology scanning and modelling tool can combine several modules, acting at different time scales, and yielding multi-purpose output information. These include:

- User needs analysis module, a part of the AIEM Graphical User Interface (GUI).
- Intelligent bots for web, bibliographic and patent databases scanning.
- Real-time data streams retrieval and analysis module, suitable for processing both, quantitative and qualitative information about AI techniques available in different social media. It performs sensitivity analysis and cascade effects for the detection of hidden technological 'black swans' in AI research data [5].
- A multi-level intelligent technological evolution model comprising quantitative expert Delphi and a group model building module [18].
- Technological evolution simulation based on various game models: cooperative, Stackelberg, and conflicting games [12].
- A set of technology ranking and recommendation algorithms, embedded in GUI.

At least one prospective technological analysis tool must be contained in the basic configuration of AIEM. The scope of database and web searches as well as the overall information processing in AIEM depends on the needs analysis outcomes. This is a supervised activity performed periodically according to the scheme: (Threat) Detection → Mitigation → Resilience Building, which is to be repeated for each class of threats. Table 1 shows a sample process referring to a real-life case of a quarry.

Table 1. Sample technological (AI and IT) needs analysis

Threat name	Detection	Mitigation	Resilience
Unauthorized intrusions	Visual monitoring, scene understanding	Supervised recommendations issued to security teams	Sound & light warning systems, optimal arrangement of fences
Rockfalls	Ground- and drone-based sensors, radars, pattern recognition	Warnings generated automatically and communicated to the staff, connecting the devices via the IIoT	Photogrammetry, rock dynamics analysis updated regularly, protection nets and fences
Rockslides and landslides	Photogrammetry with drones, remote sensing, risk maps, ground-based inspection robots	Adaptively calculating refuge areas and evacuation paths	Risk maps and rock mass dynamics analysis, earthquake warning systems
Incidental technical gas or liquid leaks	Inspection robots, gas and vapor sensors	Human-robot team actions defined based on threat assessment	Optimal coverage of the protected area by inspection robots

Out of the threats presented in Table 1, only intrusions may be related to the use of adversarial AI, such as detection of monitoring and other sensors or disturbing them. The corresponding mitigating AI tools serve as detectors and neutralizing devices which will be selected as solutions of Problem 1 and installed as DRMSS actuators. Problem 2 - related technology scanning will yield new photogrammetric techniques, image analysis, robust drones, and inspection robots. The DRMSS decision engine will be updated to include control of new devices and pattern recognition algorithms.

4.2   The Implementation of the DRMSS Operational Modules

The above-outlined DRMSS will process and fuse heterogeneous data coming from sensors, information provided in real-time by first responder teams human and robot, and from the enterprise management. The corresponding methods and algorithms will be recommended by the above AI technology scanning and modelling tool taking into account relevant uncertainty modelling techniques. The main processes performed by the DRMSS analytic engine are given below.

- Information fusion algorithms will process the sensor and response team data. This functionality will be used persistently to monitor threat mitigation activities. The data gathered will be verified when an information misrepresentation is likely.

- Intelligent DSS architecture will implement multicriteria decision making methods capable of processing validated preference information in a transparent way. These include, for example, the reference sets method [17] and direct decision consequences modelling. The DSS functionalities of the DRMSS will assist in selecting a compromise between an optimal system performance and the promptness of the response to a threat.
- The software architecture of DRMSS will include as one of its principal components a mobile, Android-, or IOS-based application to be installed on user smartphones or built in robots as an Internet of Vehicles (IoV) component. An analytic application will be installed in the crisis management headquarters.
- Depending on particular needs, the above architecture will be supplemented with robotic system supporting software as the fourth component including a SLAM module, mobile disaster knowledge base, monitoring and rescue robot coordination software, sensor and communication equipment.
- The DRMSS may include cognitive emotion recognition, such as stress measurement of the human rescue team, and its analysis, based on (i) basic physiological stress indicators measured by sensors, (ii) subjective stress assessment input to the system by the users. Stress may strongly influence the efficiency of rescue teams.

The design of the AI-based DRMSS architecture will benefit from the outcomes of the AIEM linking the threat management needs, DSS functionalities, and best-performing AI methods. The DSS is typically built with a modular architecture, while the individual modules will be interoperable and allow for easy replacements of an implemented method with an improved option recommended by the AIEM. A built-in library of algorithms assures that a replaced analytic method will be stored for an eventual later re-use. The human-DRMSS communication will be supported by GUI for back-end managers and a mobile application with front-user interface.

5   Summary and Conclusions

Full achievement of the research aims formulated in Sect. 1 of this paper, will be facilitated with an improved understanding of the interdependence of resilience needs of a complex socio-technological system, the DRMSS design and implementation constraints, and an insight into the advantages and risks related to using of AI tools.

As shown in Sect. 2, the literature on resilience support methods is fragmented and, as for now, there is no implementation of a DRMSS that would encompass a sufficiently broad spectrum of AI methods required to serving as a target reference solution for systems ensuring complex industrial plant resilience. Thus, the system architecture described in Sect. 4 and fed by the data streams from sensors should be regarded as an advent of a *"living information system"*. By definition, such systems evolve being continually aligned to recent AI and IT trends and the best business practices derived therefrom. This is why an important role will be played by system operators who will report relevant findings arising from system uses. These findings will be stored and presented as 'periodic reports' and taken into account in system updates.

AI development scanning brings a broad spectrum of methods and implementation of intelligent systems that can prevent or mitigate different threats. In case of disasters, robotic systems can rescue human lives, while multi-agent models can help to understand collective human-robot cooperation. Machine learning methods with quick fusion of big data streamed from endangered areas, processed in rule- and case-based reasoning together with decision analytics, will yield optimal decisions. Progress in AI will be scanned, verified, ranked, and finally, the best subset of AI methods will be selected and recommended for implementation in DRMSSs. Communication technologies and data science will create the background for a system's robust performance. In contradistinction to emergency management systems that use inflexible data structures and have little capacity to anticipate events that have never occurred, this disadvantage can be removed in our system design by employing intelligent anticipatory network planning [19] with artificial creativity features. Another advantage of DRMSS, compared to existing technology for emergency management, is the continual modelling of disaster characteristics and use of current updates immediately after a threat is detected. The software architecture presented in Sect. 3 builds an ecosystem of valid AI solutions to support business processes within the overall emergency management cycle: preparation, prevention, detection, response, and recovery.

The threat-related needs analysis merged with AI technology recommendations delivered by the AI evolution model, disaster data analysis and decision support engine yield a robust DRMSS with a radically novel and flexible "living system" architecture. We expect that the proposed business process management approaches oriented towards AI alignment for industrial threat resilience building and continual improvement of DRMSS architecture and design methodology can be applied to a large class of EIS that will benefit from the deployment of AI tools and techniques. This methodology is first validated and tested on two real-life cases in Poland and Spain to extend the capacity for further applications.

Acknowledgment. This research is financed in part by the Polish National Agency for Academic Exchange (NAWA), contract No. PPI/APM/2018/1/00049/U/001.

References

1. Abu Bakar, H., Mohd H., Mohd, I.: S&T converging trends in dealing with disaster: a review on AI tools. In: International Nuclear Science, Technology And Engineering Conference – Inustec 2015, AIP Conference Proceedings, vol. 1704, Article # 030001 (2016). https://doi.org/10.1063/1.4940070
2. Adadi, A., Berrada, M.: Peeking Inside the black-box: a survey on explainable artificial intelligence (XAI). IEEE Access **6**, 52138–52160 (2018). https://doi.org/10.1109/ACCESS.2018.2870052
3. Bakhshipour, M., Jabbari Ghadi, M., Namdari, F.: Swarm robotics search & rescue: a novel artificial intelligence-inspired optimization approach. Appl. Soft Comput. **57**, 708–726 (2017). https://doi.org/10.1016/j.asoc.2017.02.028
4. Bañuls, V.A., López, C., Turoff, M., Tejedor, F.: Predicting the impact of multiple risks on project performance: a scenario-based approach. Proj. Manag. J. **48**(5), 95–114 (2017). https://doi.org/10.1177/875697281704800507
5. Bécue, A., Praça, I., Gama, J.: Artificial intelligence, cyber-threats and Industry 4.0: challenges and opportunities. Artif. Intell. Rev. **54**(5), 3849–3886 (2021). https://doi.org/10.1007/s10462-020-09942-2

6. Belardo, S., Karwan, K.R., Wallace, W.: An investigation of system design considerations for emergency management decision support. IEEE Trans. Syst. Man Cybern. SMC-14(6), 795–804 (1984). https://doi.org/10.1109/TSMC.1984.6313308

7. Degener, Ph., Gösling, H., Geldermann. J.: Decision support for the location planning in disaster areas using multi-criteria methods. In: Comes, T., Fiedrich, F., Fortier, S., Geldermann, J., Müller, T. (eds.) ISCRAM 2013, pp. 278–283. KIT, Baden-Baden (2013)

8. Di Vaio, A., Palladino, R., Hassan, R., Escobar, O.: Artificial intelligence and business models in the sustainable development goals perspective: a systematic literature review. J. Bus. Res. **121**, 283–314 (2020). https://doi.org/10.1016/j.jbusres.2020.08.019

9. Domdouzis, K.: Artificial-intelligence-based service-oriented architectures (SOAs) for crisis management. In: Handbook of Research on Investigations in Artificial Life Research and Development, pp. 79–95. IGI (2018). https://doi.org/10.4018/978-1-5225-5396-0.ch005

10. Gabriel, I.: Artificial Intelligence, values, and alignment. Mind. Mach. **30**(3), 411–437 (2020). https://doi.org/10.1007/s11023-020-09539-2

11. Gerow, J.E., Thatcher, J.B., Grover, V.: Six types of IT-business strategic alignment: an investigation of the constructs and their measurement. Eur. J. Inf. Syst. **24**(5), 465–491 (2015). https://doi.org/10.1057/ejis.2014.6

12. Kar, D., et al.: Trends and applications in Stackelberg security games. In: Başar, T., Zaccour, G. (eds.) Handbook of Dynamic Game Theory, pp. 1223–1269. Springer, Cham (2018). https://doi.org/10.1007/978-3-319-44374-4_27

13. Lamnabhi-Lagarrigue, F., et al.: Systems & control for the future of humanity, research agenda: current and future roles, impact and grand challenges. Annu. Rev. Control. **43**, 1–64 (2017). https://doi.org/10.1016/j.arcontrol.2017.04.001

14. Saleh, H.A., Allaert, G.: Scientific research based optimisation and geo-information technologies for integrating environmental planning in disaster management. In: Tang, D. (ed.) Remote Sensing of the Changing Oceans, PORSEC 2008, Guangzhou, China, pp. 359–390, Springer, Heidelberg (2011). https://doi.org/10.1007/978-3-642-16541-2_19

15. Sepulveda, J., Bull, J.: A model-driven decision support system for aid in a natural disaster. In: Ahram, T., Karwowski, W., Pickl, S., Taiar, R. (eds.) IHSED 2019. AISC, vol. 1026, pp. 523–528. Springer, Cham (2020). https://doi.org/10.1007/978-3-030-27928-8_79

16. Shin, Y., Kim, S., Moon, I.: Integrated optimal scheduling of repair crew and relief vehicle after disaster. Comput. Oper. Res. **105**, 237–247 (2019). https://doi.org/10.1016/j.cor.2019.01.015

17. Skulimowski, A.M.J.: Methods of multicriteria decision support based on reference sets. In: Caballero, R., Ruiz, F., Steuer, R.E. (eds.) Advances in Multiple Objective and Goal Programming, Lecture Notes in Economics and Mathematical Systems, vol. 455, pp. 282–290. Springer, Heidelberg (1997). https://doi.org/10.1007/978-3-642-46854-4_31

18. Skulimowski, A.M.J.: Forward-looking activities supporting technological planning of AI-based learning platforms. In: Herzog, M.A., Kubincová, Z., Han, P., Temperini, M. (eds.) ICWL 2019. LNCS, vol. 11841, pp. 274–284. Springer, Cham (2019). https://doi.org/10.1007/978-3-030-35758-0_26

19. Skulimowski, A.M.J.: Multicriteria decision planning with anticipatory networks to ensuring the sustainability of a digital knowledge platform. In: Ben Amor, S., et al. (eds.) Advanced Studies in Multi-Criteria Decision Making, pp. 168–197. CRC Press, Boca Raton (2020). https://doi.org/10.1201/9781315181363-9

20. Skulimowski, A.M.J., Ćwik, A.: Communication quality in anticipatory vehicle swarms: a simulation-based model. In: Peng, S.-L., Lee, G.-L., Klette, R., Hsu, C.-H. (eds.) IOV 2017. LNCS, vol. 10689, pp. 119–134. Springer, Cham (2017). https://doi.org/10.1007/978-3-319-72329-7_11

21. Takeuchi, H., Yamamoto, S.: Business analysis method for constructing business-AI alignment model. In: Cristani, M., Toro, C., Zanni-Merk, C., Howlett, R.J., Jain, L.C. (eds.) Knowledge-Based and Intelligent Information & Engineering Systems: Proceedings of KES2020. Procedia Computer Science, vol. 176, pp. 1312–1321 (2020). https://doi.org/10.1016/j.procs.2020.09.140.
22. Turoff, M.: Past and future emergency response information systems. Commun. ACM **45**(4), 29–32 (2002). https://doi.org/10.1145/505248.505265
23. Zobel, C.W.; MacKenzie, C.A.; Baghersad, M.; Li, Y.: Establishing a frame of reference for measuring disaster resilience. Dec. Supp. Syst. **140**, 11 (2021). art.#113406. https://doi.org/10.1016/j.dss.2020.113406

A Generative Design Method Based on a Graph Transformation System

Grażyna Ślusarczyk[(✉)] [iD], Barbara Strug[iD], and Ewa Grabska[iD]

Institute of Applied Computer Science, Jagiellonian University,
Lojasiewicza 11, 30-059 Kraków, Poland
{grazyna.slusarczyk,barbara.strug,ewa.grabska}@uj.edu.pl

Abstract. This paper deals with a Computer-Aided Design (CAD) tool supporting the conceptual phase of the design process. Design drawings are created by the designer with the use of a visual editor. It is possible to interactively define design requirements and/or visual design patterns, which are important features for expected/required characteristics of the task solution. Design drawings, requirements, and specified visual patterns have their internal graph representations, which are automatically obtained during the design process. Based on design actions taken by the designer during the design process, graph transformation rules, which are used to derive graphs representing design objects, are constructed. Additional graph transformation rules are created on the ground of design patterns and requirements specified by the designer. The obtained set of graph transformation rules, called a graph transformation system, is used to automatically generate graphs corresponding to new objects. The approach is illustrated by examples of designing indoor swimming pools.

Keywords: Generative design · Design patterns · Graph structures · Graph transformation rules

1 Introduction

This paper deals with a Computer-Aided Design (CAD) tool supporting the conceptual phase of the design process by inferring graph transformation rules from initial designs, specified requirements, and design patterns. Nowadays CAD systems are widely used to support the design process by facilitating integration of its various phases and aspects [9]. However, they still lack generative methods to easily generate classes of design variants. Many models of design objects often have similar structures and differ only in sizes or the way of fitting to varying environmental conditions [2, 10].

Design is seen as a creative task because novel design problems typically are open-ended and ill-defined. Therefore CAD systems should support the creative phase of the design process by enabling the designer to experiment with candidate solutions. The proposed CAD-like interactive environment allows for generating many candidate solutions with similar structures but different geometry and parameters based on the small set of possible solutions created by the

© Springer Nature Switzerland AG 2021
L. Rutkowski et al. (Eds.): ICAISC 2021, LNAI 12855, pp. 367–378, 2021.
https://doi.org/10.1007/978-3-030-87897-9_33

designer. This type of support produces new knowledge that the designer can take into account in refining or reconsidering candidate solutions [3].

The use of graphs and graph grammars to represent and generate structures of designed objects in the initial phase of design has taken place in CAD systems since the eighties of the last century. Currently, domain-specific visual language editors encourage designers to create their initial design solutions in the form of design drawings, while graphs are internal representations of these drawings and make it possible to support the designer's actions by a CAD system. During the design process, each sequence of steps of creating a design drawing by the designer can be automatically transformed into a graph transformation rule describing the introduced changes.

The proposed approach is illustrated by examples of designing indoor swimming pools, where the designer generates design drawings using a domain-specific visual editor. In the case of designing a floor layout of a swimming pool, the designer starts by drawing an outline of a building and then divides it into areas, which are recursively divided until the level of individual rooms is reached. In this way, the designer ends with a complete floor layout showing sizes and positions of all rooms. During the design process, the architect can interactively select parts of a created design and designate them as important features for expected/required object characteristics or specify some requirements using a set of dialog windows.

In the presented approach both topological structures and semantic properties (geometrical parameters, materials, environment aspects) of designs are represented by composition graphs (CP-graphs) and their attributes, respectively [1]. A CP-graph is a labelled and attributed graph, where nodes represent components of artefacts and are labelled by names of components they represent. To each node, a number of bonds representing fragments of components and expressing potential connections between components is assigned. Each edge connects bonds of two different nodes and represents relations between fragments of components represented by these nodes and is labelled by the relation names. To nodes and bonds attributes specifying properties of components and their parts are assigned [4]. The whole design drawings generated by the designer, the specified parts (patterns), and requirements are automatically transformed into the corresponding CP-graphs.

CP-graphs representing designed objects can be generated and modified using composition graph transformation rules. Some rules of a graph transformation system are constructed on the basis of design actions taken by the designer during the design process. Additional graph transformation rules are created on the ground of design patterns and requirements specified by the designer. The obtained set of graph transformation rules is used to automatically generate CP-graphs corresponding to new objects with geometry and material properties specified by graph attributes. The possibility of relating attributes of right-hand sides of CP-graph rules to attributes of their left-hand sides enables the system to capture parametric modeling knowledge. CP-graph transformation rules are designed in the unified graphical environment delivered by the system called *GraphTool*, which supports graph transformations [11].

This paper contributes to the field of computational methods by supporting the design of a graph transformation system. Graph transformation rules are typically difficult to construct even for people with considerable background knowledge, especially when the characteristic features and design requirements should be taken into account. In our approach, graph transformation rules are inferred based on design actions taken by the designer and, design patterns and requirements specified by the designer. The proposed generative method allows the system supporting design to automatically model alternative object layouts, which can be easily adapted to different use case scenarios and environmental conditions.

2   Generating Design Drawings

In the initial stage of the design process, understanding of requirements is often associated with visualizations of early conceptual solutions. In our approach, the system supporting building design [5] enables the designer to draw building layouts with the use of a visual editor. It is also possible to interactively define design requirements and/or visual design patterns, which are important features for expected/required characteristics of the task solution. Moreover, visualizations of functional artefact models can be developed to help organise research.

In the following, the systems supporting design [5], which enable the designer to draw building layouts will be considered.

Example 1. Let us consider an example of designing an indoor swimming pool. At first, a 2D outline of the building is drawn in an orthogonal grid (Fig. 1a). In the next step, taking into consideration the type of the swimming pool (recreational, sports, learner) and an approximate number of users, the designer

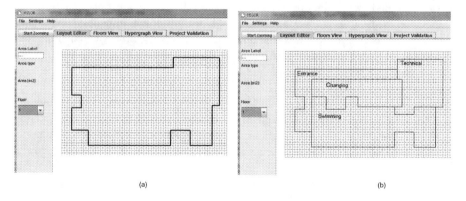

(a)   (b)

Fig. 1. a) The outline of the swimming pool b) the decomposition of the whole area into four functional areas.

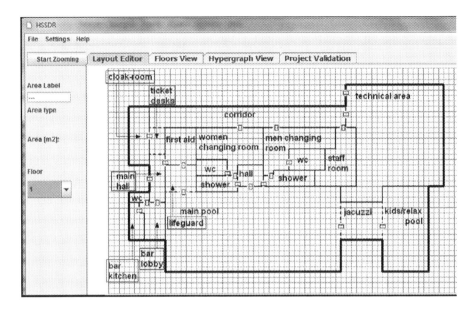

Fig. 2. The floor layout of the swimming pool.

decomposes the area determined by this contour into functional areas (Fig. 1b). Then functional areas, namely the *entrance area*, *swimming area*, and *changing area*, are decomposed into appropriate rooms. In the entrance area the *main hall*, the *ticket desks*, the *cloakroom*, the *corridor*, the *toilets*, the *bar lobby*, and the *bar kitchen* are distinguished. The changing area is decomposed into *women* and *men changing rooms*, the *toilets*, the *showers*, the *staff room*, and the *first aid room*. The swimming area is divided into the *main pool*, the *jacuzzi*, the *kids' pool*, the *lifeguard room*, and the *hall*. It is assumed that in successive design steps the designer gives labels starting from capital letters to areas that will be further divided, while labels starting from small letters are given to terminal spaces. At the last step the small rectangles representing *accessibility* relations between rooms are added and some continuous lines shared by polygons and denoting the *adjacency* relations between rooms are changed into the dashed ones, which represent the *visibility* relations between them. The whole obtained floor layout is shown in Fig. 2.

The designer has also the possibility to specify some requirements using a set of dialog windows. For example the functional areas which should be presented in the design drawing can be chosen. Each of the four areas (namely the *communication/entrance area*, *swimming area*, *changing area*, and *technical area*) together with their orientation can be selected. For each of these areas, the designer can further select rooms which it should contain. The selection process of functional areas and elements of the swimming area is shown in Fig. 3.

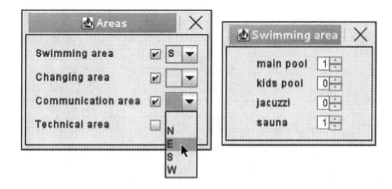

Fig. 3. Selection of functional areas and component parts of the swimming area.

Another way of specifying layout characteristic features is by indicating visual patterns of the designed drawing. In Fig. 4, which presents the layout of the first floor of a swimming pool, the visibility relation between the restaurant and the main swimming pool is indicated by the red ellipse, as the one which is important for the designer.

3   CP-Graph Representation of Designed Objects

In our approach the design drawings of building layouts are automatically transformed into their internal representations in the form of composition graphs (CP-graphs) [5–8]. A CP-graph is a labelled and attributed graph, where nodes

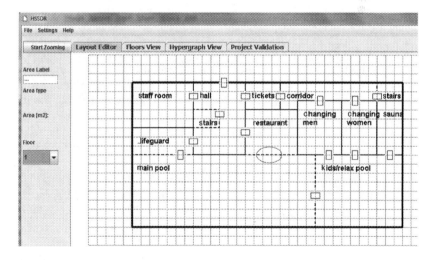

Fig. 4. Indicating the visibility relation between the restaurant and the main swimming pool.

represent components of the object being designed and are labelled by names of components they represent. To each node a number of bonds expressing potential connections between components is assigned. Nodes and bonds have attributes specifying properties of components and their parts, assigned to them, respectively [4].

Each edge of a CP-graph connects bonds of two different nodes, and it represents a relation between parts of the components represented by these bonds and is labelled by the relation name. The whole design drawings generated by the designer, the specified parts (patterns) and requirements, are automatically transformed into the corresponding CP-graphs.

The formal definition of a CP-graph is as follows.

Let Σ be a finite alphabet used to label object nodes, bond nodes and edges. Let A be a nonempty, finite set of attributes. For each attribute $a \in A$, let D_a be a fixed, nonempty set of its admissible values, known as the domain of a.

A CP-graph G is a tuple $G = (V, B, E, bd, s, t, lab, atr)$, where:

- V, B, E are pairwise disjoint finite sets, whose elements are respectively called nodes, bonds and edges,
- $bd : V \to 2^B$ is a function assigning sets of bonds to nodes in such a way that $\forall x \in B \; \exists! y \in V : x \in bd(y)$, i.e., each bond belongs to exactly one node,
- $s, t : E \to B$ are functions assigning to edges source and target bonds, respectively, in such a way that $\forall e \in E \; \exists x, y \in V : s(e) \in bd(x) \wedge t(e) \in bd(y) \wedge x \neq y$,
- $lab : V \cup B \cup E \to \Sigma$ is a labelling function,
- $atr : V \cup B \cup E \to 2^A$ is an attributing function.

It should be noted that between the same fragments of two components there can exist more than one relation. In this case, between bonds corresponding to these fragments several edges are created. However, for reasons of clarity, we draw only one edge and attach to it a set of labels representing all relations that take place between the bonds.

Example 2. A CP-graph representing the layout of the designed swimming pool building is presented in Fig. 5. This CP-graph is composed of thirteen nodes with bonds assigned to them drawn as small circles. Nodes represent rooms, while their bonds represent room walls. There are fourteen edges that represent the accessibility relation (denoted *acc*), and another fourteen representing the adjacency relation (denoted *adj*). There are also seven edges that represent the visibility relation (the label *vis* is attached to seven edges representing also the accessibility or the adjacency relations).

The selected parts of a created design and requirements specified using dialog windows also have their internal representations in the form of CP-graphs. CP-graphs representing the visibility relation between the restaurant and the main swimming pool (Fig. 4), the floor plan composed of the swimming area in the south, communication area in the east and changing area in the north, and the swimming area composed of a main pool and sauna (Fig. 3) are shown in Fig. 6.

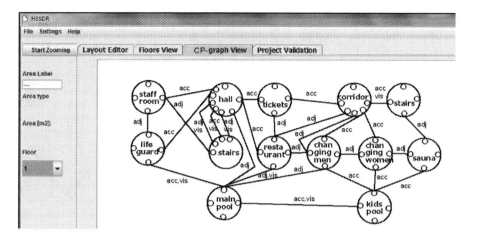

Fig. 5. A CP-graph representing the floor layout of the swimming pool from Fig. 4.

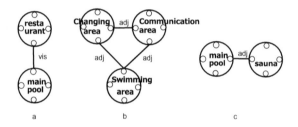

Fig. 6. CP-graphs representing design requirements.

4   CP-Graph Transformation Rule Inference

When structures of designed objects are described in terms of CP-graphs, they can be effectively generated and modified using CP-graph transformation systems. A CP-graph transformation system is composed of a set of transformation rules, which can be applied to object representations in the form of CP-graphs as long as no more of them are applicable. Based on design actions taken by the designer during the design process, CP-graph transformation rules are constructed. As successive sequences of design actions performed by the designer can be seen as steps that develop the generated solution, they correspond to various CP-graph transformation rules, which develop the corresponding CP-graph representation of the designed object. Dividing one area into several spaces in the process of designing the floor layout results in generating a rule with the left-hand side being a node labelled by the name of the divided area and the right-hand side being a CP-graph composed of nodes representing new spaces and edges representing relations between these spaces. Additional graph transformation rules are created based on design patterns and requirements specified by the designer. The obtained set of CP-graph transformation rules is used to automatically generate CP-graphs corresponding to new objects, some of which can be unexpected and innovative.

In order to apply graph transformation rules some bonds of CP-graphs occurring in these rules are specified as external ones. Let $G = (V, B, E, bd, s, t, lab, atr)$ be a CP-graph. A partial function $ext : B \rightarrow \aleph$, where $\aleph$ is a set of integers, determines external bonds of G. A CP-graph transformation rule is of the form $p = (l, r, sr)$, where l and r are attributed CP-graphs with the same number of different values of external bonds and sr is a set of functions specifying the way in which attributes assigned to nodes of l are transferred to the attributes assigned to nodes of r. Each attribute has a set of possible values, which are established during the rule application. The application of the rule p to a CP-graph G consists in substituting r for a CP-graph being an isomorphic image of l in G, replacing external bonds of the CP-graph being removed with the external bonds of r with the same numbers, and specifying values of attributes assigned to elements of r according to functions of sr. After inserting r into a host CP-graph all edges which were coming into (or out of) a bond with a given number in the CP-graph l, are coming into (or out of) bonds of r with the same number.

A CP-graph transformation system GTS is composed of a finite set of rules of the form $p = (l, r, sr)$, where l and r are attributed CP-graphs such that the set of numbers defined for bonds of l by ext is the same as the set of numbers defined by ext for r, i.e., $set(ext(B_l)) = set(ext(B_r))$.

Example 3. Selected CP-graph transformation rules inferred based on design actions taken by the designer during the generation of floor layouts of swimming pools are presented in Fig. 7. Rules *p1, p4,p6,* and *p8* are created as the result of dividing the space planned for the layout, to the entrance area, swimming area, changing area, and technical area, and then decomposing the entrance area, swimming area, and changing area into appropriate rooms, in the process of designing the floor layout presented in Fig. 2. Rules *p2, p5, p7,* and *p10* are created as the result of dividing the space planned for the layout to the communication area, swimming area, changing area, and technical area, and then decomposing these areas into appropriate rooms, in the process of designing the floor layout presented in Fig. 4. For rule *p2* the way of specifying values of the attribute *area* assigned to the nodes of the rule right-hand side CP-graph in respect to the value of this attribute defined for the node labelled *Outline* of the rule left-hand side CP-graph is shown. During this rule application, the area of the swimming space is set as half of the whole floor plan area, the changing space is set to 1/6 of the floor area, the communication space and technical space have the areas of 1/4 and 1/12 of the whole floor area, respectively.

Rules *p3* and *p9* are created based on the design requirements defined by the designer using dialog windows (Fig. 3) and represented by CP-graphs shown in Fig. 6b and Fig. 6c. The right-hand sides of these rules correspond to composition graphs from Figs. 6b and 6c. Rule *p3* allows the system to divide the floor plane into the communication area, swimming area, and changing area with establishing values of attributes assigned to the nodes representing these areas

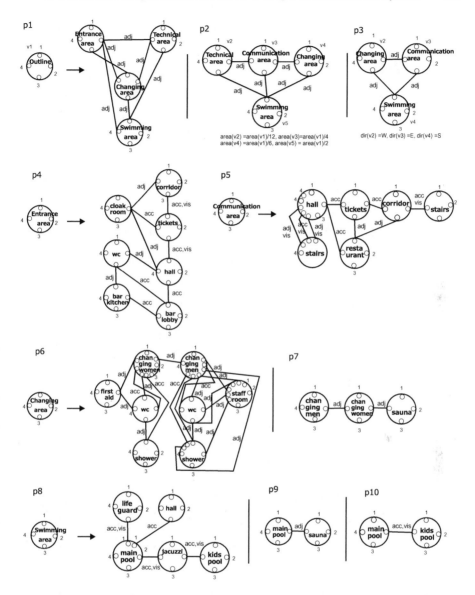

Fig. 7. Selected rules of a CP-graph transformation system inferred based on design actions taken by the designer during generation of floor layouts of swimming pools and specified design requirements

and specifying their world directions to east, south, and west, respectively. Rule *p9* enables the system to generate the swimming area composed of a main pool and sauna.

The CP-graph shown in Fig. 6a, which represents the requirement that the main pool should be visible from the restaurant, leads to creating two rules

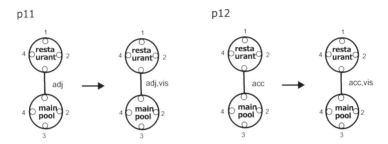

Fig. 8. Two CP-graph rules which ensure the visibility relation between the restaurant and the main pool.

presented in Fig. 8. Rule *p11* enables the system to add the visibility relation between the restaurant and the main pool, which are adjacent to each other, while rule *p12* enables to add the visibility relation between the restaurant and the main pool, which are accessible from each other.

A CP-graph transformation system, which selected rules are presented in Figs. 7 and 8, generates CP-graph representations of floor layouts of swimming pools. The embedding transformation for all these rules is such that all edges which are connected in the host CP-graph with the bond number i of the left-hand side of a production are replaced by edges connected with all bonds with number i on the right-hand side of this rule.

5 Automatic Generation of New Deigns

Combining the application of transformation rules inferred from the generation process of various designs in one derivation allows the system to generate CP-graphs representing new and sometimes innovative designs. These solutions can be treated as an inspiration for the designer, who can slightly modify them or develop further. At first, rules inferred based on the designer actions, are applied as long as it is possible starting from the CP-graph representing the initial design. Then to the obtained CP-graph, the rules changing design patterns or maintaining design requirements can be applied as long as the designer wants to or no more rules are applicable.

Example 4. Two floor layouts of the swimming pools, which correspond to CP-graphs generated using CP-graph transformation rules inferred based on design actions taken by the designer during generating swimming pool floor layouts (see: Figs. 7, 8, 10) are shown in Fig. 9. In the first case, rule *p13* which locates a swimming area at the south and a communication area at the north (Fig. 10) was used. Then the swimming area was replaced by a main pool using rule *p18* shown in Fig. 10. In the next step rule *p7* from Fig. 7 was used to divide the changing area into three rooms, and at the end rule *p15* (Fig. 10) was applied to divide the communication area into a corridor, ticket office, and a hall. In the second case the rules *p14*, *p18*, *p17*, and *p16* were used.

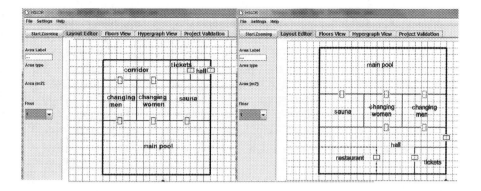

Fig. 9. Two floor layouts of swimming pools.

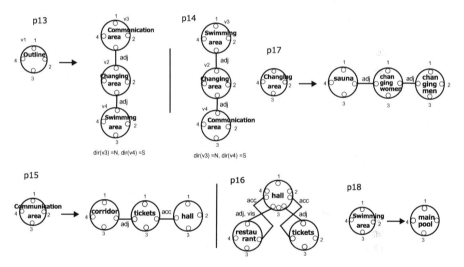

Fig. 10. The other CP-graph transformation rules inferred based on design actions taken by the designer during the generation of floor layouts of swimming pools.

6   Conclusion

In this paper, the problem of supporting the conceptual phase of the design process by inferring CP-graph transformation rules from initial designs, specified requirements, and design patterns has been considered. Initial designs are internally represented by graph structures, which are obtained automatically during the design process and can be transformed using graph rules. Based on design actions taken by the designer during the design process, CP-graph transformation rules are constructed. Additional CP-graph transformation rules are created on the ground of design patterns and requirements specified by the designer.

The obtained set of CP-graph transformation rules is used to automatically generate CP-graphs corresponding to new objects, some of which can be unexpected and innovative, and enables the designer to experiment with his/her own design ideas.

In future work, the memory of CP-graph rules will be used to ensure the presence of a predefined number of selected components within a designed object and to preserve some required characteristics of the design. We will also investigate such internal mechanisms of our design system as an ability to learn and a mechanism that would enable automatic evaluation of the obtained solutions.

References

1. Borkowski, A., Grabska, E.: Representing designs by composite graphs. In: IABSE Colloquium on Knowledge Support Systems in Civil Engineering, IABSE Reports 72, Bergamo, pp. 27–36 (1995)
2. Chung, C., Wang, S.: Computational Analysis and Design of Bridge Structures. CRC Press, Boca Raton (2014)
3. Goel, A.K., Rugaber, S.: Interactive meta-reasoning: towards a CAD-like environment for designing game-playing agents. In: Besold, T.R., Schorlemmer, M., Smaill, A. (eds.) Computational Creativity Research: Towards Creative Machines. ATM, vol. 7, pp. 347–370. Atlantis Press, Paris (2015). https://doi.org/10.2991/978-94-6239-085-0_17
4. Grabska, E.: Graphs and designing. In: Schneider, H.J., Ehrig, H. (eds.) Graph Transformations in Computer Science. LNCS, vol. 776, pp. 188–202. Springer, Heidelberg (1994). https://doi.org/10.1007/3-540-57787-4_12
5. Grabska, E., Ślusarczyk, G., Gajek, S.: Knowledge representation for human-computer interaction in a system supporting conceptual design. Fundamenta Informaticae **124**, 91–110 (2013)
6. Grabska, E., Strug, B., Ślusarczyk, G.: Cooperative design in a visual interactive environment. In: Luo, Y. (ed.) CDVE 2018. LNCS, vol. 11151, pp. 153–162. Springer, Cham (2018). https://doi.org/10.1007/978-3-030-00560-3_21
7. Grabska, E., Strug, B., Ślusarczyk, G.: A visual interactive environment for engineering knowledge modelling. In: Smith, I., Domer, B. (eds.) Advanced Computing Strategies for Engineering. EG-ICE 2018, LNCS, vol. 10863, pp. 219–230. Springer, Cham (2018). https://doi.org/10.1007/978-3-319-91635-4_12
8. Mars, A., Grabska, E., Ślusarczyk, G., Strug, B.: Design characteristics and aesthetics in evolutionary design of architectural forms directed by fuzzy evaluation. AI EDAM **34**, 147–159 (2020)
9. Pahl, G., Beitz, W., Feldhusen, J., Grote, K.-H.: Engineering Design: A Systematic Approach. Springer, London (2007). https://doi.org/10.1007/978-1-84628-319-2
10. Pipinato, A. (ed.): Innovative Bridge Design Handbook, 1st edn. Butterworth-Heinemann, Elsevier, Amsterdam (2015)
11. Ryszka, I., Grabska, E.: GraphTool - a new system of graph generation. In: Omatu, S. (ed.) ADVCOMP 2013, Porto, pp. 79–83. Curran Associates (2013)

Time Dependent Fuel Optimal Satellite Formation Reconfiguration Using Quantum Particle Swarm Optimization

K. Soyinka Olukunle[1]([✉]), Nzekwu Nwanze[2], and E. C. A. Akoma Henry[2]

[1] National Space Research and Development Agency, Obasanjo Space Center, Lugbe Abuja 900107, Nigeria
Soyinka.kunle@nasrda.gov.ng
[2] Center for Satellite Technology Development, Obasanjo Space Center, Lugbe Abuja 900107, Nigeria

Abstract. Evolutionary optimization methods have proven efficient in optimal transfer and orbit determination. This paper investigates the suitability of a Quantum Particle Swarm Optimization (QPSO) scheme for finding optimal transfer trajectories required for a fuel-efficient formation reconfiguration. It achieves this by comparing the search results for optimal transfer trajectories using QPSO, with traditional PSO. Given an initially clustered formation, a configuration, based on satellite coverage requirements, is set as the final configuration which the cluster aims to achieve, with minimal fuel consumption. Results show that QPSO had a better performance compared to traditional PSO, yielding faster convergence, higher accuracy, as well as better reliability to evade local minima by exploring a wider search spread.

Keywords: Satellite formation reconfiguration · PSO · Quantum PSO · Optimization · Lambert theorem

1 Introduction

Advances in satellite technology have fostered multi-satellite constellation and formation missions, where satellites work together to achieve a singular objective. The benefits of multiple satellite systems include, continuous global coverage, LEO operations allowing better optical resolution and lower lag time, lower cost launches, replacement and multiple angle viewing. These benefits outweigh those of larger monolithic systems, making them ideal for missions focused on research, communication, navigation, tracking and monitoring disaster as well as in-orbit activities. Consequently, there is an associated necessity for such systems' distribution not to be static, but rather be capable of reconfiguration. Reconfiguration implies conducting maneuver burns to achieve the on-orbit re-distribution of satellites in formation from their initial orbital slot to a new slot via a transfer trajectory. Lambert's theorem offers a suitable rendezvous method to

© Springer Nature Switzerland AG 2021
L. Rutkowski et al. (Eds.): ICAISC 2021, LNAI 12855, pp. 379–389, 2021.
https://doi.org/10.1007/978-3-030-87897-9_34

find time-critical intercept trajectory. Thus, Lambert's formulations present an optimization problem whose solution is sought using a numerical approach by applying swarm intelligence optimization methods.

Swarm intelligence optimization methods emerged from the study of the social interaction of organisms collectively working towards a common task. The methods have seen enormous application in spacecraft trajectory optimization problems due to their numerical accuracy and convergence rates. Algorithms such as Ant Colony Optimization (ACO), Brain Storm Optimization (BSO), Artificial Bee Colony (ABC), and Particle Swarm Optimization (PSO) have been applied to optimal spacecraft trajectory problems. A few cases include [1], where Genetic Algorithm (GA) was applied to multi-impulsive maneuvers. Also, minimum-fuel transfer orbits have been found by optimizing cost function formulations derived from Lamberts theorem [2] and [3]. Furthermore, PSO was applied in [4] and [5] to find optimal transfer trajectories for impulsive and finite burn maneuvers, the obtained results revealed PSO's effectiveness and intuitiveness towards a range of trajectory problems with varying complexity. It is possible to improve solution precision and global search ability of PSO by increasing the search particles' diversity, in order to explore a larger solution space. The desired improvements can be obtained using a variant of PSO referred to as Quantum Particle Swarm Optimization (QPSO). This is a more recent variant of PSO, based on a quantum (delta) potential well model, where particles obey principles of quantum systems such as superposition, allowing them exist simultaneously in several random states. This implies a probabilistic mechanism that ensures the algorithm's population exploits more states for search, thereby overcoming shortcomings of traditional PSO [6].

QPSO has been applied to a range of fields as found in [7–10]. The successes recorded in its application in these diverse areas further encourage its application in the field of optimal orbital spacecraft control problems. Kun et al. [11] showed the suitability of QPSO–Sequential Quadratic Programming (QPSO-SQP) hybrid for achieving a minimal fuel cooperative rendezvous between two satellites across a Far-distance, using continuous thrust. Furthermore, [12] proposed a mission planning method for On-orbit Satellite servicing involving multiple spacecraft. The paper solved a multi-objective problem, for which an improved QPSO with Non-Dominated Sorting Genetic Algorithm II (NSGA-II) was adopted to determine the optimal "Servicing Satellite to Target Satellite" assignment solution.

This paper aims to exploit a comparative study of traditional Particle Swarm Optimization (PSO), and a variant, Quantum Particle Swarm Optimization (QSPO), to achieve an optimal reconfiguration of a 3-satellite formation.

2   Lambert Theorem

Lambert's theorem provides orbit determination based on two-body equations of motion. Lambert's theorem states that given the position vectors r_1 and r_2 of two points P_1 and P_2, the transfer time Δt, from P_1 to P_2, depends only on the sum of the magnitudes of the position vectors $(r_1 + r_2)$, the semi major axis a, and the length of the chord c, joining P_1 and P_2, [14] as shown in Fig. 1.

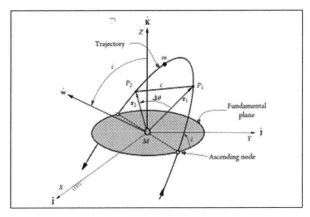

Fig. 1. Lambert's problem: transfer trajectory intersecting initial orbit at P1 and final orbit at P2 (Curtis [13])

This section shows the mathematical expressions that relate the parameters of the transfer trajectory and the total ΔV. The satellites' position vectors r_1 and r_2 are obtained from the Cartesian Orbital Elements (COE) that are predefined based on the mission requirements (mission scenario) of the satellite formation. Based on the position vectors the velocities (v_1, v_2) of the terminal points (P_1 and P_2) along the transfer orbit can be found from Eqs. (1) and (2). These equations show the relationship between the satellite's initial orbital position and velocity at a stated time to the position and velocity at a later time.

$$v_1 = \frac{1}{g}(r_2 - fr_2) \tag{1}$$

$$v_2 = \frac{1}{g}(\dot{g}r_2 - r_1) \tag{2}$$

where f and g, are Lagrange coefficients and $\dot{f}$ and $\dot{g}$ are their time derivatives. The Lagrange coefficients can be determined using the approach of Bate, Mueller and White detailed in [14]. Using mathematical formulations detailed in [14] Lambert's problem for a given transfer time can be generalized as

$$\Delta t = \frac{1}{\sqrt{\mu}}\left[\frac{y(z)}{C(z)}\right]^{\frac{3}{2}} S(z) + A\sqrt{y(z)} \tag{3}$$

where

$$A = sin\Delta\theta \sqrt{\frac{r_1 r_2}{1 - cos\Delta\theta}} \tag{4}$$

$$\Delta\theta = cos^{-1}\left(\frac{\vec{r_1}.\vec{r_2}}{r_1 r_2}\right) \tag{5}$$

$$y(z) = r_1 + r_2 + A\frac{zS(z) - 1}{\sqrt{C(z)}} \tag{6}$$

$\Delta\theta$ is the transfer angle, which is the difference between initial and final Mean Anomaly (MA), S and C are Stumpff functions which represent the Taylor series associated with the approach of applying a universal solution to Keplerian orbits. The constant z, the root of generalized Eq. (3), is found iteratively using Newton's approximation method. Using expressions (1) to (6), for a given Δt, it is possible to compute ΔV for each transfer arc. The manoeuvre is confined to one revolution thus the desired MA values are iteratively sought for within upper and lower bounds of $[0, 2\pi]$ radians.

3 QPSO

The traditional PSO algorithm simulates the behaviour of a flock of bird or school of fish, represented by n particles, in D dimensions, denoted

$$X_n^j = \left(X_n^1, X_n^2, X_n^3 \ldots X_n^D\right)^T \tag{7}$$

whose position values represent the solutions to the optimization problem. The particles' best positions, as well as the whole swarm's best position, are used to compute a velocity vector using Eq. (8) and updated to the new position in Eq. (9).

$$V_{i,n+1}^j = wV_{i,n}^j + c_i r_{i,n}^j \left(P_{i,n}^j - X_{i,n}^j\right) + c_2 R_{i,n}^j \left(G_n^j - X_{i,n}^j\right) \tag{8}$$

$$X_{i,n+1}^j = X_{i,n}^j + V_{i,n+1}^j \tag{9}$$

In Eq. (8), $r_{i,n}^j$ and $R_{i,n}^j$ are randomly generated numbers in the range $[0, 1]$, while c_1, c_2 and w, are acceleration and inertia coefficients that influence the personal best $P_{i,n}^j$, and the global best G_n^j, and ultimately the swarm's convergence. The high number of these adjustable parameters, which all influence the search behaviour, can prove problematic. This is a challenge QPSO eliminates.

Development of the algorithm's mechanism requires a look at the quantum time-space framework, where the state of a particle is described in a 3D space by the wave function $\Psi(x, t)$, and satisfies the relation:

$$|\Psi| \, dxdydz = Q \, dxdydz \tag{10}$$

where $Q \, dxdydz$ represents the probability density, which is the probability for the particle to appear in an infinitesimal volume about the point (x, y, z). Although the space around us is three dimensional, however considering that most systems change along one coordinate direction allows the assumption of a (spin-less) particle in a 1D delta potential well. The 1D particle is characterized by a time-invariant wave function $\Psi(x)$, that satisfies the stationary Schrödinger equation shown in Eq. (11).

$$\frac{d^2\Psi}{dY^2} = \frac{2m}{h^2}\left[E + \gamma\delta(Y)\right]\Psi = 0 \tag{11}$$

While there is the uncertainty about the particle position, the probability that the particle is found in the space considered is made certain (equals to 1). This is achieved by the normalization of the equation/wave function. Next, the Monte Carlo inverse transformation is applied to measure the position of the particle bound state in a 1D delta potential well. This yields a stochastic equation, which when sampled successively (n times), one can determine the particle's position as:

$$X_n = p \pm \frac{L}{2} ln\left(\frac{1}{u_n}\right), \quad u_n \sim U(0, 1) \tag{12}$$

Catering to the variation of L with each time step, as well as using the equation to measure the *ith* component of the position of particle i at each $(n + 1)th$ iteration, Eq. (13) can be written as:

$$X_{i,n+1}^j = p_{i,n}^j \pm \frac{L_{i,n}^j}{2} ln\left(\frac{1}{u_{i,n+1}^j}\right), \quad u_{i,n+1}^j \sim U(0, 1) \tag{13}$$

where

$$p_{i,n}^j = \varphi_{i,n}^j P_{i,n}^j + \left(1 - \varphi_{i,n}^j\right) G_n^j \tag{14}$$

is the local focus for the particle i at the nth iteration, $\varphi_{i,n}^j$ is a random point ($\sim U(0, 1)$) located between the personal best $P_{i,n}^j$, and global best G_n^j, in the search space, while L represents the Length of the potential well,

$$L_{i,n}^j = 2\alpha \left| X_{i,n}^j - C_n^j \right| \tag{15}$$

where C_n^j, the mean best position, is the average of the particles' personal best positions, such that

$$C_n^j = \left(\frac{1}{M}\right) \sum_{i=1}^{M} P_{i,n}^j \tag{16}$$

where, $(1 \leq j \leq N)$ and α is a positive real number known as the Contraction-Expansion (CE) coefficient. The resulting final QPSO evolution equation:

$$X_{i,n+1}^j = p_{i,n}^j \pm \alpha \left| X_{i,n}^j - C_n^j \right| \ln\left(\frac{1}{u_{i,n+1}^j}\right) \tag{17}$$

3.1 QPSO Algorithm

The steps for QPSO are enumerated in Table 1 as follows:

Table 1. Pseudocode for QPSO

	QPSO algorithm
1	Initialize parameters: population size, iteration count, particle dimension, selected α (CE) value, and establish upper and lower bounds
2	Initialize particles position $X_{i,n}^{j}$ with random positive vectors, within set bounds
3	Compute each particle's personal best $(P_{i,n}^{j})$, and swarm's global best (G_{n}^{j})
4	Compute the mean best position C_{n}^{j},
5	Generate random $\varphi_{i,n}^{j}$ and compute the local focus $p_{i,n}^{j}$
6	Update particle position $X_{i,n+1}^{j}$ (using Eq. 17),
7	Evaluate fitness function $f(X_{i,n+1}^{j})$
8	Update personal best and global best

4   Problem Formulation

The paper, considers both PSO and QPSO schemes to find optimal transfer orbit parameters which will demonstrate the time-critical reconfiguration of a 3-satellite formation. The reconfiguration applies bi-impulsive thrusts to transfer a formation from a clustered configuration in an initial orbit, to a final configuration, constrained by the requirement to increase coverage over a specific region of interest (ROI) in a single revolution, while maintaining the same orbital plane. The ROI is taken as a ground station located at (9.0765° N, 7.3986° E). To form the problem statement an initial requirement is set to determine orbital elements for the formation. The initial requirement is to:

1. Obtain a minimum of 20 passes over said ROI,
2. Achieve a minimum of 10% daily coverage.

To solve the initial problem, the Satellite Tool Kit (STK) software was used to simulate a twenty-four-hour scenario. The scenario defined the ROI using STK's Coverage Definition and Area Target objects. Computing the access between the 3 satellite Assets, and the ground grid point centered in the ROI. The minimum requirements were satisfied for the formation by spacing the satellites at 60° Mean Anomaly.

The result of the simulation in Table 2 shows the final configuration's orbital elements, as well as the Figures of Merit (FOM) that reflect its performance.

The rest of the work applies QPSO and PSO to search for the transfer time expressed in terms of the mean anomaly that minimizes the total control thrust ΔV_{Total} required for optimal transfers to the new configuration while meeting the coverage requirements defined by STK.

Table 2. Initial and final orbit elements, with figures of merit values

	Parameters	Initial configuration (cluster)	Final configuration
Orbital parameters	Orbital altitude (km)	900	1,500
	Eccentricity (e)	0.03	0.03
	Inclination (i)	40°	40°
	RAAN (Ω)	20°	20°
	Argument of perigee (ω)	10°	10°
	MA in radians	0	0–3.316
Mission scenario requirements	Satellite orbital spacing	<5°	60°
	Revisit time (hrs)	7.8	5.3
	Coverage time (%)	2.7	16.8
	Total Coverage Time (hrs)	0.6	4.0

4.1   Problem Statement

The transfer time Δt in Eq. (3), can be expressed in terms of mean anomaly in Eq. (18). This allowed the definition of an objective function J, in Eq. (19), which represents the ΔV to be minimized, for n number of chaser satellites. QPSO algorithm is used to explore the search space for a global minimum that yield particles whose components represent different mean anomaly values for each satellite.

$$\sqrt{\frac{a^3}{\mu}}(M_f - M_i) = \frac{1}{\sqrt{\mu}}\left[\frac{y(z)}{C(z)}\right]^{\frac{3}{2}}S(z) + A\sqrt{y(z)} \tag{18}$$

$$\min J = \sum_{i=1}^{n}|V_I - V_{P1}| + |V_{P2} - V_F| \tag{19}$$

Subject to

$$X = \left[M_{1f}M_{2f}M_{3f}\right]^T \in [0, 2\pi] \tag{20}$$

$$M_{2f} - M_{1f} \leq 1.0472 \tag{21}$$

$$M_{3f} - M_{2f} \leq 1.0472 \tag{22}$$

5   Results

The results of the various simulations are shown in Fig. 2 and Tables 3 and 4. Figure 2 (a) and (b) shows the result of the STK simulation indicating the revisit time and coverage over the ROI for the initial collocated satellites and the reconfigured formation. Figures 2 (c) shows the comparison of PSO and QPSO for the best run of the entire iteration Fig. 2 (d) shows the optimal transfer trajectory while Fig. 2 (e) and (f) compares the minimum ΔV_{Total} and transfer time for PSO and QPSO

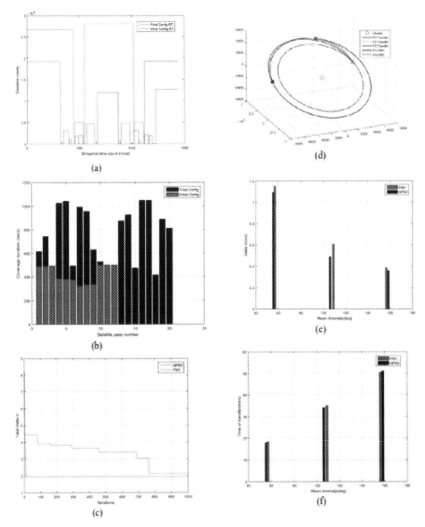

Fig. 2. (a) Comparison of Revisit Time (b) Total Coverage time comparison for both configuration (c) QPSO & PSO Minimum Delta-V result for best run (d) Optimal Transfer Trajectory (e) Comparative plot of Delta-V and Mean Anomaly for QPSO and PSO (f) Comparative plot of Transfer time and Mean Anomaly results for satellites in formation.

Table 3. Search parameters and results for PSO and QPSO

	PSO	QPSO
Best objective function value (min ΔV_{Total})	2.332 ms^{-1}	1.941 ms^{-1}
Global best positions (MA_1, MA_2, MA_3)	1.0245, 1.8745, 2.6913	0.9721, 2.0211, 3.049
Number of particles	10	10
Number of iterations (per run)	1000	1000
Dimension	3	3

A comparison of both algorithms shows QPSO performed better, and yielded a minimum ΔV_{Total} value of 1.941 ms^{-1}, to reconfigure the formation, as shown in Fig. 2(c). Additional performance metrics were employed to show an improved performance of QPSO for this transfer trajectory problem namely the mean and standard deviation in Table 4 and the convergence speed shown in Fig. 3.

Table 4. Standard deviation, mean and minimum fitness values to show performance of PSO and QPSO

	PSO	QPSO
Best ΔV_{Total} objective function value (ms^{-1})	2.3320	1.9411
Min average (or mean) ΔV_{Total} objective function value (ms^{-1})	3.9516	3.1808
Overall average (or mean) ΔV_{Total} objective function value (ms^{-1})	5.2921	3.2645
Standard deviation	2.8	2.0

The values in Table 4 are based on simulation results for 1000 iterations over 20 runs. The **Min** Average, is the minimum of all the averaged Objective function values for each run while the **Overall** Average is the average of the averages of all the Objective function values for the entire runs.

QPSO has been sparingly used for orbital transfer problems, this paper thus demonstrates its efficacy. Comparatively, QPSO obtained a better (minimum) global search final value than PSO, showing it can effectively find minimum energy bi-impulsive transfer trajectories.

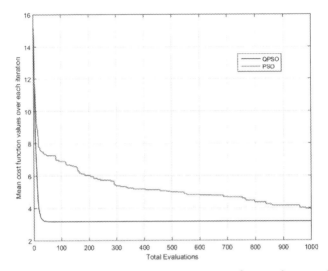

Fig. 3. Convergence history of PSO and QPSO average fitness values over 50 runs

References

1. dos Santos, D.P.S., Formiga, J.K.S.: Application of a genetic algorithm in orbital maneuvers. Computat. Appl. Math. **34**(2), 437–450 (2014)
2. Soyinka, O.K., Duan, H.: Optimal impulsive thrust trajectories for satellite formation via improved brainstorm optimization. In: Tan, Y., Shi, Y., Li, Li. (eds.) ICSI 2016. LNCS, vol. 9713. Springer, Cham (2016). https://doi.org/10.1007/978-3-319-41009-8
3. Jiashi, G., Lei, L., Yongji, W.: Spacecraft orbit design based on intelligent optimization. In: 2017 2nd International Conference on Advanced Robotics and Mechatronics (ICARM), Hefei, China. IEEE (2017). https://doi.org/10.1109/ICARM.2017.8273244
4. Mauro, P., Bruce, A.C.: Particle swarm optimization applied to space trajectories. J. Guidance Control Dynam. (2010). https://doi.org/10.2514/1.48475
5. Meireles, L., Rocco, E.M.: Study of orbital transfers with time constraint and fuel optimization. In: XVIII Brazilian Colloquium on Orbital Dynamics, 2016. Journal of Physics: Conference Series, Águas de Lindoia (2014). https://doi.org/10.1088/1742-6596/911/1/012014
6. Jun, S., Choi-Hong, L., Xiao-Jun, W.: Particle Swarm Optimisation: Classical and Quantum Perspectives. CRC Press, Taylor & Francis Group, London (2012)
7. Bagheri, A., Hamed, M., Mohsen, A.: Financial forecasting using ANFIS networks with quantum-behaved particle swarm optimization. Expert Syst. Appl. **41**(14), 6235–6250 (2014)
8. Fang, W., Sun, J., Wu, X.: Adaptive web QoS controller based on online system identification using quantum-behaved particle swarm optimization. Soft. Comput. **19**(6), 1715–1725 (2015)
9. Liu, L., Sun, J., Zhang, D., Guocheng, D., Chen, J., Wenbo, X.: Culture conditions optimization of hyaluronic acid production by Streptococcus zooepidemicus based on radial basis function neural network and quantum-behaved particle swarm optimization algorithm. Enzym. Microb. Technol. **44**(1), 24–32 (2009). https://doi.org/10.1016/j.enzmictec.2008.09.015
10. Li, S., Wang, R., Weiwei, H., Sun, J.: A new QPSO based BP neural network for face detection. In: Cao, B.-Y. (ed.) Fuzzy information and engineering, pp. 355–363. Springer, Heidelberg (2007). https://doi.org/10.1007/978-3-540-71441-5_40

11. Kun, Y., Weiming, F., Gang, L., Junfeng, Z., Piaoyi, S.: Quantum-behaved particle swarm optimization for far-distance rapid cooperative rendezvous between two spacecraft. Adv. Space Res. (2018). https://doi.org/10.1016/j.asr.2018.08.006
12. Qing, Z., Jianxin, Z., Junjie, T.: On-orbit servicing mission planning for multi-spacecraft using improved QPSO algorithm. Int. Core J. Eng. **3**, 2414–1895 (2017)
13. Curtis, H.: Lambert's Problem. In Orbital Mechanics for Engineering Students. Elsevier Butterworth-Heinemann, Oxford (2005)
14. Curtis, H.: Orbital Mechanics for Engineering Students. Elsevier, Burlington (2005)

The Efficiency of the Stock Exchange - The Case of Stock Indices of IT Companies

Paweł Trippner[1]([⊠]) and Rafał Jóźwicki[2]

[1] Department of Finance and Accounting, Institute of Finance, University of Social Sciences, Lodz, Poland
ptrippner@san.edu.pl

[2] Department of Finance and Accounting of SME's, Institute of Finance, University of Lodz, Lodz, Poland

Abstract. Since the very beginning of stock markets, investors have been looking for efficient methods to obtain high returns from their transactions. The development of technology and computational methods made it possible to use algorithms during so-called automated trading. The aim of this study was to investigate the impact of algorithmic trading on returns in the context of market efficiency theory. The research hypothesis was that the use of algorithmic trading cannot contribute to "better than market" returns, i.e. more favorable than in the case of the so-called passive control strategy. Based on the results obtained, the hypothesis adopted in the study was verified positively, as the analyzed strategies produced a worse rate of return than the passive control strategy.

Keywords: Stock market · Investment strategies · Information efficiency of the stock market · Alghorithmic trading · Neural networks

1 Introduction

The technology that is developing before our eyes is entering many areas of life and has an increasing influence on shaping human behavior. Without a shadow of a doubt, one of these areas is trading on stock exchanges and other markets that offer investors the opportunity to engage their capital. Thanks to widespread access to the Internet and the computing capabilities of computers used in the daily work of investors, the image of their work has changed significantly compared to what we observed even 10–15 years ago. Nowadays, stock exchange orders may be placed in person based on various types of brokerage investment accounts that allow the investor to view real-time quotes with various technical indicators plotted on the chart, or to simultaneously view quotes in various time frames (daily, hourly or minute charts) and the option to view order books. Another important aspect is the almost immediate access of investors to the news that may influence the course of quotations and stock exchange volatility. However, this also has certain negative effects, which can be described as market information overload, which in turn may adversely affect the investors' cool judgment and emotional composure when entering into transactions. The solution to these problems may turn out

© Springer Nature Switzerland AG 2021
L. Rutkowski et al. (Eds.): ICAISC 2021, LNAI 12855, pp. 390–402, 2021.
https://doi.org/10.1007/978-3-030-87897-9_35

to be the use of automated trading systems. By this we mean an algorithm created and implemented, which on the basis of given criteria independently makes purchases and sales of selected financial instruments. The stock exchange game takes place practically without the participation of the investor, whose role is limited only to activating the automaton. Thus, emotions accompanying the transaction are completely eliminated. Everything is done on the basis of decision algorithms, of course, previously defined by the investor.

An aspect that should also be taken into account is the so-called market efficiency, i.e. a theory stating that it is impossible to achieve a "better-than-market" rate of return by relying on historical information when deciding to buy and sell financial instruments.

Considering the possibility of eliminating emotions that interfere in the decision-making process, it is worthwhile to carry out an analysis that could settle the issues of the influence of algorithmic trading on the possible rates of return in the context of the theory of market efficiency.

Taking into account the previous considerations, the aim of the study is to investigate the influence of algorithmic trading on obtainable rates of return in the context of market efficiency theory.

The study is based on the hypothesis that the use of algorithmic trading cannot contribute to "better than market" returns, i.e. the results obtained will be less favorable than the so-called control passive strategy.

The justification for such an analysis is the fact that within the scope of the subject of this article it is difficult to find scientific studies that examine this issue in a way that takes into account the use of algorithmic trading. Stock markets have repeatedly been analyzed in terms of their efficiency, but the framing of these studies has been of a different nature than in the present study. In terms of the literature in this area, two studies in particular are worth noting. The first, was related to the evaluation of the use of so-called technical tools (averages and oscillators), and the results obtained indicated that these methods do not give above-average and, importantly, statistically significant rates of return [5]. In turn, the second study noted that the stock market in Poland was not efficient in its initial phase of development. However, in the subsequent years of its functioning, no attitudes were found to reject the poor form of information efficiency [15].

Algorithmics, on the other hand, which is a branch of computer science dealing with the design and analysis of algorithms, is an area that is very well described in the literature. An algorithm is a kind of "recipe" that leads step by step to the solution of a problem, while algorithmics is a branch of science dealing with algorithms and their properties. Neural networks, in turn, are an information processing system, in the structure of which one can find a kind of mapping of the elements of the biological nervous system. Nowadays, the application of neural networks is very wide and they are used, among others, in stock market forecasting, medical and biological research, data analysis, economic forecasting [1].

2 Algorithmic Trading

Algorithmic trading is the area of using computers to execute a specific set of instructions that enable a trade to be made with the goal of generating profits, of course. The hallmark of this is a speed and frequency that is impossible to achieve for a trader placing orders manually. Defined sets of rules, based on which transactions are executed, are based on prices, volumes and turnover for the instrument itself, or on selected indicators describing the instrument. Nothing prevents the use in this area of any algorithms based on the experience and ingenuity of their creator. For example, the trader will define a program, which will make a purchase, if the price of an instrument exceeds the average of a given value, with the oscillator level below the defined value. This means that the system will automatically track the instrument price and the average with the oscillator and will place a buy order when the predefined conditions are met. For a trader, this means that there is no need to constantly observe prices and charts, or to place orders on your own. An algorithmic trading system does this for the trader and places the orders, provided of course that the conditions defined in the system are met. Currently, the most commonly used algorithmic trading machines by investors were developed in the MQL language. Thanks to its use, a stock market player does not need to be a programming specialist, as the construction of market rules is carried out by using ready-made modules from which the entire system is created. However, the MQL language itself is based on the C++ language which is very popular among programmers [18].

The use of algorithmic trading has many advantages. The first one is the elimination of the influence of emotions on decisions. Everywhere where money is involved there are factors of human emotions that can negatively affect the decision-making process. By programming the rules of the market these emotions are eliminated, because the algorithm works only on the basis of a specific scheme. Another unquestionable advantage of algorithmic trading is the speed of implemented actions. A human being needs time to react to signals received, which combined with the aforementioned emotions can lead to a delay in placing an order, which especially for very active intra-day traders can have a significantly negative impact. Another advantage of algorithmic trading is greater processing efficiency. A computer program can analyze more data in a given unit of time and make decisions faster than a human. As a result, it is possible to build trading systems that take into account much more batch data than if a human wanted to process such data and execute transactions based on the received signals. Additionally, let us also pay attention to the aspect of the possibility of playing on foreign markets, which are active during the night in the local time. A trader can enjoy the benefits of trading while sleeping. Of course, algorithmic trading also has disadvantages. Well, there is more to trading on stock markets than just the signals themselves generated from batch data. Experienced traders in particular can point to the factor of having some sort of market feel and a broader perspective on trading than just mechanical action.

3   Efficient Market Hypotheses, Their Implications and Verification Methods

Fama presented a general Efficient Market Hypothesis (EMH), but this efficiency can take different forms depending on what type of information reaches investors [8].

The first group of information concerns the historical prices of shares, their sequences, trading volumes and obtained rates of return, i.e. all the data already available about the course of past transactions. If the price fully reflects this very information the market is efficient in the weak form. The hypothesis of weak - form EMH of the market implies that any data included in the first group are unrelated to the future returns and their use cannot contribute to obtaining above average returns for a long period of time. This means that if the market is efficient in this form, there is no point in using any methods based on the previous quotations - in particular it is not reasonable to use technical analysis.

The second group of information, which reaches the investors, is wider, because it includes those included in the first group and all publicly available information, i.e. financial reports, economic analyses, recommendations, macroeconomic data, etc. If the price fully reflects the information included in the second group it is said that the market is efficient in the sense of semi-strong hypothesis (semistrong - form EMH). This hypothesis includes in its scope the weak hypothesis, i.e. if the market is efficient in semistrong form it means that it is also efficient in the sense of weak hypothesis. It follows from the discussed hypothesis that investors using historical data and publicly available information are not able to obtain above-average rates of return in the long term - future profitability is not related to any information included in the second group. Therefore, the use of methods - especially fundamental analysis - based on this information is not justified.

The third group of information is the broadest: it includes all publicly available information and non-public information, which is known only to those who are privy to the functioning of the entity. If the price reflects all possible available information, the market is said to be efficient in the sense of strong hypothesis (EMH form). It implies that there are no investors having information which can influence future prices. Thus it is impossible to consistently obtain above-average profits on the basis of any information. Since the information included in the third group is the broadest, it means that if the market is efficient in its strong form, it is also efficient in the other two forms.

The presented efficient market hypotheses have been subject to empirical verification for many years. The results of the studies are not unequivocal: some of them confirm the hypotheses, while others raise doubts about their validity. Due to the different range of information relevant to considering the truth of individual hypotheses, their testing is carried out using different methods.

The weak hypothesis is tested using two groups of methods [4]:

1. statistical tests of independence between rates of return,
2. methods based on technical analysis.

Independence tests are designed to determine whether past quotes influence future ones. One of the basic such methods is the autocorrelation test. It consists in determining statistically whether the rate of return from period t is correlated with the rates of return from earlier periods, i.e. $t - 1, t - 2, t - 3$, etc. In the absence of such a correlation it is assumed that the capital market is efficient in the sense of the weak hypothesis.

The second group of methods for verifying the weak hypothesis is based on the tools used by technical analysis. For this purpose the so-called trading rules are created, i.e. a set of conditions that must be fulfilled in order to take a position on the market. Such a solution eliminates the "rigid" approach of statistical methods, which are not able to capture all that can be obtained on the market by combining different types of technical indicators. The results (rate of return) obtained by testing the market rules are compared with the passive "buy and hold The results (rate of return) obtained by testing the market rules are compared with a passive strategy "buy and hold policy", and on this basis it is determined whether the application of technical analysis allows to obtain better results than in the situation of buying shares at the beginning of the tested period and selling them at its end.

4   Technical Analysis Methods Used to Create Investment Strategies

Technical analysis is based on a wide range of methods related to the search for certain recurring patterns of market behavior on charts. For this purpose the following tools are used: formations created by prices, trend lines, technical indicators, and various market theories such as the Elliott wave theory, Dow theory, Carolan [11]. Many of these tools are subjective due to the analyst's individual approach. In order to be able to speak objectively about the effectiveness of technical analysis one should choose a method that does not have the above disadvantages. In this study we will use technical indicators that take specific numerical values, so that it will be possible to simulate taking specific actions in response to the values of these indicators.

Although computer programs and literature provide investors with dozens of indicators in general they can be divided into three basic categories [10]:

1. trend indicators, - among which the most important are: moving averages, MACD, Directional System, OBV.
2. Oscillators, - which include ROC, RSI, Momentum, stochastic oscillator, Williams' %R, among others.
3. Mood Indicators, - allow you to assess how optimistic or pessimistic the market crowd is. The most important among them are: NH-NL and Advance/Decline Index.

The presented use and classification of indicators make it clear that there is no single universal tool that can be used to identify the behavior of prices in the market.

Therefore, it is recommended to combine indicators in such a way that the trading system takes advantage of their advantages and at the same time rejects their disadvantages, resulting in optimal functioning of the whole, both in a flat market and during trends.

The technical indicators used to test the effectiveness of the market rules come from two groups: trend indicators and oscillators. The trend indicators most often used by investors were selected: moving average and MACD, while among the oscillators: RSI and stochastic.

Although there are many variations of moving averages used in technical analysis, the most commonly used in practice are simple, weighted and exponential moving averages. A slightly more sophisticated technique based on averages is to use the intersection of two averages of different lengths. The purpose of this approach is to eliminate the possibility of too frequent signals for transactions, in order to make them based on "more certain" trends, as they are signaled by two averages.

MACD (Moving Average Convergence/Divergence) indicator determines convergence and divergence of moving averages. To construct it, a fast and slow line is drawn and based on their mutual position investment decisions are made. The strategy for using MACD is as follows: if the fast line crosses the slow line from below it is a buy signal, if the fast line crosses the slow line from above it is a sell signal. On the MACD chart there is also an equilibrium line at the "zero" level, which causes some people to take the crossing of the fast line by the slow line as a buy or sell signal. Similarly, as in the case of averages, for which the number of periods taken into account in the calculations can be optimized, MACD can be constructed for different combinations of parameters, however, the principle of decision making obviously does not change [2, 16].

RSI (Relative Strength Index) is an oscillator developed by J. Welles and illustrates the relative strength of the given stock. It can take values from 0 to 100. The interpretation and application of RSI comes down to: determining overbought and oversold levels.

The interpretation and application of RSI comes down to: determination of overbought and oversold levels (two signal lines are plotted on the chart and their crossing by the indicator is a buy or sell signal), searching for divergences between the indicator and the Divergence between indicator and price (divergence occurs when the price movement is not confirmed by the indicator movement, for example, the price makes a new peak on the chart, while the indicator makes a peak at a lower level than the previous one), identification of formations (as it is done on the price charts) [17].

The stochastic oscillator was created by George Lane [13] and is based on the observation that during uptrends the price tends to close near its maximum values, while during downturns it tends to close near its minimums. When the existing trend is threatened, even though during the session the prices may record successive peaks or bottoms, the price may close within its previous fluctuations [7]. The application of the stochastic oscillator is similar to the principles presented for the RSI and boils down to: determining overbought and oversold levels, looking for divergence between the oscillator and the price, waiting for the moment when the %K and %D lines intersect.

5   Methodology of the Study and Results Obtained

The research results presented in the article are based on tests conducted using one of the computer programs available in this area. The analysis is based on the methodology that in the first testing period the parameters of the strategy are optimized, while in the second period the obtained parameters are used to determine the effectiveness of these strategies. Such an approach allows for an objective answer to the question of whether the optimized parameters of the algorithms underlying the strategy will allow to obtain, in the future and on their basis, a rate of return more attractive than a passive control strategy, which only assumes the purchase of a stock at the beginning of the test period and its sale at the end of the period without any active trading during that period.

The analysis covers five stock exchange indices: WIG-Informatyka index representing the Warsaw Stock Exchange, TecDax index listed on the German Frankfurt Stock Exchange, EnterNext Tech 40 index listed on Euronext, which is the integrated trading system of French, Dutch, Belgian and Portuguese markets, FTSE techMARK 100 index listed on the London Stock Exchange and finally NYSE Factset Global Robotics and

Artificial Intelligence Index representing New York Stock Exchange market. All of these indices are based on IT companies listed on their respective markets.

During the tests, optimization of investment strategies based on trend indices and oscillators was performed on the data of 2019, and the obtained parameters were used in analogous investment strategies in 2020. During the tests and optimization, several assumptions related to the terms of transactions were made:

1. total commission (buying and selling) is 1%,
2. funds that are not involved in transactions bear interest at a risk-free rate of 2%,
3. only long positions are taken on the market,
4. the trader has access to real-time quotes,
5. the position is taken on the day the signal appears.

As far as the optimization of strategy parameters for particular types of strategies is concerned, a number of further assumptions have been made.

For a single moving average the system assumes that the position is opened when the closing price is higher than the value of the average, while the position is exited when the closing price is below the moving average. Additionally a "stop" order was applied to protect the capital held against loss in case of price drop. The range of the average that was optimized was from 1 to 200, the maximum loss for the "stop" order was obtained from the range from 4 to 15%.

For two moving averages the system assumes taking the long position when the shorter average crosses the longer average from the bottom. The position is closed when the shorter average crosses the longer average from above. Both averages in the optimization process come from the range from 1 to 120. A "stop" order was also applied according to the rules discussed for the single average system.

For the MACD indicator the system assumes that a position is taken if the signal line (subject to optimization) crosses the MACD line from below. Closing takes place when the signal line crosses the MACD line from below. The range of the optimized line is from 3 to 30. The system assumes the use of a stop order according to the assumptions of a single average and two moving averages.

For the Stochastic Oscillator, the system allows a position to be opened when the Stochastic Oscillator crosses its oversold line from below and closed when the Stochastic Oscillator crosses its overbought line from above. The oversold (ranging from 5 to 45) and overbought (ranging from 55 to 95) lines are optimized. As in the previous rules the already discussed "stop" order has been used.

For the RSI the position is taken at the bottom crossing of the oversold line indicator and the close is taken at the bottom crossing of the overbought line indicator. The values that were optimized come from the ranges 5 to 45 and 55 to 95 respectively. The length of the RSI was also optimized (range 5 to 20). The system assumes the use of a defense line in the form of a "stop" order, similarly to the already discussed strategies.

Table 1. Data optimization and testing for trend indicators

Details			WIG - IT	TecDax	EnterNext Tech 40	FTSE techMARK 100	NYSE Factset Global Robotics and Artificial Intelligence Index
Data optimization	Single average	Average lentgh	37	39	41	43	31
		Annual profitability of strategy [%]	42,77	25,63	21,48	39,54	39,25
		Difference between active and passive strategy [p.p.]	**1,04**	**1,75**	**4,28**	**5,95**	**5,12**
	Two averages	Short average length	21	19	23	17	26
		Long average length	53	48	51	55	41
		Annual profitability of strategy [%]	44,12	24,69	19,21	39,14	37,58
		Difference between active and passive strategy [p.p.]	**2,39**	**0,81**	**2,01**	**5,55**	**3,45**
	MACD	Signal average lenght	19	21	18	17	23
		Annual profitability of strategy [%]	42,16	24,16	23,56	37,23	41,26

(*continued*)

Table 1. (*continued*)

Details			WIG - IT	TecDax	EnterNext Tech 40	FTSE techMARK 100	NYSE Factset Global Robotics and Artificial Intelligence Index
		Difference between antive and passive strategy [p.p.]	**0,43**	**0,28**	**6,36**	**3,64**	**7,13**
		Annual Profitability of Passive Strategy [%]	41,73	23,88	17,20	33,59	34,13
Tests on obtained parameters	Single average	Annual profitability of strategy [%]	11,23	−3,68	12,33	3,21	33,76
		Difference between active and passive strategy [p.p.]	**−24,74**	**−9,51**	**−11,31**	**−4,13**	**−11,93**
	Two averages	Annual profitability of strategy [%]	12,67	0,23	12,37	−3,79	21,03
		Difference between active and passive strategy [p.p.]	**−23,30**	**−5,60**	**−11,27**	**−11,13**	**−24,66**
	MACD	Annual profitability of strategy [%]	8,16	−6,35	2,19	−3,55	31,05
		Difference between active and passive strategy [p.p.]	**−27,81**	**−12,18**	**−21,45**	**−10,89**	**−14,64**
		Annual Profitability of Passive Strategy [%]	35,97	5,83	23,64	7,34	45,69

Source: Own calculations based on Metastock 7.0

The obtained results are summarized in Tables 1 and 2, respectively. The analysis of the obtained results allows verifying the hypothesis presented at the beginning of the research and observing other regularities. First of all, we should refer to the hypothesis underlying the research and state that the algorithmic trading system does not allow obtaining better results than the passive control strategy provides. Such conclusions are supported by the figures presented in Tables 1 and 2 in the rows defined as "Difference between active and passive strategy [p.p.]" in the section on "Tests on obtained parameters". As can be seen, in each case the difference between the active strategy (i.e., based on the optimized parameters) and the control strategy is negative - this applies to both trend indicators and oscillators. A valuable conclusion from the research is the observation that in the rows defined as "Difference between active and passive strategy [p.p.]" in the section on "Data optimization" the algorithms used allow to find such a parameter that allows obtaining an advantage of the active strategy over the passive strategy. The differences have a positive sign, but it is important to remember that these are only optimization tests on historical data.

Table 2. Data optimization and testing for oscillators

Details			WIG - IT	TecDax	EnterNext Tech 40	FTSE techMARK 100	NYSE Factset Global Robotics and Artificial Intelligence Index
Data optimization	Stochastic oscillator	Sell out	19	18	21	17	23
		Buy out	92	91	88	93	91
		Annual profitability of strategy [%]	43,26	29,66	21,29	34,57	37,82
		Difference between active and passive strategy [p.p.]	**1,53**	**5,78**	**4,09**	**0,98**	**3,69**
	RSI	RSI length	26	28	31	24	30
		Sell out	19	23	21	26	18
		Buy out	88	93	91	94	89
		Annual profitability of strategy [%]	42,36	24,17	26,31	38,92	35,76

(*continued*)

Table 2. (*continued*)

Details			WIG - IT	TecDax	EnterNext Tech 40	FTSE techMARK 100	NYSE Factset Global Robotics and Artificial Intelligence Index
		Difference between active and passive strategy [p.p.]	**0,63**	**0,29**	**9,11**	**5,33**	**1,63**
	Annual Profitability of Passive Strategy [%]		41,73	23,88	17,20	33,59	34,13
Tests on obtained parameters	Stochastic oscillator	Annual profitability of strategy [%]	16,88	−0,39	12,58	−3,48	8,96
		Difference between active and passive strategy [p.p.]	**−19,09**	**−6,22**	**−11,06**	**−10,82**	**−36,73**
	RSI	Annual profitability of strategy [%]	12,98	1,12	17,51	3,05	27,16
		Difference between active and passive strategy [p.p.]	**−22,99**	**−4,71**	**−6,13**	**−4,29**	**−18,53**
	Annual Profitability of Passive Strategy [%]		35,97	5,83	23,64	7,34	45,69

6 Conclusions

Algorithms are used in many areas of life, and for some time now the developers of trading systems have been making attempts to use them to obtain better rates of return in securities markets. The article tested the hypothesis assuming that the use of algorithmic trading will not enable an investor to obtain a better rate of return than the passive control strategy, which at the same time was part of an attempt to verify the hypothesis known since the 1970s about the efficiency of stock markets. Based on the results obtained, the hypothesis adopted in the article and assuming that algorithmic trading cannot contribute to better investment results than the passive control strategy should be verified positively, as the tested strategies yielded a worse rate of return than the control strategy.

It should be remembered, however, that stock exchange markets may be characterized by some kind of volatility related to the existence of upward or downward trends of different strength, the existence of periods of uncertainty when the market is in the so-called side trend, and all this may influence the obtained results. Taking this into account, it may turn out that changing the length of the parameter optimization and testing period would yield better results. However, these observations will be subject to further and deeper analysis by the Authors in subsequent studies. Moreover, the analysis can be enhanced using novel AI methods, e.g. fuzzy-rough sets [14], neural networks [3], particle swarm optimization [6], rough support vector machine [12] or ensemble techniques [9].

References

1. Alanis, A.Y., Arana-Daniel, N., Lopez-Franco, C. (eds.) Artificial Neural Networks for Engineering Applications. Elsevier, Amsterdam (2019)
2. Appel, G.: Technical Analysis: Power Tools for Active Investors. Pearson Education Inc., New York (2005)
3. Bilski, J., Kowalczyk, B., Marchlewska, A., Zurada, J.M.: Local Levenberg-Marquardt algorithm for learning feedforwad neural networks. J. Artif. Intell. Soft Comput. Res. **10**(4) (2020)
4. Brown, K.C., Reily, F.K.: Analiza inwestycji i zarządzanie portfelem, t. 1, (Investment analyzis and portfolio management, p. 1), PWE, Warszawa (2001)
5. Czekaj, J., Woś, M., Żarnowski, J.: Efektywność giełdowego rynku akcji w Polsce. Z perspektywy dziesięciolecia (Efficiency of the stock market in Poland. From the perspective of a decade), Wydawnictwo PWN, Warszawa (2001)
6. Dziwiński, P., Bartczuk, Ł., Paszkowski, J.: A new auto adaptive fuzzy hybrid particle swarm optimization and genetic algorithm. J. Artif. Intell. Soft Comput. Res. **10**(2) (2020)
7. Elder, A.: Trading For a Living. Psychology. Trading Tactics. Money Management. Wiley (1993)
8. Fama, E.F.: Efficient capital markets: a review of theory and empirical work. J. Finan. (2) (1970)
9. Homenda, W., Jastrzębska, A., Pedrycz, W., Fusheng, Y.: Combining classifiers for foreign pattern rejection. J. Artif. Intell. Soft Comput. Res. **10**(2) (2020)
10. Kaufman, P.J.: Trading Systems and Methods. Wiley (1998)
11. Murphy, J.J.: Technical Analysis of the Financial Markets: A Comprehensive Guide to Trading Methods and Applications. New York Institute of Finance, New York (2019)
12. Nowicki, R.K., Grzanek, K., Hayashi, Y.: Rough support vector machine for classification with interval and incomplete data. J. Artif. Intell. Soft Comput. Res. **10**(1) (2019)
13. Person, J.L.: A Complete Guide to Technical Trading Tactics: How to Profit Using Pivot Points, Candlesticks & Other Indicators. Hoboken (2004)
14. Starczewski, J.T., Goetzen, P., Napoli, Ch.: Triangular fuzzy-rough set based fuzzification of fuzzy rule-based systems. J. Artif. Intell. Soft Comput. Res. **10**(4) (2020)
15. Szyszka, A.: Efektywność giełdy papierów wartościowych w Warszawie na tle rynków dojrzałych (Efficiency of the Warsaw Stock Exchange compared to mature markets), Wydawnictwo Akademii Ekonomicznej w Poznaniu, Poznań (2003)

16. Thorp, W.A.: The MACD: A Combo of Indicators. For the Best of Both Worlds. American Association of Individual Investors Journal (2000)
17. Welles, W.: New Concepts in Technical Trading Systems, Trend Research (1978)
18. www.mql4.com

Simultaneous Contextualization and Interpretation with Keyword Awareness

Teppei Yoshino$^{(\boxtimes)}$ ⓘ, Shoya Matsumori, Yosuke Fukuchi, and Michita Imai

Keio University, Yokohama, Japan
yoshino@ailab.ics.keio.ac.jp

Abstract. Most natural-language-processing methods are designed for estimating context given an entire set of sentences at once. However, dialogue is incremental in nature. SCAIN (Simultaneous Contextualization and Interpretation) is an algorithm for incremental dialogue processing. Along with the progress of the dialogue, it can solve the interdependence problem in which the interpretation of words depends on the context, and the context is determined by the interpreted words. However, SCAIN cannot process texts that contain more words insignificant to context estimation such as in longer texts. We propose SCAIN with keyword extraction (SCAIN/KE), which extracts keywords that contribute to context estimation and eliminates the effect of insignificant words so that it can process longer texts. In the case study, SCAIN/KE updates context and interpretation better than SCAIN and obtains the keywords that contribute to context estimation better than other statistical methods. In the experiments, we evaluated SCAIN/KE on solving the ambiguity of polysemous words using the Wikipedia disambiguation pages. The results indicate that SCAIN/KE is more accurate than SCAIN.

Keywords: Dialogue context · Polysemy · Keyword extraction · SCAIN · SLAM

1 Introduction

A word can have multiple meanings, and such words are known as polysemous words. Previous studies proposed methods of processing polysemous words in natural-language processing [6,12]. A dialogue system needs to process a polysemy of word meanings and be able to identify what a word means in a conversation to interpret the speaker's intent. Context, which is composed of previous utterances, contains critical information that contributes to resolving the ambiguity of word meaning. However, most current dialogue systems only handle single-round conversation, such as a query-response pair, or predetermine the domain of a conversation and cannot take into account the context. Computational methods for context-aware word interpretation will help dialogue systems properly remove ambiguity in interpreting utterances.

© Springer Nature Switzerland AG 2021
L. Rutkowski et al. (Eds.): ICAISC 2021, LNAI 12855, pp. 403–413, 2021.
https://doi.org/10.1007/978-3-030-87897-9_36

There are many challenges in designing such a method. One important problem is the interdependence between context and word meaning: the interpretation of a word depends on the context, however, the context is determined by the interpreted words. Moreover, in a conversation, such interdependence needs to be processed sequentially. Even when the context of the dialogue is still unclear, it is necessary to interpret the meaning of an utterance and infer the context from undefined words to keep the conversation going. The dialogue system may need to withhold the interpretation of words to continue a dialogue and revise word interpretation in response to subsequent utterances. The system must carry out the following two processes simultaneously to sequentially determine the meaning of words and context in a dialogue. The first is estimating and retaining possible contexts under a certain interpretation of utterances. The second is continuously evaluating the context candidates on the basis of the interpretations of past utterances.

SCAIN [11] is an algorithm for identifying the meaning of words depending on contexts and estimating contexts depending on utterances in an incremental manner. SCAIN is based on FastSLAM [4], which is an algorithm designed for mobile robots to statistically resolve the interdependence between the robot's self-position and a map. SCAIN replaces self-position with a context and the map with a word-interpretation space to apply FastSLAM to the interdependence between context and interpretation in a sequential dialogue. In particular, the Kalman filter and a particle filter are the primary mechanisms recruited from FastSLAM. The particle filter holds multiple contexts at the same time, and word ambiguities can be clarified by selecting the interpretation with the more likely context.

However, SCAIN is not ideal with respect to processing long sentences. One of the reasons is that it uses a simple average of word vectors in estimating a context. Because of this, insignificant words that should not contribute to context estimation adversely affect the calculation of context likelihood. The more word vectors entered, the more their mean vectors converge to the center of the word-embedding space. To avoid this, it is necessary to distinguish between the words that should contribute to the context and those that should not.

We propose SCAIN with keyword extraction (SCAIN/KE). We improved upon SCAIN by introducing the idea of keywords, which are useful in estimating context. SCAIN/KE selects keywords on the basis of the assumption that the vectors of important words in a dialogue are located around a context vector that represents the entire dialogue history. A keyword extraction algorithm uses SCAIN's function in which possible context candidates are estimated in a particle-wise manner. Because SCAIN holds various possible contexts as particles, it can infer possible keywords on the basis of their possible contexts. With keyword extraction, we can reduce the effect of insignificant words and obtain more accurate context and word interpretation.

We conducted an experiment involving the Wikipedia disambiguation pages to carry out a polysemy disambiguation task and revealed that SCAIN/KE could disambiguate polysemous words more successfully than SCAIN, which indicates that by introducing the concept of keywords, SCAIN/KE can effectively resolve the problem of the interdependence between context estimation and word interpretation.

The remainder of this paper is structured as follows. In Sect. 2, we present related work regarding dialogue-context estimation and polysemy resolution and explain their challenges. In Sect. 3, we describe SCAIN/KE and investigate its effectiveness with an example dialogue. In Sect. 4, we discuss an experiment we conducted involving the Wikipedia-based polysemy resolution task used in a previous study [11], which showed that SCAIN/KE has better interpretation performance than SCAIN. Finally, we conclude the paper in Sect. 5.

2   Related Work

2.1   Multi-sense Embedding

We focus on handling the ambiguity of word meanings in dialogue. Many studies proposed word-embedding techniques that take into account polysemy. The technique word2gauss [12] uses Gaussian distribution as a representation of a word to express the ambiguity of meaning. ELMo [6] obtains context-aware word representation by concatenating a context vector with an existing word vector. BERT [2] uses masked language modeling to obtain deep bidirectional context-aware representations.

However, word2gauss acquires only the semantic field to which the word is assigned from datasets and cannot take into account the context. ELMo and BERT can interpret words from context, however, their representation of a word meaning is deterministic for each input sentence, which is problematic for sequential dialogue processing. Word meaning cannot be estimated deterministically. That is, we cannot necessarily determine the meaning of a word when it appears, and it is often the case that what a speaker intends to convey with a word is gradually clarified as the dialogue progresses. A dialogue system should retain multiple interpretations of words inferred from the current dialogue history then revise them sequentially.

2.2   Dialogue-Context Estimation

For estimating a word's meaning, it is important to infer its context, especially the long-term context, which is built from the dialogue history. HRED [9] infers a context vector using the *encoder* recurrent neural network (RNN), which embeds an utterance to the distributed representation space, and the *context* RNN which generates a context vector from the outputs of the encoder RNN. MemN2N [10] stores multiple sentence vectors in an external memory and uses them to generate a context vector using an attention mechanism [1]. Both methods can generate dialogue context representation by taking into account dialogue history, but they are not applicable to sequential dialogue processing because it is not necessarily possible to identify the exact meaning of an utterance when it is given. The interpretation of an utterance is gradually updated along with the progress in the dialogue, as we discussed in Sect. 2.1; thus, the inferred context should also be updated accordingly.

2.3 SCAIN

SCAIN [11] is an algorithm that sequentially infers context and word interpretation and based on FastSLAM [4]. SCAIN can solve the problem of the interdependence between context and interpretation, i.e., the context determines the interpretation of a word, and the context is determined by a set of words with a fixed interpretation.

In SCAIN, context x is represented by the locus of a point in a word-embedding space. The word-interpretation space m is represented by pairs of a word label and Gaussian distribution with mean μ and covariance matrix Σ in the same word-embedding space as the context vector. The m indicates possible interpretations of all words that appeared in a dialogue history. In SCAIN, the combination of x and m is represented as a particle; a single particle represents an instance of utterance comprehension, which consists of an estimated x and set of words m interpreted in that context. This enables SCAIN to sequentially interpret ambiguous words on the basis of the context while simultaneously inferring the context on the basis of the word interpretation.

However, SCAIN does not work when processing long utterance texts because of how it infers context. With SCAIN, it is assumed that the context can be simply calculated by an average of word vectors appearing in a sentence. However, all the words that appeared in an utterance are not necessarily important to infer its context. A sentence usually has words that are closely related to its context and other insignificant words. Therefore, we need to consider the connection between each word in a sentence and its context.

3 SCAIN/KE

SCAIN/KE extracts the words that represent the entire text, or keywords, by taking into account the distance between a word and its context. These keywords are used in calculating context likelihood and expected to improve context estimation. By using SCAIN's feature of holding possible sets of context and interpretation of words, SCAIN/KE can taking into account keywords even when the context and meaning of words in a dialogue remain ambiguous. Although some particles may have wrong contexts and keywords, as the dialogue continues and the context becomes clearer, appropriate particles are selected and the correct contexts and keywords gradually become dominant. SCAIN/KE consists of the following three steps.

3.1 Contextualization

SCAIN/KE updates x by using Eq. (1):

$$x_{t+1}^k = (1 - \lambda_u)\, x_t^k + \lambda_u v_{ut} + \sigma_u, \tag{1}$$

$$v_{ut} = \frac{1}{N} \sum_i^N w_i \cos\left(x_t^k, w_i\right). \tag{2}$$

where u is an utterance, λ_u is the learning rate, σ_u is the Gaussian noise corresponding to the update error of a context, and N is the number of words that appeared in utterance u. In Eq. (2), a word vector w_i is weighted by the word importance calculated from cosine similarity. This is a unique step of SCAIN/KE; SCAIN simply calculates v_{ut} as the average of w_i. Since this weight increases as the distance between a word and context decreases, the movement of the context vector is small if a word appears around the context. With this implementation, we can prevent particles that had acquired the correct context from moving to the center of the word-embedding space due to insignificant words.

3.2   Interpretation

After updating the context vectors, SCAIN/KE recalculates the interpretation distributions of words uttered by the user in each particle. This step conforms to SCAIN. First, each observation word vector z_i is calculated from pre-trained word vector w_i with $z_i = (1 - \alpha)w_i + \alpha x_{t+1}^k$, where α is the parameter indicating how much z_i is drawn to the context vector. This is explained as observation noise. To minimize this noise, the Kalman filter is applied to the interpretation distribution. These processes give each word distribution (μ_i, Σ_i).

3.3   Resampling

For each particle, the likelihood w is calculated from the pair of word distributions and estimated context vector in Eq. (3). Particles are resampled with reference to each particle w. To mitigate interference from insignificant words, the ws of the particles that detected keywords are summed with parameter λ_D.

$$w = -\log\left(\frac{1}{N}\sum_{i=1}^{N}\eta\left(l_i\right) D_M\left(<\mu_i, \Sigma_i>, x\right) + \epsilon\right) - \lambda_D d_x, \qquad (3)$$
$$d_x = \min(D_M(\mu, \Sigma)),$$

where D_M is the Mahalanobis distance representing the distance between the distribution and vector, ϵ is a minimal value to avoid division by zero, η is an attenuation term to account for the time when the interpretation is updated, l is a word label, N is the total number of uttered words, and d_x is the minimum distance between context and words. The second term in Eq. (3) implies a preference for the particles that have detected keywords, and it is one of the unique points of SCAIN/KE. The reason the weights of the particles nearing the uttered words are added under λ_D is as follows. If input sentences contain many insignificant words, the sums of the distances between all uttered words and each context vector become almost uniform. By enabling preferential treatment of the second term under a constraint λ_D, the weight of the particle is a bit dispersed even if the computation of the first term is smoothed.

Table 1. An example dialogue.

Speaker	Utterance	Time step
Human	I bought a mac	0
Agent	Where did you buy it?	1
Human	I bought it in Yokohama	1
Agent	How was it?	2
Human	There were a lot of people	2
Agent	It seems popular	3
Human	Yes, it is popular	3
	It can run heavy software	

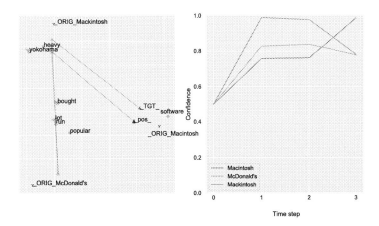

Fig. 1. (left) Visualized map of context and word vectors. (right) Confidence of interpretation of *mac* along time steps (SCAIN/KE). (Color figure online)

3.4   Case Study on SCAIN/KE

Case Study on Context and Word Sense. To examine SCAIN/KE in terms of sequential dialogue processing, we conducted a disambiguation task in a dialogue. We prepared an example dialogue in which the meaning of the polyseme *mac* is gradually revealed. We observed the transition of estimated context and word interpretation with SCAIN/KE. We also compared SCAIN/KE with SCAIN using the same dialogue. We defined the *mac* word vectors as one of the following three: *McDonald's* (hamburger), *Mackintosh* (coat), or *Macintosh* (computer). The word vectors for them were obtained from their respective Wikipedia pages. We used the pre-trained GloVe [5] 100-dimension word embeddings to define the original vectors of the words. The input sentences are listed in Table 1. The dialogue is a conversation between a person and an agent that infers the meaning of the word *mac*.

Figures 1 and 2 show the transition of the context and confidence of interpretations of *mac* with SCAIN/KE and SCAIN, respectively.

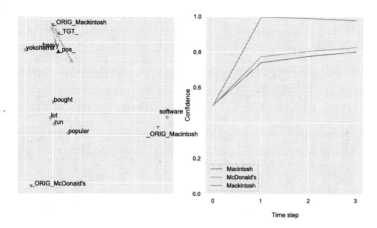

Fig. 2. (left) Visualized map of context and word vectors. (right) Confidence of interpretation of *mac* along time steps (SCAIN). (Color figure online)

The left sides of Figs. 1 and 2 show the word-embedding space that SCAIN/KE and SCAIN maintain in their particles. It is visualized by applying a principal component analysis to disperse three *mac* vectors. For visualization, we display only the context and word interpretations of the particle with the highest weight at each time step. In these figures, _pos_, represented as a blue triangle, is the position of the estimated context, _TGT_ with a red triangle is the mean of the word distribution labeled *mac*, and _ORIG_ with a red inverted triangle is a pretrained word vector of each interpretation of *mac*. Other word labels are the mean of their word distributions. The right sides of Figs. 1 and 2 show the confidence of interpretation of *mac* along time steps. The confidence was calculated from the cosine similarity between the *mac* vector of the particle with the highest weight and each candidate interpretation vector. SCAIN/KE interpreted *mac* as a computer with the highest (0.99) confidence at time step 3.

As shown on the left of Fig. 1, the context vector moved noticeably at time step 3 and reached *Macintosh*. This move occurred because SCAIN/KE estimated *software* as a keyword and recognized that the utterance was about computers. As shown on the right of Fig. 1, at time step 3 the confidence of *Macintosh* increased and decreased for the others. This is because the example dialogue does not provide any useful information on the interpretation of *mac* until time step 2, and it is not until time step 3 that we infer that it is a *Macintosh*.

As shown on the left of Fig. 2, in SCAIN, which does not take into account keywords, the context vector remained stuck to *Mackintosh* (coat). These results indicate that keyword extraction contributed to correct context inference.

Case Study on Keyword Extraction. To investigate the keyword extraction with SCAIN/KE, we compared the results of keyword extraction with three other methods: TFIDF [8], TextRank [3], and RAKE [7]. Because TFIDF requires other general documents, we used the NPS Chat Corpus, which consists of more

Table 2. Comparison of keyword extraction (with relative word-importance values).

SCAIN/KE	TFIDF	TextRank	RAKE
Computer (0.164)	Delivering (0.158)	A powerful computer (0.371)	Play latest games (0.529)
Better (0.126)	Powerful (0.158)	The latest games (0.368)	Powerful computer (0.235)
Latest (0.118)	Latest (0.137)	The power (0.262)	Power (0.059)
Games (0.115)	Games (0.137)	You (0.000)	Need (0.059)
Need (0.108)	Power (0.137)	–	Delivering (0.059)
Power (0.107)	Computer (0.088)	–	Better (0.059)
Play (0.095)	Play (0.068)	–	–
Powerful (0.088)	Better (0.060)	–	–
Delivering (0.079)	Need (0.057)	–	–

Table 3. Cosine similarities with *Macintosh.*

Word	Cosine similarity
Computer	0.756
Latest	0.444
Better	0.423
Games	0.415
Power	0.337
Need	0.328
Play	0.266
Powerful	0.265
Delivering	0.243

than 10,000 posts from chat rooms. We input the example sentence "If you will play the latest games, a powerful computer will be better for delivering the power you need." after the dialogue shown in Table 1 and compared the importance rate of each word. The results are listed in Table 2.

SCAIN/KE estimated that *computer* is important while *delivering* is not, whereas TFIDF inferred *delivering* is the keyword. TextRank assigned similar importance to *a powerful computer*, *the latest games*, and *the power*, respectively. RAKE recognized *play latest games* and *powerful computer* as idioms and regarded them as important.

The example sentence was talking about a Macintosh computer. Based on Eqs. (1) and (2), it is helpful to extract words near Macintosh as keywords to properly infer the context of this example sentence. Table 3 shows the cosine similarity between each word except for stopwords in the example sentence and

Macintosh. In accordance with Table 3, the word with the highest cosine similarity with *Macintosh* was *computer*. SCAIN/KE estimated *computer* to be significantly more important than the other words. Therefore, we can expect that the proposed keyword extraction algorithm enables SCAIN/KE to infer dialogue context more accurately than the other methods.

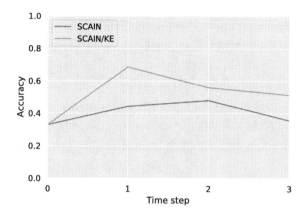

Fig. 3. Accuracy of polysemy resolutions.

4   Evaluation

4.1   Method

In a similar manner as in a previous study [11], we conducted an experiment on polysemy resolution and compared the results of SCAIN/KE with those of SCAIN. The experiment was conducted using polysemous words from Wikipedia; the disambiguation page of Wikipedia provides ambiguous word labels and descriptions of the label's interpretation candidates of the polysemous word. We randomly selected 300 disambiguation pages and extracted three candidates as possible interpretations for each page. The procedure of the experiment was as follows. First, we chose a specific topic from a disambiguation page. For each topic, we input the label of the polysemous word as a first utterance into both SCAIN and SCAIN/KE. We then input the description sentences as the following utterances. We evaluated how the meanings of the polysemous words were updated as the time steps progressed to consider the accuracy of SCAIN/KE's sequential dialogue processing. For each sentence entered, we calculated the cosine similarity between the updated polysemous word vectors and those of each correct answer. We investigated whether a candidate with the highest similarity in the particle with the highest likelihood was the correct interpretation.

4.2   Results

Figure 3 shows the results of this experiment. The horizontal axis is the number of sentences entered and the vertical axis is accuracy. At time step 0, we input only polysemous words, and from time step 1, we input one sentence per one time step. SCAIN/KE estimated the meanings of polysemous words with higher accuracy than SCAIN. In particular, SCAIN/KE had an accuracy of 0.69 in time step 1, while SCAIN had an accuracy of 0.44, indicating that SCAIN/KE could successfully update the meaning of the previous utterance when the next utterance was entered. These results suggest that, by introducing the concept of keywords, SCAIN/KE is better at solving the interdependence problem of a dialogue's context and word interpretation than SCAIN. There are possible reasons the accuracy rate did not increase as the dialogue progressed. First, some tasks generated from Wikipedia were too difficult to solve because some interpretation candidates on Wikipedia's disambiguation page were very similar to each other. Second, because Wikipedia articles are often written to reveal the topic in the first sentence, we could not fully simulate the dialogue as it gradually became clearer as the time steps progressed. A dataset that gradually reveals polysemous words as a dialogue progresses would have yielded more practical results.

5   Conclusions

We proposed SCAIN/KE, an algorithm for sequentially interpreting utterances under the problem of interdependence between context and word meaning. SCAIN/KE exploits the idea of keywords to improve the inference of context. We conducted an experiment to compare SCAIN/KE with SCAIN in a word-sense disambiguation task. The results indicate that, by using SCAIN/KE, we could estimate both the context and interpretation of utterance texts better in processing ongoing dialogue.

Acknowledgements. This work was supported by JST CREST Grant Number JPMJCR19A1, Japan.

References

1. Bahdanau, D., Cho, K., Bengio, Y.: Neural machine translation by jointly learning to align and translate. In: 3rd International Conference on Learning Representations, ICLR 2015, January 2015
2. Devlin, J., Chang, M.W., Lee, K., Toutanova, K.: BERT: pre-training of deep bidirectional transformers for language understanding. In: Proceedings of the 2019 Conference of the North American Chapter of the Association for Computational Linguistics: Human Language Technologies, Volume 1 (Long and Short Papers), Minneapolis, Minnesota, pp. 4171–4186. Association for Computational Linguistics, June 2019
3. Mihalcea, R., Tarau, P.: TextRank: bringing order into text. In: Proceedings of the 2004 Conference on Empirical Methods in Natural Language Processing, pp. 404–411 (2004)
4. Montemerlo, M., Thrun, S., Koller, D., Wegbreit, B., et al.: FastSLAM: a factored solution to the simultaneous localization and mapping problem. In: AAAI/IAAI, p. 593598 (2002)

5. Pennington, J., Socher, R., Manning, C.D.: GloVe: global vectors for word representation. In: Proceedings of the 2014 Conference on Empirical Methods in Natural Language Processing (EMNLP), pp. 1532–1543 (2014)
6. Peters, M., et al.: Deep contextualized word representations. In: Proceedings of the 2018 Conference of the North American Chapter of the Association for Computational Linguistics: Human Language Technologies, Volume 1 (Long Papers), New Orleans, Louisiana, pp. 2227–2237. Association for Computational Linguistics, June 2018. https://doi.org/10.18653/v1/N18-1202, https://www.aclweb.org/anthology/N18-1202
7. Rose, S., Engel, D., Cramer, N., Cowley, W.: Automatic keyword extraction from individual documents. Text Min. Appl. Theor. **1**, 1–20 (2010)
8. Salton, G., Buckley, C.: Term-weighting approaches in automatic text retrieval. Inf. Process. Manage. **24**(5), 513–523 (1988)
9. Serban, I.V., et al.: A hierarchical latent variable encoder-decoder model for generating dialogues. In: Proceedings of the Thirty-First AAAI Conference on Artificial Intelligence, pp. 3295–3301 (2017)
10. Sukhbaatar, S., Weston, J., Fergus, R., et al.: End-to-end memory networks. In: Advances in Neural Information Processing Systems, pp. 2440–2448 (2015)
11. Takimoto, Y., Fukuchi, Y., Matsumori, S., Imai, M.: Slam-inspired simultaneous contextualization and interpreting for incremental conversation sentences. arXiv preprint arXiv:2005.14662 (2020)
12. Vilnis, L., McCallum, A.: Word representations via gaussian embedding. In: Bengio, Y., LeCun, Y. (eds.) 3rd International Conference on Learning Representations, ICLR 2015, San Diego, CA, USA, 7–9 May 2015, Conference Track Proceedings (2015)

Aiding Long-Term Investment Decisions with XGBoost Machine Learning Model

Ekaterina Zolotareva$^{(\boxtimes)}$ (iD)

Data Analysis and Machine Learning Department, Financial University under the Government of the Russian Federation, 38 Shcherbakovskaya St., Moscow 105187, Russia
ELZolotareva@fa.ru

Abstract. The ability to identify stock market trends has obvious advantages for investors. Buying stock on the upward trend (as well as selling it in case of downward movement) results in profit. Accordingly, the start and endpoints of the trend are the optimal points for entering and leaving the market. The research concentrates on recognizing stock market long-term upward and downward trends. The key results are obtained with the use of gradient boosting algorithms, XGBoost in particular. The raw data is represented by time series with basic stock market quotes with periods labelled by experts as «Trend» or «Flat». The features are then obtained via various data transformations, aiming to catch implicit factors resulting in the change of stock direction. Modelling is done in two stages: stage one aims to detect endpoints of tendencies (i.e. "sliding windows"), stage two recognizes the tendency itself inside the window. The research addresses such issues as imbalanced datasets and contradicting labels, as well as the need for specific quality metrics to keep up with practical applicability. The model can be used to design an investment strategy though further research in feature engineering and fine calibration is required. This is the reduced version of the research, full text can be found on arxiv.org (arXiv: 2104.09341).

Keywords: XGBoost · Stock market trends · Expert opinion

1 Introduction

The ability to identify stock market trends has obvious advantages for investors. Buying stock on upward trend (as well as selling it in case of downward movement) results in profit, which makes predicting of stock markets a highly attractive topic both for investors and researchers. Sure enough, since the 1970s various methodologies have developed: fundamental analysis, technical analysis, time series econometrics, fuzzy logic, etc. are used to detect trends. Despite the long history, this field, as stated in [1], is still a promising area of research mostly because of the arising opportunities of artificial intelligence. With the rise of machine learning in the early 2010s, researchers started to take interest in applying computer science to financial market problems [1–3]. Machine learning can be thought of as an extension or even an alternative to the traditional statistical methods. Instead of using specific parametric models to explain dependencies, machine learning aims to find hidden or poorly structured dependencies in data by learning from a vast number of examples.

© Springer Nature Switzerland AG 2021
L. Rutkowski et al. (Eds.): ICAISC 2021, LNAI 12855, pp. 414–427, 2021.
https://doi.org/10.1007/978-3-030-87897-9_37

2   Literature Overview

The survey [1] provides a very thorough literature overview, which gives the historical outline of financial time series prediction, as well as the review of the main path in the literature on financial market prediction using machine learning. The authors have studied and classified 57 articles, covering the period from 1991 to 2017. In general, there are three main classes of models - artificial neural networks (ANNs), support vector machines (SVM/SVRs) and various decision tree ensembles (e.g., random forests). As of 2017, the hegemony of ANNs and SVM/SVRs has been observed - the articles based on these models accounted for 86% of articles researched in [1]. Decision trees and Random forests as the main forecasting technique were used only in 7 out of 57 selected articles. This may be explained by the relatively shorter history of scientific research in the case of decision tree ensembles: while many papers on ANNs and SVM/SVRs date back to the early 2000s [1], the papers on applying random forests to stock market forecasting appear mostly in 2010–2015[4–9].

To get a better picture of the current state of research in applying decision trees to stock market forecasting, we provide a brief overview of relevant scientific papers, published in the period between 2018 and early 2021. The classification of studies used in this paper resembles the one suggested in [1] with minor alterations. The search in the Scopus database was performed on 10.10.2020 and was tuned in order to find the most relevant articles from Q1/Q2 journals. In total, twenty papers, containing original models, from four journals were reviewed.

Same as in previous years, artificial neural networks (ANN) prevail as the method of prediction being used as the main algorithm or at least for comparison reasons (like in [10–13]). Only five groups of researchers [14–18] out of twenty do not use ANNs at all. In nine studies [19–26] ANNs were exploited as the only main algorithm. Decision tree algorithms were solely used as the main method only in three studies [11, 13, 14], as well as support vector machines/regressions [10, 17, 18]. Among other models are logistic regression [13, 14, 16], linear regression [10, 27], extreme learning machine and ARMA [12].

Besides, several researchers have exhibited model fusion – an approach discussed in [3] as a process of combining various factors (e.g. models) that can improve the performance and provide useful results. For example, in [13] the authors cascaded logistic regression onto gradient boosted decision trees for forecasting the direction of the stock market. In [28] the researchers combined ANN, SVR, random forest and boosted decision trees to predict short-term stock prices. The stacking methodology was demonstrated in [29] by fusing the outputs of four types of tree ensemble models and four types of deep learning algorithms for stock index forecasting. Model fusion appears also in combining machine learning predictions for the aims of portfolio management [11, 17].

As we can see, ANNs tend to supersede SVM/SVR (which had hegemony according to the previous literature review [1]), while the role of decision tree algorithms (decision trees/random forests/gradient boosting) remains stable, though still small. Therefore the opportunities of these types of models are not yet fully explored, especially considering the velocity of progress in the artificial intelligence field.

It should be noted that despite the common main method different researches have sufficient variations in model structure, problem formalization, and features and labels accordingly. Models also may be applied to different markets, assets and prediction horizons. A certain variation also exists in the selection of performance measures which should ensure the comparability of models.

The analysis shows, that most researchers are concentrated on predicting the direction of the stock market thereby solving a classification problem. Fewer predict prices [22, 23, 27, 28] or returns [11] by solving a regression problem. One study suggests models for both directions and prices [12] predictions introducing a hybrid variational mode decomposition and evolutionary robust kernel extreme learning machine. In all the cases the ground truth variable is extracted from the historical price series. The type of model (classification or regression) also predefines the choice of quality metrics: accuracy, AUC and F1-Score – for classification tasks and MAE, RMSE, MAPE – for regression. Also, the returns of simulated trading can be compared [18, 24].

Among the "direction type" models, the absolute majority aim to predict the next day price movement, which can be considered a short-term objective. Only in four studies the time horizon varies from one week [20, 25] to one month [29] or even several months [16]. Actually, the focus on daily basis predictions prevails within the reviewed articles for all types of target variables, except for the study [19] - but this is due to the specifics of the research problem - detecting crisis episodes.

Another important difference between the suggested models is the choice of features. Though most of the researches used market data (e.g. prices, volumes and the values derived from them - technical analysis indicators, correlations, volatilities and returns) as input variables, two studies used – and with success – only news [14] and investor sentiment [18] for the prediction of price direction. Text features were also used in [25] and [28] – along with technical analysis indicators (TA). Fundamental features (e.g. oil prices) were only found in two studies [24, 29] combined with TA and raw prices. If compared to the previous literature review, the proportion of studies that exploit fundamental variables is relatively smaller - according to [1] they were used in 26% of articles between 1991–2017. But in recent years, we again see the implementation of information fusion approach [3] in the form of feature fusion.

The geography of the research, same as in the previous years, is concentrated on the prediction of USA, Europe and China stock markets, or all of them together. This can be explained by the availability of the data, which is useful for the comparison of models, and by the boost in academic research in China, noted in [1]. Nevertheless in three studies emerging markets were explored: Malaysia [10], India [12] and Brazil [17].

In this paper we apply the gradient boosting algorithm, specifically XGBoost, to the problem of finding the start and endpoints of trends. Gradient boosting is a relatively new algorithm, suggested by [30] in 2016. It is a decision tree ensemble method that was designed to fix some of the faults of random forests. More specific information on the algorithm will be given in Sect. 7.4. Gradient boosting has several technical realizations, the most known being XGBoost, which is extensively used by machine learning practitioners to create state of art data science solutions [31].

Unlike the majority of other researchers, we aim to explore long-term trends, which last several months, not days. The start and endpoints of such trends accordingly are the optimal points for entering and leaving the market. Despite the various technical analysis algorithms and econometrics studies based solely on stock data, some market experts still argue that traders are able to see opportunities of making money (i.e. detecting trends or turning points) that cannot be formally expressed. Thus, using computer science algorithms to learn from successful traders' decisions (and not only stock data) is likely to improve financial market models. The key distinction of our model is that its ground truth vector is fully based on expert opinion data, provided by one major Russian investment company. Unlike other researchers, we do not use a mathematical formula to define a trend, instead, it is defined by an expect as a potentially profitable (or unprofitable) pattern in price dynamics. The major difficulty of this approach is that it is exposed to subjective judgements of the experts. On the other hand, if the experts are successful traders in a certain investment company, it gives the employer the chance to 'digitalize' their exceptional skills and obtain a machine learning algorithm no one else on the market can employ.

The training is performed on historical S&P stock data with additional feature engineering. The model requires only raw historical price data as input. From one point of view, it can be considered a limitation since we ignore fundamental factors and news feed. On the other hand, it makes the model unpretentious in production, since stock data is easily obtainable and can be downloaded into the company's informational systems or directly fed into the model via API.

Technically we are solving a classification problem (it is a "direction type" research), but the standard quality metrics (accuracy, AUC and F1-Score) turn out to be inapplicable because of imbalanced datasets and contradicting labels. For these reasons, the returns of simulated trading are used as the main performance indicator.

3   Problem Formulation

The research was conducted on behalf of one major Russian investment company (the Company). The Company experts have labelled historical S&P stock data, that is, they marked certain consequent periods as "Trend"," Flat" or N/A in a specially designed software with a graphical interface.

Approximately 90% of identified trends last between 40 to 600 business days, which accounts for middle- or long-term tendencies. Initially, the task was to train the model to identify the trend itself (no matter the direction) with the minimum lag from its start, for example, to recognize a 200-day-long trend 20 days after it started. Later, though, it turned out that it is also necessary to distinguish between downward or upward trends to calculate and compare the financial results of different strategies. The model should be independent of any specific stock, market, or time period, after it is properly calibrated it should be equally good for any asset and time. Another important issue is that by learning from historical patterns we aim to identify the current market situation (answer the question: what long-term tendency takes place today?) and we must always bear in mind that future data is unavailable. Breaking this condition will make the modelling results irrelevant, though minor time lag (within a couple of weeks) is quite acceptable.

4   Data Overview

The dataset to explore consists of two sources (Source I and Source II), labelled by 9 experts. The data contains quotes of 705 stocks covering the period from 2005–01–28 to 2017–09-13. The Sources have an intersection in the time period - dates from 2007–08-08 to 2017–05-24, but they have only one intersection in the list of stocks. Only 4 experts labelled both Source I and Source II data, but the second dataset is considered "cleaner" since the experts were more motivated to label data responsibly. The total number of records in both datasets amounts to 9 180 712 pieces packed in 3162 files. Each file contains on average around 2600 daily quotes (Date, Open, High, Low, Close) for a certain period and stockname, labelled by a certain expert.

5   Model Outline

Modelling is done in two stages: stage one aims to detect endpoints of tendencies ("change points", or "turning points"), stage two recognizes the tendency itself inside the window.

The performance of stage one is provided by the model which will be referred to as "ChangePoints". For each data point it returns either a value "1" ("The change of tendency occurred" or "A new window has started") or 0 ("No changepoint"). Due to various reasons, discussed later, currently the ChangePoints predictions are subject to both false positive (mainly) and false negative errors. This is why it can't be used alone and should be backed by the stage two model - "TrendOrFlat". TrendOrFlat is launched when the possible start of a new tendency is detected, i.e. "Change points" returns "1". It is important to note that once the changepoint signal occurs, we come to recognize the starting point of the new tendency, but we do not know how long it will last or when the endpoint will occur. To identify the tendency inside the new window TrendOrFlat model initially analyzes the first few, say 6, days of it, then first 7 days, then 8, etc., returning the values "1" ("Upward trend"), "-1" ("Downward trend") or "0" ("No trend/Flat") for each period. Shortly after the start, TrendOrFlat is more likely to produce incorrect predictions, but as the window widens, the tendency identification should become more and more accurate. The process continues up until a new positive signal from ChangePoints occurs, indicating the start of a new window for which the routine repeats. The final results of modelling are determined upon TrendOrFlat output. The whole process, run for a set of instruments on the chosen time period, will be referred to as 'pipeline'. The calculations were processed on Python 3.5/3.6.

6   Train and Test Sample

In order to evaluate the generalizing power of the model we traditionally divide the dataset into train and test samples. These samples should be independent, otherwise, the quality metrics would be misleadingly inflated. We considered three possible options: standard random split (inapplicable), split by source and split by date. Our final choice is to use 70% of older data as a train set and the remaining 30% as a test set. The split date is October 14th, 2014 if both Source I and Source II are analyzed, and November 6th, 2014 if only Source II is used.

7 ChangePoints Model

We start with the ChangePoints model and its features. It so happened that this model is subject to various complicated issues, while TrendOrFlat works fairly smoothly.

The list of ChangePoints features is presented below (see Table 1), totaling 22 input variables plus 1 target. The choice of input features should comply with the task conditions: the model should be independent of any specific stock, market or time period and can have only a minor time lag. That means we can't use date, stockname and absolute price or volume values as input variables, neither we can scale by such values if they are not available at a given date. On the other hand, to see the change in tendency, we need to compare data before and after the changepoint, so some time lag is inevitable. We opted for 5 business days before and after the current date, since it roughly corresponds to one business week. The use of older data (6 days before today, etc.) is also possible but will result in a number of additional variables. The use of future data determines the lag of the model and its increase is undesirable.

Table 1. ChangePoints features

Features	Description	
	Raw	Logarithmic
Close-1, … Close-5	Ratios of the 5 previous closing prices (today minus 1, minus 2 and so on) to the current closing price	The natural logarithms of the corresponding ratios, or the difference between the logarithms of numerator and denominator
Close1,… Close5	Ratios of the 5 future closing prices (today plus 1, plus 2 and so on) to the current closing price	
Volume-1, … Volume-5	Ratios of the 5 previous trading volumes (today minus 1, minus 2 and so on) to the current trading volume	
Volume1,… Volume5	Ratios of the 5 future trading volumes (today plus 1, plus 2 and so on) to the current trading volume	
High	The ratio of today's maximum price to the closing price	
Low	The ratio of today's minimum price to the closing price	
NewTrigger	Target variable, indicating whether the change in tendency has occurred today or not. It takes the value "1" the day the tendency changes and remains "0" otherwise	

First, the raw features were used for modelling. Later it appeared more appropriate to use natural logarithms instead since they are less subject to the spreads in absolute values. Furthermore, it turned out that experts were using logarithmic price scales when labelling the data.

7.1 Contradicting Labels Issue

Each expert would label a certain data point only once, but different experts can label the same data points and their expertise does not necessarily coincide. There are three major reasons for that:

1. They are different traders with different market expertise.
2. They may have misunderstood the task or completed it without due zeal.
3. As labelling was done in a GUI by mouse-clicking, technical blots could occur.

Altogether this results in several thousands of records with identical input features but different target values. Literally every positive "NewTrigger" record has a negative contradict. To suppress this issue, the following strategies have been introduced:

1 Averaging the experts' opinions by voting.
2 Triggers correction to address the technical blot issue. It forcibly pulls the start of the trend to the local (±5 days) minimum or maximum. Applying these corrections decreased the number of contradictions by 8–18% depending on the dataset.
3 Excluding irrelevant experts. The best results of the modelling were achieved after leaving only experts D and G. These experts were the main stakeholders of the research, with greater experience and motivation.
4 Ignoring the contradictions. The major pitfall of this approach is that traditional classification quality metrics (Accuracy, AUC, Precision, Recall, F-Score) will become irrelevant since there is no "ground truth" anymore [2].

7.2 Imbalanced Dataset Issue

From a formal point of view, the ChangePoints model is a binary classification model. In a perfect situation the proportion between classes should be close to 1:1, otherwise, the observations of the minority class would be "suppressed" by the majority. In highly imbalanced datasets this might lead to the total ignorance of the minority class, while this class can be especially important for the researcher. The traditional classification quality metrics, on the contrary, would perform quite well, unless you drill down into the contingency matrix or evaluate the performance on the minority and majority classes separately. Unfortunately, our dataset is highly imbalanced: records where "NewTrigger" equals "1" (positives) are the minority class with the proportion varying from 78:1 or even 331:1 depending on source filters and the application of averaging. That is natural because we seek changepoints of middle-and long-term tendencies which would happen only once in several hundred business days.

To contend with the imbalanced dataset issue, one can use special oversampling or undersampling techniques, which correct the train dataset. Another option is to alternate the algorithm performance to treat the majority or the minority class differently.

7.3 Quality Evaluation Issue

As we can see, there are at least two important groundings against traditional classification quality metrics (Accuracy, AUC, Precision, Recall, F-Score): they can be misleading both due to "ground truth" contradictions and highly imbalanced dataset. But we will find another reason to consider them irrelevant if we remember the time series issue. In a common classification model, the records are absolutely independent and if we shift the prediction of "1" to the neighboring record, we'll have a completely different result. But with the ChangePoints model, shifting one day back or forward will result in only a very minor change in terms of profit. This brings us to the conclusion that the most relevant quality metrics for such kinds of models are those profit-related.

7.4 Changepoints XGBoost Realization

There are a number of classification algorithms – e.g. logistic regression, Bayesian classifier, SVM, neural networks, decision trees and their ensembles. Out of all these diversities, it is gradient boosted ensembles that account for best modeling results during the last couple of years.

The modelling in this research was completed with XGBoost (version 0.7.post3, Python 3.6), one of the technical realizations of gradient boosting [31]. The key idea of using decision tree ensembles instead of decision trees themselves is to address the problem of overfitting: while being able to perfectly describe the known data, single trees may fail to generalize dependencies. Technically this means that the prediction error will have a relatively low bias, but quite a big variance. It can be shown, though, that ensembles (or compositions) of trees have the same low error bias as single trees, but smaller variance. Moreover, the error variance is the less the lower the correlation between the algorithms. Random forests methodology suggests the way to build decision trees in such a way to make their predictions almost independent, which is achieved by training each algorithm on a different subset of initial data. Specifically, it employs bootstrapping for samples and random subsets for features. In order to keep low error bias, the trees need to be deep, but their number must be enough to provide low variance. These requirements, obviously, result in resource intensity, especially for large datasets. Gradient boosting suggests a different approach for ensembling single algorithms: the trees are added recurrently, and each new tree is designed in a way to correct the error of the composition of previously added trees [30].

Another advantage of the XGBoost algorithm is that it allows controlling the imbalanced dataset issue by directly setting the hyperparameter "scale_pos_weight" to the proportion between the negative (majority) and positive (minority) classes in the dataset ("balance"), ensuring the parity between the classes while training the model [32].

7.5 Selecting Model Hyperparameters

There are two sets of hyperparameters to consider while training the ChangePoints model. The first set of hyperparameters concerns the data we use to train the model: we may opt for logarithmic data instead of raw, choose to average experts' opinions or not, use trigger corrections or deal without them, as well as select certain sources (use both sources or only the "cleaner" second one) and experts.

The second set of hyperparameters are the XGBoost hyperparameters themselves currently totaling a minimum of 20 items. Some of them are purely technical and some can be set to default values, while others require a more attentive approach.

In total, more than 30 ChangePoints models on different datasets were trained. The best XGBoost hyperparameters were searched by grid search/randomized grid search procedures with 5 folds cross-validation. The scoring criterion was set to 'f1_macro' (the mean of the binary F1-score metrics, giving equal weight to each class) to highlight the performance of the minority class [33]. The implementation of gridsearch can be quite time-consuming: for example, 295 fits took approximately 6 days for a 2.4 GHz 32 GB RAM machine to process. After a number of iterations, though, it became obvious that it is mainly 4 hyperparameters that matter and even they do not alter the result crucially. Increasing the maximum tree depth and the L2 regularization term generally improves the model both for 100 and 500 estimators. 500 estimators perform sufficiently better than 100 until the maximum tree depth does not exceed 10. If we set the maximum tree depth to 15, 100 estimators show higher results. Unfortunately, all the F1_marco scores vary only from 0,46 to 0,54, which means the minority class is almost suppressed by the majority. Partly it is due to the label inconsistency issue, and partly - due to the highly imbalanced dataset, and of course – the feature choice might be improper. On the other hand, we should remember, that traditional quality metrics are only rough indicators of the ChangePoints model performance and we can make final conclusions only after we run the pipeline.

The characteristics of the ChangePoints model, which demonstrated the best result on the pipeline, are the following: n_estimators = 500, max_depth = 7, reg_lambda = 3, subsample = 1, learning rate = 0.1, scale_pos_weight = 154, seed = 42, nthread = −1, other parameters set to default. The train set of the best model contained logarithmic data from both sources, but only two experts out of 9 were left. These experts were the main stakeholders of the research, with greater experience and motivation. No averaging of experts' opinions was applied, though trigger correction was.

The overall performance of the model is quite good with AUC = 86.07% and F-Score = 95% on the test set. However, drilling down we can see, that the performance on the minority subset is quite poor (F-score = 8%), mainly due to the extremely low precision (5%). This means our model creates too many "false alarms", detecting non-existent changepoints. The recall value for the minority subset is 58%, meaning that a vast amount of true changepoints is also missed.

8   TrendOrFlat Model

Again, we shall start with the choice of features. We aim to recognize the tendencies inside the windows. The model should be suitable for any stock, market, time and – ideally- window length. More than that, it should be capable of recognizing the tendency from its small patch. Below is the list of suggested features (5 + 1 target) and the intuition behind them (Table 2).

Table 2. TrendOrFlat features

Feature	Description
RegClose	The slope of the linear regression line for daily closing prices (logarithmic or not)
CloseR2	The R2 coefficient of the linear regression line for daily closing prices (logarithmic or not)
RegVol	The slope of the linear regression line for daily volumes (logarithmic or not)
VolR2	The R2 coefficient of the linear regression line for daily volumes (logarithmic or not)
LenTrend	The length of tendency (he width of the window) in business days
NewTypeBool	Target variable. The Boolean analog of the "Type" field. It takes the value "1" in case of trend and "0" otherwise

These variables do not contain absolute values, which makes them suitable for any asset.

8.1 Dataset Overview

The dataset for the TrendOrFlat model is different from that for the ChangePoints since now we are dealing with time periods, not separate trading days. The size of the dataset equals the total number of windows marked by experts. It is important for the model to recognize tendencies by their parts. To ensure this we supplemented the initial dataset, which contained full windows only, by feature vectors extracted from 5, 10, 20,....90% parts of full windows. This approach however might result in the appearance of super-short tendencies which can be considered as noise. To deal with this issue we drop off all the records with LenTrend < 6. This choice is explained by the 5-day lag of the ChangePoints model, so anyway we won't need to recognize the tendencies that last less than 6 days. The total number of records in the train set varies from 97 225 to 232 870 depending on the source filters, with 10 to 20 thousand full windows. The sample is balanced and does not contain contradictions.

8.2 TrendOrFlat Performance Overview

TrendOrFlat model is also a binary classification model. We used the XGBoost algorithm again with the following set of hyperparameters: {n_estimators = 100, max_depth = 5, reg_lambda = 3, learning_rate = 0.2, seed = 42, nthread = −1}. All the other hyperparameters were set to the algorithm default values.

In total more than 10 TrendOrFlat models were trained. Again, the train set for the best model was based on logarithmic data from both sources, but only two best experts were left. The best scores are AUC = 81.86% and F-score = 74% and, as expected, they do not vary within classes. But we will see some difference if we look at the model performance for several tendency proportions. Even for very small parts of trends, the classification quality is around 70% and sufficiently increases up to 90–95% when 80% days or more are shown to the model. That means that the TrendOrFlat model is likely to correct ChangePoints pitfalls. Also note, that if the output of the model is 1("Trend") we can easily determine the trend direction from the RegClose value (positive for upward trends and negative for downward).

9 Pipeline Results

The pipeline logic is described at the beginning of the paper. Here we shall concentrate on the discussion of the specific quality metrics.

Recalling, that the direction of the trend can be determined by the slope of the regression line, we can determine the formula to calculate the profit (in percentage) earned during the time in the short or long position for each simulated deal:

$$Profit = sign(RegClose) \cdot \frac{Close_{t1} - Close_{t0}}{Close_{t0}}, \tag{1}$$

where $Close_{t0}$– is the daily closing price at the beginning of the trend, $Close_{t1}$- the daily closing price at the end of the trend, $sign(RegClose)$ is the sign of the RegClose value, calculated for that trend. We open a long position for the upward trend, and a short – for the downward trend.

Using these statistics, we can compare the behavior of the concurrent models for the same stocks and time periods. But the results vary significantly from stock to stock, so we will need more general metrics to compare (Table 3).

Table 3. The additional pipeline performance indicators

Indicator	Description
Profit	The sum of all profits earned while in position (for all the stocks)
Days_in	The total number of business days in position (for all stocks)
Times_in	The number of times the position was opened (for all stocks)
DayProfit	The profit per one day in position, %: DayProfit = Profit/ Days_in
YearProfit	dayProfit scaled per annum, %: YearProfit = DayProfit * 250, where 250 is the average number of business days in a year
YearProfit_avg	The average annual profit, including the days not in position, %: YearProfit_avg = Profit/number of data points * 250

For the purpose of comparing models, the last two indicators - *YearProfit* and *YearProfit_avg* - are the most informative, because they are independent of the length of time period and the number of stocks in the pipeline. From a business point of view, it might also be important to see how many times we opened the positions or what was the proportion between short or long for each stock, because it influences the additional costs of trading.

The main pitfall of *YearProfit* and *YearProfit_avg* metrics is that they are not normalized, so solely by their values we can't definitely say what is a good result. But we can compare our results with those, calculated on the same datasets labelled by real experts. From simulating traders' activities, we can conclude, that experts D and G show the best results, while the performance of the "average" expert (a product of our struggle with contradictions, when experts' opinions were averaged) is comparingly poor. This actually explains, why models, based on D and G experts labels, perform better than others, and why averaging of labels didn't much improve the modelling results (though excluded contradictions).

It was mentioned previously, that our ChangePoints models produce two much "false alerts", indicating non-existent changepoints. To deal with this we can increase the classification thresholds (default 0.5) to 0.6 or even 0.8 values. A couple of such experiments was run, though, unfortunately, this approach didn't bring much improvement: as the model also produces false negatives, we start missing the true changepoints.

Totally 15 pipelines were tested. The best results are the following: *YearProfit* = 28.8% and *YearProfit_avg* = 6.9%. Of course, they are much more modest than the ground truth: the experts, who were dealing with historical data, or "saw the future", achieved at least twice more. Nevertheless, having nearly 30% on investment per annum looks quite impressive. Unfortunately, the model still did miss a lot of opportunities (76% of records are predicted as flat), so *YearProfit_avg* is not too big. Though if we add overnight interest paid on flat periods (say 7%) our average profit will reach 7% * 0.76 + 6,9% = 12,2% per annum.

10   Conclusion

The model presented in the research can be used by both individual and institutional investors. It produces "buy" and "sell" signals when starting or endpoints of trends are identified. The profit earned on days in position can reach 28.8% per annum, but, definitely, the result can be improved. Unlike in the traditional approaches, the labels (trend or flat) are not derived from prices but filled manually by experts who worked with stock data as with images. The task is quite challenging since we actually try to 'digitalize' successful traders' skills and we can only compare the performance of the model with the performance of the experts themselves. The comparison with the results of other researchers would be inadequate since we use a completely different source for ground truth.

There are several directions for this work:

- Implementing other approaches to deal with contradicting labels and the imbalanced datasets – two major issues which influence the quality of the ChangePoints model. Though the XGboost algorithm tries to correct the proportion between classes, obviously this is not enough. Perhaps certain undersampling technics might improve the situation.
- Selecting other sets of features for the ChangePoints and TrendOrFlat models. Various combinations of technical indicators should be tried for ChangePoints models and probably different time lags and threshold levels. As for TrendOrFlat, the improvement

should be concentrated on early tendency identification, i.e. recognizing trends shortly after their start.

- Experimenting with trading strategies, for example, excluding short positions as less profitable and more complicated, ignoring flat periods - that is, closing positions only when we have strong evidence of the new trend, etc.
- And finally, totally changing the model structure and using other machine learning algorithms, for example, convolutional neural networks, can also sufficiently improve the model.

References

1. Henrique, B.M., Sobreiro, V.A., Kimura, H.: Literature review: machine learning techniques applied to financial market prediction. Expert Syst. Appl. **124**, 226–251 (2019). https://doi.org/10.1016/j.eswa.2019.01.012
2. Brink, H., Richards, J.: Real-World Machine Learning Version 4 (2014)
3. Thakkar, A., Chaudhari, K.: Fusion in stock market prediction: a decade survey on the necessity, recent developments, and potential future directions. Inf. Fusion. **65**, 95–107 (2021). https://doi.org/10.1016/j.inffus.2020.08.019
4. Nair, B.B., Dharini, N.M., Mohandas, V.P.: A stock market trend prediction system using a hybrid decision tree-neuro-fuzzy system. In: Proceedings - 2nd International Conference on Advances in Recent Technologies in Communication and Computing, ARTCom 2010 (2010)
5. Nair, B.B., Mohandas, V., Sakthivel, N.R.: A decision tree- rough set hybrid system for stock market trend prediction. Int. J. Comput. Appl. (2010). https://doi.org/10.5120/1106-1449
6. Nair, B.B., Sakthivel, V.P.M.N.R.: A genetic algorithm optimized decision tree - SVM based stock market trend prediction system. Int. J. **2**(9), 2981-2988 (2010)
7. Paliyawan, P.: Stock market direction prediction using data mining classification. ARPN J. Eng. Appl. Sci. **5**, 6 (2015)
8. Patel, J., Shah, S., Thakkar, P., Kotecha, K.: Predicting stock and stock price index movement using Trend Deterministic Data Preparation and machine learning techniques. Expert Syst. Appl. **42**(1), 259–268 (2015). https://doi.org/10.1016/j.eswa.2014.07.040
9. Qin, Q., Wang, Q.-G., Li, J., Ge, S.S.: Linear and nonlinear trading models with gradient boosted random forests and application to Singapore stock market. J. Intell. Learn. Syst. Appl. **5**, 1–10 (2013). https://doi.org/10.4236/jilsa.2013.51001
10. Ismail, M.S., Md Noorani, M.S., Ismail, M., Abdul Razak, F., Alias, M.A.: Predicting next day direction of stock price movement using machine learning methods with persistent homology: evidence from Kuala Lumpur stock exchange. Appl. Soft Comput. J. **93**, 106422 (2020). https://doi.org/10.1016/j.asoc.2020.106422
11. Ma, Y., Han, R., Wang, W.: Portfolio optimization with return prediction using deep learning and machine learning. Expert Syst. Appl. **165**, 113973 (2021). https://doi.org/10.1016/j.eswa.2020.113973
12. Bisoi, R., Dash, P.K., Parida, A.K.: Hybrid variational mode decomposition and evolutionary robust kernel extreme learning machine for stock price and movement prediction on daily basis. Appl. Soft Comput. J. **74**, 652–678 (2019). https://doi.org/10.1016/j.asoc.2018.11.008
13. Zhou, F., Zhang, Q., Sornette, D., Jiang, L.: Cascading logistic regression onto gradient boosted decision trees for forecasting and trading stock indices. Appl. Soft Comput. J. **84**, 105747 (2019). https://doi.org/10.1016/j.asoc.2019.105747
14. Yang, J., Zhao, C., Yu, H., Chen, H.: Use GBDT to predict the stock market. Procedia Comput. Sci. **174**, 161–171 (2020). https://doi.org/10.1016/j.procs.2020.06.071

15. Soujanya, R., Akshith Goud, P., Bhandwalkar, A., Anil Kumar, G.: Evaluating future stock value asset using machine learning. Mater. Today Proc. **33**(7), 4808–4813 (2020)
16. Lee, T.K., Cho, J.H., Kwon, D.S., Sohn, S.Y.: Global stock market investment strategies based on financial network indicators using machine learning techniques. Expert Syst. Appl. **117**, 228–242 (2019). https://doi.org/10.1016/j.eswa.2018.09.005
17. Paiva, F.D., Cardoso, R.T.N., Hanaoka, G.P., Duarte, W.M.: Decision-making for financial trading: A fusion approach of machine learning and portfolio selection. Expert Syst. Appl. **115**, 635–655 (2019). https://doi.org/10.1016/j.eswa.2018.08.003
18. Yang, S.Y., Yu, Y., Almahdi, S.: An investor sentiment reward-based trading system using Gaussian inverse reinforcement learning algorithm. Expert Syst. Appl. **114**, 388–401 (2018). https://doi.org/10.1016/j.eswa.2018.07.056
19. Chatzis, S.P., Siakoulis, V., Petropoulos, A., Stavroulakis, E., Vlachogiannakis, N.: Forecasting stock market crisis events using deep and statistical machine learning techniques. Expert Syst. Appl. **112**, 353–371 (2018). https://doi.org/10.1016/j.eswa.2018.06.032
20. Long, J., Chen, Z., He, W., Wu, T., Ren, J.: An integrated framework of deep learning and knowledge graph for prediction of stock price trend: an application in Chinese stock exchange market. Appl. Soft Comput. J. **91**, 106205 (2020). https://doi.org/10.1016/j.asoc.2020.106205
21. Moews, B., Ibikunle, G.: Predictive intraday correlations in stable and volatile market environments: Evidence from deep learning. Phys. A Stat. Mech. its Appl. **547**, 1–59 (2020). https://doi.org/10.1016/j.physa.2020.124392
22. Vijh, M., Chandola, D., Tikkiwal, V.A., Kumar, A.: Stock closing price prediction using machine learning techniques. Procedia Comput. Sci. **167**, 599–606 (2020)
23. Zhou, F., Zhou, H., Yang, Z., Yang, L.: EMD2FNN: a strategy combining empirical mode decomposition and factorization machine based neural network for stock market trend prediction. Expert Syst. Appl. **115**, 136–151 (2019). https://doi.org/10.1016/j.eswa.2018.07.065
24. Hoseinzade, E., Haratizadeh, S.: CNNpred: CNN-based stock market prediction using a diverse set of variables. Expert Syst. Appl. **129**, 273–285 (2019). https://doi.org/10.1016/j.eswa.2019.03.029
25. Picasso, A., Merello, S., Ma, Y., Oneto, L., Cambria, E.: Technical analysis and sentiment embeddings for market trend prediction. Expert Syst. Appl. **135**, 60–70 (2019). https://doi.org/10.1016/j.eswa.2019.06.014
26. Chandrinos, S.K., Sakkas, G., Lagaros, N.D.: AIRMS: a risk management tool using machine learning. Expert Syst. Appl. **105**, 34–48 (2018). https://doi.org/10.1016/j.eswa.2018.03.044
27. Soujanya, R., Goud, P.A., Bhandwalkar, A., Kumar, G.A.: Evaluating future stock value asset using machine learning. Mater. Today Proc. **33**, 4808–4813 (2020). https://doi.org/10.1016/j.matpr.2020.08.385
28. Weng, B., Lu, L., Wang, X., Megahed, F.M., Martinez, W.: Predicting short-term stock prices using ensemble methods and online data sources. Expert Syst. Appl. **112**, 258–273 (2018). https://doi.org/10.1016/j.eswa.2018.06.016
29. Jiang, M., Liu, J., Zhang, L., Liu, C.: An improved stacking framework for stock index prediction by leveraging tree-based ensemble models and deep learning algorithms. Phys. A Stat. Mech. its Appl. **541**, 122272 (2020). https://doi.org/10.1016/j.physa.2019.122272
30. Chen, T., Guestrin, C.: XGBoost: a scalable tree boosting system. In: Proceedings of the ACM SIGKDD International Conference on Knowledge Discovery and Data Mining (2016)
31. XGBoost on GitHub Repository. https://github.com/dmlc/xgboost/tree/master/demo#machine-learning-challenge-winning-solutions
32. XGBoost Parameters. https://xgboost.readthedocs.io/en/latest/parameter.html
33. Scikit-learn Metrics and Scoring. https://scikit-learn.org/stable/modules/model_evaluation.html

Bioinformatics, Biometrics and Medical Applications

A New Statistical Iterative Reconstruction Algorithm for a CT Scanner with Flying Focal Spot

Robert Cierniak[(⊠)] and Piotr Pluta

Department of Intelligent Computer Systems, Czestochowa University of Technology,
Armii Krajowej 36, 42-200 Czestochowa, Poland
robert.cierniak@pcz.pl
http://www.iisi.pcz.pl/

Abstract. This work is related to the originally formulated 3D statistical model-based iterative reconstruction algorithm adopted to computed tomography with flying focal spot. This new reconstruction method is based on a continuous-to-continuous data model, where the forward model is formulated as a shift invariant system. The proposed approach resembles the well-known Feldkamp (FDK) algorithm, which cannot be used with a flying focal spot scanner as the paths of the X-rays used are not equi-angularly distributed. In this situation, a so-called "nutating" reconstruction algorithm is usually used, which is based on a rebinning methodology, thus transforming the reconstruction problem to that of a parallel beam system. Our approach has some significant advantages compared with the FBP methods. Moreover, although our method belongs to the category of iterative reconstruction approaches, thanks to the fact that our proposed model is derived as a shift invariant system, it is possible to use an FFT algorithm to accelerate the calculations that have to be performed. Because of this, we can obtain diagnostic images in a time comparable to that of FBP methods. Computer simulations have shown that the reconstruction method presented here outperforms referential FBP methods with regard to the image quality obtained and can be competitive in terms of time of calculation.

Keywords: Image reconstruction from projections · X-ray computed tomography · Statistical reconstruction algorithm · Flying focal spot

1 Introduction

Despite the passing of many years since the introduction of the first CT device, the search for new designs continues. At the beginning of the XXIth century,

This work was partly supported by The National Centre for Research and Development in Poland (Research Project POIR.01.01.01-00-0463/17).
The authors thank Dr. Cynthia McCoullough and the American Association of Physicists in Medicine for providing the Low-Dose CT Grand Challenge dataset.

L. Rutkowski et al. (Eds.): ICAISC 2021, LNAI 12855, pp. 431–441, 2021.
https://doi.org/10.1007/978-3-030-87897-9_38

one of these designs was for a medical spiral scanner with a flying focal spot [1]. The main aim of this new technique was to increase the sampling density of the integral lines in the reconstruction planes, and in the z-direction, to increase the sampling density in just the z-direction. Of course, this has meant that new reconstruction methods have had to be formulated that allow for the use of projections obtained from such scanners. From a practical point of view, one of the most important among these reconstruction methods is the adaptive multiple plane reconstruction (AMPR) method, (see e.g. [2]). The AMPR conception, which can be classified as a nutating reconstruction method, is a development of the advanced single slice rebinning (ASSR) method (see e.g. [1,3]). However, this rebinning approach has several drawbacks. One of these is its limited ability to suppress noise, caused by the linear nature of the signal processing. This means that it cannot be considered for systems that aim to reduce the dose of X-ray radiation absorbed by patients during examinations. On the other hand, the most interesting research directions in this area are statistical approaches, especially those belonging to the model-based iterative reconstruction (MBIR) group of methods [4]. In these conceptions, a probabilistic model of the measurement signals is formulated (a methodology based on the D-D data model). Unfortunately, up to now, the MBIR methods used commercially have some very serious drawbacks: the calculation complexity of the reconstruction problem is huge (proportional to I^4, where I is the image resolution), the statistical reconstruction procedure based on this methodology necessitates simultaneous calculations for all the voxels in the range of the reconstructed 3D image, the size of the forward model matrix $\mathbf{A}$ is huge, and the reconstruction problem is extremely ill-conditioned. These drawbacks can be reduced by using an approach that is based on a continuous-to-continuous (C-C) data model being in fact a shift-invariant system (this type of system is also used in other applications [5]).

It has been previously proposed formulations of the reconstruction problems for parallel scanner geometry [6], and for the spiral cone-beam scanner [7]. In this paper, we present the concept of a statistical reconstruction algorithm that uses spiral cone-beam projections directly, similar to FDK-type algorithms, for CT scanners with a flying focal spot.

2   Scanner Geometry

In our work, we have taken into consideration the CT scanner with flying focal spot. A general view of this system, with fundamental parameters, is shown in Fig. 1.

The measurement system is rotated around the z axis, and in each rotation the view angle α is incremented by a value Δ_α (for every focal spot position separately). At a given angle α_θ, all the detectors placed on the screen take measurements and in this way projections $p(\alpha_\theta, \beta_\psi, \dot{z}_k)$ are obtained, where the integral line runs from the focus of the screen to a given detector identified by the pair of indexes ψ and k, as depicted in Fig. 1. Other important nominal parameters of this system are the focus-isocenter distance R_f and the normal focus-detector distance R_{fd}.

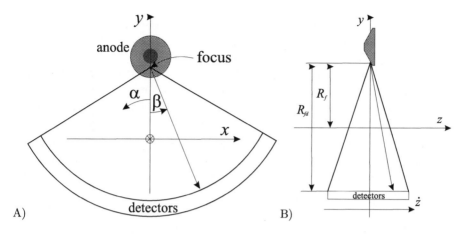

Fig. 1. General view of the measurement system of the scanner: in the reconstruction plane (A); in the z-direction (B).

In the case we analyzed, a measurement system was used that allows for deflections of the focal spot both in the reconstruction plane (αFFS) and in the z-direction (zFFS). The flying focal spot in the α-direction aims to improve the resolution of the sampling of the integral lines in the reconstruction plane, and in the z-direction aims to improve the resulting resolution of the reconstruction planes in just the z-direction. In this technique, this is possible by switching the focal spot between different places on the anode of the X-ray tube. These specific points on the anode are pointed out in Fig. 2.

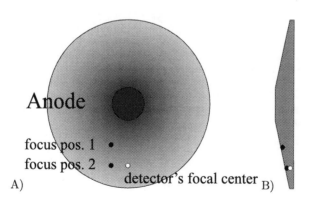

Fig. 2. Topology of the focal spots on the anode: orientation in the reconstruction plane (αFFS).

Changing the focal spot during the measurement process also affects the parameters of the projection system, i.e. the focus-isocenter distance R_f, the rotation angle α and the z position of the focal spot. In the case of focus position No. 1, only an increment in α is specified, namely δ'_α, and in the case of focus position No. 2, all three increments have to be taken into account: δ''_α, δ''_{R_f} and $\delta''_{\dot z}$. The adjusted angles of rotation can be determined using the simple relations $\alpha' = \alpha + \delta'_\alpha$ and $\alpha'' = \alpha + \delta''_\alpha$ for the focal spots No. 1 and 2, respectively. The adjusted positions of the detector with index k can be calculated using the relation $\dot z''_k = \dot z_k + \delta''_{\dot z}$ (for the focal spot No. 2), where $\dot z_k$ describes the positions of these detectors relative to the nominal situation. As for the focal spot position, the nominal focus-isocenter distance is increased easily, i.e. $R'_f = R_f + \delta'_{R_f}$. It should be noted that the integral lines are no longer equi-angularly distributed, for both focal spot positions.

3   Reconstruction Algorithm

In our approach, we utilize an optimization problem, based on the well-known maximum-likelihood (ML) estimation method, which is consistent with the following continuous-to-continuous data model:

$$\mu_{\min} = \arg \min_\mu \left(\int\limits_x \int\limits_y \left(\int\limits_{\bar x} \int\limits_{\bar y} \mu\left(\bar x, \bar y\right) \cdot h_{\Delta x, \Delta y} d\bar x d\bar y - \tilde\mu\left(x, y\right) \right)^2 dx dy \right), \quad (1)$$

where $\tilde\mu\left(x, y\right)$ is an image obtained by way of a back-projection operation, $\mu\left(x, y\right)$ is a reconstructed image, and the kernel $h_{\Delta x, \Delta y}$ is precalculated according to the formula:

$$h_{\Delta x, \Delta y} = \int\limits_0^{2\pi} int\left(\Delta x \cos \alpha + \Delta y \sin \alpha\right) d\alpha, \quad (2)$$

and $int\left(\Delta s\right)$ is a linear interpolation function used during the back-projection operation, as will be shown later.

It is clear that we have to apply the problem shown above in a discrete form, i.e. as follows:

$$\mu_{\min} = \arg \min_\mu \left(\sum_{i=1}^I \sum_{j=1}^I \left(\sum_{\bar i=1}^I \sum_{\bar j=1}^I \mu\left(x_{\bar i}, y_{\bar j}\right) \cdot h_{\Delta i, \Delta j} - \tilde\mu\left(x_i, y_j\right) \right)^2 \right), \quad (3)$$

where I is a dimension of the processed image, and the discrete kernel $h_{\Delta i, \Delta j}$ is precalculated according to the formula:

$$h_{\Delta i, \Delta j} = \frac{\Delta_\alpha}{\Delta_s^2} \sum_{\theta=0}^{\Theta-1} int\left(\Delta i \cos \theta \Delta_\alpha + \Delta j \sin \theta \Delta_\alpha\right), \quad (4)$$

and $int\left(\Delta s\right)$ is an interpolation function used during the back-projection operation.

It should be underlined that the presence of a shift-invariant system in the reconstruction problem (1) implies that this system is better conditioned than the least squares problem present in the D-D approach [8].

The reconstruction approach presented here is a full 3D reconstruction algorithm for spiral cone-beam scanner geometry. This method is based on the direct use of the projections obtained in the scanner, as in the generalized Feldkamp algorithm. The reconstruction procedure proposed by us consists of two steps: a back-projection operation and an iterative reconstruction procedure. The back-projection operation is described by the following relations:

$$
\tilde{\mu}'(x,y) \approx \frac{1}{2} \int_0^{2\pi} \int_{-\beta_m}^{\beta_m} \frac{R'_{fd}}{\sqrt{\left(R'_{fd}\right)^2 + (\dot{z}')^2}} p_{foc_1}\left(\beta, \alpha', \dot{z}\right) int_L(\Delta\beta) \, d\beta d\alpha', \quad (5)
$$

for the first focal spot position, where

$$
R'_{fd} = R_{fd}\sin\beta_\psi + \delta_{\alpha'}, \quad (6)
$$

and

$$
\tilde{\mu}''(x,y) \approx \frac{1}{2} \int_0^{2\pi} \int_{-\beta_m}^{\beta_m} \frac{R''_{fd}}{\sqrt{R''_{fd}{}^2 + (\dot{z}'')^2}} \cdot p_{foc_2}\left(\beta, \alpha'', \dot{z}\right) int_L(\Delta\beta) \, d\beta d\alpha'', \quad (7)
$$

for the second focal spot position, where:

$$
R''_{fd} = \sqrt{\left(R_{fd}\cos\beta_\psi + \delta_{R''_f}\right)^2 + (R_{fd}\sin\beta_\psi + \delta_{\alpha''})^2}. \quad (8)
$$

Having images $\tilde{\mu}'(x,y)$ and $\tilde{\mu}''(x,y)$, it is possible to perform the iterative reconstruction procedure accordingly, which is described by the formula (1). This is carried out using the sum of these two images, i.e.:

$$
\tilde{\mu}(x,y) = \tilde{\mu}'(x,y) + \tilde{\mu}''(x,y). \quad (9)
$$

There are some technical problems associated with the performing of the interpolation needed during the back-projection operation, especially regarding the second focal spot position. The distribution of the integral lines is no longer equiangular for this position and the interpolation is performed at a different resolution in every case. For details regarding the method for determining the geometrical parameters of a scanner with a flying focal spot see e.g. [10] and [11].

The complete algorithm, that is proposed here, is described schematically in Fig. 3. It is worth noting that an FFT algorithm can be implemented in this algorithm to realize the convolutions in the frequency domain, which significantly accelerates the necessary calculations.

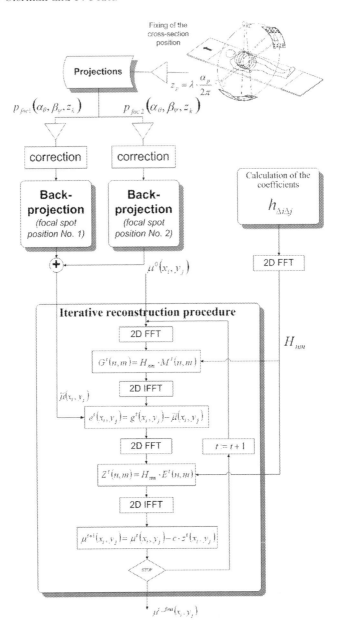

Fig. 3. Statistical iterative reconstruction algorithm.

4   Experimental Results

In our experiments, we have used projections obtained from a Somatom Definition AS+ (helical mode) scanner with the following parameters: reference tube potential 120 kVp and quality reference effective 200 mAs, $R_{fd} = 1085.6$ mm, $R_f = 595$ mm, number of views per rotation $\Psi = 1152$, number of pixels in detector panel 736, detector dimensions were 1.09 mm $\times$ 1.28 mm. We have fixed the size of the processed image at 512×512 pixels. A discrete representation of the matrix $h_{\Delta x, \Delta y}$ was established before the reconstruction process was started, and these coefficients were fixed (transformed into the frequency domain) for the whole iterative reconstruction procedure. The image obtained after the back-projection operations was then subjected to a process of reconstruction (optimization) using an iterative procedure. A specially prepared result of an FBP reconstruction algorithm was chosen as the starting point of this procedure (using projections obtained from the first focal spot position. It is worth noting that our reconstruction procedure was performed without any regularization regarding the objective function from (1).

In our experiments, we have performed calculations necessary to realize the iterative reconstruction using two approaches to hardware implementation, namely: 1) based on GPU type nVidia Titan V; 2) using a CPU with 10 cores (i.e. with an Intel i9-7900X BOX/3800 MHz processor). There are presented results obtained using the mentioned above realizations in Table 1. According to an assessment of the quality of the obtained images by a radiologist, 7000 iterations are enough to provide an acceptable image. The same results were achieved for both hardware implementations after less than 7.5 s (in the case of the GPU, it is less than 5 s), for the CPU and GPU implementations.

Table 1. Comparison of the computation times for the different realizations of the iterative reconstruction procedure.

Number of iterations	GPU	CPU
1000	850 ms	705 ms
7000	7 443 ms	4 932 ms
10000	9 945 ms	7 045 ms

One can compare the results obtained by assessing the views of the reconstructed images in Fig. 4, where the full dose projections were used, and in Fig. 5 where the quarter-dose projections were considered. In both cases, Figures (A) depict reconstructed images obtained using the standard FDK algorithm (with linear interpolation function and Shepp-Logan kernel), Figures (B) and (C) present reconstructed images where the statistical approach presented in this paper was used: only measurements performed using the first focal spot position (B), all measurements were used (C).

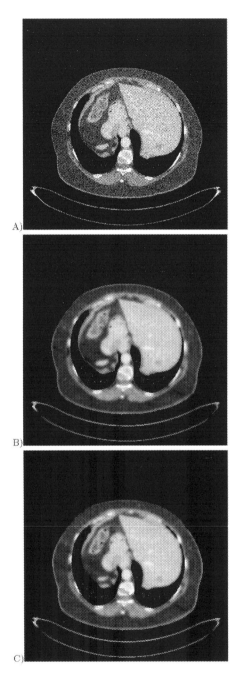

Fig. 4. View of the reconstructed image (a case with pathological changes in the liver) using full-dose projections with application of: the standard FDK algorithm (A); the statistical method presented in this paper (the first focal spot position) (B); the statistical method presented in this paper (both focal spot positions) (C).

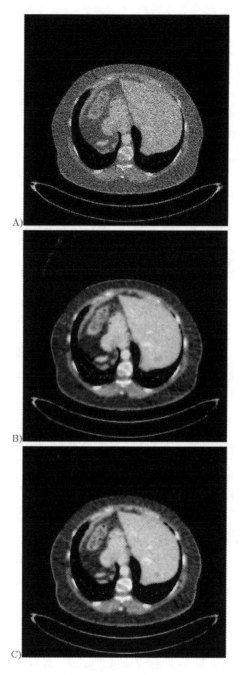

Fig. 5. View of the reconstructed image (a case with pathological changes in the liver) using quarter-dose projections with application of: the standard FDK algorithm (A); the statistical method presented in this paper (the first focal spot position) (B); the statistical method presented in this paper (both focal spot positions) (C).

5   Conclusion

An original statistical iterative reconstruction algorithm that can be used in scanners with flying focal spot has been shown here. We have carried out experiments that have proved that our reconstruction method is very fast, mainly thanks to the use of an FFT algorithm and efficient programming techniques. Regarding the complexity of our approach, it should be emphasized that if the image resolution is assumed to be $I \times I$ pixels, this complexity is proportional to $I^2 \log_2 I$. One can note that the iterative reconstruction procedure was implemented without introducing any additional regularization term, using only an early stopping regularization strategy. Our conception yields satisfactory results regarding the quality of the reconstructed images, and, to put into perspective, significantly reduces the dose of X-rays absorbed by the patients. Unfortunately, we cannot compare our method with any nutating approach because those methods give images with changed shapes, and they use projections performed within a rotation angle of 180°, whereas in our method we can use all projections from a full revolution of the measurement system. This means a halving of the dose absorbed by a patient, by the use of our reconstruction method. Our method is very easy to implement and open to the use of different modes of focal spot movement (both z and angle flying). Additionally, it should be underlined that the price of the hardware used in our experiments is relatively low (about 5000 USD in both cases). Further research will be devoted to integrating computational intelligence methods (e.g. [12–14]) into the approach presented here.

References

1. Kachelriess, M., Knaup, M., Penssel, C., Kalender, W.: Flying focal spot (FFS) in cone-beam CT. IEEE Tran. Nucl. Sci. **53**(3), 1238–1247 (2006)
2. Flohr, T., Stierstofer, K., Bruder, H., Simon, J., Polacin, A., Schaller, S.: Image reconstruction and image quality evaluation for a 16-slice CT scanner. Med. Phys. **30**(5), 832–845 (2003)
3. Cierniak, R., Pluta, P., Kaźmierczak, A.: A practical statistical approach to the reconstruction problem using a single slice rebinning method. J. Artif. Intell. Soft Comput. Res. **10**(2), 137–149 (2021)
4. Zhou, Y., Thibault, J.-B., Bouman, C.A., Hsieh, J., Sauer, K.D.: Fast model-based x-ray CT reconstruction using spatially non-homogeneous ICD optimization. IEEE Trans. Image Process. **20**, 161–175 (2011)
5. Pawlak, M., Panesar, G.S., Korytkowski, M.: A novel method for invariant image reconstruction. J.of Artif. Intell. Soft Comput. Res. **11**(1), 69–80 (2021)
6. Cierniak, R.: An analytical iterative statistical algorithm for image reconstruction from projections. Appl. Math. Comput. Sci. **24**, 7–17 (2014)
7. Cierniak, R.: Analytical statistical reconstruction algorithm with the direct use of projections performed in spiral cone-beam scanners. In: the 5th International Meeting on Image Formation in X-Ray Computed Tomography, Salt Lake City, pp. 293–296 (2018)
8. Cierniak, R., Lorent, A.: Comparison of algebraic and analytical approaches to the formulation of the statistical model-based reconstruction problem for x-ray computed tomography. Comput. Med. Imaging and Graph. **2**, 19–27 (2016)

9. Bouman, C., Sauer, K.: A unified approach to statistical tomography using coordinate descent optimization. IEEE Trans. Image Process. **5**, 480–492 (1996)

10. Flohr, T.G., Stierstofer, K., Ulzhaimer, S., Bruder, H., Promak, A.N., McCollough, C.H.: Image reconstruction and image quality evaluation for a 64-slice CT scanner with z-flying focal spot. Med. Phys. **32**(8), 2536–2547 (2005)

11. Flohr, T.G., Bruder, H., Stierstofer, K., Petersilka, M., Schmidt, B., McCollough, C.H.: Image reconstruction and image quality evaluation for a dual source CT scanner. Med. Phys. **35**(12), 5882–5897 (2008)

12. Bilski, J., Kowalczyk, B., Marchlewska, A., Zurada, J.M.: Local Levenberg-Marquardt algorithm for learning feedforwad neural networks. J. Artif. Intell. Soft Comput. Res. **10**(4), 229–316 (2020)

13. Duda, P., Jaworski, M., Cader, A., Wang, L.: On training deep neural networks using a streaming approach. J. Artif. Intell. Soft Comput. Res. **10**(1), 15–26 (2020)

14. El Zini, J., Rizk, Y., Awad, M.: An optimized parallel implementation of non-iteratively trained recurrent neural networks. J. Artif. Intell. Soft Comput. Res. **11**(1), 33–50 (2020)

A New Multi-filter Framework with Statistical Dense SIFT Descriptor for Spoofing Detection in Fingerprint Authentication Systems

Rodrigo Colnago Contreras[1]([✉])(iD), Luis Gustavo Nonato[1](iD), Maurílio Boaventura[2](iD),
Inês Aparecida Gasparotto Boaventura[2](iD), Bruno Gomes Coelho[3](iD),
and Monique Simplicio Viana[4](iD)

[1] University of São Paulo, São Carlos, SP 13566-590, Brazil
contreras@usp.br, gnonato@icmc.usp.br
[2] São Paulo State University, São José do Rio Preto, SP 15054-000, Brazil
maurilio.boaventura@unesp.br, ines@ibilce.unesp.br
[3] New York University, New York, NY 10012, USA
bgc5612@nyu.edu
[4] Federal University of São Carlos, São Carlos, SP 13565-905, Brazil
monique.viana@ufscar.br

Abstract. Fingerprint-based authentication systems represent what is most common in biometric authentication systems. Today's simplest tasks, such as unlocking functions on a personal cell phone, may require its owner's fingerprint. However, along with the advancement of this category of systems, have emerged fraud strategies that aim to guarantee undue access to illegitimate individuals. In this case, one of the most common frauds is that in which the impostor presents manufactured biometry, or spoofing, to the system, simulating the biometry of another user. In this work, we propose a new framework that makes two filtered versions of the fingerprint image in order to increase the amount of information that can be useful in the process of detecting fraud in fingerprint images. Besides, we propose a new texture descriptor based on the well-known dense Scale-Invariant Feature Transform (SIFT): the statistical dense SIFT, in which their descriptors are summarized using a set of signal processing functions. The proposed methodology is evaluated in benchmarks of two editions of LivDet competitions, assuming competitive results in comparison to techniques that configure the state of the art of the problem.

Keywords: Liveness detection · Spoofing detection · Fingerprint authentication system · Dense SIFT · Pattern recognition

1 Introduction

Currently, confirmation of a person's identity is indispensable in carrying out the simplest tasks, such as logging in to a webpage, as well as in the most important routines, such as freeing access in work environments that demand a high degree of security and employees control [1]. Thus, studies on biometric authentication systems (BAS) [18]

© Springer Nature Switzerland AG 2021
L. Rutkowski et al. (Eds.): ICAISC 2021, LNAI 12855, pp. 442–455, 2021.
https://doi.org/10.1007/978-3-030-87897-9_39

are increasingly needed, which validate the identity of users of certain services through the recognition of properties that preserve the individuality of each person. In this case, these properties are defined mainly by two types of characteristics [38]: physiological, such as fingerprints [3], faces [33], ears [5], etc.; and behavioral, such as voice [46], walking mode [26], etc. Also, systems that make use of more than one characteristic of the individual to perform authentication are not uncommon [11].

In the universe of BAS, we highlight those who use fingerprints [29] to perform the recognition of individuals, which are called fingerprint authentication system (FAS). This biometry is the most used in this context due to the ease of conducting its collection and the high amount of techniques [19] and software packages [41] available in the literature that helps in the improvement of theories involving this theme. Thus, BAS based on fingerprints are biometric systems that receive more and more attention in the academic environment and business solutions.

The convenience provided by the use of fingerprints as biometrics, combined with advances in image recognition and classification technologies, has provided a considerable expansion in the use of FAS in practical solutions. As an example, we can mention its popularization in usual applications in which it was more common to use passwords, such as access tasks involving smartphones [44]. However, the threat of fraud, that is, *spoofing attacks*, remains a disadvantage in this type of system since the security of such applications can be compromised by imposters [20]. Notably, one of the most common forms of FAS fraud is that known as *spoofing presentation attack* (SPA) [24], which consists of the improper presentation of a fingerprint manufactured using synthetic materials [7] to simulate the biometrics of a legitimate user of the system. To encourage the development of techniques to soften this situation, several competitions have been proposed in recent years [42], the first being held in 2009 [25]. The Liveness Detection (LivDet) Competition gave rise to a series of databases composed of a large volume of examples of legitimate fingerprints and synthetic fingerprints produced from different materials. These bases currently form the benchmarks that are considered for evaluating methodologies in works on this theme.

Recently, some methodologies have been proposed in an attempt to circumvent the SPA threat in FASs. Most of these strategies are based mainly on three categories of methods [30]:

C_1 methods based on texture descriptor analysis and other characteristics inherent to digital printing;
C_2 methods composed of deep learning networks,
C_3 hybrid methodologies or framework-based methods.

The first category of techniques, which is the most widely used in this theme [2], is defined by creating the artificial characteristics, or hand-crafted features (HCF) [27], extracted from the image of a fingerprint to perform the classifier training. Among these features, those obtained from the texture descriptors [16] and image quality measures [9] are recurrent. The methods that define the second category of problems are those that analyze the natural characteristics of fingerprints and generally make use of deep learning neural networks (DNN) [43], for which those particularly known as Deep Convolutional Neural Network (D-CNN) [31] present good results in the state-of-the-art for SPA treatment. Finally, the last category of methods is defined by strategies that make use of both HCF and DNNs [37,45]. This category also comprises those that define elaborated frameworks for the extraction of features, with pre-processing steps, dimensionality reduction, and training of classifiers [35].

In this work, our advances are concentrated in the categories C_1 and C_3. Specifically, we innovate on two main fronts:

- with the proposal of a new framework for the extraction of characteristics of fingerprint images,
- with a new micropattern descriptor based on measurements taken from the wellknown Dense Scale-Invariant Feature Transform (DSIFT).

The paper is organized into 5 sections. In Sect. 2, we discuss some fundamentals of the used descriptor (DSIFT). The formulation of the proposed method and the details of all its functionalities are presented in Sect. 3. Our framework is evaluated in the benchmarks of three different editions of the LivDet competition and the experimental results obtained are presented in Sect. 4. The manuscript ends with conclusions and proposals for future work in Sect. 5.

2   Dense SIFT Fundamentals

The pattern descriptor known as SIFT [23] is widely used in pattern recognition and detection tasks in images [21]. In summary, its operation is conducted from the analysis of gradient histograms present in the neighborhood of some points of interest present in the image. This measure is invariant to scale and rotation transformations, however, the characteristics represented by its descriptors are sparse, since they are dependent on the set of determined keypoints. Thus, Liu et al. [22] propose a modification of the method that takes into account all the points of an image for construct its descriptors: the DSIFT.

In recent years, many variants of this pattern descriptor have been proposed to improve its representation capacity. In this work, we will make use of one of its most robust representations: the *Pyramid Histogram Of visual Words* (PHOW) [6]. In detail, we can define this technique mathematically through 5 steps, detailed below:

1. **Step 1:** Consider the image I and a grid mesh $\mathcal{M}$ defined over I so that its nodes are spaced apart by S pixels. Also, let's assume that the $\mathcal{M}$ nodes are equally spaced representations of N pixels of I, which make up the set $\mathcal{P} = \{P_1, P_2, ..., P_N\}$.
2. **Step 2:** A set of 4 neighborhoods is made around each pixel P_i of $\mathcal{P}$, that is, $\mathcal{V} = \{V_{i,1}, V_{i,2}, V_{i,3}, V_{i,4}\}$. Each neighborhood $V_{i,j}$ is centered on the pixel P_i and is formed by grids of dimension 4×4, with each cell of these grids having dimension $\sigma_{i,j} \times \sigma_{i,j}$. Besides that, $\sigma_{i,1} < \sigma_{i,2} < \sigma_{i,3} < \sigma_{i,4}$.
3. **Step 3:** Then, the gradients [10] are calculated in each of the 16 cells of each neighborhood in $\mathcal{V}$ so that only the main 8 directions of the plan are considered, which are presented in Δ:

$$\Delta = \{(0,1); (1,0); (0,-1); (-1,0); (1,1); (1,-1); (-1,-1); (-1,1)\}.$$

In addition, gradients with a magnitude below a pre-established threshold δ are disregarded.

4. **Step 4:** With the gradients in hand, the histogram of the directions present in each cell in each neighborhood is calculated in $\mathcal{V}$. Thus, for each cell, a histogram of 8-bins is associated. Consequently, for each neighborhood $4 \cdot 4 = 16$ of these histograms are associated and, therefore, for each pixel P_i, 4 histograms of $16 \cdot 8 = 128$-bins are associated or, for simplification purposes, four vectors $\vec{d}_{i,j}$, $j = 1, 2, 3, 4$, of 128 coordinates.

5. **Step 5:** Finally, as a result of the DSIFT technique, we have a set of $4N$ descriptors in the form of 128 coordinate vectors. Thus, given an image I, the descriptors extracted with the DSIFT can be represented in their matrix form:

$$D_I = \begin{bmatrix} | & | & | & | & | & & | \\ \vec{d}_{1,1} & \vec{d}_{1,2} & \vec{d}_{1,3} & \vec{d}_{1,4} & \vec{d}_{2,1} & \cdots & \vec{d}_{N,4} \\ | & | & | & | & | & & | \end{bmatrix} \in \mathbb{R}^{128 \times 4N}. \tag{1}$$

Throughout the text, for ease of notation, we will consider D_I to be $(d_{i,j})_{i,j} \in \mathbb{R}^{128 \times 4N}$. In addition, in this work, we will follow the same parameterization of Bosch, Zisserman, and Munoz [6]. In detail, we will define on I a grid mesh with uniform spacing of $S = 5$ pixels; the dimensions of the four neighborhoods are defined by $(\sigma_{i,1}, \sigma_{i,2}, \sigma_{i,3}, \sigma_{i,4}) = (5, 7, 10, 12), \forall i$; and let's disregard gradients of magnitude less than $\delta = 10^{-6}$.

3   Proposed Multi-filter Framework and Statistical Dense SIFT for Liveness Detection in Fingerprints

In this section, we present the components that form the proposed method for detecting SPAs in FPASs. For this, we detail the operation of all strategies used through algorithms and flowcharts that facilitate the understanding and reproducibility of the developed material. Specifically, we present the following innovations obtained with the proposed work:

- A new framework for the extraction and classification of characteristics of fingerprint images to conduct the discrimination of these images into two distinct groups: the set of legitimate fingerprints and the set of fingerprints manufactured from synthetic materials;
- A new descriptor of local patterns based on measurements taken from DSIFT histograms, presented in Eq. (1),
- The proposed method is presented in the form of a generalization, and it is functional in many configurations. Therefore, we present a practical instance of it.

3.1   Multi-filter Framework

Pattern descriptors, especially those dedicated to representing textures [34], may be dependent on illumination conditions [40]. In this way, a correction step conducted by histogram equalization strategies should be used to enhance the ability to represent the pattern descriptor used. This being one of the initial steps of the proposed framework.

It is known that, in the analysis of fingerprint images, some natural phenomena associated with the human finger may occur, which end up compromising the image collection performance by the sensor used [28]. As an example, we can mention the cases in which the fingers are too wet or too dry, are dirty, are excessively oily, among others. Thus, the characterization of the collected fingerprint is impaired in these situations and the use of a smoothing filter can be used to mitigate these difficulties. However, the use of this type of technique can make it difficult to detect important features of a fingerprint, which can be crucial for the classification step, since they are dissolved in the image using these filters. This problem is intensified in cases where the image captured by the sensor does not fit into any of the problem situations mentioned and, therefore, does not have any noise class in its composition. For this reason, using the original image together with its smoothed version is a powerful strategy in the task of representing the texture. Also, we propose the use of a sharpening filter so that the characteristics that are not very outstanding can be highlighted in the image and, consequently, are used in the task of representing the image together with the features extracted from the original image and the smoothed image.

In the special case of the synthetic fingerprint recognition problem, some authors [32] have already highlighted classes of patterns that are inherent to the counterfeiting process, such as artifacts in the form of "holes" present inside of the finger and in the form of extensive homogeneous regions that have lost many details through the manufacturing process. For these situations, the filtering can also be useful in highlighting substantial differences compared to the original image, since the effect of the filtering can be more intense in images of synthetic fingers. For example, in Fig. 1, three versions of the fingerprint image from the same individual, whose code is "002_4_0" in the database LivDet 2015 [15], is presented.

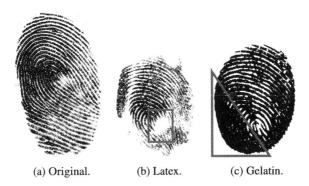

(a) Original.   (b) Latex.   (c) Gelatin.

Fig. 1. Fingerprint of index "002_4_0" from "Hi_Scan 2015" train database. Highlighted in red are the patterns and artifacts generated in the fingerprints during the production process.

Specifically, in Fig. 1b, we see an artifact inside the fingerprint manufactured with latex, highlighted by a red rectangle, which defines a region of high frequencies. In this image, a smoothing would be much more efficient, and therefore more significant, than it would be in the real fingerprint, shown in Fig. 1a. On the other hand, when analyzing the finger construction made with gelatin, shown in Fig. 1c, we notice the presence of a very homogeneous region, highlighted by the red triangle, in which the use of sharpening filtering would be more effective than it would be in the real image. In summary, extracting the descriptor from three different versions of the same fingerprint should increase the representational capacity of the descriptor, since the difference between the features extracted from each of the three versions of the same image must be more intense in manufactured fingerprints, due to the presence of artifacts, than in legitimate fingerprints.

Thus, the proposed framework for SPA detection in FPAS consists of conducting the following steps:

- **Parameter initialization**: to use the framework, it is necessary to define a pattern descriptor that will be used to represent the images. Then, a histogram equalization technique is defined. Also, the used filtering techniques are defined, one being smoothing and one sharpening.
- **Definition of the database**: the method demands the use of a set of images of fingerprints known to be legitimate and of fingerprints known to be manufactured since it is necessary to carry out the training of a classifier.
- **Illumination unbalance correction:** In this phase of the method, the histogram of the fingerprint image is equalized.
- **Filtering:** A smoothed version and a sharpened version of the original image are calculated.
- **Pattern extraction:** A feature vector is extracted from the three versions considered in the image using the selected pattern descriptor.
- **Construction of feature vector:** In this step, the feature vector of each of the three versions of the fingerprint image is concatenated to compose only one feature vector that represents the image. Then, according to the descriptor used, it may be necessary to carry out the normalization of the characteristic vector, as well as to conduct some feature selection process.
- **Definition of the recognition model:** After extracting a feature vector from each of the considered fingerprint images, it is necessary to define a classifier, usually based on some machine learning technique, using the fingerprints in which its category is known. In other words, a training base is used to make the selected classifier capable of separating the feature vectors between vectors extracted from legitimate fingerprints and vectors extracted from manufactured fingerprints. Finally, after training, the spoofing-detection model is defined based on the classifier.

In summary, in Fig. 2, a flowchart of the proposed framework is presented.

3.2 Statistical Dense SIFT

In this work, we propose a new way of summarizing the descriptors extracted from the fingerprint image using DSIFT: the Statistical DSIFT (sDSIFT). In other words, given an image I, we propose to represent its DSIFT descriptors D_I, presented in Eq. (1), by a vector of real coordinates, which must be obtained using simple measures extracted from the 128 lines of D_I. A preliminary strategy, which proved to be very effective

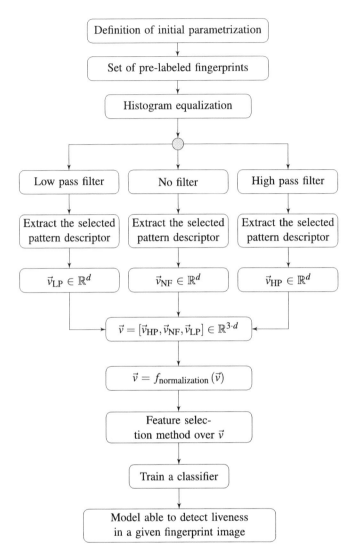

Fig. 2. Flowchart of the proposed framework. In this case, d is the dimension of the feature extracted from the image.

in the task of representing texture images, was proposed by Erpenbeck et al. [13], in which the authors summarize the SIFT descriptors of 36 keypoints considered using the mean and standard deviation. Our technique consists of a generalization in which, specifically, given a matrix of descriptors D_I in the form presented in Eq. (1), we extract from this matrix a set of vectors in $\mathbb{R}^{128}$ using a pre-established set of K functions $\mathcal{F}_{SP}$, as defined in Eq. (2), and finally, we concatenate all these vectors to form the feature vector that represents I.

$$\mathcal{F}_{\text{SP}} := \{f_1, f_2, f_3, \ldots, f_K\}, \tag{2}$$

in which, f_j is, for every j, a function defined in the form of the Eq. (3):

$$f_j : \mathbb{R}^{128 \times N'} \longrightarrow \mathbb{R}^{128}$$
$$D \longmapsto f_j(D) := \vec{v}_j, \tag{3}$$

where $N' = 4N$, and N is the pixel number considered in the DSIFT calculation.

Therefore, the proposed descriptor associates to a given image I, a vector $\vec{v} \in \mathbb{R}^{128K}$, formed by statistics or measurements of the descriptors D_I of the considered image. In Algorithm 1, we present in detail, the operation of the proposed technique.

Algorithm 1. Proposed Statistical DSIFT.

Input:	I	A given image.
	$\mathcal{F}_{\text{SP}}$	The set with K measures to be extracted from the DSIFT descriptors.

1: $D := \text{DSIFT}(I)$ ▷ Extract DSIFT descriptor from I.
2: **for** $f_j \in \mathcal{F}_{\text{SP}}$ **do**
3: $\quad \vec{v}_j := f_j(D)$ ▷ Calculate the statistics of the descriptors (D) using f_j.
4: **end for**
5: $\vec{v}_{\text{sDSIFT}} := [\vec{v}_1, \vec{v}_2, \ldots, \vec{v}_K]$ ▷ Join the vectors.

Output:	$\vec{v}_{\text{sDSIFT}}$	The Statistical DSIFT feature vector extracted from I.

3.3 Proposed Instance

The proposed method, described in the two previous Sects. 3.1 and 3.2, is a generalization that allows several configurations, since our framework allows the use of any histogram equalization technique, smoothing filtering, sharpening filtering, and even pattern descriptor. The same goes for our sDSIFT, as the definition of this descriptor is dependent on the set of measures $\mathcal{F}_{\text{SP}}$. Thus, it is necessary to establish a specific instance to make use of the proposed method in the detection of SPAs in FPASs. Below, we indicate in detail the parameterization proposed and used in this paper:

- **Framework:**
  - We use as histogram equalization technique, the *automatic contrast-limited adaptive histogram equalization* (ACLAHE) [8], which is a technique based on the adaptive histogram equalization that makes use of textural information from blocks of the Image.
  - As a smoothing filter, we are using a Gaussian filter defined by a standard deviation kernel equal to 1.
  - To perform the sharpening, we use a Laplace filter, in which the mask that defines it is a matrix 5×5 with a central coordinate equal to 24 and other coordinates equal to -1.

- The pattern descriptor that we use in the framework is the proposed sDSIFT.
- We evaluate two normalization functions used in isolation in the framework. In this case, the functions $f_{\text{normalization},1}$ and $f_{\text{normalization},2}$ used are, respectively, the well-known normalizations Min-Max and z-score [17].
- The feature selection is done with a recent technique based on the meta-heuristic genetic algorithm [36].
- To classify the images, we use a linear support vector machine (SVM).

- **sDSIFT parameters**: we use five measures ($K = 5$) to summarize the DSIFT descriptors of the images. In this case, these used measures are [12]: the average (f_1), the standard deviation (f_2), the maximum value (f_3), the average energy (f_4) and the entropy (f_5) between the columns of the descriptor matrix. Mathematically, considering the matrix $D = (d_{i,j})_{i,j} \in \mathbb{R}^{128 \times N'}$, the functions of $\mathcal{F}_{\text{SP}}$ are presented in Eq. (4):

$$f_1(D) := \frac{1}{N'} \left[\sum_{j=1}^{N'} d_{1,j}, \ldots, \sum_{j=1}^{N'} d_{128,j} \right], \tag{4a}$$

$$f_2(D) := \frac{1}{N'} \left[\sum_{j=1}^{N'} \left(d_{1,j} - \sum_{r=1}^{N'} \frac{d_{1,r}}{N'} \right)^2, \ldots, \sum_{j=1}^{N'} \left(d_{128,j} - \sum_{r=1}^{N'} \frac{d_{128,r}}{N'} \right)^2 \right], \tag{4b}$$

$$f_3(D) := \left[\max_{j \in \{1,2,\ldots,N'\}} \{d_{1,j}\}, \ldots, \max_{j \in \{1,2,\ldots,N'\}} \{d_{128,j}\} \right], \tag{4c}$$

$$f_4(D) := \frac{1}{N'} \left[\sum_{j=1}^{N'} d_{1,j}^2, \ldots, \sum_{j=1}^{N'} d_{128,j}^2 \right], \tag{4d}$$

$$f_5(D) := - \left[\sum_{j=1}^{N'} p_{1,j} \cdot \log_2 (p_{1,j} + \varepsilon), \ldots, \sum_{j=1}^{N'} p_{128,j} \cdot \log_2 (p_{128,j} + \varepsilon) \right], \tag{4e}$$

in which, $p_{i,j} := \dfrac{d_{i,j}}{\sum_{r=1}^{N'} d_{i,r}}$ and $\varepsilon := 10^{-10}$.

4   Experiments and Results

To validate the proposed material, we conduct practical assessments on the most well-known benchmarks on the topic. The results show that the framework is able to increase the fraud detection efficiency in FPASs with the use of sDSIFT. In this case, the two most used editions of the Liveness Detection competition were considered for evaluations. In detail, we consider the LivDet 2013 base [14], which consists of three sensors[1]: Biometrika, Italdata, and Swipe; and we consider the LivDet 2015 base [15], consisting

---

[1] The CrossMatch sensor has a cataloging error, so we do not consider it in the evaluations.

of four sensors: CrossMatch, GreenBit, Digital, and Hi_Scan. To carry out the training, we apply the proposed methodology in the form of the framework to each sensor in each base, making no exchange of information between sensors or bases.

To conduct our experiments, we developed a computational prototype of our framework in MATLAB R2018a. To implement the proposed sDSIFT descriptor, we used the VL_Feat library [39], widely used in image processing and recognition tasks.

For our analysis, we consider three different versions of the proposed method:

V1 (sDSIFT + SVM): In this version, we are evaluating only the proposed descriptor without considering any stage of the framework. At the end of the extraction of the descriptor of each sensor, we conduct training and classification using a linear SVM.

V2 (FW + sDSIFT + $f_{normalization,1}$): In this version, we consider the proposed descriptor and all the steps of the defined instance of the framework, and the step of feature vector normalization is done by the normalization function $f_{normalization,1}$.

V3 (FW + sDSIFT + $f_{normalization,2}$): A version similar to the previous one, with the exception that we are using in this the normalization function $f_{normalization,2}$.

In Table 1, we present the results obtained using the three proposed versions of the method and compare them with the performance of methods that define the state of the art. In detail, for comparison, we consider: the winners of each edition of fingerprint liveness detection competition [14,15]; three different techniques by Tan et al. [35]; and the Alshdadi et al. [4] method. These last two were chosen because they are recent techniques that are similar to the proposed methodology.

Table 1. Comparison of accuracy (Acc), in percentage (%), obtained by three versions of the proposed method. The best values are highlighted in bold. **Avg** represents the average accuracy for each method considering all the sensors in each year of the competition. **AVG** represents the average accuracy of each method considering all the sensors and all the years of competition.

Edition	2013				2015					
Method	Biometrika	Italdata	Swipe	Avg	Hi_Scan	CrossMatch	Digital	GreenBit	Avg	**AVG**
V1	95.85	90.3	92.34	92.83	86.88	90.74	83.68	90.26	87.89	90.01
V2	98.8	**99.85**	97.3	**98.65**	**99.20**	**99.49**	92.40	91.50	95.65	**96.93**
V3	**99.6**	96.6	96.5	97.57	98.40	99.35	**98.10**	94.40	**97.56**	**97.56**
CoALBP [35]	97	99.4	95.8	97.40	92.16	97.29	93.24	92.79	93.87	95.38
CoALBP-GIF [35]	96.7	98.6	95.2	96.83	90.16	97.18	93.24	94.83	93.85	95.13
Guided filter [35]	98.1	**99.85**	96.3	98.08	93.36	98.77	94.08	94.27	95.12	96.39
Winner [14,15]	95.3	96.5	85.93	92.58	94.36	98.1	93.72	95.4	95.40	94.19
Q-FFF [4]	98.7	98.8	96.5	98.00	96.4	96.73	91	**97.37**	95.38	96.5

According to the obtained results, we can see that, even though it is a very simple descriptor, the proposed sDSIFT, used in isolation for the training of a linear SVM and represented by **V1**, was able to overcome the winner's results of the 2013 edition of the LivDet competition. However, the other results presented by this technique are inferior to the majority of the results that configure the state of the art in the specialized literature. As an example, we can note that this technique was the only one considered to have an average accuracy of less than 90% in the 2015 edition. This fact is useful to highlight the importance of using the framework, represented by the other two versions of the proposed technique (**V2** and **V3**) that use the framework together with the sDSIFT descriptor since the results presented by these are much better than the results presented with the isolated use of sDSIFT. Mathematically, the use of the framework compared to the isolated use of sDSFIT improved 6.92% the accuracy obtained by the method in **V2** and 7.55% in **V3**. Indeed, these techniques have the two best average accuracy results (**AVG**) considering all sensors from the 2013 and 2015 competitions.

When we use the proposed descriptor together with the framework with normalization performed by $f_{\mathrm{normalization},1}$ (**V2**), we obtain the best accuracy value in most of the considered sensors. In addition, this version of the technique presents, on average, the best accuracy when considering only the 2013 edition of the competition. In the case of the 2015 competition, **V2** has the second-best average accuracy value among the considered techniques, having shown accuracy greater than 99% in the classification of two sensors from this base.

The use of sDSFIT together with the normalization function $f_{\mathrm{normalization},2}$ in the framework (**V3**) seems to add greater stability to the method since this technique presented the best overall average accuracy (**AVG**) among all the other techniques. Being its worst performance presented in the GreenBit sensor of LivDet 2015, in which it presented 94.4% accuracy, which configures a result similar to those presented by the techniques of Tan et al. [35] and only 1% less compared to the result presented by the winner of the respective edition of the competition. Furthermore, when considering only the 2015 edition, **V3** presents more than 2% of better average accuracy compared to the methods that represent the state of the art.

Thus, we can see that the results obtained by the proposed descriptor, although adequate in some instances, have been considerably improved with the use of two different versions of the proposed framework. Thus, the versions **V2** and **V3** of the proposed method differ slightly from each other, presenting competitive results to those that represent the state of the art.

5   Conclusion

In this work, we propose a new texture descriptor, sDSIFT, and a new framework that intends to improve the ability of descriptors to detect SPAs in FPASs. The method is very wide and, therefore, it is necessary to define a specific instance of it to conduct its use and, finally, classify fingerprints as being spoofings or legitimate.

Three different versions of the method were evaluated, one composed solely of the proposed texture descriptor (**V1**) and two other versions composed of different configurations of the framework used in conjunction with sDSIFT (**V2** and **V3**). In this case, the results presented by these last two proved to be much superior to the results presented by the isolated use of sDSIFT in solving the problem, which serves as an indication for the proof that the use of the proposed framework can improve the representation capacity of a pattern descriptor. Besides, the results presented by **V2** and **V3** compare, or even surpass, the techniques that represent the state of the art in the problem.

We intend to analyze each of the stages of the proposed framework in isolation and how they influence the ability to improve representation in the considered descriptor. Also, we will evaluate more instances of the framework considering several different configurations. In detail, we will evaluate the performance of other meta-heuristics and other techniques, such as those based on auto-encoders, in the feature selection stage. Finally, we will extend the proposed material theory to make it possible to fusion more than one texture descriptor into the framework.

Acknowledgments. This study was financed in part by the São Paulo Research Foundation (FAPESP), process #15/14358-0, by the Brazilian National Council for Scientific and Technological Development (CNPq), process #381991/2020-2, and by the *"Coordenação de Aperfeiçoamento de Pessoal de Nível Superior - Brasil"* (CAPES) - Finance Code 001.

References

1. Afandi, F., Sarno, R.: Android application for advanced security system based on voice recognition, biometric authentication, and internet of things. In: 2020 International Conference on Smart Technology and Applications (ICoSTA), pp. 1–6. IEEE (2020)
2. Agarwal, R., Jalal, A., Arya, K.: A review on presentation attack detection system for fake fingerprint. Mod. Phys. Lett. B **34**(05), 2030001 (2020)
3. Ali, S.S., Baghel, V.S., Ganapathi, I.I., Prakash, S.: Robust biometric authentication system with a secure user template. Image Vis. Comput. **104**, 104004 (2020)
4. Alshdadi, A., Mehboob, R., Dawood, H., Alassafi, M.O., Alghamdi, R., Dawood, H.: Exploiting level 1 and level 3 features of fingerprints for liveness detection. Biomed. Sig. Process. Control **61**, 102039 (2020)
5. Annapurani, K., Sadiq, M., Malathy, C.: Fusion of shape of the ear and tragus-a unique feature extraction method for ear authentication system. Expert Syst. Appl. **42**(1), 649–656 (2015)
6. Bosch, A., Zisserman, A., Munoz, X.: Image classification using random forests and ferns. In: 2007 IEEE 11th International Conference on Computer Vision, pp. 1–8. IEEE (2007)
7. Cappelli, R., Maio, D., Maltoni, D.: Synthetic fingerprint-database generation. In: Object Recognition Supported by User Interaction for Service Robots, vol. 3, pp. 744–747. IEEE (2002)
8. Chang, Y., Jung, C., Ke, P., Song, H., Hwang, J.: Automatic contrast-limited adaptive histogram equalization with dual gamma correction. IEEE Access **6**, 11782–11792 (2018)
9. Chugh, T., Cao, K., Jain, A.K.: Fingerprint spoof buster: use of minutiae-centered patches. IEEE Trans. Inf. Forensics Secur. **13**(9), 2190–2202 (2018)
10. Dalal, N., Triggs, B.: Histograms of oriented gradients for human detection. In: 2005 IEEE Computer Society Conference on Computer Vision and Pattern Recognition (CVPR 2005), vol. 1, pp. 886–893. IEEE (2005)
11. Dinca, L.M., Hancke, G.P.: The fall of one, the rise of many: a survey on multi-biometric fusion methods. IEEE Access **5**, 6247–6289 (2017)
12. Djebbar, F., Ayad, B.: Energy and entropy based features for wav audio steganalysis. J. Inf. Hiding Multimedia Sig. Process. **8**(1), 168–181 (2017)
13. Erpenbeck, D., et al.: Basic statistics of SIFT features for texture analysis. In: Tolxdorff, T., Deserno, T.M., Handels, H., Meinzer, H.-P. (eds.) Bildverarbeitung für die Medizin 2016. I, pp. 98–103. Springer, Heidelberg (2016). https://doi.org/10.1007/978-3-662-49465-3_19
14. Ghiani, L., et al.: LivDet 2013 fingerprint liveness detection competition 2013. In: 2013 International Conference on Biometrics (ICB), pp. 1–6. IEEE (2013)

15. Ghiani, L., Yambay, D.A., Mura, V., Marcialis, G.L., Roli, F., Schuckers, S.A.: Review of the fingerprint liveness detection (LivDet) competition series: 2009 to 2015. Image Vis. Comput. **58**, 110–128 (2017)
16. Gragnaniello, D., Poggi, G., Sansone, C., Verdoliva, L.: An investigation of local descriptors for biometric spoofing detection. IEEE Trans. Inf. Forensics Secur. **10**(4), 849–863 (2015)
17. Jain, A., Nandakumar, K., Ross, A.: Score normalization in multimodal biometric systems. Pattern Recogn. **38**(12), 2270–2285 (2005)
18. Jain, A.K., Flynn, P., Ross, A.A.: Handbook of Biometrics. Springer, Cham (2007). https://doi.org/10.1007/978-0-387-71041-9
19. Jain, A.K., Nandakumar, K., Ross, A.: 50 years of biometric research: accomplishments, challenges, and opportunities. Pattern Recogn. Lett. **79**, 80–105 (2016)
20. Kiefer, R., Stevens, J., Patel, A., Patel, M.: A survey on spoofing detection systems for fake fingerprint presentation attacks. In: Senjyu, T., Mahalle, P.N., Perumal, T., Joshi, A. (eds.) ICTIS 2020. SIST, vol. 195, pp. 315–334. Springer, Singapore (2021). https://doi.org/10.1007/978-981-15-7078-0_30
21. Leng, C., Zhang, H., Li, B., Cai, G., Pei, Z., He, L.: Local feature descriptor for image matching: a survey. IEEE Access **7**, 6424–6434 (2018)
22. Liu, C., Yuen, J., Torralba, A.: SIFT flow: dense correspondence across scenes and its applications. IEEE Trans. Pattern Anal. Mach. Intell. **33**(5), 978–994 (2010)
23. Lowe, D.G.: Distinctive image features from scale-invariant keypoints. Int. J. Comput. Vis. **60**(2), 91–110 (2004)
24. Marcel, S., Nixon, M.S., Fierrez, J., Evans, N.: Handbook of Biometric Anti-Spoofing: Presentation Attack Detection. Springer, Cham (2019). https://doi.org/10.1007/978-3-319-92627-8
25. Marcialis, G.L., et al.: First international fingerprint liveness detection competition—LivDet 2009. In: Foggia, P., Sansone, C., Vento, M. (eds.) ICIAP 2009. LNCS, vol. 5716, pp. 12–23. Springer, Heidelberg (2009). https://doi.org/10.1007/978-3-642-04146-4_4
26. Medikonda, J., Madasu, H., Panigrahi, B.K.: Information set based gait authentication system. Neurocomputing **207**, 1–14 (2016)
27. Nanni, L., Ghidoni, S., Brahnam, S.: Handcrafted vs. non-handcrafted features for computer vision classification. Pattern Recogn. **71**, 158–172 (2017)
28. Patil, M.S., Patil, S.S.: Wet and dry fingerprint enhancement by using multi resolution technique. In: 2016 International Conference on Global Trends in Signal Processing, Information Computing and Communication (ICGTSPICC), pp. 188–193. IEEE (2016)
29. Prasad, P.S., Sunitha Devi, B., Janga Reddy, M., Gunjan, V.K.: A survey of fingerprint recognition systems and their applications. In: Kumar, A., Mozar, S. (eds.) ICCCE 2018. LNEE, vol. 500, pp. 513–520. Springer, Singapore (2019). https://doi.org/10.1007/978-981-13-0212-1_53
30. Raja, K.B., Raghavendra, R., Venkatesh, S., Gomez-Barrero, M., Rathgeb, C., Busch, C.: A study of hand-crafted and naturally learned features for fingerprint presentation attack detection. In: Marcel, S., Nixon, M.S., Fierrez, J., Evans, N. (eds.) Handbook of Biometric Anti-Spoofing. ACVPR, pp. 33–48. Springer, Cham (2019). https://doi.org/10.1007/978-3-319-92627-8_2
31. Samma, H., Suandi, S.A.: Transfer learning of pre-trained CNN models for fingerprint liveness detection. In: Biometric Systems. IntechOpen (2020)
32. Sharma, R.P., Dey, S.: Fingerprint liveness detection using local quality features. Vis. Comput. **35**(10), 1393–1410 (2018). https://doi.org/10.1007/s00371-018-01618-x
33. Silva, E., Boaventura, M., Boaventura, I., Contreras, R.: Face recognition using local mapped pattern and genetic algorithms. In: Proceedings of the International Conference on Pattern Recognition and Artificial Intelligence, pp. 11–17 (2018)

34. Susan, S., Hanmandlu, M.: Difference theoretic feature set for scale-, illumination-and rotation-invariant texture classification. IET Image Process. **7**(8), 725–732 (2013)
35. Tan, G., Zhang, Q., Hu, H., Zhu, X., Wu, X.: Fingerprint liveness detection based on guided filtering and hybrid image analysis. IET Image Process. **14**(9), 1710–1715 (2020)
36. Too, J., Abdullah, A.R.: A new and fast rival genetic algorithm for feature selection. J. Super-comput., 1–31 (2020). https://doi.org/10.1007/s11227-020-03378-9
37. Toosi, A., Bottino, A., Cumani, S., Negri, P., Sottile, P.L.: Feature fusion for fingerprint liveness detection: a comparative study. IEEE Access **5**, 23695–23709 (2017)
38. Tripathi, K.: A comparative study of biometric technologies with reference to human inter-face. Int. J. Comput. Appl. **14**(5), 10–15 (2011)
39. Vedaldi, A., Fulkerson, B.: VLFeat: an open and portable library of computer vision algo-rithms (2008). http://www.vlfeat.org/
40. Veerashetty, S., Patil, N.B.: Novel LBP based texture descriptor for rotation, illumination and scale invariance for image texture analysis and classification using multi-kernel SVM. Mul-timedia Tools Appl. **79**(15), 9935–9955 (2020). https://doi.org/10.1007/s11042-019-7345-6
41. Velapure, A., Talware, R.: Performance analysis of fingerprint recognition using machine learning algorithms. In: Proceedings of the Third International Conference on Computational Intelligence and Informatics, pp. 227–236 (2020)
42. Yambay, D., Ghiani, L., Marcialis, G.L., Roli, F., Schuckers, S.: Review of fingerprint pre-sentation attack detection competitions. In: Marcel, S., Nixon, M.S., Fierrez, J., Evans, N. (eds.) Handbook of Biometric Anti-Spoofing. ACVPR, pp. 109–131. Springer, Cham (2019). https://doi.org/10.1007/978-3-319-92627-8_5
43. Yuan, C., Xia, Z., Sun, X., Wu, Q.J.: Deep residual network with adaptive learning frame-work for fingerprint liveness detection. IEEE Trans. Cogn. Dev. Syst. **12**(3), 461–473 (2019)
44. Zafar, M.R., Shah, M.A.: Fingerprint authentication and security risks in smart devices. In: 2016 22nd International Conference on Automation and Computing (ICAC), pp. 548–553. IEEE (2016)
45. Zhang, Y., Zhou, B., Wu, H., Wen, C.: 2D fake fingerprint detection based on improved CNN and local descriptors for smart phone. In: You, Z., et al. (eds.) CCBR 2016. LNCS, vol. 9967, pp. 655–662. Springer, Cham (2016). https://doi.org/10.1007/978-3-319-46654-5_72
46. Zheng, T.F., Li, L.: Robustness-Related Issues in Speaker Recognition. Springer, Singapore (2017). https://doi.org/10.1007/978-981-10-3238-7

Application of a Neural Network to Generate the Hash Code for a Device Fingerprint

Marcin Gabryel[(⊠)] and Milan Kocić

Spark Digitup, Plac Wolnica 13 lok. 10, 31-060 Kraków, Poland
marcin.gabryel@sparkdigitup.pl

Abstract. Device fingerprint (also known as browser fingerprint) is a tool that makes a user identifiable online without the use of cookies. Many organisations believe that device fingerprinting could compromise user privacy. However, it does allow for a number of positive applications, including facilitating detection of online fraud and abuse. The device fingerprint is created from the values of a number of parameters which are collected only from a web browser. Due to the very large number of parameters, the values are fed into the hash function input to create a unique, short hash code, which makes them easy to compare with each other. This paper describes an efficient method for generating hash codes using a deep neural network with an autoencoder structure. Two types of the fingerprint have been examined: a stable one with a low uniqueness, and an unstable one with a high accuracy and uniqueness. The results were compared with fingerprint hash codes obtained using well-known hash functions such as MD5 or SHA.

Keywords: Browser fingerprint · Device fingerprint · Autoencoder

1 Introduction

One of the problems of detecting fraud in the online advertising industry is related to the problem of identifying the source of fraudulent traffic (returning visitors' devices/browsers, bot generated traffic). The simplest solution seems to be to track entries made from a computer by a repeating IP address. However, this can be easily avoided by using different types of proxy servers, TOR networks or by using a dynamically assigned IP address, which is quite common in mobile networks. Many research papers provide descriptions of different ways how to uniquely identify a user's browser/device [1] using other mechanisms. Such identification makes it possible, despite attempts to hide subsequent visits by deleting cookies and changing the IP address, to recognise subsequent visits to a website made from the same device/browser. In literature, this type of identifier can be found under the name "browser fingerprint" or "device fingerprint", and it is obtained by using the fact that each of the computers accessing a given website differs in terms of some minor hardware parameters and software configuration. A set of such parameter values gives a unique fingerprint. However, there are some disadvantages which prevent flawless identification of devices. One of them is that some of these parameters are not very stable. Browsers are often updated, which often

© Springer Nature Switzerland AG 2021
L. Rutkowski et al. (Eds.): ICAISC 2021, LNAI 12855, pp. 456–463, 2021.
https://doi.org/10.1007/978-3-030-87897-9_40

results in changes in system settings, thus affecting the values of collected parameters. Another problem is maintaining the uniqueness of a given fingerprint. Different devices with a similar configuration may, in some cases, receive the same identifier. A particular problem here is distinguishing devices of the same model, especially if they very often also have the same software configuration (e.g. smartphones). In this paper, a solution is proposed that improves the efficiency of user recognition by using a deep neural network to generate the hash code for a device fingerprint. This effectiveness will be compared with obtaining the hash code of the fingerprint using a hash function (MD5, SHA or a similar one). The described solution will be implemented using a neural network in the structure of an autoencoder performing semantic hashing. Such network will enable obtaining the hash code from the units in one of the hidden layers of the network.

It is claimed that the device fingerprint is quite unstable, because the values of its parameters may change over time [2, 5]. Therefore, two types of fingerprints can be used: stable and unstable ones. A stable fingerprint consists of parameters whose values are stable in time. There are only a few of these, so although the value of such parameter can be assigned to a browser user for a long period of time, it is of low variability. Many browsers can also have the same fingerprint. The unstable fingerprint, on the other hand, is made up of the largest possible number of elements. Its value is more unique and there is little likelihood that another browser will have the same value. The disadvantage is, however, that such a fingerprint is valid in a very short period of time. Some parameter values are unstable and change quickly over time.

This paper presents a solution that allows the generation of a fingerprint with a significantly improved variability and uniqueness compared to the standard fingerprint resulting from implementation of a simple hash function. The paper is divided into several sections. Section 2 introduces the device fingerprint and its applications. Section 3 describes the deep neural network working in an autoencoder system with two encoders used in the presented research study. The next section describes how the research study was carried out and the obtained results of the experiments conducted. The paper concludes with a short summary and conclusions.

2 Device Fingerprint

The device fingerprint (also known as browser fingerprint) first appeared in the literature in 2009 in paper [6], where it was first noted that by using various parameters collected from a browser, we can identify its user with a high degree of probability. This was confirmed in subsequent studies [7], which were conducted on a larger number of users. There are papers describing a variety of extracted browser parameters, their type and effectiveness in creating a fingerprint [9]. Some works also focus on the analysis of the stability of individual parameters [2, 10], the study of different types of browsers [8] or applied security features that limit the possibility of feature extraction [5].

Creating and using a device fingerprint for the purpose of user identification is widely criticized because of a certain degree of likelihood of invasion of user privacy. In fact, the need for privacy in the Internet has been recognized by the European Parliament, and for several years now, websites have included information about the use of cookies [11]. While privacy is an obvious necessity, unambiguous identification of the device or browser is extremely valuable when it comes to ensuring security and preventing online fraud. Among the many frauds that are committed online are fraudulent credit card payments, artificial generation of Internet traffic, so-called abusive traffic, generation of click fraud or automated collection of web content. Practically, the device fingerprint can also be used in web tracking, bot and fraud prevention, web crawler detection and an augmented authentication. The possibility of using device fingerprinting is accessible to everyone, as it has been commercialized [12]. The paper [18] presents the possibilities offered by browser fingerprinting to verify the software and hardware stack of a mobile or desktop client. The presented system can, for example, distinguish between traffic sent by an original smartphone running an original browser from an emulator or desktop client spoofing the same configuration.

In this paper, the parameters directly retrieved from the browser are divided into two sets. According to [5], stable parameters included font information, WebGL support and provider, number of colours, some parameters retrieved from the User-agent header, plug-ins, touchscreen information, time zone, memory and processor information, browser-specific information (related to JavaScript implementation). The following parameters were used as unstable: canvas fingerprint, all screen and audio information, and some parameters from the User-agent header. A stable fingerprint consists of 40 parameters, while an unstable fingerprint is made up of 56 parameters.

3   Autoencoder

An autoencoder is a neural network with at least one hidden layer. The output and input of the network have the same size. The autoencoder is trained to try to copy at its output the values given at its input. It consists of two parts: an encoder and a decoder. The encoder is supposed to compress the data into a low-dimensional representation, and the decoder is designed to reconstruct the input data from the low-dimensional representation generated by the encoder [3].

A practical example of using an autoencoder is the so-called symmetric hashing published in [17]. An encoded object is fed to the input of a neural network, which generates a low-dimensional binary vector. Two similar objects will have two identical or very similar hash codes. A similar application may be applied to encoding the information comprising a device fingerprint.

For the encoding of stable and unstable features of the fingerprint, it is necessary to acquire two low-dimensional representations. The autoencoder will therefore consist of two encoder-decoder pairs for stable and unstable parameters respectively. In this way the autoencoder will have two hidden layers h_1 and h_2. After the learning process, the decoder part is disconnected from the network. Layers h_1 and h_2 then become the output layers, which will return the hash code values of the fingerprint fed to the input of the network. The diagrams of the two examined autoencoder structures are shown in Fig. 1. The two networks differ in the way in which hash code h_2 is obtained. In network a) h_2 is generated only from the values of unstable parameters, while in network b) from the values of all parameters given to the network input. The values obtained at the output of the hidden layer h_1 will form the stable fingerprint hash code, because it is created only from the stable parameters of vector x_i. In contrast, the hash code of the unstable fingerprint will be obtained for a) as a combination of h_1 and h_2, and for b) as h_2.

a)   b)

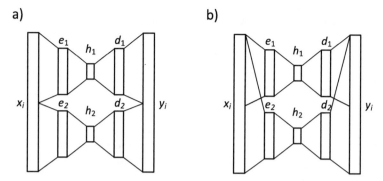

Fig. 1. The two structures of the autoencoders under this study.

For autoencoder a) training set $x_i \in \{x_{1j}, x_{2k}\} \in R^m$ is considered, where $x_{1j} \in R^{m_1}$, $x_{2k} \in R^{m_2}$ are the stable and unstable input data, respectively, m_1 is the number of the stable data, m_2 is the number of the unstable data and $m = m_1 + m_2$. When the structures are simplified to one hidden layer, the values obtained at the output of these layers will then be of the form:

$$h_{1j} = f_{\theta_1}(x_{1j}) = s(W_1 x_{sj} + b_1), \tag{1}$$

$$h_{2k} = f_{\theta_2}(x_{2k}) = s(W_2 x_{nk} + b_2), \tag{2}$$

where: $W_1 \in R^{m_1 \times n_1}$, $W_2 \in R^{m_2 \times n_2}$ are the weights of hidden layers h_1 and h_2, respectively, n_1 and n_2 are the number of units in hidden layers h_1 and h_2, $b_1 \in R^{m_1}$ and $b_2 \in R^{m_2}$ – bias vectors, $\theta_1 = \{W_1, b_1\}$ i $\theta_2 = \{W_2, b_2\}$. The autoencoder output value $y_i \in \{y_{1j}, y_{2k}\} \in R^m$ will require the following calculations:

$$y_{1j} = g_{\theta_1'}(x_{1j}) = s\left(W_1' h_{1j} + b_1'\right), \tag{3}$$

$$y_{2k} = g_{\theta_1'}(x_{2k}) = s\left(W_2' h_{2k} + b_2'\right), \tag{4}$$

where: $W_1' \in R^{n_1 \times m_1}$, $W_2' \in R^{n_2 \times m_2}$ are the weights, $b_1' \in R^{m_1}$, $b_2' \in R^{m_2}$ are bias vectors and $\theta_1' = \{W_1', b_1'\}$ and $\theta_2' = \{W_2', b_2'\}$. The purpose of learning is therefore to find optimal values for the parameters $\theta_1, \theta_2, \theta_1'$ and θ_2', which minimize the error between input and output for the whole training set:

$$\theta_1, \theta_2, \theta_1', \theta_2' = \underset{\theta_1, \theta_2, \theta_1', \theta_2'}{\arg\min} \; L(x, y) \tag{5}$$

For our autoencoder the loss function $L(x, y)$ will be expressed by the following formula:

$$L(x_i, y_i) = \frac{L(x_{1i}, y_{1i}) + L(x_{2i}, y_{2i})}{2} = \frac{\|x_{1i} - y_{1i}\|^2 + \|x_{1i} - y_{1i}\|^2}{2}$$

$$= \frac{\left\|x_{1i} - s\left(W_1' s(W_1 x_{1i} + b_1) + b_1'\right)\right\|^2 + \left\|x_{2i} - s\left(W_2' s(W_2 x_{2i} + b_2) + b_2'\right)\right\|^2}{2} \quad (6)$$

After the learning process has been carried out, only the encoders are used for further research and these are treated as hash functions. The fingerprint feature values are given as one-hot encode data to the encoders' input. The output values obtained by the two layers h_1 and h_2 are rounded to integers. If the sigmoid function is used as the activation function defined by formula:

$$s(t) = \frac{1}{1 + exp^{-t}} \quad (7)$$

and after rounding the results, a sequence of bits 0–1 is obtained at the encoders' outputs.

For autoencoder b), the structure of the neural network is identical to autoencoder a), both networks differ only in terms of the size of layers e_2 and d_2. However, changes occur in the training set, which is defined as $x_i \in \{x_{1j}, x_{2k}\} \in R^m$, where $x_{1j} \in R^{m_1}$, $x_{2k} \in R^{m_1} \cup R^{m_2}$. The weights, on the other hand, are as follows: $W_1 \in R^{m_1 \times n_1}$, $W_2 \in R^{m_1 \times n_2} \cup R^{m_2 \times n_2}$ for the encoder and $W_1' \in R^{n_1 \times m_1}$, $W_2' \in R^{n_2 \times m_1} \cup R^{n_2 \times m_2}$ for the decoder. The bias vectors take the values: $b_1 \in R^{m_1}$, $b_2 \in R^{m_1} \cup R^{m_2}$ oraz $b_1' \in R^{m_1}$, $b_2' \in R^{m_1} \cup R^{m_2}$ for the encoder and decoder, respectively.

4   Experimental Research Study

In order to conduct the research, data was collected from 213,000 entries to the websites of 43 different online stores. The data was collected over 90 days. The entries of 33,000 unique users were selected, who for the purposes of the research were also identified with a unique identifier stored in a cookie. The selected users were those who visited the website at least 5 times. The data was divided into two parts. All the visits of a given user except the last visit went in the learning set. The final visits of the users went in the testing set.

The recognition accuracy of the test users based on the stable and unstable fingerprints was analyzed. The measures of precision and recall were used as a measure of the effectiveness of the conducted tests [4]. The parameter combining both above parameters is $F1$ score which is the harmonic mean of precision and recall. The recognition accuracy of the test users based on the stable and unstable fingerprints was analyzed. The measures of precision and recall were used as a measure of the effectiveness of the conducted tests [4]. The parameter combining both the above parameters is F1 score, i.e. the harmonic mean of precision and recall.

In this research study, the experiments were performed on a number of different autoencoder structures. The input data were encoded by a one-hot-encoder. The network had 3160 inputs and the corresponding (i.e. the same) number of outputs. The most promising experimental results are shown in Table 1. The best results were selected from various combinations of hyperparameters of both types of networks. The networks were learnt by stochastic gradient descent with momentum (learning rate = 0.0001, momentum = 0.9) over 20 epochs. The following columns show the parameters of the tested algorithms (type of algorithm, network structure, number of hash code bits) and the precision, recall and F1 scores for the test data. As can be seen, the best results were achieved using network a), where two network fragments are learnt separately with different - stable and unstable - parameters.

Table 1. Experimental results for different network structures when compared to using the hash function SHA1.

Algorithm	Network structure	Number of hash code bits: stable unstable	Stable fingerprint: precision recall F1 score	Unstable fingerprint: precision recall F1 score
SHA1	–	h1: 160 h2: 160	0.13972 0.95434 0.24376	0.00040 0.00038 0.00039
Network a)	512-256-128-256-512 512-256-128-256-512	h1: 128 h1 + h2: 256	0.20727 0.91568 **0.33802**	0.12992 0.33564 **0.18733**
Network b)	512-256-128-256-512 512-256-128-256-512	h1: 128 h2: 128	0.09340 0.92111 0.16960	0.01149 0.12639 0.02106

5   Conclusions

This paper presents a method for obtaining the hash code for a device fingerprint using a neural network structure running in an autoencoder structure. The autoencoder allows to obtain two hash codes during one learning cycle. This makes it possible to obtain two types of fingerprint simultaneously, i.e. a stable one and an unstable one. The presented experimental results show that the efficiency of searching for similar fingerprints is much higher in comparison to the use of hash codes generated with the use of an ordinary hash function.

A major advantage of the proposed method is that the encoding results obtained by the h_1 and h_2 encoder layers can be easily stored in a database. This provides a possibility for a practical application of the algorithm. Both methods can also be applied to create device fingerprints for different browsers used by the same user [1]. In such case it is advisable to carry out beforehand an accurate analysis of changes in the values of fingerprint features and create relevant groups of stable and unstable features.

Browser fingerprinting is a tool which can reduce fraud on the Internet. The algorithm proposed in this paper can increase detection rate of fraudulent user activities on the web by identifying them more accurately. Further development of the research is planned towards the use of the recurrent neural network [13, 15, 19], which may be applicable for better classification of repeat visits of the same user to a given website. A more efficient and faster learning process will be possible by using a different optimization algorithm [14]. It may also be an interesting idea to use an evolutionary algorithm in order to select relevant parameters for stable and unstable fingerprints [16].

Acknowledgments. The presented results are obtained within the realization of the project "Traffic Watchdog 2.0 – verification and protection system against fraud activities in the on-line marketing (ad frauds) supported by artificial intelligence and virtual finger-print technology" financed by the National Centre for Research and Development; grant number POIR.01.01.01-00-0241/19-01.

References

1. Cao, Y., Li, S., Wijmans, E.: (Cross-) browser fingerprinting via OS and hardware level features. In: NDSS (2017, March)
2. Kobusińska, A., Pawluczuk, K., Brzeziński, J.: Big Data fingerprinting information analytics for sustainability. Futur. Gener. Comput. Syst. **86**, 1321–1337 (2018)
3. Goodfellow, I., Bengio, Y., Courville, A.: Deep Learning. MIT Press, Cambridge (2016)
4. Leskovec, J., Rajaraman, A., Ullman, J.D.: Mining of Massive Datasets. Cambridge University Press, Cambridge (2014)
5. Gabryel, M., Grzanek, K., Hayashi, Y.: Browser fingerprint coding methods increasing the effectiveness of user identification in the web traffic. J. Artif. Intell. Soft Comput. Res. **10**(4), 243–253 (2020)
6. Mayer, J.R.: "Any person... a pamphleteer": internet anonymity in the age of Web 2.0. Undergraduate Senior Thesis, Princeton University (2009)
7. Eckersley, P.: How unique is your web browser? In: Atallah, M.J., Hopper, N.J. (eds.) PETS 2010. LNCS, vol. 6205, pp. 1–18. Springer, Heidelberg (2010). https://doi.org/10.1007/978-3-642-14527-8_1
8. Laperdrix, P., Bielova, N., Baudry, B., Avoine, G.: Browser fingerprinting: a survey (2019). https://arxiv.org/abs/1905.01051
9. Steven, E., Arvind, N.: Online tracking: a 1-million-site measurement and analysis. In: Proceedings of the 2016 ACM SIGSAC Conference on Computer and Communications Security (CCS 2016). ACM, New York (2016), pp. 1388–1401
10. Alaca, F., Van Oorschot, P.C.: Device fingerprinting for augmenting web authentication: classification and analysis of methods. In: Proceedings of the 32nd Annual Conference on Computer Security Applications, December 2016, pp. 289–301 (2016)
11. Low, C.: Cookie law explained (2016). https://www.cookielaw.org/the-cookie-law/. Accessed Mar 2021
12. FingerprintJS. Fraud detection API. https://fingerprintjs.com/. Accessed Mar 2021
13. El Zini, J., Rizk, Y., Awad, M.: An optimized parallel implementation of non-iteratively trained recurrent neural networks. J. Artif. Intell. Soft Comput. Res. **11**(1), 33–50 (2020)
14. Bilski, J., Kowalczyk, B., Marchlewska, A., Zurada, J.M.: Local Levenberg-Marquardt algorithm for learning feedforward neural networks. J. Artif. Intell. Soft Comput. Res. **10**(4), 299–316 (2020)

15. Shewalkar, A., Nyavanandi, D., Ludwig, S.A.: Performance evaluation of deep neural networks applied to speech recognition: RNN, LSTM and GRU. J. Artif. Intell. Soft Comput. Res. **9**(4), 235–245 (2019)
16. Łapa, K., Cpałka, K., Laskowski, Ł, Cader, A., Zeng, Z.: Evolutionary algorithm with a configurable search mechanism. J. Artif. Intell. Soft Comput. Res. **10**(3), 151–171 (2020)
17. Salakhutdinov, R., Hinton, G.: Semantic hashing. Int. J. Approx. Reason. **50**(7), 969–978 (2009)
18. Bursztein, E., Malyshev, A., Pietraszek, T., Thomas, K.: Picasso: lightweight device class fingerprinting for web clients. In: Proceedings of the 6th Workshop on Security and Privacy in Smartphones and Mobile Devices, October, 2016, pp. 93–102 (2016)
19. Niksa-Rynkiewicz, T., Szewczuk-Krypa, N., Witkowska, A., Cpałka, K., Zalasiński, M., Cader, A.: Monitoring regenerative heat exchanger in steam power plant by making use of the recurrent neural network. J. Artif. Intell. Soft Comput. Res. **11**(2), 143–155 (2021). https://doi.org/10.2478/jaiscr-2021-0009

Fingerprint Device Parameter Stability Analysis

Marcin Gabryel[(⊠)] and Milan Kocić

Spark Digitup, Plac Wolnica 13 lok. 10, 31-060 Kraków, Poland
marcin.gabryel@sparkdigitup.pl

Abstract. Web-based device fingerprints (also known as browser fingerprints) are designed to identify users without leaving a cookie trail. Device fingerprints are a set of different parameter values that are collected through JavaScript from a user's web browser when they visit a website. A combination of many different parameter values makes it possible to create a unique device identifier. Unfortunately, most of the values of collected parameters change during subsequent visits to the website, which means that a new fingerprint is then created. As a result, a subsequent entry of a given user to the website is treated as a new entry. The paper presents an analysis of the data collected during a study on the variability of parameters included in the fingerprint device. The parameters studied can be classified into stable and unstable parameters on the basis of two criteria, i.e. variability and diversity of parameter values during subsequent visits. The paper offers a helpful insight into selecting appropriate values for the parameters comprising the device fingerprint.

Keywords: Web-based device fingerprint · Browser fingerprint

1 Introduction

Browser fingerprint (or device fingerprint) is a set of device-specific information about the browser, device, operating system, and environmental and location settings of a given user. Data is collected by a JavaScript script placed on the monitored web page, and by a web server that receives HTTP requests for the displayed page [1]. From these collected parameter values, a unique fingerprint is created, which allows for a unique identification of the browser and its user. The fingerprint is created from parameters retrieved from the web browser. The ideal situation occurs when the values of the parameters are identical during subsequent visits of the user to the website. Unfortunately, as shown in the research study presented in this article, there are a number of parameters that are very unstable and should be omitted when creating the fingerprint.

The information comprising the browser fingerprint is ultimately stored in a database. In order to be able to quickly compare and search for identical devices or browsers, the parameter values are compressed, for example, using a hash function. One such function is, for example, SHA1, which returns alphanumeric strings of fixed length. The disadvantage of this type of solution is its susceptibility to changes in the values of the parameters that make up the fingerprint. Changing even one value completely changes the value of the alphanumeric fingerprint string. Despite this drawback, using

© Springer Nature Switzerland AG 2021
L. Rutkowski et al. (Eds.): ICAISC 2021, LNAI 12855, pp. 464–472, 2021.
https://doi.org/10.1007/978-3-030-87897-9_41

a hash function is the easiest and fastest method for comparing such a large number of fingerprint parameter values. There are other methods of fingerprint device encoding available, which for example, include those given in paper [2], yet they require additional computationally complex algorithms.

Various analyses of fingerprint parameters have been discussed in the literature on the subject [4, 5]. The paper [3], in particular, provides information about the stability of the parameters and their suitability for creating a unique identifier. The authors of the paper noted that parameter stability is maintained for an average of 6 days. This paper reports slightly different results, but a much larger number of parameters were analyzed. In general, other works usually present comprehensive information about individual features included in the fingerprint [1, 3, 6, 7] and their research on different types of browsers [1, 3] as well as the possibility of obtaining different parameter values from them. A lot of information on the practical possibilities of using the fingerprint can be found in several contexts, such as web tracking [6, 7], bot and fraud prevention [8] and an augmented authentication [9].

This paper is organized as follows. In Sect. 2 a definition of the browser fingerprint is given and the characteristics of the features chosen for its creation are described. Section 3 presents descriptions and results of two stages of the study. Section 4 concludes the paper and offers suggestions for future work.

2   Definition of the Device Fingerprint

The browser (or device) fingerprint is a set of data relating to a user's device and browser. It contains information about the hardware, the operating system and the browser and its settings [1]. The information is collected only directly from the browser by the Javascript script and by a web server. The user remains unaware that they are being identified because the use of the browser fingerprint leaves no trace. For the purposes of a quick search and comparison, a set of collected data is processed by a hash function. The resulting hash becomes the unique identifier of the given browser. Assuming that such browser is used by the same person, the hash will become a unique identifier of the user.

Approximately 162 parameters were extracted as part of the study conducted, which include: parameters related to the hardware, parameters related to the browser type, font list, time zone, supported language, screen and browser size, parameters related to the operating system, parameters related to JavaScript implementation, multimedia playback functionality (recognised codecs), checking access to popular TCP/IP ports, examining characteristics of HTML5 (non-existent image, new features of canvas properties/elements, coordinates of non-existent rectangles), WebGL information, audio playback properties, i.e. the so-called AudioCanvas, identification of logins on social media sites, and parameters extracted from the HTTP User-Agent header.

3   The Research and Its Results

The research study discussed in this section consisted in the analysis of changes in the values of the device fingerprint parameters occurring during subsequent visits to the monitored website. The research has been aimed at finding the parameters that change most over time and those that are the most stable over time.

The data extracted from the users' browsers were collected while monitoring visits to five different Polish online shop websites. The monitoring of these websites took place from 29 September to 16 December 2020. Over 2.5 million visits made by 1272445 different anonymous users were recorded. For the purposes of the study, a fingerprint-independent unique user identifier was additionally stored in the cookie. The data obtained from the browser was collected by a specially prepared script written in JavaScript, activated at the moment of entering the web page. The operation of the script consisted in collecting the values of 162 parameters, creating a device fingerprint, assigning it to the unique identifier stored in the cookie, and sending it to a database located on a third-party server.

The study was conducted in two stages. The first stage was aimed at determining the overall number of the changes of all values of the parameters included in the fingerprint during the subsequent visits. In the second stage, the changes in the values of individual parameters comprising the device fingerprint were analysed. From among all 1.2 million users taking part in the study, 21625 persons who visited the monitored web sites on the first day of the study were selected. The selected users were those from whom complete sets of data were successfully collected. Collected data may be incomplete when the execution of the script is unexpectedly interrupted by the user clicking another link that takes them to another page, closing the browser or a bookmark, moving back to a previous page, or it may simply be caused by an error. In such cases, the script does not have enough time to extract the values of all parameters and send them to the database. The number of entries on subsequent days is presented in the form of a graph in Fig. 1. It can be seen that many users, despite a significant lapse of time, appeared repeatedly on the following days. When looking at the graph, we can notice a rather distinct "peak" that occurred on day 59 of the study. This day was the so-called "Black Friday", i.e. the time of sales. It resulted in a dramatic increase in online shop visits. The said group of 21625 users who appeared on the first day of the study were further analysed.

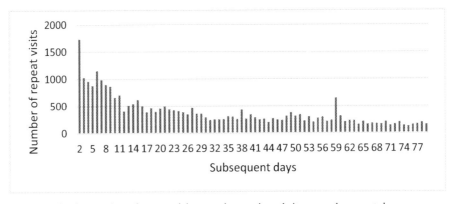

Fig. 1. Number of repeat visitors to the monitored sites on subsequent days.

The next graph (Fig. 2) presents the number of changes in parameter values on specific days recorded during successive visits to the websites. The trendline with squares presents the number of changes of the parameter values occurring on a given day in comparison with the parameter values which were recorded on the first day of the study. On the other hand, the trendline with triangles presents the number of changes of the parameter values made on a given day in comparison with the parameter value from the previous visit. Additionally, the number of user visits on a given day is shown using the trendline with circles. The graph shows that the number of changes occurring is significant. The number of recorded changes is several times greater than the number of user visits to the website. It is worth noting that each change of the parameter has an impact on the resulting device fingerprint. In the case of the fingerprints generated on the first day of the study, it may turn out that subsequent visits generate completely different fingerprints.

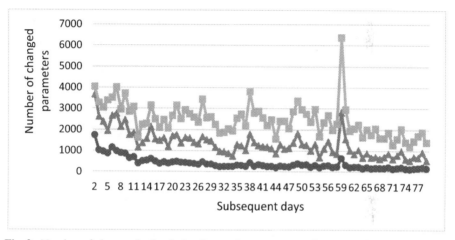

Fig. 2. Number of changes in the device fingerprint parameter values occurring on subsequent days: compared to the first visit (square), compared to the previous visit (triangle).

The graph in Fig. 3 presents the number of repeat visits on subsequent days, where at least one change in the parameter value defining the fingerprint was recorded. The trendline with squares presents the number of visitors on a given day with at least one change of the fingerprint parameter in comparison to the values registered on the first day of the study. The trendline with triangles presents the number of users on a given day with at least one change in comparison with the data extracted from the previous visit. Additionally, the trendline with triangles presents the number of all repeat visits on a given day. As can be observed, over the first days of the study, only some users were noted to have had some changes in the values of fingerprint parameters. However, on subsequent days, all the lines start to converge, which means that each entry to the website is accompanied by a number of changes. This means that each entry to the website shows at least one changed parameter in comparison to the first entry or the changes occurring during the previous visit.

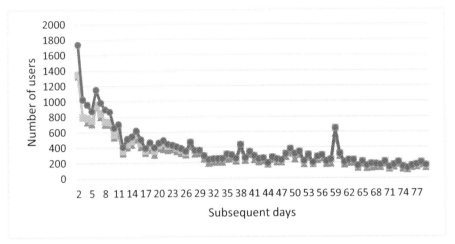

Fig. 3. Number of repeat visits to the monitored websites with changes in the values of the fingerprint parameters: compared to the first visit (square), compared to the previous visit (triangle).

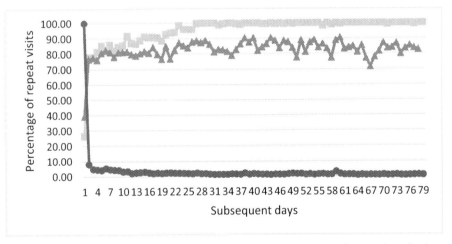

Fig. 4. Percentage of visitors returning to the website that have a change in the value of at least one parameter in comparison with the first visit (squares) and with the previous visit (triangles).

We can observe the rate of changes of the fingerprint parameter values by analysing the trendlines presented in Fig. 4. The line with circles shows the percentage of repeat visits (users returning in subsequent days) in comparison to the number of visitors accounted for on the first day of the study. The other trendlines show the percentage of visitors to the site on a given day where there was a change in the value of at least one parameter in comparison to the first visit (the squares) and to the previous visit (triangles). As can be observed, the changes actually take place from the first days of the study. As early as the second day of the study, most of the returning visitors have at least one parameter changed. However, the scale of changes in the number of parameters is not large. The average number of changed parameters of individual users can be seen in the graph in Fig. 5. At the end of the study it can be seen that the average number of changes of parameters that make up the fingerprint of one user when compared to the first visit is ca. 5 parameters (the squares), while in the case of the changes when compared to the previous visit it is ca. 2 parameters (the triangles). It is therefore possible that there are only a few unstable parameters, which show to be extremely variable during successive visits and might result in generating a completely new fingerprint.

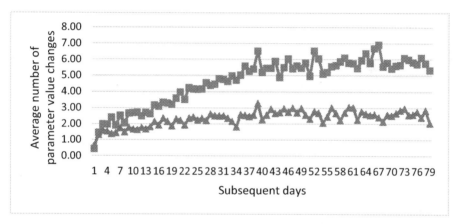

Fig. 5. Average number of changed parameter values in repeat visitors on subsequent days.

Thus, in order to identify the most unstable parameters contributing to the generation of the device fingerprint in the second stage of the study, changes in each parameter among returning users were analyzed separately. During subsequent visits we recorded the changes in the values of each parameter in comparison to the visit on the first day of the study as well as to the previous visit. The summary of the analysis of the collected data is presented in Tables 1 and 2. Table 1 lists the parameters with the least changes during the repeat visits, while Table 2 shows the parameters with the most changes in the values. In addition, the entropy value is given for each parameter. The entropy value allows us to identify those parameters which have a high diversity in values. The most valuable parameters are those characterised by a high stability with a concurrent high diversity in values. They are the most suitable parameters to be used for generating the device fingerprint.

Table 1. List of parameters with the least changes during the subsequent visits.

Changes compared to the visit on the first day			Changes compared to the previous visit		
Parameter name	Average number of changes per visit	Entropy	Parameter name	Average number of changes per visit	Entropy
ua_android_device_type	0.0003	0.51	oscpu	0	0.29
ua_device_brand_name	0.0003	2.62	ua_android_device_type	0.0002	0.51
ua_device_brand	0.0003	2.15	ua_device_brand_name	0.0002	2.62
ua_os_name	0.0004	1.47	ua_device_brand	0.0002	2.15
ua_os_short_name	0.0004	1.47	ua_os_name	0.0003	1.47
ua_device_type	0.0005	1.23	ua_os_short_name	0.0003	1.47
ua_client_type	0.0006	0.66	ua_device_type	0.0004	1.23
eval_str_len	0.0006	0.80	ua_device_model	0.0004	5.25
logic_cores	0.0006	1.08	ua_device_code	0.0004	5.45
sessionStorage_enabled	0.0006	0.01	ua_secondary_client_type	0.0005	0.36
indexedDB_enabled	0.0006	0.01	ua_secondary_client_name	0.0005	0.69
localStorage_enabled	0.0007	0.09	eval_str_len	0.0005	0.8
webgl_support	0.0007	0.01	logic_cores	0.0005	1.08
ua_device_model	0.0008	5.25	ua_client_type	0.0005	0.66
ua_device_code	0.0008	5.45	sessionStorage_enabled	0.0006	0.01
color_depth	0.0008	0.71	indexedDB_enabled	0.0006	0.01
device_memory	0.0009	1.90	localStorage_enabled	0.0006	0.09
platform	0.0010	1.86	color_depth	0.0007	0.71
ua_client_name	0.0010	2.87	webgl_support	0.0007	0.01
ua_client_short_name	0.0010	2.43	platform	0.0007	1.86

Table 2. List of parameters that have shown the most changes during subsequent visits.

Changes compared to the visit on the first day			Changes compared to the previous visit		
Parameter name	Average number of changes per visit	Entropy	Parameter name	Average number of changes per visit	Entropy
languages	0.0486	2.89	languages	0.0308	2.89
wout_width	0.0762	3.82	ac-outputLatency	0.0316	0.7
win_width	0.0796	3.94	extensions	0.0340	4.26
ua_uses_mobile_brow	0.0821	0.40	win_width	0.0541	3.94
rects	0.0857	5.30	timezone	0.0551	0.9
extensions	0.0919	4.26	rects	0.0576	5.3
canvas_2d_fingerprint	0.0985	6.95	wout_width	0.0589	3.82
logs	0.0995	2.28	canvas_2d_fingerprint	0.0664	6.95
ports	0.1112	1.64	logs	0.0785	2.28
placeholder_img_h	0.1278	1.25	br_version	0.0912	4.1
placeholder_img_w	0.1325	1.24	placeholder_img_h	0.1067	1.25
timezone	0.1537	0.90	placeholder_img_w	0.1106	1.24
ac-state	0.1722	1.00	ports	0.1107	1.64
wout_height	0.1994	7.39	ua_client_version	0.1214	5.53
ai2_nt_vc_result	0.2201	5.13	app_version	0.1367	12
win_height	0.2259	7.74	wout_height	0.1553	7.39
br_version	0.2371	4.10	ac-state	0.1636	1
ua_client_version	0.3082	5.53	win_height	0.1737	7.74
app_version	0.3327	12.00	ai2_nt_vc_result	0.2027	5.13
ai2_cc_result	0.5848	5.33	ai2_cc_result	0.5830	5.33

4 Conclusion

The conducted study has shown that developing a unique device fingerprint that would explicitly and clearly identify a given user should not only consist in collecting the values of the largest possible number of parameters. It is also important to consider the stability of their values during subsequent visits made by users. The analysis carried out during the research study shows that some of the parameters, despite having a high entropy, also have a very large number of changes (see Table 2). However, removing these parameters from the fingerprint generation process reduces the variety of parameters and may result in the generated fingerprint being identical to a fingerprint obtained from another user. Therefore one of the possibilities of using all the parameters can be the generation of two types of fingerprints:

- short-term fingerprint – featuring a high variety and concurrently a high instability. It can be created using as many parameters as possible, which may have the same values over a short period of time.
- long-term fingerprint – typically with a low variety, but at the same time a low variation of parameter values, which guarantees that the device remains the same over time, but at the same time it allows for the possibility of duplicates.

However, the development of a set of parameters assigned to both types of the fingerprint will require a number of additional studies. The algorithm could also be extended to allow encoding using artificial neural networks [13] with an appropriate structure [11, 15], big data algorithms [14], fuzzy methods [10] or others [12].

Acknowledgements. The presented results are obtained within the realization of the project "Traffic Watchdog 2.0 – verification and protection system against fraud activities in the online marketing (ad frauds) supported by artificial intelligence and virtual finger-print technology" financed by the National Centre for Research and Development; grant number POIR.01.01.01-00-0241/19-01.

References

1. Laperdrix, P., Bielova, N., Baudry, B., Avoine, G.: Browser fingerprinting: a survey. arXiv preprint arXiv:1905.01051 (2019)
2. Gabryel, M., Grzanek, K., Hayashi, Y.: Browser fingerprint coding methods increasing the effectiveness of user identification in the web traffic. J. Artif. Intell. Soft Comput. Res. **10**(4), 243–253 (2020)
3. Kobusińska, A., Pawluczuk, K., Brzeziński, J.: Big data fingerprinting information analytics for sustainability. Future Gener. Comput. Syst. **86**, 1321–1337 (2018)
4. Acar, G., Eubank, C., Englehardt, S., Juarez, M., Narayanan, A., Diaz, C.: The web never forgets: persistent tracking mechanisms in the wild. In: Proceedings of the 2014 ACM SIGSAC Conference on Computer and Communications Security, pp. 674–689 (2014)
5. Laperdrix, P., Rudametkin, W., Baudry, B.: Beauty and the beast: diverting modern web browsers to build unique browser fingerprints. In: 2016 IEEE Symposium on Security and Privacy (SP), pp. 878–894 (2016)

6. Steven, E., Arvind, N.: online tracking: a 1-million-site measurement and analysis. In: Proceedings of the 2016 ACM SIGSAC Conference on Computer and Communications Security (CCS '16), pp. 1388–1401, ACM, New York (2016)

7. Eckersley, P.: How unique is your web browser? In: Atallah, M.J., Hopper, N.J. (eds.) PETS 2010. LNCS, vol. 6205, pp. 1–18. Springer, Heidelberg (2010). https://doi.org/10.1007/978-3-642-14527-8_1

8. Vastel, A., Rudametkin, W., Rouvoy, R., Blanc, X.: FP-crawlers: studying the resilience of browser fingerprinting to block crawlers. In: NDSS Workshop on measurements, attacks, and defenses for the web (MADWeb'20) (2020)

9. Mouawi, R., Elhajj, I.H., Chehab, A., Kayssi, A.: Crowdsourcing for click fraud detection. EURASIP J. Inf. Secur. **2019**(1), 1–18 (2019). https://doi.org/10.1186/s13635-019-0095-1

10. Starczewski, J.T., Goetzen, P., Napoli, C.: Triangular fuzzy-rough set based fuzzification of fuzzy rule-based systems. J. Artif. Intell. Soft Comput. Res. **10**(4), 271–285 (2020). https://doi.org/10.2478/jaiscr-2020-0018

11. Bilski, J., Kowalczyk, B., Marchlewska, A., Zurada, J.M.: Local levenberg-marquardt algorithm for learning feedforwad neural networks. J. Artif. Intell. Soft Comput. Res. **10**(4), 299–316 (2020). https://doi.org/10.2478/jaiscr-2020-0020

12. Starczewski, A., Goetzen, P., Er, M.J.: A New method for automatic determining of the DBSCAN parameters. J. Artif. Intell. Soft Comput. Res. **10**(3), 209–221 (2020). https://doi.org/10.2478/jaiscr-2020-0014

13. El Zini, J., Rizk, Y., Awad, M.: An optimized parallel implementation of non-iteratively trained recurrent neural networks. J. Artif. Intell. Soft Comput. Res. **11**(1), 33–50 (2020). https://doi.org/10.2478/jaiscr-2021-0003

14. Koren, O., Hallin, C.A., Perel, N., Bendet, D.: Decision-making enhancement in a big data environment: application of the k-means algorithm to mixed data. J. Artif. Intell. Soft Comput. Res. **9**(4), 293–302 (2019). https://doi.org/10.2478/jaiscr-2019-0010

15. Niksa-Rynkiewicz, T., Szewczuk-Krypa, N., Witkowska, A., Cpałka, K., Zalasiński, M., Cader, A.: Monitoring regenerative heat exchanger in steam power plant by making use of the recurrent neural network. J. Artif. Intell. Soft Comput. Res. **11**(2), 143–155 (2021). https://doi.org/10.2478/jaiscr-2021-0009

RNA Folding Codes Optimization Using the Intel SDK for OpenCL

Mateusz Gruzewski and Marek Palkowski[✉]

Faculty of Computer Science and Information Systems, West Pomeranian University
of Technology in Szczecin, Zolnierska 49, 71210 Szczecin, Poland
mpalkowski@zut.edu.pl
http://www.wi.zut.edu.pl

Abstract. RNA folding algorithms are challenging dynamic programming tasks to optimize because they are resource-intensive and have
a large number of non-uniform dependences. Fortunately, these bioinformatics problems can be represented within the polyhedral model. It
allows us to use well-known source-to-source optimizing compilers. In
this paper, we applied cache efficient strategies developed within the
TRACO and Pluto compilers as well as manual codes of transpose and
classical loop skewing, and implemented these codes using Intel SDK
for OpenCL. This OpenCL platform allows running the unified code on
various computing units like multicore CPUs and GPUs including the
Intel graphic accelerators. We proposed the Intel OpenCL code generator using parallel implementations. We compared performance on chosen
HPC platforms using the basic Nussinov's RNA folding algorithm as an
example of non-serial uniform dynamic programming. Experiments were
carried out to achieve significant locality improvement and speed-up for
cross-platform unified code. We outlined related and future work.

Keywords: RNA folding · High performance computing · Intel
OpenCL · Loop tiling · Parallelization · Unified programming

1 Introduction

In this paper, we present an approach to generate the OpenCL for Nussinov's RNA folding. The Nussinov RNA secondary structure prediction is a
computing-heavy bioinformatics task. There are a lot of parallel implementations based mainly on the OpenMP pragmas, Intel TBB or C++17 Parallel STL
for CPUs, and CUDA framework for GPUs [3,4,6,7,14,17,18]. Unfortunately,
the approaches are often manual implementations or do not offer cross-platform
and unified programming.

We designed a tool which translates the OpenMP parallel codes [10] for Intel
OpenCL SDK [1]. The toolkit allows us to run codes for many target devices,
including multicore CPUs and GPUs. We chose Intel OpenCL SDK to achieve
access, especially to Intel GPUs. These cards cannot be used with NVIDIA CUDA
or OpenMP frameworks. We consider their performance, stability, and usage.

© Springer Nature Switzerland AG 2021
L. Rutkowski et al. (Eds.): ICAISC 2021, LNAI 12855, pp. 473–482, 2021.
https://doi.org/10.1007/978-3-030-87897-9_42

Nussinov's RNA folding is a useful dynamic programming benchmark that makes opportunities for cache-efficient scheduling on multi-core processors. Nussinov's two-dimensional arrays are accessed in a triply nested program loop. For chosen Intel CPUs and GPUs, we developed six cross-platform kernels to fold RNA strands. The optimized codes are calculated by compilers based on affine transformation framework [2], tile correction [11], time-space tiling [12] and present classical loop skewing and array transpose techniques. We carried experimental study to check the performance of Nussinov's RNA folding OpenCL codes and compare the performance of the kernels. We observed also that the efficiency of Intel embedded GPUs is significant and outperforms original serial execution on CPUs.

The rest of the paper is organized as follows: Sect. 2 introduces Nussinov's RNA folding algorithm, Sect. 3 presents related work, Sect. 4 describes the Intel SDK for OpenCL implementations details, Sect. 5 outlines the source-to-source tool for Intel OpenCL code generation, Sect. 6 shows the experimental study and the last section concludes the paper.

2   Nussinov's RNA Folding Algorithm

Nussinov made one of the first attempts at folding RNA in a computationally efficient way as the base pair maximization approach in 1978 [9]. An RNA sequence is a chain of nucleotides from the alphabet G (guanine), A (adenine), U (uracil), C (cytosine). Given an RNA sequence, the Nussinov algorithm solves the problem of RNA non-crossing secondary structure prediction through computing the maximum number of base pairs for subsequences, starting with subsequences of length 1 and building upwards, storing the result of each subsequence in a dynamic programming array.

Let N be a $n \times n$ Nussinov matrix and $\sigma(i, j)$ be a function which returns 1 if RNA[i], RNA[j] are a pair in the set (AU, UG, GC) and $i < j-1$, or 0 otherwise. Then the following recursion $N(i, j)$ is defined over the region $1 \leq i \leq j \leq n$ as

$$N(i, j) = max(N(i + 1, j - 1) + \sigma(i, j), \max_{1 \leq j \leq n} (N(i, k) + N(k + 1, j))) \quad (1)$$

and zero elsewhere [17].

The equation leads directly to the C/C++ code with triple-nested loops presented in Listing 1 [8].

Listing 1. Nussinov loop nest.

```
for (i = N-1; i >= 0; i--) {
 for (j = i+1; j < N; j++) {
  for (k = 0; k < j-i; k++) {
   S[i][j] = MAX(S[i][k+i] + S[k+i+1][j], S[i][j]); //s0
  }
  S[i][j] = MAX(S[i][j], S[i+1][j-1] + bond(RNA,i,j));//s1
 }
}
```

3   Related Work

There are many manual and automatic approaches implementing parallel and cache-efficient Nussinov's RNA folding [3,4,6,7,11,14,17,18] on CPUs, GPUs, co-processors, and FPGA platforms. Most of them are manual or dedicated only to one target hardware.

Li and et al. [5] proposed an interesting manual solution for Nussinov's RNA folding algorithm. Using lower and unused parts of Nussinov's, they changed column reading to more efficient row reading. Diagonal scanning exposes parallelism in the output code. The authors presented separated codes for CPUs and GPUs using OpenMP and CUDA, respectively. We chose their implementation for our approach.

Zhao et al. [19] improved the *Transpose* method and performed the experimental study of the energy-efficient codes. The approach based on the LRU cache model requires about half as much memory as does Li's Transpose. Unfortunately, the authors do not present parallel codes.

Pluto [2] is the most popular state-of-the-art source-to-source polyhedral code generator that transforms C programs to parallel coarse-grained code with enhanced data locality. The compiler is limited to code for CPUs with the OpenMP pragmas. Based on ATF compiler PPCG [16] produces code for only NVIDIA cards.

The idea of tiling presented in paper [11] is to transform (correct) original rectangular tiles so that all target tiles are valid under lexicographical order. Tile correction is performed by applying the transitive closure to loop nest dependence graphs. The paper presents very efficient codes for Nussinov's loop nests (the code is presented in Listing 2). Another cache-efficient is a space-time tiling technique for bioinformatics dynamic programming kernels [12]. Both optimization approaches are implemented within the TRACO compiler.

Listing 2. The Nussinov RNA folding loop nest implemented within the tile correction strategy.

```
for (c1 = 1; c1 < N + floord((N - 2),128); c1 += 1)
// parallel c3
for(c3 = (c1 - 1) / 129; c3<= max(0, -N + c1 + 1); c3++){
  r1 = N - c1 + 129 * c3 + 127;
  r2 = (N-c1+c3-1);
  for( c5 = 0; c5 <= 8 * c3; c5 += 1)
    for( c9 = N - c1 + 129 * c3; c9 <= min(N - 1, r1); c9 += 1)
      for( c11 = 16 * c5; c11 <= min(128 * c3, 16 * c5 + 15); c11 += 1)
        S[r2][c9] = max(S[r2][c11+r2] + S[c11+r2+1][c9], S[r2][c9]);
  for( c9 = N - c1 + 129 * c3; c9 <= min(N - 1, r1); c9 += 1)
    for( c10 = max(0, N - c1 + 129 * c3 - c9 + 1); c10 <= 1; c10 += 1) {
      if (c10 == 1) {
        S[r2][c9] = max(S[r2][c9], S[r2+1][c9-1]) + can_pair(RNA, r2, c9);
      } else
        for( c11 = 128 * c3 + 1; c11 <= -N + c1 - c3 + c9; c11 += 1)
          S[r2][c9] = max(S[r2][c11+r2] + S[c11+r2+1][c9], S[r2][c9]);
    }
}
```

Paper [13] presents an automated transpose technique for Nussionv's RNA folding. This is a pre-processing approach for optimizing compilers like TRACO or Pluto. The solution is not limited only to dynamic programming kernels, however, it can be applied only to two-dimensional arrays like Nussinov's array.

The approach presented in this paper allows us to produce the unified code for CPUs and GPUs using OpenCL. Furthermore, it is possible to profile the method for the target device in the same environment choosing between discussed approaches like loop tiling or transpose kernels. The code is generated automatically, portable and unified.

4   The Intel OpenCL Nussinov RNA Folding Algorithm Implementation

The Intel OpenCL is a software development kit for general-purpose parallel programming of heterogeneous systems. It provides a uniform programming environment that's used to write portable code for client personal computers, high-performance computing servers, hybrid, and embedded systems. It can be installed as a package for Intel Studio or Microsoft Visual Studio IDEs.

The Nussinov RNA folding code using Intel OpenCL implements two files: the host code and kernel file with .cl extension. Listings 1 and 4 present fragments of code executed by the host. Listing 5 presents kernel code executed at the target device. The codes are built based on Listing 2.

First, the host code makes OpenCL initialization, device, context, queue, and kernel creation (Listing 3). These operations reside in the OpenCLBasic (variable *oclobjects*) and OpenCLProgramOneKernel classes with function *NussinovDeviceKernelTileCor* implemented in kernel file *Nussinov.cl*. The outermost loop is executed in serial. The second loop is parallel and implemented with OpenCL. The function *NussinovKernel* collects pointers to arrays S (Nussinov's array) and RNA (sequence), OpenCL classes, lower and upper bounds of the second loop and value of index variable of the outermost loop, *c1*.

Listing 3. The OpenCL initialization and the Nussinov's outermost tiled loop nest.

```
OpenCLProgramOneKernel executable(oclobjects, L"Nussinov.cl", "",
"NussinovDeviceKernelTileCor");
for (c1 = 1; c1 < N + floord((N - 2),128); c1 += 1)
{
    cl_int ub = (c1 - 1) / 129;
    cl_int lb = max(0, -N + c1 + 1);
    NussinovKernel(S\_par, RNA\_par, N, oclobjects, executable, lb, ub, c1);
}
```

Listing 4 presents fragments of the *NussinovKernel* function. Arrays S and RNA are allocated in the memory of target device. Their content are copied. Next, we prepare arguments of the kernel with the OpenCL function *clSetKernelArg*. *Global_wor_size* represents number of threads. It is defined as *ub-lb+1*. Hence, each thread executes the body for one value of the index variable of the second parallel loop. Kernel is queued and executed with *ocl.queue*. The code of Nussinov.cl is executed on the target device. The host waits with the *clFinish* method. Next, the content of the S array is copied back to the host memory.

Listing 4. The part of NussinovKernel function body.

```
cl_mem cl_S_buffer = clCreateBuffer(
ocl.context, CL_MEM_READ_WRITE | CL_MEM_USE_HOST_PTR,
zeroCopySizeAlignment(sizeof(cl_int) * N * N),
S_par, &err
);

cl_mem cl_RNA_buffer = clCreateBuffer(
ocl.context, CL_MEM_READ_ONLY | CL_MEM_USE_HOST_PTR,
zeroCopySizeAlignment(sizeof(cl_char) * N),
RNA_par, &err
);
...
err = clSetKernelArg(exec.kernel, 0, sizeof(cl_mem), (void*) &cl_S_buffer);
err = clSetKernelArg(exec.kernel, 1, sizeof(cl_mem), (void*) &cl_RNA_buffer);
err = clSetKernelArg(exec.kernel, 2, sizeof(cl_mem), (void*) &cl_lb_buffer);
err = clSetKernelArg(exec.kernel, 3, sizeof(cl_mem), (void*) &cl_c1_buffer);

size_t global_work_size[1] = {(size_t)ub-lb+1};
clEnqueueNDRangeKernel(ocl.queue, exec.kernel, 1, NULL, global_work_size,
    NULL, 0, NULL, NULL);
err = clFinish(ocl.queue);

void* tmp_ptr = NULL;
tmp_ptr = clEnqueueMapBuffer(ocl.queue, cl_S_buffer, true, CL_MAP_READ, 0,
                            sizeof(cl_int) * N * N, 0, NULL, NULL, NULL);
if(tmp_ptr!=S_par)
    throw Error("clEnqueueMapBuffer failed to return original pointer");
```

Listing 5 represents the kernel function implemented in the Nussinov.cl file. Variable x represents a number of thread. The index variable of the third loop is computed as a sum of x and lower bound lb of the index variable of the second parallel loop. The tiled loops in the kernel are executed in serial order. The code is rewritten from the source parallel code. The listings present the optimized Nussinov code with applied the tile correction approach. The rest of the parallel codes discussed in this paper are built in the same manner. The kernel functions can be placed in the same Nussinov.cl file with different function names.

Listing 5. The kernel code executed on the target device.

```
__kernel void NussinovDeviceKernelTileCor(__global int S[N][N], __global char*
  RNA, __global int* lb_arr, __global int *c1_arr)
{
    const int x = get_global_id(0);
    int lb = lb_arr[0];
    int c1 = c1_arr[0];
    int c3,c5,c9,c11,c10;
    c3 = lb + x;

    r1 = N - c1 + 129 * c3 + 127;
    r2 = (N-c1+c3-1);
    for( c5 = 0; c5 <= 8 * c3; c5 += 1)
        for( c9 = N - c1 + 129 * c3; c9 <= min(N - 1, r1); c9 += 1)
            for( c11 = 16 * c5; c11 <= min(128 * c3, 16 * c5 + 15); c11 += 1)
                S[r2][c9] = max(S[r2][c11+r2] + S[c11+r2+1][c9], S[r2][c9]);
    for( c9 = N - c1 + 129 * c3; c9 <= min(N - 1, r1); c9 += 1)
        for( c10 = max(0, N - c1 + 129 * c3 - c9 + 1); c10 <= 1; c10 += 1) {
            if (c10 == 1) {
                S[r2][c9] = max(S[r2][c9], S[r2+1][c9-1]) + can_pair(RNA, r2, c9);
            } else
                for( c11 = 128 * c3 + 1; c11 <= -N + c1 - c3 + c9; c11 += 1)
                    S[r2][c9] = max(S[r2][c11+r2] + S[c11+r2+1][c9], S[r2][c9]);
}    }
```

5   Code Generator

As part of scientific research, a code transformer from OpenMP to optimized code using OpenCL was created. The code generator analyzes the loops that are marked for parallelization and uses them to prepare the output code in OpenCl. The algorithm used by the generator is presented below.

Input: The parallel and optimized code in the OpenMP format
Output: The parallel and optimized code in the Intel OpenCL format

1. Rewrite the input code to the line with the parallel directive and add it to the host code. The outer loops index variables are shared.
2. Detect shared arrays, their sizes, and describe their usage (write, read, or read-write).
3. Store lower and upper bounds of the parallel loop to the variables *lb*, *ub*, respectively. Calculate them using an expression from the parallel loop.
4. Declare arrays and copy them to the device memory.
5. Calculate number of threads as *ub-lb+1*.
6. Run kernel.
7. Calculate the index variable of the parallel loop inside the kernel as *lb* + *x*, where *x* is a number of the kernel instance executed on the device.
8. Rewrite the rest code to the kernel and execute it in the serial order.
9. After the kernel execution on the target device, the written arrays are copied to the host memory.

The tool is developed in the Python language. Below is an example of the result of the loop analysis through the OpenCL code generator. It describes a parallel loop, an index variable of this loop and its bounds, shared array and indexes, the statements of the loop, and the original code of the loop from the input code. Using this information, the tool creates the output code of Nussinov's algorithm in the OpenCL format.

Listing 6. Loop analysis example.

```
for_n(
init={'name': 'j', 'value': 'i+1'},
end_condition={'var': 'j', 'value': 'N-1'},
increment={'var': 'j', 'inc': 1},
 instructions=[<P....0>,
 {
'var': 'S',
'index': '[i][j]',
'val': 'MAX(S[i][j],S[i+1][j-1]+bond(RNA,i,j))',
'original_line': 'S[i][j]=MAX(S[i][j],S[i+1][j-1]+bond(RNA,i,j));'
}], variables={}, is_parallel=True, original_line='for(j=i+1;j<N;j++)'
}
```

6   Experimental Study

To carry out experiments, we used two Intel machines: Intel Core i5-4440 CPU 3.10GHz (4 threads, 8GB RAM, 6MB Cache) with GPU Intel HD 4600 (24

Shading Units, 12 Execution Unit, Boost Clock 900 MHz), and Intel Core i7-8700 CPU 3.20 GHz (12 threads, 16 GB RAM, 12 MB Cache) with GPU Intel UHD 630 (192 Shading Units, 24 Execution Unit, Boost Clock 1050 MHz). The project is compiled with the Visual Studio C++ 2019 on Windows 10 with the Intel OpenCL SDK 18.1 package.

We converted five OpenMP implementations of the Nussinov RNA folding to the Intel OpenCL format, executed them on CPUs, and compared them to the original code. We considered three tiled codes produced with Pluto (ATF) and TRACO (Time-spacing tiled code and tile correction). The Pluto is not able to tile the innermost loop nest of the three Nussinov's code loop nests [11] and this fact limits the efficiency of this approach. We also examined two cache-efficient codes implementing array transpose [5,13]. Additionally, we applied the classical loop skewing for the Nussinov RNA folding code. This technique is applied to a nested loop iterating over a Nussinov's array, where each iteration of the inner loop depends on previous iterations, and rearranges its array accesses so that the only dependencies are between iterations of the outer loop. For Nussinov's RNA folding code iterations [i,j,k] are skewed to [i+j,j,k] to parallelize the second loop (the outermost loop is serial). The schedule of this code can be calculated using the ISL framework [15].

Table 1. Time results for Intel Core i5-4440 CPU 3.10 GHz.

N	Original	Pluto ATF	Tile correction	Time-space tiling	Traco transpose	Li transpose	Loop skewing
1000	1,28	0,54	0,55	0,54	0,93	0,72	0.75
2500	30,98	4,81	3,24	2,33	3,98	2,44	4.30
5000	250,25	53,72	18,7	14,52	18,33	12,36	37.12
7500	1020,18	213,38	67,02	60,61	57,47	46,91	117.60
10000	2052,15	522,26	121,25	124,57	119,65	84,92	299.60

Table 2. Time results for Intel Core i7-8700 CPU 3.20 GHz.

N	Original	Pluto ATF	Tile correction	Time-space tiling	Traco transpose	Li transpose	Loop skewing
1000	0,12	0,65	0,9	0,66	0,75	0,68	0.71
2500	4,63	1,95	2,15	1,12	1,36	1,21	1.62
5000	69,67	14,08	9,609	5,47	7,47	7,37	10.39
7500	301,66	53,32	34,49	27,21	21,33	24,12	37.06
10000	771,68	137,08	61,94	49,75	49,8	56,2	97.14

The time results in seconds are presented in Tables 1 and 2 for i5 CPU and i7 CPU, respectively.

Table 3. Time results for Intel HD 4600.

N	Pluto ATF	Tile correction	Time-space tiling	Traco transpose	Li transpose	Loop skewing
1000	3,52	14,67	3,09	3,12	1,94	2,19
2500	22,96	88,43	34,67	34,98	15,42	9,98
5000	143,6	455,54	163,32	287,65	118,56	52,33
7500	460,55	2076,55	449,303	1145,35	383,94	173,12
10000	1117,09	4012,11	978,901	2688,88	852,34	419,06

Table 4. Time results for Intel UHD 630.

N	Pluto ATF	Tile correction	Time-space tiling	Traco transpose	Li transpose	Loop skewing
1000	2,69	10,33	4,07	2,39	1.43	1,03
2500	16,909	44,88	23,86	15,39	8,48	3,07
5000	100,83	248,66	104,86	130,99	62.33	16,7
7500	291,888	789.54	254,94	335,78	121,96	46,23
10000	657,07	1563,99	455,156	981,44	319,05	116,62

Next, we executed the optimized codes on the GPUs (Intel HD 4600 and UHD 630). The graphic processors solve Nussinov's RNA folding without any code modifications (only a target machine is changed in the user's settings). The time results in seconds are presented in Tables 3 and 4 for HD 4600 and UHD 630, respectively.

Analyzing time results in tables, we conclude that: i) the best results for i7 CPU are achieved with the time-space tiled and transposed codes produced by compiler TRACO, for older i5 the most efficient code is the implementation of the Li's Transpose approach; ii) GPUs produces faster code than original code executed on one CPU core; iii) the best time results for both Intel GPUs is classical loop skewing; iv) tiled and transposed code is not so efficient for Intel GPUs as well as for CPUs (processors have more complex cache structure and it gives chance for better locality); vi) skewed loops on GPUs are faster than the Pluto code performed on CPUs (the innermost RNA Nussinov's loop tiling is crucial to achieving satisfactory performance).

Figure 1 depicts the top 10 time results (below 125 s) for RNA strands with ten thousand nucleotides. We can find the Nussinov's RNA folding skewed loop nests executed on UHD 630 in this set on the 7th position.

The Nussinov RNA folding Intel OpenCL implementation is available on the github repository https://github.com/lshadown/nussinov-opencl. The code generator sources are available on the github repository https://github.com/lshadown/opencl-generator.

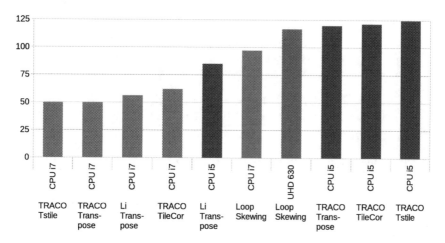

Fig. 1. Time results below 125 s for RNA sequences with 10000 length.

7 Conclusion

The paper presents the approach to generate the Intel OpenCL codes for Nussinov's RNA Folding. We applied parallel cross-platform programming and run the same code for CPUs and GPUs. We chose multi-core Intel processors for the experimental study. The approach can be used as a post-processor tool for state-of-art optimizing compilers like Pluto or TRACO. Loop tiling and array transposition improves code locality and are very efficient for CPU units. The performance of Intel graphic processors is significant for the Nussinov RNA folding skewed loop nests.

In the future, we are going to consider using the presented approach for other cross-platform frameworks like SyCL or oneApi to produce optimized code for Nussinov's RNA folding. We will extend our research to achieve OpenMP code with two parallel loops to rewrite them to the 2D OpenCL kernels for GPUs. We plan also to apply the approach for other bioinformatics dynamic programming kernels to minimize free energy or probabilistic models for RNA secondary structure prediction.

References

1. Intel SDK for OpenCL applications (2021). https://software.intel.com/content/www/us/en/develop/tools/opencl-sdk.html
2. Bondhugula, U., et al.: A practical automatic polyhedral parallelizer and locality optimizer. SIGPLAN Not. **43**(6), 101–113 (2008). http://pluto-compiler.sourceforge.net
3. Chang, D.J., Kimmer, C., Ouyang, M.: Accelerating the Nussinov RNA folding algorithm with CUDA/GPU. In: The 10th IEEE International Symposium on Signal Processing and Information Technology, pp. 120–125, December 2010

4. Jacob, A., Buhler, J., Chamberlain, R.D.: Accelerating Nussinov RNA secondary structure prediction with systolic arrays on FPGAs. In: Proceedings of the 2008 International Conference on Application-Specific Systems, Architectures and Processors, pp. 191–196, ASAP 2008. IEEE Computer Society, Washington, DC, USA (2008)

5. Li, J., Ranka, S., Sahni, S.: Multicore and GPU algorithms for Nussinov RNA folding. BMC Bioinform. **15**(8), S1 (2014). https://doi.org/10.1186/1471-2105-15-S8-S1

6. Liu, L., Wang, M., Jiang, J., Li, R., Yang, G.: Efficient nonserial polyadic dynamic programming on the cell processor. In: IPDPS Workshops, Anchorage, Alaska, pp. 460–471. IEEE (2011)

7. Mathuriya, A., Bader, D.A., Heitsch, C.E., Harvey, S.C.: GTfold: a scalable multicore code for RNA secondary structure prediction. In: Proceedings of the 2009 ACM Symposium on Applied Computing, New York, NY, USA, pp. 981–988, SAC 2009. ACM (2009)

8. Mullapudi, R.T., Bondhugula, U.: Tiling for dynamic scheduling. In: Rajopadhye, S., Verdoolaege, S. (eds.) Proceedings of the 4th International Workshop on Polyhedral Compilation Techniques, Vienna, Austria, January 2014

9. Nussinov, R., Pieczenik, G., Griggs, J.R., Kleitman, D.J.: Algorithms for loop matchings. SIAM J. Appl. Math. **35**(1), 68–82 (1978)

10. OpenMP Architecture Review Board: OpenMP application program interface version 4.0 (2012). http://www.openmp.org/mp-documents/OpenMP4.0RC1_final.pdf

11. Palkowski, M., Bielecki, W.: Parallel tiled Nussinov RNA folding loop nest generated using both dependence graph transitive closure and loop skewing. BMC Bioinform. **18**(1), 290 (2017)

12. Palkowski, M., Bielecki, W.: Tiling Nussinov's RNA folding loop nest with a space-time approach. BMC Bioinform. **20**(1), 208:1–208:11 (2019)

13. Palkowski, M., Bielecki, W., Gruzewski, M.: Automatic generation of parallel cache-efficient code implementing Zuker's RNA folding. In: Rutkowski, L., Scherer, R., Korytkowski, M., Pedrycz, W., Tadeusiewicz, R., Zurada, J.M. (eds.) ICAISC 2020. LNCS (LNAI), vol. 12415, pp. 646–654. Springer, Cham (2020). https://doi.org/10.1007/978-3-030-61401-0_60

14. Rizk, G., Lavenier, D.: GPU accelerated RNA folding algorithm. In: Allen, G., Nabrzyski, J., Seidel, E., van Albada, G.D., Dongarra, J., Sloot, P.M.A. (eds.) ICCS 2009. LNCS, vol. 5544, pp. 1004–1013. Springer, Heidelberg (2009). https://doi.org/10.1007/978-3-642-01970-8_101

15. Verdoolaege, S.: Integer set library - manual. Technical report (2011). www.kotnet.org/~skimo//isl/manual.pdf

16. Verdoolaege, S., Janssens, G.: Scheduling for PPCG (2017)

17. Wonnacott, D., Jin, T., Lake, A.: Automatic tiling of "mostly-tileable" loop nests. In: IMPACT 2015: 5th International Workshop on Polyhedral Compilation Techniques, At Amsterdam, The Netherlands (2015)

18. Xia, F., Dou, Y., Zhou, X., Yang, X., Xu, J., Zhang, Y.: Fine-grained parallel RNAalifold algorithm for RNA secondary structure prediction on FPGA. BMC Bioinform. **10**(S1), S37 (2009)

19. Zhao, C., Sahni, S.: Cache and energy efficient algorithms for Nussinov's RNA folding. BMC Bioinform. **18**(15), 518 (2017)

A Comparison of Machine Learning Techniques for Diagnosing Multiple Myeloma

Luveshan Marimuthu[1], Nelishia Pillay[1(✉)] [iD], Rivak Punchoo[2,3] [iD],
and Sachin Bhoora[3] [iD]

[1] Department of Computer Science, University of Pretoria, Gauteng, South Africa
`nelishia.pillay@up.ac.za`
[2] National Health Laboratory Service, Tshwane Academic Division,
Gauteng, South Africa
[3] Department of Chemical Pathology, University of Pretoria, Gauteng, South Africa
`rivak.punchoo@up.ac.za`

Abstract. Multiple myeloma is a type of bone marrow cancer. Patient's blood samples are analysed from protein gel strips and densitometer graphs which are then interpreted by a pathologist to diagnose multiple myeloma. This manual process of diagnosis is slow which is problematic as patients need to be diagnosed as soon as possible in order to prevent the condition from worsening, hence the need to automate this process. Given the success of machine learning in diagnosing diseases from images, this study investigates the use of machine learning approaches for the diagnosis of multiple myeloma. This is the first study investigating the automation of this process and hence presents a novel application of machine learning. The study compares machine learning approaches, artificial neural networks, convolutional neural networks, random forests and support vector machines, for the diagnosis of multiple myeloma from gel strips and densitometer graphs. The study has revealed that convolutional neural networks, specifically VGG16, is the most suitable approach for the detection of multiple myeloma.

Keywords: Machine learning · Multiple myeloma · Classification

1 Introduction

Multiple myeloma is a type of bone marrow cancer that affects the plasma cells in the blood. Plasma cells originate in the bone marrow and play an important role in the immune system and the secretion of antibodies that help fight off foreign viruses and bacteria. The diagnosis of multiple myeloma involves analyzing a patient's blood serum samples in the form of gel strips and densitometer graphs and is a relatively slow and sometimes inconsistent process. This process is also important as a follow-up examination for patients who are undergoing treatment

© Springer Nature Switzerland AG 2021
L. Rutkowski et al. (Eds.): ICAISC 2021, LNAI 12855, pp. 483–494, 2021.
https://doi.org/10.1007/978-3-030-87897-9_43

[7]. Given the importance of this procedure, having a faster, more consistent method of diagnosis will definitely be of benefit to both patients and pathologists.

Since the diagnostic process is time consuming and sometimes prone to inconsistency between interpretations, it can be an issue for patients because delayed diagnosis can negatively affect the disease's development. There is currently no existing automation for the diagnosis of multiple myeloma using gel strips and/or densitometer graphs. This study aims at automating the process of diagnosing multiple myeloma from gel strips and densitometer graphs using machine learning. The effectiveness of machine learning for such medical diagnosis from images is evident from the study conducted by Wang et al. [13] on classifying mediastinal lymph node metastasis. The authors conduct a comparative study of machine learning approaches and doctor diagnosis. Majority of the machine learning approaches produced a better accuracy than that of the doctor.

Chemical pathologists diagnose multiple myeloma from gel strips and densitometer graphs. The research presented in this paper compares different machine learning approaches generally successful for disease diagnosis, namely, artificial neural networks (ANNs), convolutional neural networks (CNNs), support vector machines (SVMs) and random forests (RFs), for automating this process. The study revealed that CNNs are the most effective for multiple myeloma diagnosis, giving the best accuracy.

The following section describes previous work investigating machine learning techniques for multiple myeloma diagnosis and highlights how the research presented in this paper differs from this. Section 3 describes the manual process of diagnosing multiple myeloma that this study aims to automate using machine learning. The machine learning approaches employed for automation are presented in Sects. 4.1 through to 4.4. The experimental setup used to evaluate the performance of the different machine learning techniques in diagnosing multiple myeloma is provided in Sect. 5. The performance of the approaches in diagnosing multiple myeloma is compared in Sect. 6. Finally, Sect. 7 summarizes the findings of the research and presents future extensions of the work.

2   Machine Learning for Multiple Myeloma Diagnosis

This section provides an overview of machine learning approaches applied for multiple myeloma diagnosis and illustrates how the research presented in this paper differs from these studies.

Previous work applying machine learning techniques to detect multiple myeloma include the study by Xu et al. [15] which used machine learning techniques to detect bone lesions from images. The study compared various machine learning approaches for this purpose and convolutional neural networks were found to produce the best accuracy. In the study conducted by Bhattacharyya et al. [2] bone lesions were detected by applying machine learning to identify potential biomarkers indicative of bone lesions in blood serum samples of patients. Both partial least squares discriminant analysis and random forests were found to be effective. Turki et al. [10] used convolutional neural networks with transfer learning, trained on data for other types of cancer, to predict drug sensitivity of multiple myeloma patients. Waddell et al. [12] used machine learning to predict the susceptibility of early-onset (before the age of 40) of multiple myeloma from single-nucleotide polymorphism data using support vector machines.

The research presented in this paper differs from previous work in that machine learning is employed to perform the diagnosis of multiple myeloma traditionally performed by chemical pathologists from gel strips and densitometer graphs. This is the first study conducting such an investigation. From the literature it is evident that machine learning approaches generally successful for disease diagnosis include artificial neural networks (ANNs), convolutional neural networks (CNNs), support vector machines (SVMs) and random forests RFs).

ANNs have been used in numerous studies to diagnose diseases from different images such as blood serum samples [11], ultrasonic images [14] and PET/CT scans [13]. These studies report highly accurate models for diagnosis. CNNs are excellent feature extractors and can avoid the complicated and expensive feature engineering that is often needed for medical images [16].

In a study using PET/CT images to detect lung cancer, support vector machines had the fastest training time when compared to ANN, RF and CNN [13]. The ability of SVMs to map input features to higher dimensions could lead to more accurate learning of the patterns in the data. Sopharak et al. [9] applied SVM classification to automatic micro-aneurysm detection for diabetic retinopathy. The algorithm analyzed retinal micro-aneurysms with a diameter of 10–$100\,\mu m$ and the results showed that the SVM performed better than Naïve Bayes and K-Nearest Neighbor. Ullah et al. [11] compared a SVM, random forest and neural networks for the risk prediction of asthma disease using blood serum samples and the results showed that the SVM was the most suitable technique. The SVM also showed the highest accuracy in a study using histopathological images to classify head and neck tumours [1].

In a study on the prediction of fatty liver disease by Wu et al. [14], the random forest classifier showed a higher performance than ANNs and Naïve Bayes.

The research presented in this paper differs from previous work in that it attempts to automate the diagnosis performed by chemical pathologist from gel strips and densitometer graphs. This is the first study investigating machine learning for this purpose. From the literature surveyed, it is evident that different machine learning techniques produced better results for the diagnosis of different diseases. The most commonly used techniques include ANNs, CNNs, RF and SVMs. A comparative study of these techniques for diagnosing multiple myeloma from gel strips and densitometer graphs is conducted. The following section describes the manual process employed by chemical pathologists to diagnose multiple myeloma from gel strips and densitometer graphs.

3   Diagnosing Multiple Myeloma

Chemical pathologists use gel strips and densitometer graphs to diagnosis multiple myeloma. Figure 1 illustrates examples of two gel strips that are typically used for multiple myeloma diagnosis. The first gel strip is that of a patient that does not have multiple myeloma and the second of a patient that does. The corresponding densitometer graphs are depicted in Fig. 2.

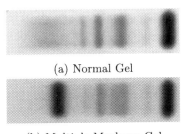

(a) Normal Gel

(b) Multiple Myeloma Gel

Fig. 1. Example gel strips

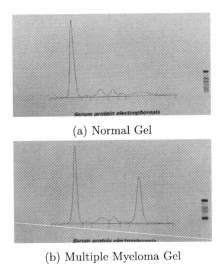

(a) Normal Gel

(b) Multiple Myeloma Gel

Fig. 2. Example graphs

Chemical pathologists analyse either the gel strips or the densitometer graphs or both to determine the following in order to diagnose multiple myeloma:

– M-Protein - whether M-Protein is present or not.
– Albumin - whether the Albumin concentration is low, normal or high.
– Alpha1 - whether the Alpha1 concentration is low, normal or high.
– Alpha2 - whether the Alpha2 concentration is low, normal or high.

- Beta1 - whether the Beta1 concentration is low, normal or high.
- Beta2 - whether the Beta2 concentration is low, normal or high.
- Gamma - whether the gamma concentration is low, normal or high.

In an attempt to automate this process this study uses machine learning techniques to determine the presence of M-Protein and the concentration of Albumin, Alpha1, Alpha2, Beta1, Beta2 and Gamma from images of the gel strips and densitometer graphs. The techniques are applied separately to the gel strip and densitometer graph images. Supervised learning is used and each image in the training set has labels indicating the presence of M-Protein (yes or no) and the concentration of Albumin, Alpha1, Alpha2, Beta1, Beta2 and Gamma (low, normal or high).

4   Machine Learning Approaches

The machine learning approaches employed for multiple myeloma diagnosis are described in the sections that follow.

4.1   Artificial Neural Network (ANN)

This section describes the ANN. The architecture is comprised of one input layer, three hidden layers and one output layer. The size of the input layer is determined by the size of the input image. The three hidden layers have 256, 128 and 64 nodes respectively. The output layer has 2 or 3 nodes depending on the target values of the problem (M-protein target uses 2 output nodes and the region targets use 3 output nodes). The three hidden layers use the ReLU activation function and the output layer uses the Softmax activation function. The weights in the hidden layers are initialized using the He Normal initialization method [3] which helps avoid saturation and works well for rectified activations. L2 regularization is also incorporated into the hidden layers with a lambda value of 0.001. The Adam optimizer [4] is used for its popularity and overall performance and speed in practice. The loss function used is categorical cross-entropy for multi-class classification.

4.2   Convolutional Neural Network (CNN)

In this case the architecture of the CNN is the VGG16 model proposed by K. Simonyan and A. Zisserman [8] which was used in the ImageNet ILSVRC2014 competition. A smaller, custom-made CNN was tested initially but was outperformed by the VGG16 network in both speed and accuracy.

The VGG16 model was downloaded from the Keras Applications API and used with the provided pre-trained weights. The architecture of the VGG16 network requires the shape of the input images to be 224×224. The 1000 node output layer is replaced with an output layer with 2 or 3 nodes depending on the target classes for each problem (M-protein uses 2 output nodes and the region targets use 3 output nodes). Only the output layer and the last fully connected layer have trainable weights and all previous layers have their weights fixed to the pre-trained values. If only the output layer's weights are freed the model does not perform as well and if more fully connected layers weights are freed the number of parameters becomes large enough to start over-fitting/'memorizing' the current dataset. If the dataset grows much larger in the future, it could be feasible for more weights to be freed in the model for better results.

The output layer uses the softmax activation function. All specifics of the convolutional, pooling and fully connected layers remain unchanged from the original VGG16 network. The Adam optimizer [4] is used for its overall performance and speed. The loss function used is categorical cross-entropy for multi-class classification.

4.3   Random Forest (RF)

The Scikit-Learn library was used to implement the random forest in the study. The RF classifier is initialized with the feature split criterion, a random state and the number of estimator/trees to be used. For the split criterion, the Gini impurity index is used.

4.4   Support Vector Machine (SVM)

The support vector machine used is scikit-learn's C-Support Vector Classification (SVC) model. The model is initialized with a kernel type, degree (for polynomial kernel), gamma value, class weight and decision function shape. The radial basis function(rbf) kernel was compared to the polynomial kernel with varying degrees. Testing showed that the polynomial kernel is able to find decision boundaries more effectively than rbf for all problem cases. The gamma value is set to scale which calculates gamma as the inverse of the number of features multiplied by the variance of the data i.e. $1/(n\_features * X.var())$. The class weight is set to balanced which uses the target values to adjust weights inversely proportional to the class frequencies. The decision function shape is one-vs-one which is the required function for multi-class classification.

5   Experimental Setup

This section describes the experimental setup employed to compare the machine learning approaches for diagnosing multiple myeloma.

5.1   ANN Parameters

The parameters that are varied for each problem is the learning rate(LR) and number of epochs(Ep) and is provided in Table 1 below. These values were determined empirically by performing trial runs.

Table 1. Artificial neural network parameters

	Gel		Graph	
	Ep	LR	Ep	LR
Mprotein	130	0.0000002	75	0.0000001
Albumin	50	0.0005	60	0.0001
Alpha1	100	0.00005	50	0.000002
Alpha2	100	0.00005	50	0.000005
Beta1	120	0.00005	100	0.00005
Beta2	100	0.00005	75	0.00001
Gamma	175	0.0001	75	0.00002

As seen in the table, a very small learning rate is needed across all cases. The number of epochs was determined by looking at the training accuracy as well as the processing time required. The gel images trained over more epochs because the gel images themselves were relatively small and did not take up a lot of training time. The graph images however were much larger and took a substantial while longer to train.

Another parameter that was tuned for cases is the batch size used for mini-batch training. This was tuned based on either accuracy or training time as a slightly bigger batch size usually achieves slightly faster training times. A batch size of 6 was used for the M-protein model for both gels and graphs and the Beta2 model for graphs. All other models use a batch size of 4. The Alpha1, Alpha2 and Beta2 models incorporate early stopping because the accuracy observed during training at times showed no improvement over a certain number of epochs. The parameters for early stopping were a minimum change of 0.0001 and a patience of 5, 10, 15 for the Alpha1, Alpha2 and Beta2 regions respectively.

5.2   CNN Parameters

The parameters that are varied for each problem is the learning rate(LR) and number of epochs(Ep) as shown in Table 2 below.

Table 2. Convolutional neural network parameters

	Gel		Graph	
	Ep	LR	Ep	LR
Mprotein	10	0.0001	10	0.0001
Albumin	6	0.0001	10	0.0001
Alpha1	6	0.0001	10	0.0001
Alpha2	6	0.0001	10	0.0001
Beta1	10	0.00005	10	0.00005
Beta2	10	0.0002	15	0.0001
Gamma	12	0.00005	12	0.00005

5.3   RF Parameters

Table 3 below provides the values for the number of estimators/trees (n-Est) used which were fine-tuned for each problem case.

Table 3. Random forest parameters

	Gel	Graph
	n-Est	n-Est
Mprotein	10	12
Albumin	17	21
Alpha1	23	11
Alpha2	39	23
Beta1	99	7
Beta2	7	7
Gamma	45	73

5.4   SVM Parameters

Table 4 below shows the degrees of the polynomial kernels used for each problem case which were determined through trail and error.

Table 4. Support vector machine parameters

	Gel	Graph
	degree	degree
Mprotein	1	2
Albumin	2	2
Alpha1	4	4
Alpha2	2	2
Beta1	4	2
Beta2	2	2
Gamma	4	3

The training times for the SVM were fairly quick, much faster than the neural networks but not as fast as random forest.

5.5   Statistical Tests

Statistical tests were performed to determine the significance of the performance comparisons of the machine learning approaches. If the results met the assumptions of normal distribution and homogeneity of variances, the parametric t-test and one-way analysis of variance (ANOVA) test was used [6]. If not, the non-parametric Man-Whitney U [5] test and the Kruskal-Wallis H test was used. To determine which model to choose the post hoc Dunn test was performed. All tests were performed at a 5% significance level.

6   Results and Discussion

This section compares the performance of the machine learning approaches for determining the presence of M-Protein and the concentration of Albumin, Alpha1, Alpha2, Beta1, Beta2 and Gamma from gel strips and densitometer graphs. The performance of the approaches are compared according to the average and best accuracy. In the tables in the following sections, 'GEL' refers to the model using the gel strip images and 'SPEP' refers to the model using the densitometer graph images.

Table 5, 6, 7, 8, 9, 10 and 11 present the performance comparisons for M-Protein, Albumin, Alpha1, Alpha2, Beta1, Beta2 and Gamma respectively. The processing time is presented in seconds. As is evident from the tables VGG-16 performs better than the other approaches. This result was found to be statistically significant for Beta 1 for graphs and not for the other comparisons. The processing time of the VGG-16 is also reasonable in comparison to the other the approaches.

Table 5. Performance comparison for M-Protein

	ANN		VGG16		RF		SVM	
	GEL	SPEP	GEL	SPEP	GEL	SPEP	GEL	SPEP
Average training accuracy	83.82	81.47	99.97	99.91	98.57	97.97	78.27	100.0
Average test accuracy	78.19	77.22	87.68	92.83	78.86	78.1	73.42	74.68
Best training accuracy	84.66	81.47	100.0	100.0	99.68	99.36	78.27	100.0
Best test accuracy	81.01	77.22	89.87	93.67	82.28	81.01	73.42	74.68
Processing time(s)	100.14	176.49	65.73	66.99	0.38	1.42	21.61	43.23

Table 6. Performance comparison for Albumin

	ANN		VGG16		RF		SVM	
	GEL	SPEP	GEL	SPEP	GEL	SPEP	GEL	SPEP
Average training accuracy	82.02	81.47	97.58	98.7	99.5	97.97	88.82	100.0
Average test accuracy	89.75	89.87	83.8	84.39	85.32	78.1	75.95	75.95
Best training accuracy	86.9	81.47	100.0	100.0	100.0	99.68	88.82	100.0
Best test accuracy	92.41	89.87	87.34	87.34	89.87	91.14	75.95	75.95
Processing time(s)	57.12	204.67	39.91	66.01	0.44	1.42	3.69	43.23

Table 7. Performance comparison for Alpha1

	ANN		VGG16		RF		SVM	
	GEL	SPEP	GEL	SPEP	GEL	SPEP	GEL	SPEP
Average training accuracy	75.41	69.65	96.28	98.33	99.89	98.2	100.0	100.0
Average test accuracy	75.23	74.68	75.11	78.9	75.82	65.99	79.75	64.56
Best training accuracy	78.59	69.65	99.04	100.0	100.0	99.36	100.0	100.0
Best test accuracy	79.75	74.68	79.75	82.28	82.28	73.42	79.75	64.56
Processing time(s)	116.74	47.38	40.34	14.21	0.56	1.67	4.71	42.63

Table 8. Performance comparison for Alpha2

	ANN		VGG16		RF		SVM	
	GEL	SPEP	GEL	SPEP	GEL	SPEP	GEL	SPEP
Average training accuracy	83.91	52.29	97.68	99.12	99.96	99.64	91.69	100.0
Average test accuracy	80.17	46.41	78.86	74.14	69.75	52.62	79.75	55.7
Best training accuracy	89.46	59.11	100.0	100.0	100.0	100.0	91.69	100.0
Best test accuracy	87.34	46.84	87.34	77.22	74.68	59.49	79.75	55.7
Processing time(s)	113.31	75.76	53.25	13.93	0.85	2.66	3.53	53.3

Table 9. Performance Comparison for Beta1

	ANN		VGG16		RF		SVM	
	GEL	SPEP	GEL	SPEP	GEL	SPEP	GEL	SPEP
Average training accuracy	83.31	78.27	98.54	98.0	99.99	96.3	100.0	100.0
Average test accuracy	74.39	70.89	75.65	80.38	71.22	67.3	75.95	67.09
Best training accuracy	86.58	78.27	100.0	100.0	100.0	98.4	100.0	100.0
Best test accuracy	77.22	70.89	77.22	82.28	73.42	72.15	75.95	67.09
Processing time(s)	136.07	341.26	25.03	13.89	3.0	1.29	3.78	49.77

Table 10. Performance Comparison for Beta2

	ANN		VGG16		RF		SVM	
	GEL	SPEP	GEL	SPEP	GEL	SPEP	GEL	SPEP
Average training accuracy	72.92	62.02	96.05	94.25	96.9	96.55	86.9	100.0
Average test accuracy	60.72	58.65	62.07	61.39	57.43	52.49	63.29	44.3
Best training accuracy	77.0	68.69	99.36	99.68	98.08	98.08	86.9	100.0
Best test accuracy	67.09	60.76	67.09	68.35	64.56	63.29	63.29	44.3
Processing time(s)	115.26	76.87	60.3	13.93	0.21	1.11	4.63	49.48

Table 11. Performance comparison for Gamma

	ANN		VGG16		RF		SVM	
	GEL	SPEP	GEL	SPEP	GEL	SPEP	GEL	SPEP
Average training accuracy	81.02	66.42	99.56	98.26	99.93	99.97	100.0	100.0
Average test accuracy	74.22	66.92	73.84	81.69	72.7	76.12	72.15	63.29
Best training accuracy	88.5	75.72	100.0	100.0	100.0	100.0	100.0	100.0
Best test accuracy	81.01	77.22	78.48	86.08	78.48	79.75	72.15	63.29
Processing time(s)	198.74	259.71	66.65	16.56	0.95	9.22	4.92	47.67

7 Conclusion

The research presented examined automating the process traditionally performed manually by chemical pathologists for diagnosing multiple myeloma from gel strips and densitometer graphs. The diagnosis is performed by determining the presence of M-Protein and the concentration of Albumin, Alpha1, Alpha2, Beta1, Beta2 and Gamma from the gel strips and densitometer graphs. This study compared the performance of an ANN, VGG-16, random forest and support vector machine for automating this diagnosis process from images of gel strips and densitometer graphs. The VGG-16 was found to perform the best, with reasonable processing times, followed by the ANN and random forest and the SVMs giving the worst performance.

The study also revealed that the process of tuning the parameters for each of the processes is a time consuming task that requires expert knowledge. Future work will look at automating this process and providing a tool for multiple myeloma diagnosis for non-artificial intelligence experts.

Acknowledgements. This work is based on the research supported wholly/in part by the National Research Foundation of South Africa (Grant Numbers 46712). Opinions expressed and conclusions arrived at, are those of the author and are not necessarily to be attributed to the NRF. The authors also acknowledge the Sebia Group Cartridge for sponsoring the printing of the gel strips and graphs.

References

1. Al Zorgani, M., Ugail, H.: Comparative study of image classification using machine learning algorithms. Technical report (2018)
2. Bhattacharyya, S., Epstein, J., Suva, L.J.: Biomarkers that discriminate multiple myeloma patients with or without skeletal involvement detected using SELDI-TOF mass spectrometry and statistical and machine learning tools. Dis. Markers **22**(4), 245–255 (2006)
3. He, K., Zhang, X., Ren, S., Sun, J.: Delving deep into rectifiers: surpassing human-level performance on imagenet classification. In: Proceedings of the IEEE International Conference on Computer Vision, pp. 1026–1034 (2015)
4. Kingma, D., Adam, B.J.: A method for stochastic optimization. arxiv [cs. lg]. 2014 (2017)
5. Lund, A., Lund, M.: Kruskal-Wallis H test using SPSS statistics (2018). https://statistics.laerd.com/spss-tutorials/kruskal-wallis-h-test-using-spss-statistics.php (2018)
6. Lund, A., Lund, M.: One-way anova (2018). https://statistics.laerd.com/statistical-guides/one-way-anova-statistical-guide.php (2018)
7. Santo, L., Vallet, S., Raje, N.: Multiple myeloma. BMJ best practice. https://bestpractice.bmj.com/topics/en-us/179 (2018)
8. Simonyan, K., Zisserman, A.: Very deep convolutional networks for large-scale image recognition. arXiv preprint arXiv:1409.1556 (2014)
9. Sopharak, A., Uyyanonvara, B., Barman, S.: Comparing SVM and Naive Bayes classifier for automatic microaneurysm detections. Int. J. Comput. Electr. Autom. Control Inf. Eng. **8**(5), 797–800 (2014)

10. Turki, T., Wei, Z., Wang, J.T.: Transfer learning approaches to improve drug sensitivity prediction in multiple myeloma patients. IEEE Access **5**, 7381–7393 (2017)
11. Ullah, R., Khan, S., Ali, H., Chaudhary, I.I., Bilal, M., Ahmad, I.: A comparative study of machine learning classifiers for risk prediction of asthma disease. Photodiagn. Photodyn. Ther. **28**, 292–296 (2019)
12. Waddell, M., Page, D., Shaughnessy Jr, J.: Predicting cancer susceptibility from single-nucleotide polymorphism data: a case study in multiple myeloma. In: Proceedings of the 5th International Workshop on Bioinformatics, pp. 21–28 (2005)
13. Wang, H., Zhou, Z., Li, Y., Chen, Z., Lu, P., Wang, W., Liu, W., Yu, L.: Comparison of machine learning methods for classifying mediastinal lymph node metastasis of non-small cell lung cancer from 18 F-FDG PET/CT images. EJNMMI Res. **7**(1), 1–11 (2017)
14. Wu, C.C., Yeh, W.C., Hsu, W.D., Islam, M.M., Nguyen, P.A.A., Poly, T.N., Wang, Y.C., Yang, H.C., Li, Y.C.J.: Prediction of fatty liver disease using machine learning algorithms. Comput. Meth. Prog. Biomed. **170**, 23–29 (2019)
15. Xu, L., et al.: Automated whole-body bone lesion detection for multiple myeloma on 68Ga-Pentixafor PET/CT imaging using deep learning methods. Contrast Media Mol. Imaging **2018**, 2391925 (2018)
16. Yadav, S.S., Jadhav, S.M.: Deep convolutional neural network based medical image classification for disease diagnosis. J. Big Data **6**(1), 1–18 (2019). https://doi.org/10.1186/s40537-019-0276-2

Fuzzy Granulation Approach to Face Recognition

Danuta Rutkowska[1]([✉]), Damian Kurach[2], and Elisabeth Rakus-Andersson[3]

[1] Information Technology Institute, University of Social Sciences,
90-113 Lodz, Poland
drutkowska@san.edu.pl
[2] Czestochowa University of Technology, 42-201 Czestochowa, Poland
[3] Department of Mathematics and Natural Sciences,
Blekinge Institute of Technology, 37179 Karlskrona, Sweden

Abstract. In this paper, a new approach to face description is proposed. The linguistic description of human faces in digital pictures is generated within a framework of fuzzy granulation. Fuzzy relations and fuzzy relational rules are applied in order to create the image description. By use of type-2 fuzzy sets, fuzzy relations, and fuzzy IF-THEN rules, an image recognition system can infer and explain its decision. Such a system can retrieve an image, recognize, and classify – especially a human face – based on the linguistic description.

Keywords: Fuzzy granulation · Face recognition · Linguistic description · Type-2 fuzzy sets · Fuzzy rules · Fuzzy relations · Explainable AI

1 Introduction

There are many publications concerning face recognition, including top biometric technologies, and projects realized by Google, Apple, Facebook, Amazon, Microsoft, in the field of Artificial Intelligence. A list of highly cited papers, journals and books, can be found on the Internet, e.g.: face-rec-org. Deep learning, in particular deep convolutional neural networks, are very popular, with increasing contributions and spectacular applications; see survey papers, e.g. [8]. However, as mentioned in [11], there is a need of linguistic description of the facial features, based on fuzzy approach to image recognition and retrieval [9]. This is still reasonable to study and apply despite the great success of the deep learning methods in image recognition (see e.g. [28]).

It seems obvious that faces - with regard to recognition or classification with explanation - should be considered using fuzzy sets. According to Lotfi Zadeh - who introduced fuzzy sets [32] and fuzzy logic [35] - it is not possible to determine precise borders between particular parts of a human face, such as the nose, cheek, etc. Thus, the regions of a face corresponding to the specific parts (e.g. nose, cheek) should be viewed as fuzzy areas of every face. An illustration of this

© Springer Nature Switzerland AG 2021
L. Rutkowski et al. (Eds.): ICAISC 2021, LNAI 12855, pp. 495–510, 2021.
https://doi.org/10.1007/978-3-030-87897-9_44

concept is presented in [17, 25]. This idea is also applied and developed by other authors, e.g. [10–15].

In the next section, the results presented in [26] are used and ideas introduced in this paper are developed.

2   Face Description by Use of Fuzzy Sets

In [17] fuzzy sets with their membership functions for selected facial features of particular parts of a human face have been created and fuzzy IF-THEN rules formulated for a fuzzy classifier. Different features have been distinguished, including eye colors. In [26], our attention is focused on two attributes: height and width, with regard to different parts of a face, as well as the face as a whole.

The regions corresponding to the specific parts of a face can be defined by membership functions of two arguments associated with the height and width attributes (see [25] and [17]). In this way, the face is partitioned by fuzzy granulation into particular fuzzy regions, the areas of nose, eyes, cheeks, etc.

Let us imagine that a face, and its particular parts are detected and indicated by use of rectangle frames around them (Fig. 4, in Sect. 4, portrays selected main parts.) These frames are crisp areas that characterize the face. Using the fuzzy approach, these crisp areas of the face are compared with fuzzy sets defined by membership functions of two arguments (associated with the height and width attributes). The fuzzy sets correspond to linguistic values, such as: *long* and *narrow* nose.

Fuzzy IF-THEN rules that include linguistic variables, representing particular parts of a face, and the face as a whole, allow to infer a linguistic description of the face, e.g. in the following form: *This is a rectangle shape face, with wide and short forehead, big eyes, long and narrow nose, wide and thin mouth, etc.*

The problem is – how to define the membership functions? We can assume Gaussian, triangular, or trapezoidal shapes of these functions but we do not know exact values of parameters, e.g. centers and widths of the Gaussian functions.

Usually, in fuzzy expert systems, the fuzzy IF-THEN rules are formulated based on expert knowledge. However, in many cases it is difficult to acquire such a domain knowledge. On the other hand, in the cases when a sufficient amount of data is available, the knowledge represented by the rules can be gathered by a learning procedure, like in a neuro-fuzzy system; see e.g. [24].

When a face is considered, with regard to the linguistic description, we can use the anthropological data and the statistical knowledge applied to the results of measurement of a large number of human faces. For example, the results concerning measurements of human heads can be found on the Internet: antropologia-fizyczna.pl.

In [26], the anthropological data are employed in order to define membership functions of fuzzy sets refering to linguistic values of face attributes, first of all - the height and width of a face. The data of this kind, also with regard to other attributes, summarize the measurements of a large number of different human heads, conducted by anthropologists.

The data from Wikipedia Anthropometry pages, provided by Mark Steven Cohen [3], had served as a source of the average values of measurements of human heads. Values of the vertical distance from the bottom of the chin to the level of the top of the head, in inches, for 1st, 5th, 50th, 95th, 99th pecentiles, are presented in [26]; different ones for men and women.

The membership functions of different shapes, e.g. Gaussian, triangular, or trapezoidal, can be applied. In [26] triangular membership functions represent linguistic values of height and width of human faces, such as *short*, *medium*, *long*. Now, we employ trapezoidal membership functions of the fuzzy sets instead of the triangular ones.

Figure 1 illustrates the membership functions of fuzzy sets *very short*, *short*, *medium*, *long*, *very long*, with regard to the height of men and women faces, constructed based on the anthropological data presented in [26].

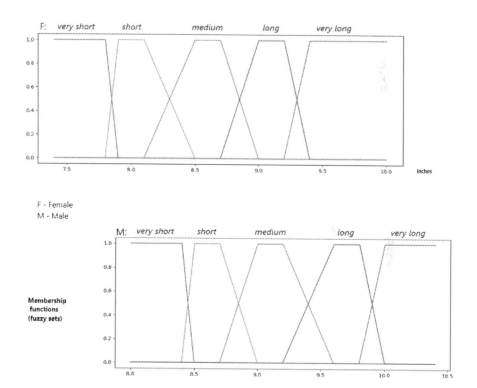

Fig. 1. Membership functions of fuzzy sets representing linguistic values of the height of men and women faces; denoted, respectively, as M (Male) and F (Female).

Apart from the height of human faces, many other results of face measurements are presented in [3]. Among others, there are values of measurements of the head breadth (usually above and behind the ears) and the bitragion breadth (from the right tragion to the left). The former head breadth is wider because it includes ears while the latter one concerns the distance between the tragions, i.e. anthropometric points situated in the notch just above the tragus of each ear. The lower values of the head breadth are more suitable when the shape of human faces is considered. Speaking more precisely, the proportion of the height of a face to the width of the face (breadth of the head) is analyzed in such a case.

Analogously like for the height of faces, fuzzy sets can be created, with regard to linguistic values concerning the width of faces, e.g. *very narrow, narrow, medium wide, wide, very wide*. For details concerning values of the anthropological measurements, for 1st, 5th, 50th, 95th, 99th pecentiles, see [26].

In the situation when human faces are classified based on their height and/or width, separately within the groups of men and women faces, we can use the fuzzy sets presented in Fig. 1 for the height parameter (and analogous, for the width) in fuzzy inference rules. However, the problem of face classification or recognition can be considered without such a differentiation. This means that the fuzzy IF-THEN rules can also be formulated in more general forms, related to human faces, not distinguishing men and women groups.

3   Type-1 and Type-2 Fuzzy Sets for Face Description

In [26], type-2 fuzzy sets are introduced in order to generate the face description. The membership functions of fuzzy sets that represent linguistic values of the height of men and women faces are considered. In [26], triangular membership functions are employed. It is worth mentioning that triangular functions are often used in type-2 fuzzy systems; see e.g. [20,29]. Now, we apply trapezoidal membership functions, instead of the triangular ones, and create type-2 fuzzy sets in a little different way.

Let us start from type-1 fuzzy sets, and focus our attention on the example of the fuzzy set *medium*. Figure 2 portrays the membership functions of two fuzzy sets – F: *medium* and M: *medium* – presented in Fig. 1 for F (Female) and M (Male), respectively.

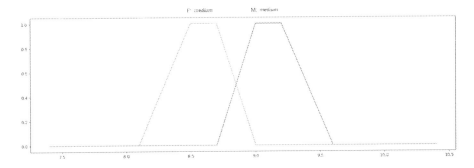

Fig. 2. Membership functions of fuzzy sets representing linguistic values *medium* of the height of human faces, for F (Female) and M (Male), respectively.

Figure 3 a) illustrates the type-2 fuzzy sets *medium*, created based on the type-1 fuzzy sets portrayed in Fig. 2. Two different secondary membership functions are depicted in Fig. 3 a). The decreasing (red color) function refers to F: *medium* while the increasing one (green) refers to M: *medium*.

The secondary membership functions (for the type-2 fuzzy set), shown in Fig. 3 a), are linear functions defined on the interval [0,1] and taking values from the interval [0,1]. However, the functions of this kind are also used as the secondary membership functions of fuzzy sets having shorter intervals as their support, and taking values from 0 to 1. This is illustrated in Fig. 3 b).

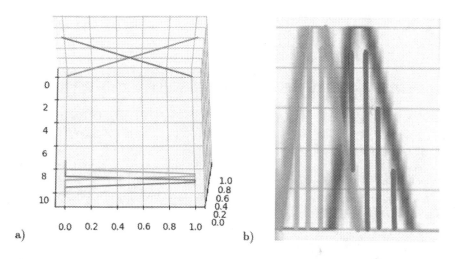

Fig. 3. Membership functions of type-2 fuzzy sets, representing linguistic values *medium* of the height of human faces, for F (Female) and M (Male), respectively: a) 3D, b) 2D. (Color figure online)

Figure 3 b) focuses on the projection of the secondary membership functions on the 2D space with the primary membership functions that represent the F: *medium* and M: *medium* fuzzy sets shown in Figs. 2 and 3 a).

By use of the linear secondary membership functions, at first we create F: *medium* and M: *medium* fuzzy sets of type-2. Let us notice that Fig. 3 b) represents the type-2 fuzzy set F: *medium*. In this case, the secondary membership functions (green lines) take values 1 for the primary membership values of fuzzy set F: *medium* and 0 for the primary membership values of the fuzzy set M: *medium*. The second part of this figure refers to the fuzzy set M: *medium* and the decreasing lines (red color). This also means that the secondary membership functions take values 0 for the primary membership values of the fuzzy set M: *medium* and 1 for the primary membership values of the fuzzy set F: *medium*.

The type-2 fuzzy set M: *medium* is created in the similar way. In order to illustrate this fuzzy set, we should apply the decreasing (red) linear secondary membership functions to the first part of Fig. 3 b) that refers to the F: *medium* type-1 fuzzy set. The increasing (green) secondary membership functions have to be applied to the M: *medium* type-1 fuzzy set.

Based on both type-2 fuzzy sets F: *medium* and M: *medium*, by use of the union operation on type-2 fuzzy sets, and the extension principle introduced by Zadeh [34], the type-2 fuzzy set *medium* has been created.

In the same way, type-2 fuzzy sets *very short*, *short*, *long*, and *very long* have been constructed.

Thus, for particular crisp value of the height of a human face, the type-2 membership value is calculated, and the result is obtained in the form of the appropriate linear function defined on the interval determined by this crisp value (the heigh in inches).

Analogously, type-2 fuzzy sets representing the width of human faces, as well as other attributes, referring to particular parts of the faces, can be created.

4   Fuzzy Granulation Approach to Face Description

Values of measurements of the height and width of a human face can be used in order to recognize and/or classify the shape of the face. As emphasized in [21], it is commonly believed that shape defines the most important feature of objects that we perceive. The authors of this paper explain that shape is the most discriminative property that allows to infer and classify real-world objects.

With regard to shape recognition, usually an algorithm for edge detection is employed to determine regions of interest; see e.g. [6]. However, when fuzzy granulation [36] is considered the exact shape is not so crucial. Figure 4 illustrates main parts of a face that can be viewed as fuzzy granules, inside the face granule that is also a fuzzy granule. Figure 5 presents four lines that determine the shape of the face.

The ratio of the face width to the face height describes the proportional relationship between these two measurements of a head. Thus, this provides an information about the shape, e.g. narrow, rectangle or round head. Of course, in our approach we consider the shape in terms of the fuzzy concept.

There are several different shapes of a face distinguished in the literature, e.g. oblong, oval, round, rectangular, square, triangular, diamond, inverted triangle, heart; for details see [19].

The proportions of the four line sections, portrayed in Fig. 5 in yellow color, inform about the shape of the face. For example, if a face is oval the top half and the bottom half of the face are very similar, so the vertical centerline segment is longer than the middle centerline segment (which is the cheekbone line section). In the case of a square face, the forehead line section and the jaw line section are of similar length, like in Fig. 5. For a round face, the length of the vertical line segment and the horizontal line segment are very similar. The difference between the round face and square face concerns the jawline area (circular versus more angular squared). If a face is of the heart shape the top forehead line section is longer than the jaw line section.

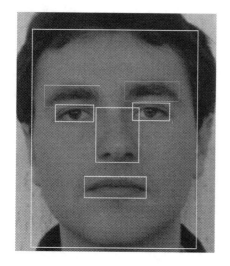

Fig. 4. Main parts of a face.

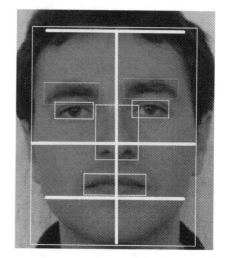

Fig. 5. Lines determining shape.

The two attributes considered in Sects. 2 and 3, the height and width of a human face, are the most important. The propotions of these two attributes allow to give an information concerning the shape of a face. For example, if the height of a face is short then probably the shape is round. It is easy to formulate a rule with the conclusion that the shape is oval, infered based on the ratio of the face width to the face height.

It should be mentioned that there are some computer programs that recognize shapes of human faces based on a photo. However, they do not use fuzzy rules that allow to explain the result with regard to grades of membership.

Of course, the length of the four line segments, portrayed in Fig. 5, are considered as fuzzy sets. In accordance to the height and width of a human face, these line segments can also be viewed as type-2 fuzzy sets.

As mentioned earlier, particular parts of a human face can be considered as fuzzy granules. The main parts are depicted in Fig. 4, as well as in Fig. 5. In addition to those portrayed in the figures, the forehead, chin, cheeks, ears, and other parts can be viewed as fuzzy granules. Two-dimensional membership functions can easily be defined for the parts of a human face, within the rectangular frame that includes the face (2D space). All the fuzzy granules can be viewed as type-2 fuzzy sets.

The fuzzy granulation is hierarchical which means that smaller granules are included in the bigger one, like the particular parts of a face are considered as granules in the face granule. Further, a face granule can be a part of a human figure in an image, and so on.

With regard to the shapes of human faces some of them are very similar, such as round and square faces, as described earlier. Therefore, at first, both can be viewed as one shape granule, and then two different shapes can be considered within this one. In this way, a hierarchy of different face shapes can be created.

Thus, Fuzzy IF-THEN rules formulated in order to decide about a face shape can also be constructed in the hierarchical manner.

5   Fuzzy Relations in Face Description

Fuzzy relations play a very important role in image description, and particularly with regard to face description. Fuzzy relations, as a generalization of classic relations, describe associations between two or more objects. A crisp relation represents the presence or absence of association, interaction, or interconnect-edness between elements of two or more crisp sets. In the concept of fuzzy rela-tions, various degrees of strengths of relation or interaction beetween elements are allowed. Degrees of association can be represented by membership grades in a fuzzy relation in the same way as degrees of set membership are represented in fuzzy sets. For details, see [32] and [24].

A relation matrix is very useful to represent a fuzzy relation in the case of a finite universe of discourse. Elements of the matrix are values of the membership function. These elements correspond to (x, y) where $x \in X$ and $y \in Y$. A fuzzy relation from a crisp set X to a crip set Y is a fuzzy subset of the Cartesian product $X \times Y$. The fuzzy relation is a mapping from the Cartesian space $X \times Y$ to the interval [0,1] where the strength of the mapping is expressed by the mem-bership function of the relation. The strength of the relation between ordered pairs of the two universes is measured with the membership function expressing various degree of strength within [0,1].

Fuzzy relations are mapping elements of one universe to those of another universe through the Cartesian product of two universes, X, Y. A fuzzy relation represents the strength of association between elements of two sets.

For example, let us consider the rectangle around the mouth in Fig. 4. The area inside this rectangle constitutes a crisp set in the universe of discourse that is the whole area of the image. This crisp set can be generalized to the fuzzy set that represents the specific part of the face called "mouth". In this case, X corresponds to the horizontal line at the bottom of the image while Y to the vertical line at the left side of the image. Thus, the image is considered within the $X \times Y$ Cartesian space.

The fuzzy set "mouth", in Fig. 4, can also be viewed as a fuzzy relation between points $x \in X$ and $y \in Y$ that belong to the crisp intervals referring to the width and height sides of the "mouth" rectangle, respectively.

In the same way, every part of a human face can be considered as a fuzzy relation defined within the 2D space of the image that is viewed as the Cartesian product of two one-dimensional spaces corresponding to the axes that refer to the width and height attributes of the image, respectively.

When all the fuzzy relations (fuzzy granules) that represent particular parts of a face are defined then relationships between the granules (objects – parts of the face), such as "mouth", "nose", etc., are analysed.

It is very important to distinguish what relations and relationships mean. However, in this paper, with regard to the face description, both fuzzy relations (in the mathematical sense) as well as relationships between the relations (fuzzy sets) are considered.

Since fuzzy relations are fuzzy sets, then operations on fuzzy sets can also be applied to fuzzy relations. For example, we can obtain the fuzzy relation "eyes" as the union of relations "left eye" and "right eye" both defined in the similar way as the fuzzy relation "mouth".

Figure 6 illustrates projections (from 2D space) of the membership function of the type-1 fuzzy set (fuzzy relation) "mouth". In the case of this part of the face, the membership function is created based on the triangular and trapezoidal functions.

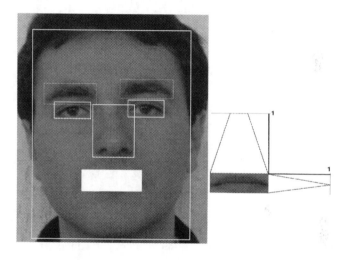

Fig. 6. Membership functions of the fuzzy granule "mouth".

In the same way, by use of the union operation of fuzzy relations "eyes", "nose", "mouth", "forehead", "cheeks", "chin", and "ears", the fuzzy relation "face" is determined. Membership functions of the particular fuzzy sets are defined analogously to the membership function of the fuzzy granule "mouth". In [25] and [17], an illustration of the membership function of fuzzy granule "eye" is portrayed. Of course, different shapes of the membership functions can be applied.

When fuzzy relations of particular parts of a human face are type-1 fuzzy sets, then the type-1 fuzzy set "face" (fuzzy relation) can be created for the "male face" and "female face" separately. However, a type-2 fuzzy set "face" can be defined as the type-2 fuzzy relation to describe the more general granule that include both "male face" and "female face".

The type-2 fuzzy relation "face" can be treated as a pattern of a face, and apply in a process of face recognition in digital images. By use of such a pattern, the face recognition is realized in the same way as human beings perceive faces. This is very important from artificial intelligence point of view.

At first glance, a human being sees a face as a fuzzy granule, and then recognizes details of particular parts of the face, so he/she can distinguish different faces. It also happens that sometimes we see patterns of a fuzzy face granule, for example inside a cloud in the sky, in grass, on a floor, etc. This means that we use such a fuzzy pattern of a face to recognize real human faces. Of course, people remember patterns (as fuzzy granules) of various objects and use them to recognize the objects.

In the process of face recognition within the framework of fuzzy granulation approach, the inclusion relations as well as similarity relations are applied; see [32,33]. The similarity relations are fuzzy counterparts of crisp equivalence relations. The concept of generalized equivalence relations has been extended on type-2 fuzzy sets [4]. This can be useful in the comparison between type-2 fuzzy sets (fuzzy relations).

6   Fuzzy Relational Rules

A fuzzy relational rule is a fuzzy IF-THEN rule that includes a fuzzy relation in its antecedent part; see [30,31].

The classic form of a fuzzy rule (see e.g. [24]), usually employed in fuzzy systems, is presented as follows:

$$\textbf{IF } x_1 \text{ is } A_1 \text{ AND } x_2 \text{ is } A_2 \text{ AND } \ldots x_n \text{ is } A_n \textbf{ THEN} \ldots \qquad (1)$$

where $x_1, x_2, \ldots, x_n$ are linguistic variables defined on the universes of discourse $X_1, X_2, \ldots, X_n$, and $A_1, A_2, \ldots, A_n$ are fuzzy sets in $X_1, X_2, \ldots, X_n$, respectively.

A fuzzy relational rule, as explained above, can be formulated including fuzzy relations to the IF part of rule (1), in the following way:

$$\textbf{IF } x_1 \text{ is } A_1 \text{ AND } x_2 \text{ is } A_2 \text{ AND } \ldots x_n \text{ is } A_n \text{ AND } (A_i, A_j) \text{ is } R_k \qquad (2)$$

for each k, where R_k is a fuzzy relation defined on the Cartesian product $X_i \times X_j$.

Of course, more than one fuzzy relation may be included in (2). For details, see [22]. Let us also present an example referring to the face granules.

In our case of the face description by use of fuzzy relations, linguistic variables, $x_1, x_2, \ldots, x_n$, in (2), correspond to the height and width of particular parts of a human face (see the rectangles in Fig. 6). The fuzzy relations in this rule refer to the "eyes", "nose", "mouth", "forehead", "cheeks", "chin", and "ears" fuzzy granules. Some of these fuzzy relations can be considered as separate fuzzy sets, e.g. "left eye" and "right eye" granules. For example, the following rule concerns the "left eye" and "right eye" granules, denoted as G_{Leye} and G_{Reye}, respectively,

$$\textbf{IF } x_1 \text{ is } A_1 \text{ AND } x_2 \text{ is } A_2 \text{ AND } x_3 \text{ is } A_3 \text{ AND } x_4 \text{ is } A_4$$
$$\text{AND } (A_1, A_2) \text{ is } G_{Leye} \text{ AND } (A_3, A_4) \text{ is } G_{Reye} \textbf{ THEN } \textit{small eyes} \qquad (3)$$

where x_1, x_3 are linguistic variables that refer to the "width" and x_2, x_4 to the "height" parameters of the eyes.

The fuzzy relational IF-THEN rules contain the fuzzy relations within a human face in the antecedent parts, and moreover the relationships between particular parts of a human face are comprised. Therefore, the relational rules are very suitable for application in a face recognition system.

When fuzzy IF-THEN rules are applied in fuzzy systems, including decision support systems, degrees of activation of the rules, called rule firing levels, are determined. This is very important because the systems produce decisions based on the conclusions (consequent parts) of the activated rules. Thus, the antecedent matching degrees are employed in the inference process. The rule firing levels are usually calculated by use of the *t-norm* operation, that is product or min.; for details, see e.g. [24].

It is much easier to determine the rule firing levels for classic IF-THEN rules, presented as (1) than for the fuzzy relational rules, with antecedent parts formulated as (2). It should be emphasized that in both cases it is easy to calculate the degree of rule activation when input data (corresponding to linguistic values) are singletons. It is obvious because singletons, as a matter of fact, are single crisp values (with membership to the singleton fuzzy set equal to 1).

However, in our case, when a human face is described by use of the fuzzy relational rules, the input values are not always singletons. As we see in Fig. 6, fuzzy sets – with triangular or trapezoidal membership functions – are employed. But, consequent parts of the rules (as well as the inference process) are much simpler for classification tasks than in fuzzy control systems. Thus, the antecedent matching degrees are not so difficult to determine, also when the input fuzzy sets are interval type instead of singletons. As mentioned at the end of Sect. 5, the inclusion relations can be applied. In addition, the hierarchical approach to fuzzy granulation helps to cope this problem.

It should be emphasized that with regard to the linguistic description of human faces, in addition to the particular fuzzy granules, relationships between the granules are also taken into account. For example, let us consider the rule formulated as (3). The antecedent part of this rules refers to fuzzy granules that represent two eyes (left and right) but treated as separate parts of a human face. This is sufficient to produce the conclusion such as *small eyes*. This means that we take it for granted that the eyes are parts of the face.

The relationships between particular fuzzy granules are very important in the case when a human face should be detected from an image. The relationships that describe fuzzy distances and locations of the granules within the area of the face help to distinguish the face from other objects in the image. The relationships inform, for example, that the eyes are located between nose and forehead, and moreover that the nose is below the eyes while the forehead is above the eyes. Other relationships concern the mouth – below the nose, and the chin – below the mouth, etc. Knowing the relationships, the system can recognize a human face based on this knowledge, and use this knowledge in order to explain the recognition process.

7   Face Recognition Based on Linguistic Description

While fuzzy sets of particular parts of a face are defined, fuzzy IF-THEN rules can be formulated in order to classify human faces. Then, the system for image recognition and face classification produces an output for input images that are digital pictures of human faces.

Having data concerning measurements of the parts of human faces, we can define suitable fuzzy sets (type-1), and formulate fuzzy IF-THEN rules with conclusions referring to linguistic descriptions of the face attributes.

Fuzzy IF-THEN rules with type-2 fuzzy sets can also be employed in the system for face classification (including the particular parts of faces). The rules allow to produce linguistic description of classified faces. The system also generates an explanation that supports its decision.

It should be emphasized that the main goal is to recognize and classify faces (pictures of faces) with explanation. The system has to produce a linguistic description that explains the decision concerning recognized, retrieved, and classified pictures. Such an explanation is possible based on the knowledge represented by fuzzy IF-THEN rules. If attribute values of the input picture match the antecedent (IF part) of a rule, it is easy to explain the conclusion – included in THEN parts of the rule (or rules). It is also important that the decision is inferred by use of fuzzy logic, so grades of membership – with linguistic labels, such as *little, medium, much* – can be applied in the explanation.

The interpretability and explainability of a model or system is significant from the XAI (Explainable Artificial Intelligence) point of view; see [7]. Unlike many machine learning (ML) methods, including neural networks (see e.g. [37]) and deep learning (see e.g. [1]) that are data-driven and viewed as "black boxes" (with hidden knowledge), rule-based methods are knowledge-driven and hence they are interpretable and explainable. Therefore, some researchers try to extract knowledge (in the form of rules) from "black box models" like neural networks (see e.g. [27]), and also deep networks, e.g. [2]. Many authors emphasize the significance of rule representation, e.g. [18], and apply fuzzy rules for classification (e.g. [23]). Fuzzy rules are also employed by other researchers in problems of image retrieval, e.g. [16].

Another advantage of using the fuzzy granular approach to the face description is the possibility to formulate different queries to the system by use of the linguistic description. For example, with regard to the shape, the following queries can be applied: "Find oval faces". In addition, with regard to particular parts of a face, the examplary queries "Find faces with big nose" or "Find faces with wide mouth" or "Find faces with narrow nose and height forehead" can be employed. Of course, a full linguistic description of a face can be provided to the system. Thus, in contrary to neural networks (that are "black boxes"), this system does not require complete data concerning image attributes, and is explainable (rule-based system).

It is worth emphasizing that the system can produce the linguistic description of input images of human faces without any information whether the pictures present male or female faces. Therefore, the type-2 fuzzy sets are employed in the rules. However, in the case when such an information is introduced to the system, the type-1 fuzzy sets can be sufficient, like those in Fig. 1, different for men and women.

Let us notice that the shape of a human face is determined, according to the approach described in Sect. 4, regardless of the information whether the face is masculine or feminine.

Based on the fuzzy sets of type-1 and type-2, applied in order to formulate fuzzy IF-THEN rules, it is possible to generate the linguistic description of a face, for example, in the following form: *This is small, round-shaped face, with long and narrow nose, small eyes, medium mouth, big ears, and wide forehead.*

8   Conclusions and Final Remarks

The main aim of this paper is to present the fuzzy granular approach to face recognition and the concept of a system that generates the linguistic description of a face portrayed in a digital picture (photo). The system can retrieve a picture or classify pictures based on this description.

The linguistic description of a human face is created by use of fuzzy IF-THEN rules that include fuzzy sets of type-1 and type-2. The fuzzy rules can be employed to provide an explanation of the system's decision concerning the face recognition, retrieval, and/or classification.

A very important aspect of the approach proposed in this paper is the utilization of the knowledge coming from the anthropological data. Therefore, the system can infer decisions, and generate the linguistic description, based on a small collection of digital pictures of faces, without learning from examples of face images in large datasets.

The linguistic description of human faces is very important in the process of image retrieval from databases that consist of thousands or million of face images; see e.g. [5].

Although the system could be employed in order to detect human faces in images, it is intended to analyse the already detected faces, by use of the linguistic description that refers to particular parts of the face, such as forehead, eyes, eyebrows, nose, cheeks, mouth (lips), chin, ears. The face detection outputs the rectangle on the face. Then, the fuzzy granules of the parts of the face are analysed by the system.

The most important advantage of this approach is the possibility to create a system that "understands" its decision – description of the image (face). Image understanding is a challenging research area in Computer Vision and Artificial Intelligence. Deep neural networks try to understand every pixel of an image, and needs a large number of well-labeled images in order to learn to recognize an object presented at a picture. The fuzzy granular approach, proposed in this paper, tries to understand particular areas of a picture within the whole image. In contrary to deep learning methods, the fuzzy granular approach allows to perceive a picture in the similar way as humans do. This makes the Artificial Intelligence to be closer to natural intelligence when performs face recognition.

References

1. Bengio, Y.: Learning deep architectures for AI. Found. Trends Mach. Learn. **2**(1), 1–127 (2009)
2. Bologna, G., Hayashi, Y.: Characterization of symbolic rules embedded in deep DIMLP networks: a challange to transparency of deep learning. J. Artif. Intell. Soft Comput. Res. **7**(4), 265–286 (2017)
3. Cohen, M.S.: Table of average head dimensions based on data from Wikipedia Anthropometry pages. File:AvgHeadSizes.png. Wikipedia (2017). commons. wikimedia.org/wiki/File:AvgHeadSizes.png
4. Dutta, D., Sen, M., Deshpande, A.: Generalized type-2 fuzzy equivalence relation. In: Proceedings of the National Academy of Sciences. India Sect. A Phys. Sci. **107**, 2411–2502 (2020)
5. Grycuk, R., Najgebauer, P., Kordos, M., Scherer, M.M., Marchlewska, A.: Fast image index for database management engines. J. Artif. Intell. Soft Comput. Res. **10**(2), 113–123 (2020)
6. Grycuk, R., Wojciechowski, A., Wei, W., Siwocha, A.: Detecting visual objects by edge crawling. J. Artif. Intell. Soft Comput. Res. **10**(3), 223–237 (2020)
7. Gunning, D., Aha, D.: DARPA's explainable artificial intelligence (XAI) program. AI Mag. **40**(2), 44–58 (2019)
8. Guo, G., Zhang, N.: A survey on deep learning based face recognition. Comput. Vis. Image Underst. **189**, 102805 (2019)
9. Iwamoto, H., Ralescu, A.: Towards a multimedia model-based image retrieval system using fuzzy logic. In: Proceedings of SPIE, vol. 1827, pp. 177–185. Model-based Vision (1992)
10. Kaczmarek, P., Pedrycz, W., Reformat, M., Akhoundi, E.: A study of facial regions saliency: a fuzzy measure approach. Soft Comput. **18**, 379–391 (2014)
11. Kaczmarek, P., Kiersztyn, A., Rutka, P., Pedrycz, W.: Linguistic descriptors in face recognition: a literature survey and the perspectives of future development. In: Proceedings of SPA 2015 (Signal Processing: Algorithms, Architectures, Arrangements, and Applications), Poznań. Poland, pp. 98–103 (2015)
12. Kaczmarek, P., Pedrycz, W., Kiersztyn, A., Rutka, P.: A study in facial features saliency in face recognition: an analytic hierarchy process approach. Soft Comput. **21**, 703–7517 (2016). Springer
13. Karczmarek, P., Kiersztyn, A., Pedrycz, W., Dolecki, M.: Linguistic descriptors and analytic hierarchy process in face recognition realized by humans. In: Rutkowski, L., Korytkowski, M., Scherer, R., Tadeusiewicz, R., Zadeh, L.A., Zurada, J.M. (eds.) ICAISC 2016. LNCS (LNAI), vol. 9692, pp. 584–596. Springer, Cham (2016). https://doi.org/10.1007/978-3-319-39378-0_50
14. Karczmarek, P., Kiersztyn, A., Pedrycz, W.: An evaluation of fuzzy measure for face recognition. In: Rutkowski, L., Korytkowski, M., Scherer, R., Tadeusiewicz, R., Zadeh, L.A., Zurada, J.M. (eds.) ICAISC 2017. LNCS (LNAI), vol. 10245, pp. 668–676. Springer, Cham (2017). https://doi.org/10.1007/978-3-319-59063-9_60
15. Kiersztyn, A., Kaczmarek, P., Dolecki, M., Pedrycz, W.: Linguistic descriptors and fuzzy sets in face recognition realized by humans. In: Proceedings of 2016 IEEE International Conference on Fuzzy Systems (FUZZ), pp. 1120–1126 (2016)

16. Korytkowski, M., Senkerik, R., Scherer, M.M., Angryk, R.A., Kordos, M., Siwocha, A.: Efficient image retrieval by fuzzy rules from boosting and metaheuristic. J. Artif. Intell. Soft Comput. Res. **10**(1), 57–69 (2020)
17. Kurach, D., Rutkowska, D., Rakus-Andersson, E.: Face classification based on linguistic description of facial features. In: Rutkowski, L., Korytkowski, M., Scherer, R., Tadeusiewicz, R., Zadeh, L.A., Zurada, J.M. (eds.) ICAISC 2014. LNCS (LNAI), vol. 8468, pp. 155–166. Springer, Cham (2014). https://doi.org/10.1007/978-3-319-07176-3_14
18. Liu, H., Gegov, A., Cocea, M.: Rule based networks: an efficient and interpretable representation of computational models. J. Artif. Intell. Soft Comput. Res. **7**(2), 111–123 (2017)
19. Medlej, J.: Human anatomy fundamentals. Design & Illustration. https://design.tutsplus.com
20. Niewiadomski, A., Kacprowicz, M.: Type-2 fuzzy logic systems in applications: managing data in selective catalytic reduction for air pollution prevention. J. Artif. Intell. Soft Comput. Res. **11**(2), 85–97 (2021)
21. Pawlak, M., Panesar, G.S., Korytkowski, M.: A novel method for invariant image reconstruction. J. Artif. Intell. Soft Comput. Res. **11**(1), 69–80 (2021)
22. Pierrard, R., Poli, J-P., Hudelot, C.: Learning fuzzy relations and properties for explainable Artificial Intelligence. In: 2018 IEEE International Conference on Fuzzy Systems (FUZZ-IEEE), Rio de Janeiro, Brasil (2018)
23. Riid, A., Preden, J.-S.: Design of fuzzy rule-based classifiers through granulation and consolidation. J. Artif. Intell. Soft Comput. Res. **7**(2), 137–147 (2017)
24. Rutkowska, D.: Neuro-Fuzzy Architectures and Hybrid Learning, vol. 85. Springer, Heidelberg (2002). https://doi.org/10.1007/978-3-7908-1802-4
25. Rutkowska, D.: An expert system for human personality characteristics recognition. In: Rutkowski, L., Scherer, R., Tadeusiewicz, R., Zadeh, L.A., Zurada, J.M. (eds.) ICAISC 2010. LNCS (LNAI), vol. 6113, pp. 665–672. Springer, Heidelberg (2010). https://doi.org/10.1007/978-3-642-13208-7_83
26. Rutkowska, D., Kurach, D., Rakus-Andersson, E.: Face recognition with explanation by fuzzy rules and linguistic description. In: Rutkowski, L., Scherer, R., Korytkowski, M., Pedrycz, W., Tadeusiewicz, R., Zurada, J.M. (eds.) ICAISC 2020. LNCS (LNAI), vol. 12415, pp. 338–350. Springer, Cham (2020). https://doi.org/10.1007/978-3-030-61401-0_32
27. Setiono, R.: Extracting rules from neural networks by prunning and hidden-unit splitting. Neural Comput. **9**, 205–225 (1997)
28. Singh, H.: Practical Machine Learning and Image Processing: For Facial Recognition, Object Detection, and Pattern Recognition Using Python. Apress, India (2019)
29. Starczewski, J.T., Goetzen, P., Napoli, C.: Triangular fuzzy-rough set based fuzzification on fuzzy rule-based systems. J. Artif. Intell. Soft Comput. Res. **10**(4), 271–285 (2020)
30. Yager, R.: The representation of fuzzy relational production rules. Appl. Intell. **1**(1), 35–42 (1991)
31. Yager, R.R., Filev, D.P.: Relational partitioning of fuzzy rules. Fuzzy Sets Syst. **80**(1), 57–69 (1996)
32. Zadeh, L.A.: Fuzzy sets. Inf. Control **8**, 338–353 (1965)
33. Zadeh, L.A.: Similarity relation and fuzzy ordering. Inf. Sci. **3**, 177–200 (1971)
34. Zadeh, L.A.: The concept of a linguistic variable and its application to approximate reasoning-1. Inf. Sci. **8**, 199–249 (1975)

35. Zadeh, L.A.: Fuzzy logic = computing with words. IEEE Trans. Fuzzy Syst. **4**, 103–111 (1996)
36. Zadeh, L.A.: Toward a theory of fuzzy information granulation and its centrality in human reasoning and fuzzy logic. Fuzzy Sets Syst. **90**, 111–127 (1997)
37. Żurada, J.M.: Introduction to Artificial Neural Systems. West Publishing Company, St. Paul (1992)

Dynamic Signature Vertical Partitioning Using Selected Population-Based Algorithms

Marcin Zalasiński[1]($\boxtimes$) (iD), Tacjana Niksa-Rynkiewicz[2] (iD),
and Krzysztof Cpałka[1] (iD)

[1] Department of Computational Intelligence, Czestochowa University of Technology,
Częstochowa, Poland
{marcin.zalasinksi,krzysztof.cpalka}@pcz.pl
[2] Department of Marine Mechatronics, Gdańsk University of Technology, Gdańsk,
Poland
tacniksa@pg.edu.pl

Abstract. The dynamic signature is a biometric attribute used for identity verification. It contains information on dynamics of the signing process. There are many approaches to the dynamic signature verification, including the one based on signature partitioning. Partitions are the regions created on the basis of signals describing the dynamics of the signature. They contain information on the shape of the signature characteristic of a given individual. In this paper, we focus on so-called vertical partitioning and different population-based algorithms which are used to determine partition division points. In the verification process we use an authorial one-class classifier.

Keywords: Dynamic signature verification · Population-based algorithms · Fuzzy systems · Biometrics

1 Introduction

The dynamic signature is a biometric attribute which is acquired using a digital input device, e.g. graphics tablet. This type of signature is described by signals changing over time which contain information on the signing process of an individual. This biometric attribute is commonly used in the identity verification process.

In the literature there are many approaches to the dynamic signature verification. In this paper we focus on the one based on signature partitioning. Signature partitions are created on the basis of the division of signals describing dynamics of the signing process, e.g. pen velocity or pressure. These partitions contain information on the signature shape characteristic of a given individual.

In this paper we propose partitioning of a signature into vertical partitions using different population-based algorithms (PBAs). Vertical partitions are associated with the time moment of the signing process and contain information on

© Springer Nature Switzerland AG 2021
L. Rutkowski et al. (Eds.): ICAISC 2021, LNAI 12855, pp. 511–518, 2021.
https://doi.org/10.1007/978-3-030-87897-9_45

the shape of the signature in different signing phases, e.g. initial and final. Verification of the signature proposed in this paper is realized using an authorial one-class classifier with the simulations performed using the BioSecure dynamic signature database DS2 [16].

This paper is organized into four sections. Section 2 describes the proposed population-based approach for vertical signature partitioning. Section 3 contains information on the simulations carried out in the study presented in this paper. Conclusions are drawn in Sect. 4.

2   Population-Based Approach for Vertical Signature Partitioning

Partitioning is one of the methods used for creating characteristic regions of the signature. These regions contain information about the signature which can be used to create signature descriptors. Descriptors are used in the identity verification process. Vertical partitions are created on the basis of the value of the time moment of the signing process. Due to this, descriptors created in partitions contain information on the similarity of the signature trajectories in the regions associated with, for example, the initial, middle and final time moments of the signing process. In this paper we investigate signature partitioning using different population-based algorithms (see e.g. [2,8,11–13,21,26,34–36]). The details of the proposed method are presented in this section.

Partitions of the signature determined in J training signatures of user i should be pre-processed using commonly known methods [15] to, among others, match their lengths. After that, we use a PBA to select division points $div_i^{\{s\}}$ used for creating the partitions, where s is a signal used for determining a partition. We assume that for each signal s, the signature is divided into two partitions. In our method each individual $\mathbf{X}_{i,ch}$ from the population encodes division points determined for user i:

$$
\begin{aligned}
\mathbf{X}_{i,ch} &= \left\{ div_i^{\{s_1\}}, div_i^{\{s_2\}}, \dots, div_i^{\{s_{Ns}\}} \right\} \\
&= \left\{ X_{i,ch,s_1}, X_{i,ch,s_2}, \dots, X_{i,ch,s_{Ns}} \right\},
\end{aligned}
\tag{1}
$$

where ch is the number of the individual, $\{s_1, s_2, \dots, s_{Ns}\}$ is a set of signals used for the signature alignment and associated with the created partitions, Ns is the number of signals used for the signature alignment and associated with the created partitions.

The algorithm presented in this paper uses a specially designed fitness function (FF) whose value is dependent on the values of descriptors $d_{i,j,r}^{\{s,a\}}$ (where j is the training signature number, r is the partition number, and a is a trajectory signal used for creating the descriptor) created in the partitions determined using division points encoded in the individuals and ratio $Rp_{i,j}^{\{s\}}$ between the number of the discretization points in the signature partitions. Value of $Rp_{i,j}^{\{s\}}$ is determined as follows:

$$Rp_{i,j}^{\{s\}} = \begin{cases} 1 - \dfrac{L_{i,1}^{\{s\}}}{L_{i,0}^{\{s\}}} \text{ for } L_{i,0}^{\{s\}} \geq L_{i,1}^{\{s\}} \\ 1 - \dfrac{L_{i,0}^{\{s\}}}{L_{i,1}^{\{s\}}} \text{ otherwise,} \end{cases} \tag{2}$$

where j is the index of the reference signature, r is the index of the partition, $L_{i,r}^{\{s\}}$ is the number of discretization points in partition r of user i created on the basis of signal s.

Values of the descriptors represent similarity of the training signatures to the template. Detailed information on how they are determined can be found in [9]. In order to be taken into account in the evaluation function, descriptors' values should be normalized. In our paper the normalization process is performed by a properly designed membership function $\mu\left(d_{i,j,r}^{\{s,a\}}\right)$ which is defined as follows:

$$\mu\left(d_{i,j,r}^{\{s,a\}}\right) = \frac{1}{1 + \exp\left(5 - 2 \cdot d_{i,j,r}^{\{s,a\}}\right)}. \tag{3}$$

In the proposed method the values of parameters $Rp_{i,j}^{\{s\}}$ and $\mu\left(d_{i,j,r}^{\{s,a\}}\right)$ are averaged in the context of each reference signature j of the individual. The average value of descriptors $avgD_{i,j}$ in the context of signature j of user i and the average value $avgR_{i,j}$ of parameters $Rp_{i,j}^{\{s\}}$ are determined as follows:

$$\begin{cases} avgD_{i,j} = \dfrac{\left(\begin{array}{c} \mu\left(d_{i,j,0}^{\{v,x\}}\right) + \mu\left(d_{i,j,1}^{\{v,x\}}\right) + \ldots \\ \ldots + \mu\left(d_{i,j,0}^{\{z,y\}}\right) + \mu\left(d_{i,j,1}^{\{z,y\}}\right) \end{array} \right)}{4 \cdot r} \\ avgR_{i,j} = \frac{1}{2} \cdot \left(Rp_{i,j}^{\{v\}} + Rp_{i,j}^{\{z\}} \right). \end{cases} \tag{4}$$

The value of the FF is calculated using weighted algebraic triangular norm $T^* \{\cdot\}$ [7]:

$$\begin{aligned} F\left(\mathbf{X}_{i,ch}\right) &= T^* \left\{ \begin{array}{l} avgD_{i,1}, \ldots, avgD_{i,J}, avgR_{i,1}, \ldots, avgR_{i,J}; \\ wD_{i,1}, \ldots, wD_{i,J}, wR_{i,1}, \ldots, wR_{i,J} \end{array} \right\} \\ &= \left((1 - wD_{i,1} \cdot (1 - avgD_{i,1})) \cdot \ldots \cdot (1 - wR_{i,J} \cdot (1 - avgR_{i,J})) \right), \end{aligned} \tag{5}$$

where t-norm $T^* \{\cdot\}$ is a generalization of the usual two-valued logical conjunction considered in the classical logic, $wD_{i,j} \in [0,1]$ and $wR_{i,j} \in [0,1]$ are the weights of importance of arguments $avgD_{i,j}$ and $avgR_{i,j}$.

In the proposed algorithm, the value of the FF in form (5) is maximized. The values of the best individual in the population are division points $div_i^{\{v\}}$ and $div_i^{\{z\}}$ which are used for creating vertical partitions. The partitions indicated by the elements of vector $\mathbf{r}_i^{\{s\}} = \left[r_{i,l=1}^{\{s\}}, r_{i,l=2}^{\{s\}}, \ldots, r_{i,l=L_i}^{\{s\}} \right]$ are determined as follows:

$$r_{i,l}^{\{s\}} = \begin{cases} 0 \text{ for } l < div_i^{\{s\}} \\ 1 \text{ for } l \geq div_i^{\{s\}}, \end{cases} \tag{6}$$

where l is the index of the signal sample. This step is performed only in the learning phase for the reference signatures.

When the partitions of the signature are determined, we can create templates $\mathbf{tc}_{i,r}^{\{s,a\}} = \left[tc_{i,r,l=1}^{\{s,a\}}, ..., tc_{i,r,l=Lc_{i,r}^{\{s\}}}^{\{s,a\}} \right]$ in the partitions, where $Lc_{i,r}^{\{s\}}$ is the number of discretization points in partition r of user i associated with signal s. They are average values of reference signatures' trajectories and are determined as follows:

$$tc_{i,r,l}^{\{s,a\}} = \frac{1}{J} \sum_{j=1}^{J} a_{i,j,r,l}^{\{s\}}, \tag{7}$$

where $\mathbf{a}_{i,j,r}^{\{s\}} = \left[a_{i,j,r,l=1}^{\{s\}}, ..., a_{i,j,r,l=Lc_{i,r}^{\{s\}}}^{\{s\}} \right]$ is a trajectory (x or y) of reference signature j of user i created on the basis of signal s, which belongs to partition r. Templates are determined only in the training phase of the algorithm.

When we have the templates, descriptors $d_{i,r}^{\{s,a\}}$ can be determined as follows:

$$d_{i,j,r}^{\{s,a\}} = \sqrt{\sum_{l=1}^{Lc_{i,r}^{\{s\}}} \left(tc_{i,r,l}^{\{s,a\}} - a_{i,j,r,l}^{\{s\}} \right)^2}. \tag{8}$$

The descriptors are determined in the training phase of the algorithm and also in the test phase for the signatures which need to be verified.

At the end of the algorithm operation, identity verification is performed. The signature of the test user is classified as genuine or a forgery. The classification process (see e.g. [29,30]) is realized by the flexible fuzzy one-class classifier proposed in our previous works (see e.g. [10]). More information about fuzzy systems can be found, e.g., in [3,7,14,20,33].

3   Simulations

Our simulations were performed using an authorial test environment implemented in C# and the BioSecure dynamic signature database DS2 [16], which contains signatures of 210 users. In the training phase we used 5 randomly selected genuine signatures of each signer, and in the test phase we used 10 genuine signatures and 10 so-called skilled forgeries [19] of each signer.

We selected 4 PBAs to compare their performance in our algorithm: the differential evolution algorithm (DE, [28]), the imperialist competitive algorithm (ICA, [1]), the golden ball algorithm (GB, [27]), and the grey wolf optimizer algorithm (GWO, [25]).

We adopted the following assumptions about the PBAs: **a)** the number of individuals in the population: 100, **b)** the method of selecting individuals in the

Table 1. Comparison of the accuracy of the method using the on-line signature vertical partitioning with the PBA to the other methods for the DS verification using the BioSecure database. The best results are given in bold.

Id.	Method	Average FAR	Average FRR	Average error
1.	Different methods presented in [18]	-	-	3.48%–30.1 %
2.	Method using vertical partitioning presented in [9]	3.13 %	4.15 %	3.64 %
3.	Our method using DE	2.80 %	3.04 %	2.92 %
4.	Our method using ICA	2.84 %	3.36 %	3.10 %
5.	Our method using GB	2.90 %	3.34 %	3.12 %
6.	Our method using GWO	**2.78 %**	**2.94 %**	**2.86 %**

DE and GB (also the selection method of players that face each other to score a goal): the roulette wheel method, **c)** parameter CR in the DE: 0.5, parameter F in the DE: 0.75, **d)** the number of empires in the ICA: 10, parameter ϵ in the ICA: 0.1, parameter β in the ICA: 2.0, parameter γ in the ICA: 0.15, **e)** the number of goal chances in the GB: 20, the number of teams in the GB: 10, and the number of matches in the league competition in the GB: *number of teams·number of teams* (each team plays with each other).

The results of the simulations are presented in Table 1. We can see, that the accuracy of the proposed method is higher than the accuracy of the other methods using the BioSecure database collected in [18]. Effectiveness of the verification method using population-based partitioning is also higher than in the case of the method using fixed-size vertical partitions proposed in [9]. Moreover, the accuracy of the proposed method is relatively high regardless of the PBA type used, which confirms correctness of the adopted assumptions. In Table 1 we can also see that the GWO algorithm was the best in the context of creating vertical partitions in comparison to the DE, ICA, and GB algorithms.

4 Conclusions

In this paper, we propose a method for dynamic signature vertical partitioning using selected PBAs. The proposed algorithm selects division points of the signature which allow to create partitions containing characteristic information about the individual. The simulations conducted using the BioSecure database have proved that the use of PBAs in the partitioning process improves the effectiveness of the signature verification. This is due to the fact that partitions created by the proposed method are better suited for the individual than the partitions created in a classic way. The simulations have also shown that the GWO algorithm in this case works the best of all the tested PBAs.

In the future, we are planning to use deep learning methods (see e.g. [17, 31, 32]) based on artificial neural networks (see e.g. [4–6, 22–24]) in the dynamic signature verification process.

Acknowledgment. This paper was financed under the program of the Minister of Science and Higher Education under the name 'Regional Initiative of Excellence' in the years 2019–2022, project number 020/RID/2018/19 with the amount of financing PLN 12 000 000.

References

1. Atashpaz-Gargari, E., Lucas, C.: Imperialist competitive algorithm: an algorithm for optimization inspired by imperialistic competition. In: Proceedings of the IEEE Congress on Evolutionary Computation, vol. 7, pp. 4661–4666 (2007)
2. Bartczuk, Ł., Dziwiński, P., Goetzen, P.: Nonlinear fuzzy modelling of dynamic objects with fuzzy hybrid particle swarm optimization and genetic algorithm. In: International Conference on Artificial Intelligence and Soft Computing, pp. 315–325. Springer, Cham (2020)
3. Bartczuk, Ł, Przybył, A., Cpałka, K.: A new approach to nonlinear modelling of dynamic systems based on fuzzy rules. Int. J. Appl. Math. Comput. Sci. (AMCS) **26**(3), 603–621 (2016)
4. Bilski, J., Kowalczyk, B., Żurada, J.M.: Application of the givens rotations in the neural network learning algorithm. In: Rutkowski, L., Korytkowski, M., Scherer, R., Tadeusiewicz, R., Zadeh, L.A., Zurada, J.M. (eds.) ICAISC 2016. LNCS (LNAI), vol. 9692, pp. 46–56. Springer, Cham (2016). https://doi.org/10.1007/978-3-319-39378-0_5
5. Bilski, J., Kowalczyk, B., Żurada, J.M.: Parallel implementation of the givens rotations in the neural network learning algorithm. In: Rutkowski, L., Korytkowski, M., Scherer, R., Tadeusiewicz, R., Zadeh, L.A., Zurada, J.M. (eds.) ICAISC 2017. LNCS (LNAI), vol. 10245, pp. 14–24. Springer, Cham (2017). https://doi.org/10.1007/978-3-319-59063-9_2
6. Bilski, J., Rutkowski, L., Smoląg, J., Tao, D.: A novel method for speed training acceleration of recurrent neural networks. Inf. Sci. **553**, 266–279 (2021). https://doi.org/10.1016/j.ins.2020.10.025
7. Cpałka, K.: Design of Interpretable Fuzzy Systems. Springer, Cham (2017)
8. Cpałka, K., Łapa, K., Przybył, A.: Genetic programming algorithm for designing of control systems. Inf. Technol. Control **47**(4), 668–683 (2018)
9. Cpałka, K., Zalasiński, M.: On-line signature verification using vertical signature partitioning. Expert Syst. Appl. **41**, 4170–4180 (2014)
10. Cpałka, K., Zalasiński, M., Rutkowski, L.: New method for the on-line signature verification based on horizontal partitioning. Pattern Recogn. **47**, 2652–2661 (2014)
11. Dziwiński, P., Bartczuk, Ł, Paszkowski, J.: A new auto adaptive fuzzy hybrid particle swarm optimization and genetic algorithm. J. Artif. Intell. Soft Comput. Res. **10**, 95–111 (2020)
12. Dziwiński, P., Bartczuk, Ł, Przybyszewski, K.: A population based algorithm and fuzzy decision trees for nonlinear modeling. In: Rutkowski, L., Scherer, R., Korytkowski, M., Pedrycz, W., Tadeusiewicz, R., Zurada, J.M. (eds.) ICAISC 2018. LNCS (LNAI), vol. 10842, pp. 516–531. Springer, Cham (2018). https://doi.org/10.1007/978-3-319-91262-2_46
13. Dziwiński, P., Bartczuk, Ł: A new hybrid particle swarm optimization and genetic algorithm method controlled by fuzzy logic. IEEE Trans. Fuzzy Syst. **28**(6), 1140–1154 (2019)

14. Ferdaus, M.M., Anavatti, S.G., Garratt, M.A., Pratamam, M.: Development of C-Means clustering based adaptive fuzzy controller for a flapping wing micro air vehicle. J. Artif. Intell. Soft Comput. Res. **9**, 99–109 (2020)
15. Fierrez, J., Ortega-Garcia, J., Ramos, D., Gonzalez-Rodriguez, J.: HMM-based on-line signature verification: feature extraction and signature modeling. Pattern Recogn. Lett. **28**, 2325–2334 (2007)
16. Homepage of Association BioSecure. http://biosecure.it-sudparis.eu. Accessed 13 Nov 2020
17. Hou, Y., Holder, L.B.: On graph mining with deep learning: introducing model R for link weight prediction. J. Artif. Intell. Soft Comput. Res. **9**, 21–40 (2019)
18. Houmani, N., et al.: BioSecure signature evaluation campaign (BSEC 2009): evaluating online signature algorithms depending on the quality of signatures. Pattern Recogn. **45**, 993–1003 (2012)
19. Jain, A.K., Ross, A.: Introduction to biometrics. In: Jain, A.K., Flynn, P., Ross, A.A. (eds.) Handbook of Biometrics. Springer, Heidelberg (2008)
20. Korytkowski, M., Senkerik, R., Scherer, M.M., Angryk, K.M., Siwocha, A.: Efficient image retrieval by fuzzy rules from boosting and metaheuristic. J. Artif. Intell. Soft Comput. Res. **10**, 57–69 (2020)
21. Krell, E., Sheta, A., Balasubramanian, A.P.R., King, S.A.: Collision-free autonomous robot navigation in unknown environments utilizing PSO for path planning. J. Artif. Intell. Soft Comput. Res. **9**, 267–282 (2019)
22. Laskowski, Ł, Laskowska, M., Jelonkiewicz, J., Boullanger, A.: Molecular approach to Hopfield neural network. In: Rutkowski, L., Korytkowski, M., Scherer, R., Tadeusiewicz, R., Zadeh, L.A., Zurada, J.M. (eds.) ICAISC 2015. LNCS (LNAI), vol. 9119, pp. 72–78. Springer, Cham (2015). https://doi.org/10.1007/978-3-319-19324-3_7
23. Laskowski, Ł.: Objects auto-selection from stereo-images realised by self-correcting neural network. In: Rutkowski, L., Korytkowski, M., Scherer, R., Tadeusiewicz, R., Zadeh, L.A., Zurada, J.M. (eds.) ICAISC 2012. LNCS (LNAI), vol. 7267, pp. 119–125. Springer, Heidelberg (2012). https://doi.org/10.1007/978-3-642-29347-4_14
24. Laskowski, Ł.: Hybrid-maximum neural network for depth analysis from stereo-image. In: Rutkowski, L., Scherer, R., Tadeusiewicz, R., Zadeh, L.A., Zurada, J.M. (eds.) ICAISC 2010. LNCS (LNAI), vol. 6114, pp. 47–55. Springer, Heidelberg (2010). https://doi.org/10.1007/978-3-642-13232-2_7
25. Mirjalili, S., Mirjalili, S.M., Lewis, A.: Grey wolf optimizer. Adv. Eng. Softw. **69**, 46–61 (2014)
26. Nasim, A., Burattini, L., Fateh, M.F., Zameer, A.: Solution of linear and non-linear boundary value problems using population-distributed parallel differential evolution. J. Artif. Intell. Soft Comput. Res. **9**, 205–218 (2019)
27. Osaba, E., Diaz, F., Onieva, E.: Golden ball: a novel meta-heuristic to solve combinatorial optimization problems based on soccer concepts. Appl. Intell. **41**, 145–166 (2014)
28. Price, K.V., Storn, R.M., Lampinen, J.A.: Differential Evolution: A Practical Approach to Global Optimization. Springer, Berlin (2005)
29. Rutkowski, L.: On Bayes risk consistent pattern recognition procedures in a quasi-stationary environment. IEEE Trans. Pattern Anal. Mach. Intell. **PAMI-4**(1) 84–87 (1982)
30. Rutkowski, L.: Sequential pattern recognition procedures derived from multiple Fourier series. Pattern Recogn. Lett. **8**(4), 213–216 (1988)

31. Shewalkar, A., Nyavanandi, D., Ludwig, S.A.: Performance evaluation of deep neural networks applied to speech recognition: RNN, LSTM and GRU. J. Artif. Intell. Soft Comput. Res. **9**, 235–245 (2019)

32. Souza, G.B., Silva Santos, D.F., Pires, R.G., Marananil, A.N., Papa, J.P.: Deep features extraction for robust fingerprint spoofing attack detection. J. Artif. Intell. Soft Comput. Res. **9**, 41–49 (2019)

33. Starczewski, J.T., Goetzen, P., Napoli, C.: Triangular fuzzy-rough set based fuzzification of fuzzy rule-based systems. J. Artif. Intell. Soft Comput. Res. **10**, 271–285 (2020)

34. Szczypta, J., Przybył, A., Cpałka, K.: Some aspects of evolutionary designing optimal controllers. In: Rutkowski, L., Korytkowski, M., Scherer, R., Tadeusiewicz, R., Zadeh, L.A., Zurada, J.M. (eds.) ICAISC 2013. LNCS (LNAI), vol. 7895, pp. 91–100. Springer, Heidelberg (2013). https://doi.org/10.1007/978-3-642-38610-7_9

35. Tambouratzis, G., Vassiliou, M.: Swarm algorithms for NLP-the case of limited training data. J. Artif. Intell. Soft Comput. Res. **9**, 219–234 (2019)

36. Zalasiński, M., Cpałka, K., Hayashi, Y.: New fast algorithm for the dynamic signature verification using global features values. In: Rutkowski, L., Korytkowski, M., Scherer, R., Tadeusiewicz, R., Zadeh, L.A., Zurada, J.M. (eds.) ICAISC 2015. LNCS (LNAI), vol. 9120, pp. 175–188. Springer, Cham (2015). https://doi.org/10.1007/978-3-319-19369-4_17

Author Index